T0215740

Classical Feedback Control with Nonlinear Multi-Loop Systems

With MATLAB® and Simulink®

Third Edition

Automation and Control Engineering

*Series Editors - Frank L. Lewis, Shuzhi Sam Ge,
and Stjepan Bogdan*

Networked Control Systems with Intermittent Feedback
Domagoj Tolic and Sandra Hirche

Analysis and Synthesis of Fuzzy Control Systems: A Model-Based Approach
Gang Feng

Subspace Learning of Neural Networks
Jian Cheng Lv, Zhang Yi, Jiliu Zhou

Synchronization and Control of Multiagent Systems
Dong Sun

System Modeling and Control with Resource-Oriented Petri Nets
MengChu Zhou, Naiqi Wu

Deterministic Learning Theory for Identification, Recognition, and Control
Cong Wang and David J. Hill

Optimal and Robust Scheduling for Networked Control Systems
Stefano Longo, Tingli Su, Guido Herrmann, and Phil Barber

Electric and Plug-in Hybrid Vehicle Networks
Optimization and Control
Emanuele Crisostomi, Robert Shorten, Sonja Stüdli, and Fabian Wirth

Adaptive and Fault-Tolerant Control of Underactuated Nonlinear Systems
Jiangshuai Huang, Yong-Duan Song

Discrete-Time Recurrent Neural Control
Analysis and Application
Edgar N. Sánchez

Control of Nonlinear Systems via PI, PD and PID
Stability and Performance
Yong-Duan Song

Multi-Agent Systems: Platoon Control and Non-Fragile Quantized Consensus
Xiang-Gui Guo, Jian-Liang Wang, Fang Liao, Rodney Swee Huat Teo

For more information about this series, please visit: https://www.crcpress.com/
Automation-and-Control-Engineering/book-series/CRCAUTCONENG

Classical Feedback Control with Nonlinear Multi-Loop Systems

With MATLAB® and Simulink®

Third Edition

Boris J. Lurie and Paul J. Enright

CRC Press
Taylor & Francis Group
Boca Raton London New York

CRC Press is an imprint of the
Taylor & Francis Group, an **informa** business

MATLAB® and Simulink® are trademarks of the MathWorks, Inc. and are used with permission. The MathWorks does not warrant the accuracy of the text or exercises in this book. This book's use or discussion of MATLAB® and Simulink® software or related products does not constitute endorsement or sponsorship by the MathWorks of a particular pedagogical approach or particular use of the MATLAB® and Simulink® software.

CRC Press
Taylor & Francis Group
6000 Broken Sound Parkway NW, Suite 300
Boca Raton, FL 33487-2742

First issued in paperback 2021

© 2020 by Taylor & Francis Group, LLC
CRC Press is an imprint of Taylor & Francis Group, an Informa business

No claim to original U.S. Government works

ISBN 13: 978-1-03-224056-5 (pbk)
ISBN 13: 978-1-1385-4114-6 (hbk)

DOI: 10.1201/9781351011853

This book contains information obtained from authentic and highly regarded sources. Reasonable efforts have been made to publish reliable data and information, but the author and publisher cannot assume responsibility for the validity of all materials or the consequences of their use. The authors and publishers have attempted to trace the copyright holders of all material reproduced in this publication and apologize to copyright holders if permission to publish in this form has not been obtained. If any copyright material has not been acknowledged, please write and let us know so we may rectify in any future reprint.

Except as permitted under U.S. Copyright Law, no part of this book may be reprinted, reproduced, transmitted, or utilized in any form by any electronic, mechanical, or other means, now known or hereafter invented, including photocopying, microfilming, and recording, or in any information storage or retrieval system, without written permission from the publishers.

For permission to photocopy or use material electronically from this work, please access www.copyright.com (http://www.copyright.com/) or contact the Copyright Clearance Center, Inc. (CCC), 222 Rosewood Drive, Danvers, MA 01923, 978-750-8400. CCC is a not-for-profit organization that provides licenses and registration for a variety of users. For organizations that have been granted a photocopy license by the CCC, a separate system of payment has been arranged.

Trademark Notice: Product or corporate names may be trademarks or registered trademarks, and are used only for identification and explanation without intent to infringe.

Publisher's Note
The publisher has gone to great lengths to ensure the quality of this reprint but points out that some imperfections in the original copies may be apparent.

Library of Congress Cataloging-in-Publication Data

Names: Lurie, B. J., author. | Enright, Paul J., author.
Title: Classical feedback control with nonlinear multi-loop systems : with MATLAB and Simulink / Boris J. Lurie and Paul J. Enright.
Other titles: Classical feedback control with MATLAB
Description: Third edition. | Boca Raton : Taylor & Francis, CRC Press, [2020] | Series: Automation and control engineering | Original edition published under title: Classical feedback control with MATLAB. | Includes bibliographical references and index.
Identifiers: LCCN 2019016601| ISBN 9781138541146 (hardback : alk. paper) | ISBN 9781351011853 (e-book)
Subjects: LCSH: Feedback control systems. | Nonlinear systems. | MATLAB. | SIMULINK.
Classification: LCC TJ216 .L865 2020 | DDC 629.8/3028553--dc23
LC record available at https://lccn.loc.gov/2019016601

Visit the Taylor & Francis Web site at
http://www.taylorandfrancis.com

and the CRC Press Web site at
http://www.crcpress.com

Contents

Preface

Classical Feedback Control describes the design and implementation of high-performance feedback controllers for engineering systems. The book emphasizes the frequency-domain approach, which is widely used in practical engineering. It presents design methods for linear and nonlinear high-order controllers for single-input, single-output and multi-input, multi-output analog and digital feedback systems. Although the title word *classical* refers to the frequency-domain approach, this book goes well beyond the over-simplified classical designs that are typically included in contemporary texts on automatic control.

Modern technology allows very efficient design and implementation of high-performance controllers at a very low cost. Conversely, several analysis tools that were previously considered an inherent part of control system courses limit the design to low-order (and therefore low-performance) compensators. Among these are the root-locus method, the detection of right half-plane polynomial roots using the Routh–Hurwitz criterion, and extensive algebraic calculations using the Laplace transform. These methods are obsolete and are granted only a brief treatment in this book, making room for loop-shaping, Bode integrals, structural simulation of complex systems, multi-loop systems, and nonlinear controllers, all of which are essential for good design practice.

In the design philosophy adopted in *Classical Feedback Control*, Bode integral relations play a key role. These integrals are employed to estimate the available system performance and to determine the ideal frequency responses that maximize the disturbance rejection and the feedback bandwidth. This ability to quickly analyze the attainable performance, before detailed synthesis and simulation, is critical for efficient system-level trades in the design of complex engineering systems, of which the controller is one of many subsystems. Only at the final design stage do the compensators need to be designed in detail, using high-order approximations of the ideal frequency responses. Nonlinear dynamic compensators are used to provide global stability and to improve transient responses. The controllers are then economically implemented using analog and digital technology.

The first six chapters support a one-semester introductory course in control systems. Later chapters consider nonlinear control, robustness, global stability, and complex system simulation. Throughout the book, MATLAB© with Simulink and SPICE are used for simulation and design—no preliminary experience with this software is required. The student should be familiar with frequency responses and have some knowledge of the Laplace transform. The required theory is reviewed in Appendix 2. The first appendix is an elementary treatment of feedback control, which can be used as an introduction to the course.

It was the authors' intention to make *Classical Feedback Control* not only a textbook but also a reference for students becoming engineers, enabling them to design high-performance controllers and easing the transition from school to the competitive industrial environment. The methods described in this book were used by the authors and their colleagues as the major design tools for feedback loops of aerospace and telecommunication systems. Most examples are from space systems developed by the authors at the Jet Propulsion Laboratory in Pasadena, California.

The third edition adds the presentation of a nonlinear multi-loop system with substantially increased feedback for systems whose performance is limited by uncertainties in the high-frequency area.

We would be grateful for any comments, corrections, and criticism—please take the trouble to communicate with us by e-mail at bolurie@gmail.com and enrightcontrols@att.net.

Acknowledgments

We thank Alla Roden for technical editing and also Asif Ahmed. We appreciate previous discussions on many control issues with Professor Isaac Horowitz, and collaboration, comments, and advice of our colleagues at the Jet Propulsion Laboratory.

In particular, we thank Drs Alexander Abramovich, John O'Brien, David Bayard (who helped edit the chapter on adaptive systems), Dhemetrio Boussalis, Daniel Chang, Gun-Shing Chen (who contributed Appendix 7), Ali Ghavimi (who contributed to the design example A13.14), Fred Hadaegh (who co-authored papers on multiwindow control), John Hench, Mehran Mesbahi, Jason Modisette (who suggested many changes and corrections), Gregory Neat, Samuel Sirlin, and John Spanos. We are grateful to Edward Kopf, who told the authors about the jump resonance in the attitude control loop of the Mariner 10 spacecraft. Suggestions and corrections made by Profs Randolph Beard, Arthur Lanne, Roy Smith, and Michael Zak allowed us to improve the manuscript. Alan Schier contributed the example of the control of a mechanical snake control. To all of them we extend our sincere gratitude.

Boris J. Lurie
Paul J. Enright

MATLAB® is a registered trademark of the MathWorks, Inc. For product information, please contact:

The MathWorks, Inc.

3 Apple Hill Drive

Natick, MA 01760-2098 USA

Tel: 508 647 7000

Fax: 508-647-7001

E-mail: info@mathworks.com

Web: www.mathworks.com

To Instructors

The book presents the techniques that the authors found the most useful for designing control systems for industry, telecommunications, and space programs. It also prepares the reader for research in the area of high-performance nonlinear controllers.

Plant and compensator. In classical control, the plant is characterized by its frequency response over the range of effective feedback. Classical design does not utilize the plant's internal variables nor estimates of them. The appropriate loop response is achieved by a stand-alone high-order compensator. For these reasons this book begins with the concepts of feedback, disturbance rejection, loop shaping, and compensator design, rather than an extensive discussion of plant modeling. It is a goal of feedback to greatly reduce the sensitivity of the closed-loop system to plant characteristics.

Book architecture. The material contained in this book is organized as a sequence of, roughly speaking, four design layers. Each layer considers linear and nonlinear systems, feedback, modeling, and simulation, including the following:

1. Control system *analysis*: Elementary linear feedback theory, a short description of the effects of nonlinearities, and elementary simulation methods (Chapters 1 and 2).

2. Control system *design*: Feedback theory and design of linear single-loop systems developed in depth (Chapters 3 and 4), followed by implementation methods (Chapters 5 and 6).

This completes the first one-semester course in control.

3. Integration of linear and nonlinear subsystem models into the system model, utilization of the effects of feedback on impedances, various simulation methods (Chapter 7), followed by a brief survey of alternative controller design methods and of adaptive systems (Chapters 8 and 9).

4. Nonlinear systems study with practical design methods (Chapters 10 and 11), the elimination or reduction of process instability (Chapter 12), and composite nonlinear controllers (Chapter 13), as well as nonlinear multi-loop control (Chapter 14).

Each consecutive layer is based on the preceding layers. For example, the introduction of absolute stability and Nyquist-stable systems in the second layer is preceded by a primitive treatment of saturation effects in the first layer; global stability and absolute stability are then treated more precisely in the fourth layer. Treatment of the effects of links' input and output impedances on the plant uncertainty in the third layer is based on the elementary feedback theory of the first layer and the effects of plant tolerances on the available feedback developed in the second layer.

This architecture reflects the multifaceted character of design, and allows the illustration of system theory using realistic examples without excessive idealization.

Design examples. Examples were chosen among those designed by the authors for various space robotic missions at the Jet Propulsion Laboratory, including the following:

A prototype for the controller of a retroreflector carriage on the Chemistry spacecraft (Section 4.2.3). The plant has flexible modes, and the controller is high order with a Bode step and nonlinearities.

A nonlinear digital controller for the pointing of the Mars Pathfinder high-gain antenna (Section 5.10.5).

A switched-capacitor controller for the STRV spacecraft cryogenic cooler vibration rejection (Section 6.4.2).

Thermal control for the Cassini Narrow Angle Camera (Section 7.1.2 and Example 13.6).

Vibration damping for a space stellar interferometer (Examples 7.2 and 7.20).

An analog feedback loop for a microgravity accelerometer (Example 11.12).

Thrust vector control for the Cassini spacecraft during main engine burns (A13.1), which is depicted on the front cover.

An optical receiver for multichannel space system control (A13.15).

Software. Many design examples in the book use MATLAB with Simulink from MathWorks, Inc. The MATLAB functions are described as they are used for analysis and plotting. Additional MATLAB functions written by the authors and described in Appendix 14 may complement the control design methods.

Several examples are given using SPICE, which is particularly useful in electrical engineering for combining conventional design and control system simulation. Also, some simple examples of coding in C are described in Chapter 5.

Undergraduate course: The first six chapters, which may constitute a first course in control, do include some material that is better suited for a graduate course. This material can be omitted from a one-semester course—the sections to be bypassed are listed at the beginning of each chapter.

Digital controllers: The best way to design a digital controller is to design a high-order continuous controller, break it properly into several links of low order, and then convert each of these to a digital link. This conversion can be performed using the two small tables included in Chapter 5 or by the MATLAB function c2d. The accuracy of the Tustin transform is adequate for this purpose and prewarping is unnecessary.

Analog controller implementation: Because the inputs and outputs of many sensors and actuators are electrical, the analog controllers of Chapter 6 can be easier and cheaper to design than their digital counterparts, and it remains important for all engineers to understand them. However, this chapter does not need to be fully covered in a one-semester course—it will be useful for self-study or as a reference.

Second course in advanced classical control: Chapters 7–14 can be used for a second one-semester course (it may be a graduate course), and as a reference for practicing engineers.

Chapter 7 describes structural design and simulation systems with drivers, motors, and sensors. In particular, it shows that tailoring the output impedance of the actuator is important to reduce the plant tolerances and to increase the feedback in the outer loop.

Chapter 8 gives a short introduction to quantitative feedback theory and H_∞ control and time-domain control based on the system's state variables.

Adaptive systems are described briefly in Chapter 9. Although these systems are often not necessary, designers should be aware of the major concepts, advantages, and limitations of adaptive control, and be able to recognize the need for such control and, at the same time, not waste time trying to achieve the impossible. The material in Chapter 9 will guide engineers to design their own adaptive systems, or to understand the specialized literature in the area of adaptive control research.

The design of high-order nonlinear controllers is covered in Chapters 10–13. These design methods have been proven very effective in practice but are far from being perfect—further research needs to be done to advance these methods. Chapter 14 introduces nonlinear multi-loop control, which has the potential to double the achievable bandwidth compared to the single-loop ideal Bode system.

Authors

Boris J. Lurie received his PhD degree from the Institute of Transportation Engineering, St. Petersburg, and the DSc. degree from the Higher Attestational Commission, Moscow. The author or co-author of over 100 publications, including six books and nine patents, Dr Lurie worked for many years in the telecommunication and aerospace industries, and taught at Russian, Israeli, and American universities. After many years as a senior staff member of the Jet Propulsion Laboratory, California Institute of Technology, Dr Lurie is retired and now lives in San Diego.

Paul J. Enright received his MS in Aeronautics and Astronautics from Stanford University, and his PhD from the University of Illinois at Urbana-Champaign where he pioneered the use of direct transcription methods for the optimization of spacecraft trajectories. As a member of the technical staff at the Jet Propulsion Laboratory, California Institute of Technology, he designed attitude control systems for interplanetary spacecraft and conducted research in the area of nonlinear control. Dr Enright currently works in the field of quantitative finance in Chicago.

1

Feedback and Sensitivity

Chapter 1 introduces the basics of feedback control. The purpose of feedback is to make the output insensitive to plant parameter variations and disturbances. Negative, positive, and large feedback are defined and discussed along with sensitivity and disturbance rejection. The notions of frequency response, the Nyquist diagram, and the Nichols chart are introduced. (The Nyquist stability criterion is presented in Chapter 3.)

Feedback control and block diagram algebra are explained at an elementary level in Appendix 1, which can be used as an introduction to this chapter. Laplace transfer functions are described in Appendix 2.

1.1 Feedback Control System

It is best to begin with an example. Figure 1.1a depicts a *servomechanism* regulating the elevation of an antenna. Figure 1.1b shows a block diagram for this control system made of cascaded elements, i.e., *links*. The capital letters stand for the signals' Laplace transforms and also for the transfer functions of the linear links.

There is one input command U_1, which is the commanded elevation angle, and just one output U_2, which is the actual elevation of the antenna, so the system is said to be *single-input single-output* (SISO). Evidently there is one feedback loop, and so the system is also referred to as *single-loop*.

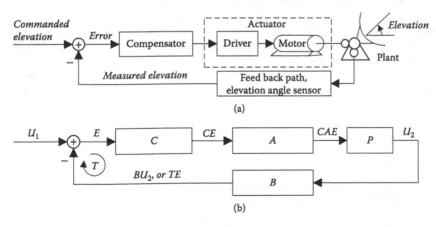

(a)

(b)

FIGURE 1.1
Single-loop feedback system.

The *feedback path* contains some sort of sensor for the output variable and has the transfer function B. Ideally, the measured output value BU_2 equals the commanded value U_1, and the *error* $E = U_1 - BU_2$, at the output of the summer, is zero. In practice, most of the time the error is nonzero but small.

The error is amplified by the *compensator* C and applied to the *actuator* A, in this case a motor regulator (driver) and a motor, respectively. The motor rotates the *plant* P, the antenna itself, which is the object of the control. The compensator, actuator, and plant make up the *forward path* with the transfer function CAP.

The *return signal*, which goes into the summer from the feedback path, is $BU_2 = TE$, where the product $T = CAPB = BU_2/E$, is called the *loop transfer function* or the *return ratio*.

The output of the summer is:

$$E = U_1 - ET \tag{1.1}$$

so that the error can be expressed as

$$E = \frac{U_1}{T+1} = \frac{U_1}{F} \tag{1.2}$$

where $F = T + 1$ is the *return difference*. Its magnitude $|F|$ is the *feedback*. It is seen that when the feedback is large, the error is small.

If the feedback path was not present, the output U_2 would simply equal the product $CAPU_1$, and the system would be referred to as *open-loop*.

Example 1.1

A servomechanism for steering a toy car (using wires) is shown in Figure 1.2. The command voltage U_1 is regulated by a joystick potentiometer. Another identical potentiometer (angle sensor) placed on the shaft of the motor produces voltage U_{angle} proportional to the shaft rotation angle. The feedback makes the error small, so that the sensor voltage approximates the input voltage, and therefore the motor shaft angle tracks the joystick-commanded angle.

This arrangement of a motor with an angle sensor is often called servomotor, or simply *servo*. Similar servos are used for animation purposes in movie production.

The system of regulating aircraft-control surfaces using joysticks and servos was termed "fly by wire" when it was first introduced to replace bulky mechanical gears and cables. The required high reliability was achieved by using four independent parallel analog electrical circuits.

The telecommunication link between the control box and the servo can certainly also be wireless.

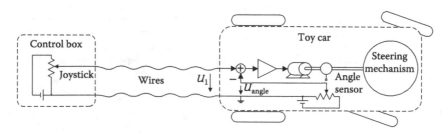

FIGURE 1.2
Joystick control of a steering mechanism.

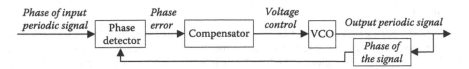

FIGURE 1.3
Phase-locked loop.

Example 1.2

A *phase-locked loop* (PLL) is shown in Figure 1.3. The plant here is a *voltage-controlled oscillator* (VCO).

The VCO is an ac generator whose frequency is proportional to the voltage applied to its input. The *phase detector* combines the functions of phase sensors and input summer: its output is proportional to the phase difference between the input signal and the output of the VCO.

Large feedback makes the phase difference (phase error) small. The VCO output therefore has a small phase difference compared with the input signal. In other words, the PLL synchronizes the VCO with the input periodic signal.

It is very important to also reduce the noise that appears at the VCO output.

PLLs are widely used in telecommunications (for tuning receivers and for recovering the computer clock from a string of digital data), for synchronizing several motors' angular positions and velocities, and for many other purposes.

1.2 Feedback: Positive and Negative

The output signal in Figure 1.1b is $U_2 = ECAP$, and from Equation 1.2, $U_1 = EF$. The input-to-output transfer function of the system U_2/U_1, commonly referred to as the *closed-loop transfer function*, is:

$$\frac{U_2}{U_1} = \frac{ECAP}{EF} = \frac{CAP}{F} \tag{1.3}$$

so the feedback reduces the input–output signal transmission by the factor $|F|$.

The system is said to have *negative* feedback when $|F| > 1$ (although the expression $|F|$ is certainly positive). This definition was developed in the 1920s and has to do with the fact that negative feedback reduces the error $|E|$ and the output $|U_2|$, i.e., produces a negative increment in the output level when the level is expressed in logarithmic values (dB, for example) preferred by engineers.

The feedback is said to be *positive* if $|F| < 1$, which makes $|E| > |U_1|$. Positive feedback increases the error and the level of the output.

We will adhere to these definitions of negative and positive feedback since very important theoretical developments, to be studied in Chapters 3 and 4, are based on them.

Whether the feedback is positive or negative depends on the amplitude and phase of the return ratio (and not only on the sign at the feedback summer, as is sometimes stated in elementary treatments of feedback).

Let's consider several numerical examples.

Example 1.3

The forward path gain coefficient *CAP* is 100, and the feedback path coefficient *B* is −0.003. The return ratio *T* is −0.3. Hence, the return difference *F* is 0.7, the feedback is positive, and the closed-loop gain coefficient (100/0.7 = 143) is greater than the open-loop gain coefficient.

Example 1.4

The forward path gain coefficient is 100, and the feedback path coefficient is 0.003. The return ratio T is 0.3. Hence, the return difference F is 1.3, the feedback is negative, and the closed-loop gain coefficient (100/1.3=77) is less than the open-loop gain coefficient.

It is seen that when T is small, the sign of the transfer function about the loop determines whether the feedback is positive or negative.

When $|T| > 2$, then $|T+1| > 1$ and the feedback is negative. That is, when $|T|$ is large, the feedback is always negative.

Example 1.5

The forward path gain coefficient is 1,000 and $B=0.1$. The return ratio is therefore 100. The return difference is 101, the feedback is negative, and the closed-loop gain coefficient is 9.9.

Example 1.6

In the previous example, the forward path transfer function is changed to −1,000, and the return ratio becomes −100. The return difference is −99, the feedback is still negative, and the closed-loop gain coefficient is 10.1.

1.3 Large Feedback

Multiplying the numerator and denominator of Equation 1.3 by B yields another meaningful formula:

$$\frac{U_2}{U_1} = \frac{1}{B}\frac{T}{F} = \frac{1}{B}M \tag{1.4}$$

where

$$M = \frac{T}{F} = \frac{T}{T+1} \tag{1.5}$$

Equation 1.4 indicates that the closed-loop transfer function is the inverse of the feedback path transfer function multiplied by the coefficient M. When the feedback is *large*, i.e., when $|T| \gg 1$, the return difference $F \approx T$, the coefficient $M \approx 1$, and the output becomes

$$U_2 \approx \frac{1}{B}U_1 \tag{1.6}$$

One result of large feedback is that the closed-loop transfer function depends nearly exclusively on the feedback path, which can usually be constructed of precise components. This feature is of fundamental importance since the parameters of the actuator and the plant in the forward path typically have large uncertainties. In a system with large feedback, the effect of these uncertainties on the closed-loop characteristics is small. The larger the feedback, the smaller the error expressed by Equation 1.2.

Manufacturing an actuator that is sufficiently powerful and precise to handle the plant without feedback can be prohibitively expensive or impossible. An imprecise actuator may be much cheaper, and a precise sensor may also be relatively inexpensive. Using feedback, the cheaper actuator and the sensor can be combined to form a powerful, precise, and reasonably inexpensive system.

According to Equation 1.6, the antenna elevation angle in Figure 1.1 equals the command divided by B. If the elevation angle is required to be q, then the command should be Bq.

If $B=1$, as shown in Figure 1.4a, then the closed-loop transfer function is just M and $U_2 \approx U_1$; i.e., the output U_2 follows (tracks) the commanded input U_1. Such *tracking systems* are widely used. Examples are a telescope tracking a star or a planet, an antenna on the roof of a vehicle tracking the position of a knob rotated by the operator inside the vehicle, and a cutting tool following a probe on a model to be copied.

Example 1.7

Figure 1.4b shows an amplifier with unity feedback. The error voltage is the difference between the input and output voltages. If the amplifier gain coefficient is 10^4, the error voltage constitutes only 10^{-4} of the output voltage. Since the output voltage nearly equals the input voltage, this arrangement is commonly called a *voltage follower*.

Example 1.8

Suppose that $T=100$, so that $M=T/(T+1)=0.9901$. If P were to deviate from its nominal value by +10%, then T would become 110. This would make $M=0.991$, an increase of 0.1%, which is reflected in the output signal. Without the feedback, the variation of the output signal would be 10%. Therefore, introduction of negative feedback in this case reduces the output signal variations 100 times. Introducing positive feedback would do just the opposite—it would increase the variations in the closed-loop input–output transfer function.

Example 1.9

Consider the voltage regulator shown in Figure 1.5a with its block diagram shown in Figure 1.5b. Here, the differential amplifier with transimpedance (ratio of output current I to input voltage E) 10 A/V and high input and output impedances plays the dual role of compensator and actuator. The power supply voltage is VCC. The plant is the load resistor R_L. The potentiometer with the voltage division ratio B constitutes the feedback path.

The amplifier input voltage is the error $E=U_1-TE$, and the return ratio is $T=10BR_L$. Assume that the load resistor is 1 kΩ and the potentiometer is set to $B=0.5$. Consequently, the return ratio is $T=5,000$.

The command is the 5 V input voltage (when the command is constant, as in this case, it is commonly called a *reference* and the control system is called a *regulator*). Hence, the

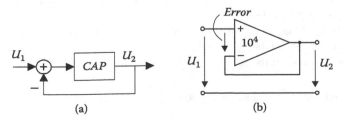

(a)   (b)

FIGURE 1.4
(a) Tracking system and (b) voltage follower.

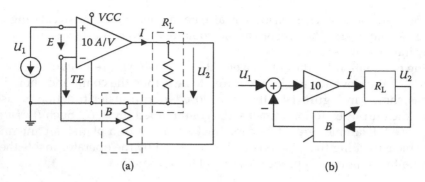

FIGURE 1.5
Voltage regulator: (a) schematic diagram, (b) block diagram.

output voltage according to Equation 1.4 is $10\times5,000/5,001 = 9.998$ V. (The *VCC* should be higher than this value; 12 V to 30 V would be more appropriate.)

When the load resistance is reduced by 10%, without the feedback the output voltage would be 10% less. With the feedback, T decreases by 10% but the output voltage is equal to $10\times4,500/4,501 \approx 9.99778$, i.e., only 0.002% less. The feedback reduces the output voltage variations 10%/0.002% = 5,000 times.

This example also illustrates another feature of feedback. Insensitivity of the output voltage to the loading indicates that the regulator output resistance is very low. The feedback dramatically alters the output impedance from very high to very low. (The same is true for the follower shown in Figure 1.4b.) The effects of feedback on impedance will be studied in detail in Chapter 7.

1.4 Loop Gain and Phase Frequency Responses

1.4.1 Gain and Phase Responses

The sum (or the difference) of sinusoidal signals of the same frequency is a sinusoidal signal with the same frequency. The summation is simplified when the sinusoidal signals are represented by vectors on a complex plane (*phasors*). The modulus of the vector equals the signal amplitude and the phase of the vector equals the phase shift of the signal.

Signal $u_1 = |U_1|\sin(\omega t + \phi_1)$ is represented by the vector $U_1 = |U_1|\angle\phi_1$, i.e., by the complex number $U_1 = |U_1|\cos\phi_1 + j\,|U_1|\sin\phi_1$.

Signal $u_2 = |U_2|\sin(\omega t + \phi_2)$ is represented by the vector $U_2 = |U_2|\angle\phi_2$, i.e., by the complex number $U_2 = |U_2|\cos\phi_2 + j\,|U_2|\sin\phi_2$.

The sum of these two signals is:

$$u = |U|\sin(\omega t + \phi) = \left(|U_1|\cos\phi_1 + |U_2|\cos\phi_2\right)\sin\omega t + \left(|U_1|\sin\phi_1 + |U_2|\sin\phi_2\right)\cos\omega t$$

That is, $\text{Re}U = \text{Re}U_1 + \text{Re}U_2$ and $\text{Im}U = \text{Im}U_1 + \text{Im}U_2$, i.e., $U = U_1 + U_2$. Thus, the vector for the sum of the signals equals the sum of the vectors for the signals.

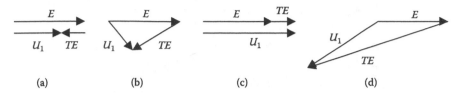

FIGURE 1.6
Examples of phasor diagrams for (a and b) positive and (c and d) negative feedback.

Example 1.10

If $u_1 = 4\sin(\omega t + \pi/6)$, it is represented by the vector $4\angle\pi/6$, or $3.464 + j2$. If $u_2 = 6\sin(\omega t + \pi/4)$, it is represented by the vector $6\angle\pi/4$, or $4.243 + j4.243$. The sum of these two signals is represented by the vector (complex number) $7.707 + j6.243 = 9.92\angle0.681$ rad or $9.92\angle39.0°$.

Example 1.11

Figure 1.6 shows four possible vector diagrams $U_1 = E + TE$ of the signals at the feedback summer at some frequency. In cases (a) and (b), the presence of feedback signal TE makes $|E| > |U_1|$; therefore, $|F| < 1$ and the feedback at this frequency is positive. In cases (c) and (d), $|E| < |U_1|$, and the feedback at this frequency is negative.

Replacing the Laplace variable s by $j\omega$ in a Laplace transfer function, where ω is understood to be the (real) frequency in rad/s, results in a frequency-dependent transfer function. The transfer function is the ratio of the signals at the link's output and input, which depend on the frequency. Plots of the gain and phase of a transfer function versus frequency are referred to generically as the *frequency response*.

The *loop frequency response* is defined by the complex function $T(j\omega)$. The magnitude $|T(j\omega)|$ expressed in decibels (dB) is referred to as the *loop gain*. The angle of $T(j\omega)$, $\angle T$ or arg T, is the *loop phase shift*, which is usually expressed in degrees. Plots of the gain (and sometimes the phase) of the loop transfer function with logarithmic frequency scale are often called *Bode diagrams*, in honor of Hendrik Bode, who, although he did not invent the diagrams, did develop an improved methodology of using them for feedback-system design [6] to be explained in detail in Chapters 3 and 4. The plots can be drawn using the angular frequency ω in rad/s or $f = \omega/(2\pi)$ in Hz.

Example 1.12

Let

$$T(s) = \frac{num}{den} = \frac{5,000}{s(s+5)(s+50)} = \frac{5,000}{s^3 + 55s^2 + 250s + 0}$$

where *num* and *den* are the numerator and denominator polynomials.

The frequency responses for the loop gain in dB, $20\log|T(j\omega)|$, and the loop phase shift in degrees, $(180/\pi)$ arg $T(j\omega)$, can be plotted over the 0.1 to 100 rad/s range with the software package MATLAB® from MathWorks, Inc. using the following script:

```
w = logspace(-1, 2);
% log scale of angular
% frequency w
num = 5000;
den = [1 55 250 0];
bode(num, den, w)
```

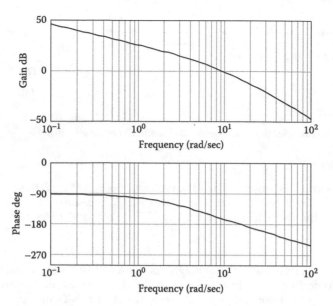

FIGURE 1.7
MATLAB plots of gain and phase loop responses.

The plots are shown in Figure 1.7. The loop gain rapidly decreases with frequency, and the slope of the gain response gets even steeper at higher frequencies; this is typical for practical control systems.

The loop gain is 0 dB at 9 rad/s, i.e., at $9/(2\pi) \approx 1.4$ Hz. The phase shift gradually changes from $-90°$ toward $-270°$, i.e., the phase lag increases from 90° to 270°.

The MATLAB function conv can be used to multiply the polynomials s, $(s+5)$, and $(s+50)$ in the denominator:

```
a = [1 0]; b = [1 5]; c = [1 50];
ab = conv(a,b); den = conv(ab,c)
```

results in:

```
den = 1   55   250   0
```

More information about the MATLAB functions used above can be obtained by typing help bode, logspace, or conv in the MATLAB working window, and from the MATLAB manual.

Conversions from one form of a rational function to another can also be done using the MATLAB functions tf2zp (transfer function to zero-pole form) and zp2tf (zero-pole form to transfer function).

Example 1.13

The return ratio from Example 1.12, explicitly expressed as a function of jω, is:

$$T(j\omega) = \frac{5{,}000}{j\omega(j\omega + 5)(j\omega + 50)}$$

At the frequency $\omega = 2$, $T \approx 10\angle-110°$, and $F \approx 9\angle-115°$, so the feedback is negative. $|T|$ reduces with frequency. At the frequency $\omega = 9$, $T \approx 1\angle-160°$, and $F \approx 0.2\angle-70°$, so the feedback is positive.

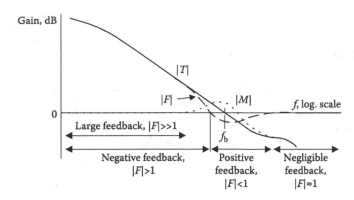

FIGURE 1.8
Typical frequency responses for T, F, and M.

The Bode plot of $|T|$ for a typical tracking system is shown in Figure 1.8 by the solid line. The loop gain decreases with increasing frequency. The diagram crosses the 0-dB line at the *crossover frequency* f_b where, by definition, $|T(f_b)| = 1$.

The frequency response of the feedback $|F|$ is shown by the dashed line. It can be seen that the feedback is negative (i.e., $20 \log|F| > 0$) up to a certain frequency, becomes positive in the neighborhood of f_b, and then becomes negligible at higher frequencies, where $F \to 1$ and $20 \log|F| \to 0$.

The input–output closed-loop system response is shown by the dotted line. The gain is 0 dB (i.e., the gain coefficient is 1) over the entire bandwidth of large feedback. The hump near the crossover frequency is a result of the positive feedback. This hump, as will be demonstrated in Chapter 3, results in an oscillatory closed-loop transient response, and should be bounded.

In general, feedback improves the tracking system's accuracy for commands whose dominant Fourier components belong to the area of negative feedback, but degrades the system's accuracy for commands whose frequency content is in the area of positive feedback.

1.4.2 Nyquist Diagram

To visualize the transition from negative to positive feedback, it is helpful to look at the plot of T as the frequency varies from 0 to ∞, which is referred to as the *Nyquist diagram* and is shown in Figure 1.9. Either Cartesian (ReT, ImT) or polar coordinates ($|T|$ and arg$T$) can be used. We refer to this as the *T-plane* Nyquist diagram to distinguish it from the logarithmic diagram, which is described below.

The Nyquist diagram is a major tool in feedback-system design and will be discussed in detail in Chapter 3. Here, we use the diagram only to show the locations of the frequency bands of negative and positive feedback in typical control systems.

At each frequency, the distance to the diagram from the origin is $|T|$, and the distance from the –1 point is $|F|$. It can be seen that $|F|$ becomes less than 1 at higher frequencies, which means the feedback is positive there.

The Nyquist diagram should not pass too close to the *critical point at* –1 or else the closed-loop gain at this frequency will be unacceptably large.

In practice, Nyquist diagrams are commonly plotted with rectangular coordinate axes for the phase and the gain of T, as shown in Figure 1.10a. This is referred to as the *L-plane* diagram, because the real and imaginary parts of the complex logarithm of T are the logarithmic gain and phase. Notice that the critical point –1 of the T-plane maps to point (–180°, 0 dB) on the L-plane. The Nyquist diagram should avoid this point by a certain margin.

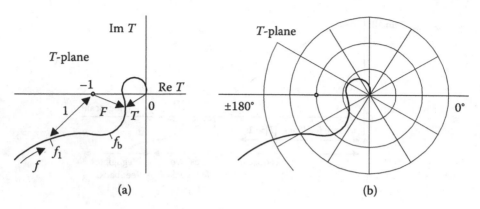

FIGURE 1.9
Nyquist diagram with (a) Cartesian and (b) polar coordinates; the feedback is negative at frequencies up to f_1.

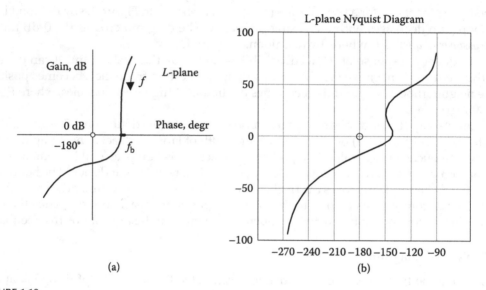

FIGURE 1.10
Nyquist diagrams on the *L*-plane, (a) typical for a well-designed system and (b) MATLAB generated for Example 1.14.

Example 1.14

Figure 1.10b shows the *L*-plane Nyquist diagram for $T = (20s + 10)/(s^4 + 10s^3 + 20s^2 + s)$ plotted with the following MATLAB script:

```
num= [20 10]; den= [1 10 20 1 0];
[mag, phase] =bode(num, den);
plot(phase, 20*log10(mag), 'r', -180, 0, 'wo')
title('L-plane Nyquist diagram')
set(gca,'XTick',[-270 -240 -210 -180 -150 -120 -90])
grid
```

It is recommended for the reader to run this program for modified transfer functions and to observe the effects of the polynomial coefficient variations on the shape of the Nyquist diagram.

1.4.3 Nichols Chart

The *Nichols chart* is an *L*-plane template for the mapping from *T* to *M*, according to Equation 1.5, and is shown in Figure 1.11. When the Nyquist diagram for *T* is drawn on this template, the curves indicate the tracking system gain: 20 log |*M*|.

It is seen that the closer the Nyquist diagram gets to the critical point (−180°,0 dB), the larger is |*M*|, and therefore the higher the peak of the closed-loop frequency response in Figure 1.8. The limiting case has |*M*| approaching infinity, indicating that the system goes unstable. Typically, |*M*| is allowed to increase no more than two times, i.e., not to exceed 6 dB. Therefore, the Nyquist diagram should not penetrate into the area bounded by the line marked 6 dB.

Consider several examples that make use of the Nichols chart.

Example 1.15

The loop gain is 15 dB, the loop phase shift is −150°. From the Nichols chart, the closed-loop gain is 14 dB. The feedback is 15−1.4=13.6 dB. The feedback is negative.

Example 1.16

The loop gain is 1 dB, the loop phase shift is −150°. From the Nichols chart, the closed-loop gain is 6 dB. The feedback is 1−6=−5 dB. The feedback is positive.

Example 1.17

The loop gain is −10 dB, the loop phase shift is −170°. From the Nichols chart, the closed-loop gain is −7 dB. The feedback is −10−(−7)=−3 dB. The feedback is positive.

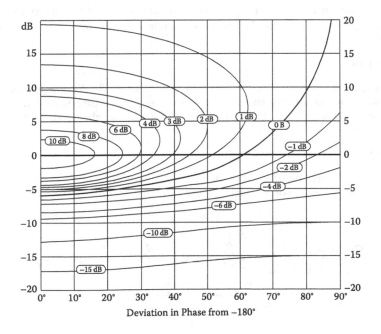

FIGURE 1.11
Nichols chart.

1.5 Disturbance Rejection

Disturbances are signals that enter the feedback system at the input or output of the plant or actuator, as shown in Figure 1.12, and cause undesirable signals at the system output. In the antenna pointing control system, disturbances might be due to wind, gravity, temperature changes, and imperfections in the motor, the gearing, and the driver. The disturbances can be characterized either by their time history or, in the frequency domain, by the disturbance *spectral density*.

The frequency response of the effect of a disturbance at the system's output can be calculated in the same way that the output frequency response to a command is calculated: it is the open-loop effect (D_1AP, for example) divided by the return difference F.

In Figure 1.12, three disturbance sources are shown. Because in linear systems the combined effect at the output of several different input signals is the sum of the effects of each separate signal, the disturbances produce the output effect:

$$\frac{D_1AP + D_2P + D_3}{F}$$

The effects of the disturbances on the output are reduced when the feedback is negative and increased when the feedback is positive. Disturbance rejection is the major purpose for using negative feedback in most control systems.

For some systems there is no real command and disturbance rejection is the only purpose of introducing feedback. Such systems are called *homing systems*. A typical example is a homing missile, which is designed to follow the target. No explicit command is given to the missile. Rather, the missile receives only an error signal, which is the deviation from the target. The feedback causes the vehicle's aerodynamic surfaces to reduce the error. This error can be considered a disturbance, and large feedback reduces the error effectively. Another popular type of system without an explicit command is active suspension, in which motors or solenoids are used to attenuate the vibration propagating from the base to the payload.

Example 1.18

The feedback in a temperature control loop of a chamber is 100. Without feedback, when the temperature outside the chamber changes, the temperature within the chamber changes by 6°. With feedback, the temperature within the chamber changes by only 0.06°.

Example 1.19

Gusty winds disturb the orientation of a radio telescope. The winds contain various frequency components, some slowly varying in time and others rapidly oscillating.

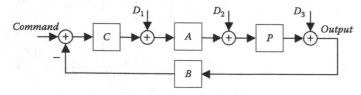

FIGURE 1.12
Disturbance sources in a feedback system.

The feedback in the antenna attitude control loop is 200 at very low frequencies, but drops with frequency (since motors cannot move the huge antenna rapidly), and at 0.1 Hz the feedback is only 5. The disturbance components are attenuated by the feedback accordingly, 200 times for the effect of steady wind, and 5 times for the 0.1 Hz gust components. Detailed calculations for a similar example will be given in the next section.

To further reduce the higher-frequency disturbances, an additional feedback loop might be introduced that would adjust the position not of the entire antenna dish, but of some smaller mirror in the optical path from the antenna dish to the front end of the receiver (or from the power amplifier of the transmitter).

1.6 Example of System Analysis

We proceed now with the analysis of the simplified antenna elevation control system that was shown in Figure 1.1. Assume that the elevation angle sensor function is 1 V/ rad, the feedback path coefficient is $B = 1$, the actuator transfer function (the ratio of the output torque to the input voltage) is $A = 5{,}000/(s+10)$ Nm/V, and the antenna is a rigid body with a moment of inertia $J = 5{,}000$ kgm^2. The plant's input variable is torque, and the output variable is the elevation angle; i.e., the plant is a double integrator with gain coefficient $1/J$. Since the Laplace transform of an integrator is $1/s$, the transfer function of the plant is:

$$P(s) = \frac{1}{(Js^2)}$$

As shown in the block diagram Figure 1.13, the torque τ applied to the antenna is the sum of the torque produced by the actuator and the disturbance wind torque, τ_w.

It is known that for large antennas, the wind torque can be modeled by the following noise shaping filter:

$$\frac{1}{(s+0.1)(s+2)}$$

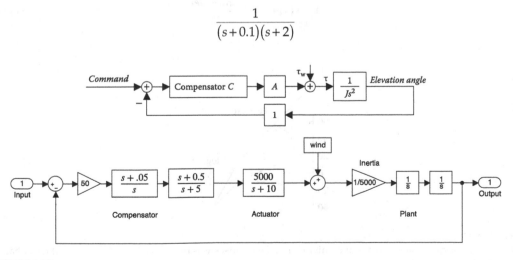

FIGURE 1.13
Elevation control system and Simulink block diagram.

The root mean square disturbance in the antenna elevation angle is therefore proportional to

$$\frac{1}{(s+0.1)(s+2)}\frac{1}{s^2}=\frac{1}{s^4+2.1s^3+0.2s^2} \tag{1.7}$$

The frequency response of the disturbance can be plotted using MATLAB with:

```
w = logspace(-2,1);
% frequency range
% 0.01 to 10 rad/sec
num = 1;
den = [1 2.1 0.2 0 0];
bode(num, den, w)
```

The plot is shown in Figure 1.14 but normalized to 100 dB at $\omega = 0.01$.

The spectral density is larger at lower frequencies. Large feedback must be introduced at these frequencies to reject the disturbance.

This simple compensator makes the system work reasonably well, although not optimally:

$$C(s) = \frac{50(s+0.05)(s+0.5)}{s(s+5)}$$

The loop transfer function is:

$$T(s) = CAP = \frac{50(s+0.05)(s+0.5)}{s(s+5)} \times \frac{5{,}000}{s+10} \times \frac{1}{5{,}000s^2}$$

i.e.,

$$T(s) = \frac{num}{den} = \frac{50(s+0.05)(s+0.5)}{s^3(s+5)(s+10)} = \frac{50s^2+27.5s+1.25}{s^5+15s^4+50s^3}$$

The return difference is:

$$F(s) = T(s)+1 = \frac{(num+den)}{den} \tag{1.8}$$

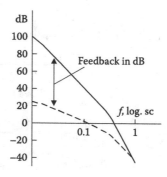

FIGURE 1.14
Elevation angle disturbance, in relative units: before the feedback was introduced, solid line; with the feedback, dashed line.

The closed-loop transfer function $M(s) = T/F = num/(num + den)$.

The plots of the gain and phase for the loop transfer function $T(j\omega)$, for $F(j\omega)$, and for $M(j\omega)$ can be made in MATLAB with:

```
w=logspace(-1,1);               % frequency range 0.1 to 10 rad/sec
den=[1 15 50 0 0 0];
num=[0 0 0 50 27.5 1.25];       % equal length of the vectors
g=num+den;                      % makes the addition allowable
bode(num, den, w)               % for T
hold on
bode(g, den, w)                 % for F
bode(num, g, w)                 % for M
hold off
```

The plots are shown in Figure 1.15a. The labels are placed with mouse and cursor, one at a time, using the MATLAB command gtext('label'). Alternatively, the plots can be made using the Linear Analysis tool. The feedback is large at low frequencies and is negative up 0.8 rad/s.

The closed-loop gain response 20 log $|M|$ is nearly flat up to 1.4 rad/s, i.e., up to 0.2 Hz. The gain peaks at 0.8 rad/s. The hump on the gain response does not exceed 6 dB, which satisfies the design rule mentioned in Section 1.4.3. More precise design methods will be studied in the following chapters.

The plot for disturbances in the system with feedback, the dashed line in Figure 1.14, is obtained by subtracting the feedback response (in dB) from the disturbance response, or directly by dividing Equation 1.7 by Equation 1.8. The disturbances are greatly reduced by the feedback.

The mean square of the output error is proportional to the integral of the spectral density with linear scales of the axes. The plots required to calculate the reduction in the mean square error can be generated with MATLAB, and the areas under the responses found graphically, or directly calculated using MATLAB functions.

The *L*-plane Nyquist diagram is shown in Figure 1.15b. The diagram avoids the critical point by significant margins: by 20 dB from below, by 40 dB from above, and by 42° from the right.

The compensated system has the required feedback for disturbance rejection, but the Nyquist diagram reaches −270° at lower frequencies, so, for large input signals, saturation of the actuator can lead to instability by effectively reducing the gain, and causing the Nyquist plot to slide down and encompass the critical point. To be stable under these conditions, the system needs a compensator with a nonlinear dynamic element, as will be shown in Sections 11.7–11.9.

The Linear Analysis tool can be used to obtain the Bode diagram of the (linearized) system with saturation, or the MATLAB function linmod can be used on the nonlinear model as shown in Figure 1.15d.

Example 1.20

If the compensator gain coefficient in the system with the Nyquist diagram shown in Figure 1.15a is increased five times, i.e., by 14 dB, the Nyquist diagram shifts up by 14 dB and the margin from below decreases from 20 dB to 6 dB.

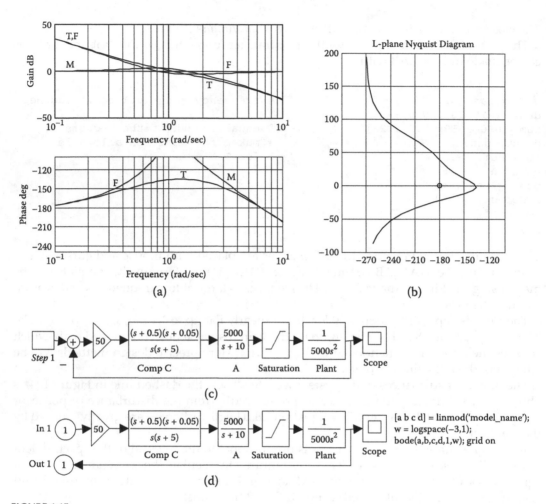

FIGURE 1.15
(a) Loop frequency response for the elevation control system: at lower frequencies, T and F overlap; at higher frequencies, T and M overlap; (b) L-plane Nyquist diagram; (c) Simulink for step response; (d) Simulink for Bode diagram.

If the loop gain is increased by (approximately) 20 dB, the Nyquist diagram shifts up by 20 dB, the return ratio becomes 1 at a certain frequency, and the closed-loop gain at this frequency therefore becomes infinite. As we already mentioned in Section 1.4.3, this is a condition for the system to become unstable. Similarly, if the loop gain is reduced by 40 dB, the Nyquist diagram shifts down, at some frequency the return ratio becomes 1, and the system becomes unstable. Using the Nyquist diagram for stability analysis will be discussed in detail in Chapter 3.

The transient response of the closed-loop system to a 1 radian *step command* (instantly increase the elevation angle by 1 radian) can be found using the MATLAB commands:

```
num = [0 0 0 50 27.5 1.25];
den = [1 15 50 0 0 0];
g = num + den; step(num, g)
grid
```

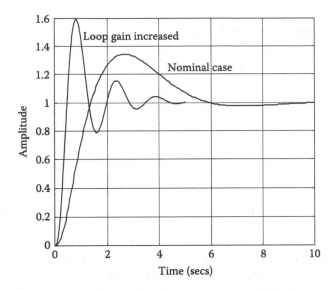

FIGURE 1.16
Output response to 1 radian step command.

Alternatively, the response can be found by hitting the "Run" button and then using the Simulation Data Inspector. The step response is shown in Figure 1.16 for the nominal gain case. The output doesn't rise instantly by 1 radian as would be ideal: it rises by 1 radian in less than 2 seconds, but then overshoots by 30%, then slightly undershoots, and settles to 1 radian with reasonable accuracy in about 10 seconds. This closed-loop transient response may be acceptable, but as mentioned above, global stability will require the introduction of a nonlinear dynamic element in the compensator, and this will improve the response for large signals.

Example 1.21

The effect of the Nyquist diagram passing closer to the critical point can be seen on the closed-loop transient response. The response generated with den = 0.2*den is shown Figure 1.16 by the curve labeled "Loop gain increased." (This was discussed in Example 1.20—the loop gain is increased by a factor of 5 which is 14 dB.) The response is faster, but it overshoots much more and is quite oscillatory.

1.7 Effect of Feedback on the Actuator Nondynamic Nonlinearity

Actuators can be relatively expensive, bulky, heavy, and power hungry. Economy requires that the actuator be as small as possible. However, these actuators will not be able to reproduce signals of relatively large amplitudes without distortion. We will consider these issues in Chapter 3 and later in Chapters 9–13.

Because the output power of any actuator is limited, *saturation* limits the amplitude of the output signal. The *hard* saturation in Figure 1.17a is shown by the dashed line. In many actuators, the saturation is *soft*, as shown by the solid curve.

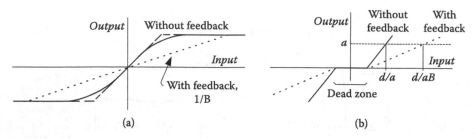

FIGURE 1.17
Input–output characteristic of the actuator with (a) soft saturation and (b) dead zone.

Both hard and soft saturations are *nondynamic*, i.e., they do not depend on the frequency of the input signal. When the amplitude of the input increases, the ratio of the output to the input decreases. Loosely speaking, the gain coefficient of the saturation link decreases. This will be studied in detail in Section 11.1. (Sections 11.8 and 11.9 also describes the use of nonlinear *dynamic* links to provide for global stability.)

Large feedback about the actuator changes the shape of the input–output characteristic. If the input signal level is such that the slope (differential gain) of the saturation curve is not yet too flattened out, then the differential feedback may remain large and the closed-loop differential transfer function can be quite close to B^{-1}. The closed-loop amplitude characteristic, shown by the dotted line, is therefore a segment of a nearly straight line. (The slope of the line is shallow since the feedback reduces the input–output differential gain.) Therefore, in a system with soft saturation, the input–output curve appears as hard saturation when the feedback is large.

The *dead-zone* characteristic is shown in Figure 1.17b. Large feedback reduces the differential input–output gain coefficient and therefore makes the input–output characteristic shallower, as shown by the dotted curve. Therefore, for any large amplitude a of the output signal, increasing the feedback will cause the input signal amplitude to change from d to $2d$. The value, d, that causes no response in the output, a, decreases. In other words, the feedback reduces the relative width of the dead zone. This feature allows the achievement of high resolution and linearity in control systems that use actuators and drivers with rather large dead zones (such actuators and drivers may be less expensive or consume less power from the power supply line, such as push-pull class B amplifiers or hydraulic spool valve amplifiers, briefly described in Section 7.1.3, or motors with mechanical gears).

Next, consider the output signal distortions caused by a small deviation of the actuator from linearity. In response to a sinusoidal input with frequency f, the output of the nonlinear forward path consists of a fundamental component with amplitude U_2 and additional Fourier components called *nonlinear products*. The ratio of the amplitude of a nonlinear product to the amplitude of the fundamental is the *nonlinear product coefficient*.

Consider one of these products having frequency nf and amplitude $U_{2(n)}$. If the forward path is approximately linear, the nonlinear product can be viewed as a disturbance source added to the output, as shown in Figure 1.18.

Now, compare two cases: (1) the system without feedback and (2) the system with feedback and with the input signal increased so that the output signal amplitude U_2 is preserved. In the second case, the disturbance, i.e., the amplitude of the nonlinear product

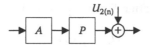

FIGURE 1.18
Equivalent representation of nonlinear distortions in the actuator.

at the system's output, is reduced by the value of the feedback. Therefore, the dynamic nonlinear product coefficient for a closed-loop system is reduced by the value of feedback at the frequency of the product.

Example 1.22

Without feedback, the nonlinear distortion coefficient in an audio amplifier is 5%. When feedback of 100 is introduced over the entire frequency band of interest, the coefficient becomes 0.05%.

Example 1.23

An amplifier is used to amplify signals of several TV channels. The feedback in the amplifier is very large at lower frequencies but drops to 5 at 200 MHz. A third harmonic of the 67 MHz signal produces an undesirable nonlinear product with frequency 200 MHz that falls within the band of a higher-frequency channel. The feedback reduces the amplitude of this product five times.

Low nonlinear distortion in the telecommunication feedback amplifiers invented by Harold Black at the Bell Laboratories in the 1920s made possible the development of multichannel telecommunication systems. The Bell Laboratories' scientists Harry Nyquist and Hendrik Bode established the basis for the frequency-domain design of feedback systems. Later, these methods were applied to other feedback control systems to maximize the accuracy and speed of operation.

1.8 Sensitivity

Sensitivity functions are generally used to quantify the undesirable effects on a transfer function when some parameters deviate from their nominal values. Sensitivity is not a transfer function; although, numerically it can happen to be equal to one. For control engineers, of particular interest is the sensitivity of the closed-loop transfer function to plant parameter variations.

An infinitesimal relative change in the plant transfer function dP/P causes an infinitesimal relative change $d\,(U_2/U_1)/(U_2/U_1)$ in the closed-loop system transfer function U_2/U_1. The feedback-system *sensitivity* is defined as the ratio of these changes:

$$S = \frac{d(U_2/U_1)/(U_2/U_1)}{dP/P} = \frac{d\left[(\log(U_2/U_1)\right]}{d(\log P)} \tag{1.9}$$

The smaller the sensitivity, the better. From Equation 1.3,

$$\frac{U_2}{U_1} = \frac{CAP}{CAPB+1}$$

then

$$S = \frac{P}{U_2/U_1} \frac{d(U_2/U_1)}{dP} = \frac{P(CAPB+1)}{CAP} \frac{\left[CA(CAPB+1) - C^2A^2PB\right]}{(CAPB+1)^2} = \frac{1}{(CAPB+1)}$$

Therefore, as was established by Harold Black,

$$S = \frac{1}{F} \tag{1.10}$$

The sensitivity is small when the feedback is large, and the feedback reduces small variations in the output variables, expressed either in percent or in logarithmic values, $|F|$ times.

Example 1.24

When the feedback is 10 and the plant magnitude $|P|$ changes by 10%, the closed-loop transfer function changes by only 1%.

Example 1.25

When the feedback is 10 and the plant magnitude $|P|$ changes by 3 dB, the closed-loop gain changes by 0.3 dB.

When the plant transfer function depends on some parameter q (temperature, pressure, power supply voltage, etc.), this dependency can be characterized by the sensitivity $S_{Pq} = [dP/P] / [dq/q]$. The chain rule can then be used to find the sensitivity of the closed-loop transfer function to q, i.e., the product of the sensitivities: $S \times S_{Pq}$.

Example 1.26

The feedback is 10, and the plant's sensitivity to one of its elements is 0.5. When the value of this element changes by 20%, the closed-loop transfer function changes by 1%.

Since the actuator and compensator transfer functions enter Equation 1.3 in the same way as P, similar formulas describe the effects of variations in the actuator and compensator transfer functions. Since the sensitivity of a large-feedback closed-loop response to the feedback path is nearly 1, the accuracy of the compensator and actuator is much less important than the accuracy of the feedback path.

Example 1.27

The feedback is 100, and the plant gain uncertainty is 3 dB. When the actuator and compensator implementation accuracy (together) is 0.2 dB, the total uncertainty of the forward path gain is 3.2 dB. Due to the feedback, the closed-loop uncertainty will be 0.032 dB.

If the feedback path uncertainty is, say, 0.01 dB, the total closed-loop response uncertainty is 0.042 dB.

1.9 Effect of Finite Plant Parameter Variations

Plant parameters often vary widely. For example, payloads in a temperature-controlled furnace can be quite different. The inertial properties of a rocket change significantly while the propellant is being used up. The mass of a cart whose position or velocity must be controlled depends on what is placed in the cart. The moment of inertia of an antenna about one axis might depend on the angle of rotation about another axis.

The sensitivity analysis described in Section 1.8 is a convenient tool for calculation of closed-loop response error and provides sufficient accuracy when plant parameter variations are small. When the plant P deviates from the nominal plant P_0 by ΔP, which is more than 50% of P_0, the *Horowitz sensitivity* [26] can be used for better accuracy:

$$S_H = \frac{(\Delta U_2/U_{20})}{(\Delta P/P_0)}$$

This is a ratio of finite relative changes, where the plant perturbed value $P = P_0 + \Delta P$. With the input U_1 kept the same for both cases and U_{20} the nominal output, the perturbed output $U_2 = U_{20} + \Delta U_2$. It can be shown that the Horowitz sensitivity is the inverse of the feedback for the perturbed plant (see Problem 31):

$$S_H = \frac{1}{(CAPB)} = \frac{1}{F}$$

Again it is seen that large negative feedback renders the closed-loop transfer function insensitive to plant parameter variations.

Example 1.28

If the nominal value of the plant is $P_0 = 100$, the perturbed plant is $P = 200$, and the feedback is $F = 50$ (with this P), then the relative output variation is found to be:

$$\frac{\Delta U_2}{U_{20}} = \frac{\Delta P}{P_0} F = \frac{100}{100} \frac{1}{50} = 0.02$$

1.10 Automatic Signal Level Control

The block diagram of automatic volume control in an AM receiver is shown in Figure 1.19. The goal for the control system is to maintain the carrier level constant at the AM detector, in spite of variations of the signal strength at the antenna.

The antenna signal is amplified and applied to the multiplier M. The signal from the multiplier is further amplified 1,000 times and applied to the AM detector. The signal that appears at the output of the detector is the sum of the audio component and the very low-frequency component that is proportional to the carrier. The low-pass filter B with corner frequency 0.5 Hz removes the audio signal, and the amplitude of the carrier multiplied by B is then compared with the reference voltage.

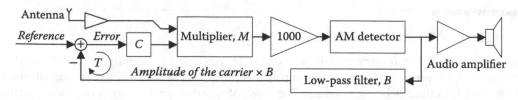

FIGURE 1.19
Automatic carrier level control.

The error signal processed by the compensator is applied to the second input of the multiplier, changing its transfer function for the RF signal. When the feedback is large, the error gets small and the output carrier level approximates *Reference / B*.

Example 1.29

If, for example, *Reference* = 0.5 V, *B* = 0.5, *C* = 100, and the RF carrier amplitude at the modulator input is 0.001 V, the return ratio is *T* = 50. The carrier amplitude at the detector equals *Reference* × *T*/[(*T* + 1)*B*] = 0.98 V, which is very close to the desired carrier amplitude *Reference / B* = 1 V.

Since RF signals can vary over a very large range, the gain in the feedback loop can change accordingly. The feedback must be large even when the RF signal is the smallest acceptable (when the signal level is only slightly larger than the level of the always-present noise); on the other hand, the system must perform well when the loop gain increases by an additional 60 to 80 dB due to the increase of the signal at the antenna.

1.11 Lead and *PID* Compensators

Compensator design will be discussed in Chapters 3–6. However, to make this chapter a short, self-contained course on servo design, we present below simple design rules for two of the most frequently used types of compensators. We assume ω_b has been already chosen (it must be, typically, at least 10 times lower than the frequency of the plant structural mode, and the sensor imperfection and noise at ω_b must not be excessive).

A *lead compensator* $C(s) = k(s + a) / (s + b)$ is often used when the plant transfer function is close to a double integrator. The zero is placed at $a \approx 0.3\omega_b$ and the pole at $b \approx 3\omega_b$. The coefficient k is adjusted for the loop gain to be 0 dB at ω_b, $k \approx b / [\omega_b \times Plant(\omega_b)]$. Lead compensators are used in Problems 42–44.

A *PID compensator* transfer function $C(s) = P + I / s + Dqs / (s + q)$ is tuned by adjusting the three real coefficients: *P* (proportional), *I* (integral), and *D* (derivative). The proportional coefficient $P \approx 1 / Plant(\omega_b)$. The integral coefficient $I \leq 0.2P\omega_b$. (If $I = 0$, the compensator becomes simply a lead.) The derivative coefficient *D* is 0 for a single-integrator plant and approximately $0.3P/\omega_b$ for a double-integrator plant. The pole at $q \approx 4\omega_b$ provides a realizable differentiation over the required bandwidth.

The compensator parameters are fine-tuned either experimentally or using a mathematical plant model and plotting computer-simulated open-loop and closed-loop frequency responses and the closed-loop step response. The loop phase shift at ω_b (i.e., arg $T(\omega_b)$) must be kept between $-120°$ and $-150°$ (as shown in Figure 1.15). The compensator hardware and software implementation is described in Chapters 5 and 6.

A *TID* controller (Section 6.6.2) is better than *PID* but remains suboptimal relative to a Nyquist/Bode design, which is the best choice when the loop gain uncertainty is modest (up to ±6 dB).

Larger loop gain variation, say by ±15 dB, which typically occurs at higher frequencies, does not permit the implementation of stable linear feedback, nor can linear feedback compensate for transmission line delay. In these cases a nonlinear multi-loop system may improve the feedback bandwidth by 1.3 octaves, as described in Chapter 14.

1.12 Conclusion and a Look Ahead

The material presented in this chapter enables some analysis of single-loop linear control systems. The reader can probably even design some control systems and also demonstrate that feedback vastly improves the system performance. The examples in Appendix 13 should start to become interesting and comprehensible. Still, in order to design systems with performance close to the best possible, the following topics need to be mastered:

Addition of prefilters and feedforward paths to improve the closed-loop responses.

Analysis and design of multi-loop control systems, since, for example, in a real antenna attitude control system there also exists a feedback loop stabilizing the motor rate that is "nested" within the main loop.

Design of feedback systems with maximum available accuracy, since the feedback is limited, as we will learn, by some fundamental laws.

Implementation of controllers with analog and digital technology.

Building of mathematical models of plants and control systems to evaluate system performance.

Analysis of the effects of the links' nonlinearities, and the design of nonlinear controllers. (The controller for the system having the Bode and Nyquist diagrams shown in Figure 1.15 must include nonlinear elements to ensure system stability when the actuator becomes overloaded.)

PROBLEMS

1. Using proper names for the actuators, plants, and sensors, draw block diagrams for feedback systems controlling the following:
   a. Temperature in a chamber
   b. Pressure in a chamber
   c. Angular velocity of a rotating machine element
   d. Luminescence of an illuminated surface

   e. Frequency of an oscillator

   f. Pitch, yaw, and roll of an airplane

2. When a control system is being designed, conversions from numbers to dB and back need to be performed fast. To do this, one must memorize the following table:

Decibel	0	1	3	6	10	20	20n
Number	1	1.12	1.4	2	3.16	10	10^n

Then, say, what is 30 dB? 30 dB $= 10$ dB $+ 20$ dB, i.e., $31.6 \approx 30$ times.

Or, what is 54 dB? 54 dB $= 60$ dB $- 6$ dB, i.e., $1{,}000/2 = 500$ times.

Convert to dB:

  a. 100

  b. 200

  c. 3,000

  d. 5

  e. 8

Convert to numbers:

  f. 110 dB

  g. 63 dB

  h. 12 dB

  i. 18 dB

  j. 30 dB

3. T is equal to (a) 0.01, (b) −0.01, (c) 0.1, (d) −0.1, (e) 2.72, (f) −0.9, (g) 10, (h) −10, (i) $1.5 \angle 150°$.

    Calculate F and M, and conclude whether the feedback is positive or negative, large or negligible.

4. For $T = 99$ and $B =$ (a) 0.01, (b) 0.1, (c) 1, (d) 0.05, (e) 2.72, and (f) 3, calculate the closed-loop transfer functions, and find the error E and the command U_1 to make the output $U_2 = 10$.

5. The open-loop gain coefficient is 3,000, the closed-loop gain coefficient is (a) 100, (b) 200, (c) 3,000, (d) 5, and (e) 2.72. What are the feedback, return difference, and return ratio? Is this a case of positive or negative feedback?

6. The open-loop gain coefficient is 5,000, and the closed-loop gain coefficient is (a) 100, (b) 200, (c) 3,000, (d) 5, (e) 2.72. Is the feedback large? How much will the closed-loop gain coefficient change when, because of changes in the plant, the open-loop gain coefficient becomes 6,000?

7. In an antenna elevation control system, the feedback is large, the antenna moment of inertia is 430 kgm², the three-phase 10 kW motor with gear ratio 1,200:1 is painted green, and the angle sensor gain coefficient $B = 0.1$ V/degree. What must be the command for the elevation angle to be the following:

  a. 30.5°

  b. 15.5°

    c. $3.5°$

    d. $1.5°$

    e. $2.72°$

    f. 30 mrad

8. Use the MATLAB commands `conv` or `zp2tf` to calculate the coefficients of the numerator and the denominator polynomials for the functions having the following coefficient k, zeros, and poles:

    a. $k=10$; zeros: $-1, -3, -8$; poles: $-4, -35, -100, -200$

    b. $k=20$; zeros: $-3, -3, -9$; poles: $-4, -65, -100, -400$

    c. $k=13$; zeros: $-1, -5, -8$; poles: $-6, -35, -100, -600$

    d. $k=25$; zeros: $-3, -7, -9$; poles: $-5, -65, -300, -400$

    e. $k=2.72$; zeros: $-1, -1, -2.72$; poles: $-1, -10, -100, -100$

    f. $k=20$; zeros: $-3, -4, -12$; poles: $-4, -165, -150, -500$

    g. $k=1,300$; zeros: $-1, -50, -80$; poles: $-6, -35, -300, -500$

    h. $k=150$; zeros: $-3, -70, -90$; poles: $-5, -150, -300, -400$

9. Use the MATLAB command `root` or `tf2zp` to calculate the poles and zeros of the following functions:

    a. $(20s^2+30s+40)/(2s^4+10s^3+100s^2+900)$

    b. $(s^2+3s+4)/(s^4+2s^3+20s^2+300)$

    c. $(10s^2+10s+40)/(2s^4+20s^3+100s^2+2,000)$

    d. $(s^2+20s+200)/(s^4+5s^3+50s^2+300)$

    e. $(2.72s^2+27.2s+200)/(s^4+2.72s^3+50s^2+272)$

    f. $(s^2+10s+8)/(s^4+2s^3+12s^2+150)$

10. Use the MATLAB command `bode` to plot the frequency response for the first-order, second-order, and third-order functions:

    a. $10/(s+10)$

    b. $100/(s+10)^2$

    c. $1,000/(s+10)^3$

Describe the correlation between the slope of the gain frequency response and the phase shift.

  Plot the step time response. Describe the correlation between the slope of the gain response at higher frequencies and the curvature of the time response at small times.

11. Use MATLAB to plot the frequency response and step time response for the first- and second-order functions:

    a. $10/(s+10)$

    b. $100/(s^2+4s+100)$

    c. $100/(s^2+2s+100)$

    d. $100/(s^2+s+100)$

Describe the correlation between the step time responses and the shapes of the frequency responses.

Find the denominator roots. Describe the correlation between the denominator polynomial roots, the shapes of the frequency responses, and the step responses.

12. Use MATLAB to convert the function to a ratio of polynomials and plot the frequency response for the function:

   a. $50(s+3)(s+12)/[(s+30)(s+55)(s+100)(s+1,000)]$

   b. $60(s+3)(s+16)/[(s+33)(s+75)(s+200)(s+2,000)]$

   c. $10(s+2)(s+22)/[(s+40)(s+65)(s+150)]$

   d. $-20(s+2)(s+26)/[(s+43)(s+85)(s+250)(s+2,500)]$

   e. $2.72(s+7)(s+20)/[(s+10)(s+100)(s+1,000)]$

   f. $-25(s+2)(s+44)/[(s+55)(s+66)(s+77)(s+8,800)]$

13. For Example 2 in Section 1.12, find $F(s)$ and $M(s)$, and plot their frequency responses using MATLAB.

14. When $|T|$ monotonically decreases in relation to frequency (as in most feedback control systems), what happens to F and M?

15. With the return ratio equal to the function in Problem 12, and $B = 1$, plot the closed-loop frequency responses for (a) to (f).

16. With the return ratio equal to the function in Problem 12, and $B = 4$, plot the closed-loop frequency responses for (a) to (f).

17. Considering functions in Problem 12 as the return ratio, and $B = 10$, plot the closed-loop frequency responses and the return ratio responses for (a) to (f).

18. Plot Nyquist diagrams on the L-plane for the following functions:

   a. $(20s^2+30s+40)/(2s^4+s^3+s^2+3)$

   b. $(s^2+30s+4)/(s^4+2s^3+2s^2+3)$

   c. $(10s^2+10s+40)/(2s^4+2s^3+s^2+3)$

   d. $(s^2+20s+5)/(s^4+5s^3+s^2+3)$

19. What feedback needs to be introduced in an amplifier with harmonic coefficient 5%, for the resulting harmonic coefficient to be 0.02%?

20. Find the third harmonic coefficient in a feedback amplifier with an open-loop harmonic coefficient of 5%, if the feedback at the frequency of the fundamental is (a) 100, (b) 200, (c) 300, (d) 400, or (e) 272, and the return ratio is inversely proportional to the frequency.

21. Before introduction of feedback, the actuator dead zone was (a) 5 N and (b) 200 mrad. What is the dead zone after the feedback of 50 was introduced?

22. Before introduction of feedback, the maximum actuator output signal was 100 m/s. What is this parameter after the feedback of 30 was introduced?

23. The open-loop gain is (a) 80 dB, (b) 100 dB, and (c) 120 dB. The closed-loop gain is 20 dB. Because of plant parameter variations, the open-loop gain is reduced by 1 dB. What is the change in the closed-loop gain?

24. The feedback is 80 dB. Plant gain is uncertain within ±1.5 dB. What is the uncertainty in the closed-loop gain?

25. In an amplifier, differential gain variations constitute 0.1 dB. What will these variations be when 40 dB of feedback is introduced?

26. The open-loop gain coefficient is (a) 1,000, (b) 2,000, and (c) 50,000. The closed-loop gain coefficient is 20. Because of plant parameter variations, the open-loop gain is reduced by 5%. What is the change in the closed-loop gain?

27. The feedback is (a) 200, (b) 100, (c) 1, and (d) 0.5. Plant gain is uncertain within ±15%. What is the uncertainty in the closed-loop gain coefficient?

28. The feedback is (a) 10, (b) 100, (c) 1, and (d) 0.5. The plant's sensitivity to the temperature is 0.1 dB/degree. When the temperature changes by −6°, by how much will the output change?

29. Tubes were expensive in the early days of tube radio receivers, and positive (regenerative) feedback was used to increase the amplifier gain. The positive feedback also improved the selectivity of the regenerative receiver (but narrowed the bandwidth of the received signal).

    Illustrate the above using the following example. The forward path consists of a resonance contour tuned at the signal frequency 100 kHz and an amplifier. The forward path transfer function is $10,000/(s^2 + 12.5s + 628^2)$, where the angular velocity is expressed in krad/s. The feedback path coefficient $B = -0.0077$.

    Plot the gain response of the receiver without and with regenerative feedback. By how much does the feedback increase the receiver gain? What is the sensitivity at the frequency of the resonance? How do small deviations in the amplifier's gain affect the output signal? What needs to be done in the feedback path to keep the closed-loop gain constant?

    What, in your opinion, will happen when the amplifier gain coefficient increases by 2% but the feedback path is not adjusted? By 2.5%?

30. In the previous example, introduce negative feedback (degenerative feedback) with $B = 0.001$ and then $B = 0.01$. Explain the effect of the feedback on the gain and the selectivity.

31. Prove that Horowitz sensitivity equals $1/F$.

32. The nominal value of the plant is 30, and the perturbed value is 50. The perturbed value of the feedback is 20. With the nominal plant, the value of the output would be 10. Using Horowitz sensitivity, find the change in the value of the output when the plant changes from nominal to perturbed.

33. The perturbed value of the plant is 30, the nominal value is 50. The nominal value of the feedback is 20. With the perturbed plant, the value of the output is 10. Using Horowitz sensitivity, find the change in the value of the output when the plant changes from nominal to perturbed.

34. Derive the expression for the Bode sensitivity for the following links: P, A, C. Give numerical examples. What is the required implementation accuracy for these links?

    For example, the plant sensitivity (Equation 1.10) can be derived by using Equation 1.3 while keeping U_1 constant:

$$S = \frac{dU_2/U_2}{dP/P} = \frac{PdU_2}{U_2dP} = \frac{P}{\dfrac{CAPBU_1}{1+CAPB}} \frac{d}{dP} \frac{U_1CAPB}{1+CAPB} = \frac{1}{F}$$

Therefore, the plant implementation accuracy can be $|F|$ times worse than the required accuracy for the closed-loop system transfer function.

35. Derive an expression for the Bode sensitivity of the output to variations in the transfer function of the feedback path link B. Give numerical examples, including cases where the feedback is large (negative), positive, and negligible.

36. From the frequency hodograph of T plotted on the Nichols chart in Figure 1.20, find the closed-loop gain frequency response of a tracking system (assuming $B=1$) and plot this frequency response with the logarithmic frequency scale.

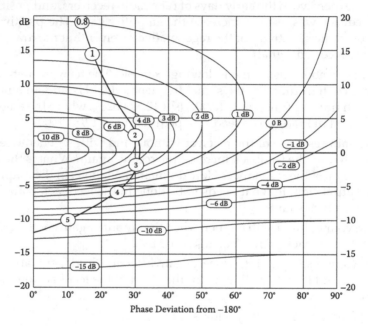

FIGURE 1.20
Locus of T on the Nichols chart.

37. Prove that if T is plotted upside-down on the Nichols chart, the curvilinear coordinates give $20 \log |F|$.

38. A periodic disturbance with frequency 0.05 Hz and amplitude 2 mrad in pointing of a spacecraft is caused by magnetometer boom oscillation. Find the value of feedback in the pointing control loop at this frequency required for reducing the disturbance to less than:

a. 0.1 mrad

b. 34 μrad

c. 12 nrad

d. 0.2 mrad

e. 2.72 arcsec

39. Consider the following noise shaping filters: (a) $k/[(s+1)s]$, (b) $k/[(s+2)s]$, (c) $k/[(s+5)s]$, (d) $k/[(s+10)s]$, and (e) $k/[(s+2.72)s]$.

The return ratio is $1{,}000(s+20)/[(s+1)s]$. Plot the relative noise with and without the feedback.

40. For the voltage regulator depicted in Figure 1.4, derive the dependence of the feed-back and the output voltage on the load resistor R_L. By comparison with the known formula for the voltage at a source terminal: $V = \text{emf} \times R_L / (R_L + R_S)$, find the equivalent output impedance R_s of the voltage regulator.

41. Consider a current regulator with the schematic diagram shown in Figure 1.21. Here, B is the current sensing resistor, i.e., the system has current feedback at the output. Therefore, the output current in the load R_L is stabilized by the feedback, and the regulator output represents a current source. Hence, the output impedance of the regulator must be high.

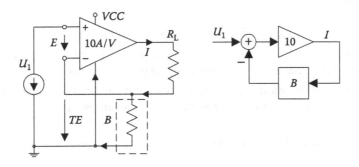

FIGURE 1.21
Current regulator.

The reference voltage is 0.5 V. The output impedance of the amplifier is much larger than $R_L + B$. Set B for the output current to be:

a.   0.1 A

b.   0.25 A

c.   0.5 A

d.   0.4 A

e.   2.72 A

f.   0.66 A

In each case find the return ratio and the output impedance of the regulator.

42. A temperature control loop is shown in Figure 1.22. The dimensionality of the signals is shown at the block joints. The heater is a voltage-controlled power source with the gain coefficient 2 kW/V. The transfer function (in °C/kW) of the loaded furnace was measured and the experimental response was approximated by transfer function $P(s) = 500/[(s+0.1)(s+25)]$. (At dc, when $s = 0$, $P(0) = 200°C/kW$.) The compensator transfer function $C = 8(s+1.6)/(s+0.2)$. The thermometer transfer function is 0.01 V/°C. The command is 4 V.

   a.   Find the return ratio and the feedback at dc (in a stationary regime).

   Find the loop transfer function and input–output transfer function. Plot Bode diagrams for the loop transfer function and for the input–output transfer function. Plot the L-plane Nyquist diagram. Check whether the diagram enters the area surrounded by the 6 dB line on the Nichols diagram.

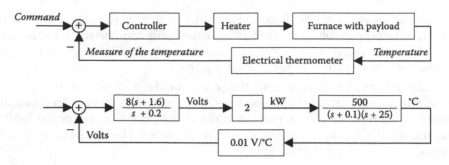

FIGURE 1.22
Temperature regulator.

Plot the output time response for the 0.01 V step command.

Plot the output time response to the step disturbance −100 W applied to the input of the furnace that represents some undesired cooling effects.

b.   Create a Simulink model. Include in the heater model a saturation with threshold 2 kW. Plot output responses to command steps of different amplitudes. Explain the results.

43. A scanning mirror with an angular velocity drive is shown in Figure 1.23.

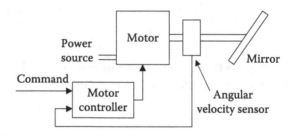

FIGURE 1.23
Mirror drive.

The mirror angular velocity control loop is shown in Figure 1.24. The controller transfer function is $C(s) = 10(s + 30)/[(s + 300)s]$. The motor (with driver) is a voltage-controlled velocity source with transfer function 300 (rad/s)/V (the actuator might have an internal angular velocity control loop). The mirror angular velocity differs from the angular velocity of the motor because of flexibility of the motor shaft and the mirror inertia. The plant transfer function (the ratio of the mirror angular velocity to the motor angular velocity) is $P(s) = 640,000/(s^2 + 160s + 640,000)$. At dc, the plant transfer function is 1. The gain coefficient of the angular velocity sensor is 0.01 V/(rad/s).

a.   Find the loop transfer function and input–output transfer function. Plot Bode diagrams for the loop transfer function and for the input–output transfer function. Plot an *L*-plane Nyquist diagram. Determine whether the diagram enters the area surrounded by the 6 dB line on the Nichols chart.

b.   Plot the output time response for the 1 V step command.

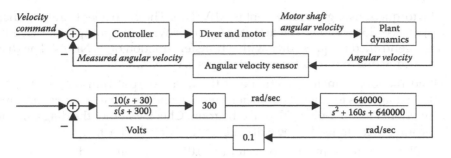

FIGURE 1.24
Angular velocity control.

Plot the output time response to step disturbance—0.1 rad/s applied to the input of the plant that represents the inaccuracy of the dc permanent magnet motor caused by switching between the stator windings.

44. A pulley with a torque drive is shown in Figure 1.25. The goal is to maintain the prescribed profile of the torque of the pulley so that the force in the cable lifting a load will be as desired. The motor torque control loop is shown in Figure 1.26. The plant dynamics, i.e., the transfer function from torque to the angle, is out of the feedback loop since the torque sensor is measuring the pulley torque directly at the motor.

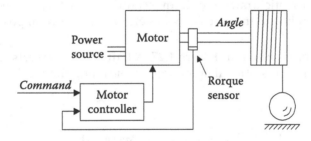

FIGURE 1.25
Torque regulation in a pulley.

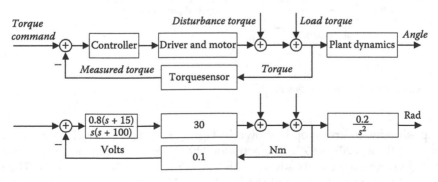

FIGURE 1.26
Torque control.

The torque sensor gain coefficient is 0.1 V/Nm. The controller transfer function is $C(s) = 0.8(s+15)/(s+100)$. The actuator (a motor with an appropriate driver) is a voltage-controlled torque source with a transfer function of 30 Nm/V. The disturbance torque due to the motor imperfections is 0.1 Nm.

a. Find the loop transfer function and the input–output transfer function. Plot Bode diagrams for the loop transfer function and for the input–output transfer function. Plot an *L*-plane Nyquist diagram. Check whether the diagram enters the area surrounded by the 6 dB line on the Nichols chart.

    Plot the torque time response for the 0.01 V step command.

    Plot the torque time response to the disturbance torque as a step function.

b. Assuming that the pulley radius is 0.1 m and the load mass is 2 kg, i.e., the plant moment of inertia is 0.02 kg×m², i.e., the plant is a double integrator and its transfer function is $50/s^2$, plot the position time history in response to a step torque command (assume that at zero time the load is on the ground).

c. Create a Simulink model. Include in the driver model a saturation with threshold 2 kW. Plot output responses to command steps of different amplitude. Explain the results.

d. Using Simulink, plot the position time history in response to the following command: 1 V for the duration of 2 s, 0 V for the duration of the next 2 s, and −1 V for the duration of the next 2 s.

e. Make a schematic drawing of torque control in a drilling rig.

f. Make a schematic drawing of torque control in a lathe, for keeping constant the force applied to the cutting tool (there could be several possible kinematic schemes).

45. An *x*-positioner is shown in Figure 1.27. A ball screw converts the motor rotational motion to translational motion of a table along the *x*-axis, with gear ratio 0.5 mm/rad.

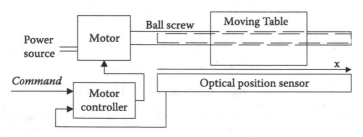

FIGURE 1.27
A positioner with an optical position sensor.

The positioner control loop is shown in Figure 1.28. The controller transfer function $C(s) = 0.03(s+20)/(s+70)$. The actuator (the driver with the motor) is a voltage-controlled angular velocity source with transfer function 500 (rad/s) / V. The plant transfer function is the ratio of the position to the velocity, i.e., $(1/s)0.5$. The gain

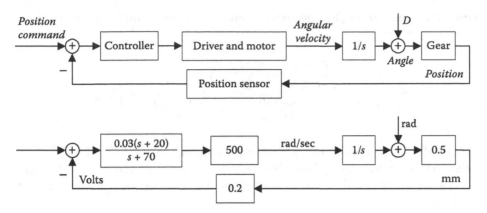

FIGURE 1.28
Position control.

coefficient of the optical position sensor is 0.2 V/mm. The disturbance in the angle represents the ball screw imperfections.

a. Find the loop transfer function and input–output transfer function. Plot Bode diagrams for the loop transfer function and for the input–output transfer function. Plot an *L*-plane Nyquist diagram. Check whether the diagram enters the area surrounded by the 6 dB line on the Nichols diagram.

  Plot the output time response for the 0.01 V step command.

  Plot the output time response to step disturbance −0.1 mm.

  If the gear has a dead zone of 0.02 mm, what is the resulting dead zone of the entire closed-loop system?

b. Create a Simulink model. Include a saturation with threshold 3,000 rpm in the driver and motor model. Plot output responses to command steps of different amplitude. Explain the results.

46. The signal at an antenna of an AM receiver varies 100 times, from station to station and due to changing reception conditions. The linearized (differential) loop gain of automatic level control at the largest signal level is 1,000. What is the range of the output signal carrier variations?

47. The dependence of M on T and many other formulas of this chapter are linear fractional functions, i.e. ratios of linear functions. It is known that a linear fractional function maps a circle (or a straight line) from the complex plane of the variable onto a circle (or a straight line) in the complex plane of the function. Why then are the coordinate curves in the Nichols chart not circles?

48. What is the reason for using voltage followers?

49. A common electric heating pad has thermal control. What would happen if it were left under the blankets? What would happen if there were no thermal control?

ANSWERS TO SELECTED PROBLEMS

1a. Heater, furnace with payload, thermometer.

3b. $F = 0.99$, $M = -0.01$, the feedback is positive but negligible (i.e., $|F| \approx 1$).

21a. 0.1 N

31. Given the values P and P_0 for the nominal and perturbed plant transfer functions, from Equation 1.9:

$$S_H = \frac{P_0}{P - P_0} \frac{\left[\dfrac{CAPB}{1+CAPB} - \dfrac{CAP_0B}{1+CAP_0B} \right]}{\dfrac{CAP_0B}{1+CAP_0B}} = \frac{1}{1+CAPB} = \frac{1}{F}$$

Notice the peculiarity of this formula: the changes in the left side (in S_H) are relative to the *nominal* plant value, while the feedback in the right side is calculated for the *perturbed* plant value. The opposite is also true, since the nominal and perturbed values can be swapped.

48. Due to the feedback, the voltage follower has high-input and low-output impedances, and it relieves the preceding signal source from its voltage being reduced when the load is connected.

49. The temperature will be kept safely low by reducing the consumed power. Without the thermal control, the consumed power will be nearly constant, and the temperature under the blanket may reach unsafe levels.

2

Feedforward, Multi-Loop, and MIMO Systems

The command feedforward scheme is equivalent to using a prefilter, or using a feedback path with a specified transfer function. All of these methods enable one to obtain the desired input–output closed-loop transfer function with any arbitrary compensator transfer function. Therefore, the compensator transfer function can be chosen as required to maximize the disturbance rejection, i.e., the feedback, while the desired closed-loop transfer function is obtained by using an appropriate feedback path, a command feedforward, or a prefilter.

While the command feedforward scheme and its equivalents do not affect disturbance rejection and plant sensitivity, the error feedforward scheme and Black's feedforward increase disturbance rejection and reduce plant sensitivity.

Multi-loop feedback systems are defined, following Bode, as those having a nonlinear (saturation) element in each loop. This definition reflects the importance of taking into account the large uncertainty in the signal transmission of the saturation link caused by the signal amplitude changes.

Major practical types of multi-loop systems are studied: local and common loops, nested loops, crossed loops, and the main/vernier loop configuration.

The methods for equivalent transformation of block diagrams are described.

The chapter ends with an introduction of multi-input multi-output (MIMO) systems, coupling, and decoupling matrices in the forward and the feedback path. Typical MIMO systems are discussed, with examples.

Example 2.1 and Sections 2.3 and 2.4 can be omitted from an introductory control course.

2.1 Command Feedforward

The accuracy of the system transfer function can be improved not only by the feedback but also by feeding certain signals forward and combining the signals at the load, or by some combination of feedback and *feedforward*. There are several feedforward schemes, each having decisive advantages for specific applications; the schemes are briefly described below.

The first type is referred to as *command feedforward* and is shown in Figure 2.1. The transfer function of the command feedforward path is:

$$(AP_0)^{-1}$$

where P_0 is the *nominal plant transfer function*.

The input–output transfer function of the system with command feedforward is found by summing the transmission functions of parallel paths of the input-to-output signal propagation. In each path, the feedback reduces the transmission F times:

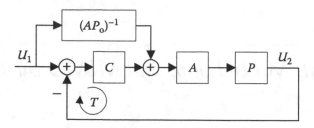

FIGURE 2.1
Command feedforward.

$$\frac{U_2}{U_1} = \left(\frac{1}{AP_o}AP\right)\frac{1}{F} + \frac{T}{F} = \frac{\dfrac{P}{P_o}+T}{F} = \frac{1+T_o^{-1}}{1+T^{-1}}$$

where $T_o = CAP_o$. The transfer function approaches 1 in two cases: (1) when $C \to \infty$ and both T^{-1} and T_o^{-1} vanish, and (2) when the plant transfer function does not deviate much from the known nominal plant transfer function P_o, i.e., $P_o \approx P$, and therefore $T_o \approx T$. For example, if $C = 0$ (open-loop case), it follows directly from the block diagram (and certainly from the formula) that the input–output transfer function is $(P_o/P) \approx 1$.

The input–output transfer function can also be expressed as:

$$\frac{T(T_0+1)}{T_0(T+1)} = \frac{M}{M_0}$$

where $M_o = T_o/(T_o+1)$. If the deviations of the plant P from the nominal value P_o are small, then $M \approx M_o$ and the system transfer function approaches 1, even when the feedback is not large.

The command feedforward can substantially improve the accuracy of the output response to the command, especially over the frequency band where the feedback cannot be made large. There are three limitations, however, on the use of command feedforward:

1. The plant must be known pretty well. If the uncertainty in the plant transfer function is large, then P_o cannot be made close to P, and the advantages of the command feedforward decrease drastically.

2. The power of the numerator polynomial of the feedforward path transfer function should be smaller than the power of the denominator, for the transfer function to be feasible.

   There is also a limit on the bandwidth: at higher frequencies $|P|$ typically decreases; however, making $|P_0|$ too small, and therefore making the feedforward path gain too large, would produce an excessively large signal at the input of the actuator, and the actuator would become saturated.

3. A plant with substantial pure delay would imply using feedforward with substantial phase advance, which is not feasible.

Command feedforward does not change the feedback about the plant, the sensitivity to the plant parameter variations, or the disturbance rejection.

Example 2.1

Biological, robotic, and many other engineering systems must perform well in both slow and fast modes of operation. In the slow mode, high accuracy must be available, as when surgery is performed. The same actuators (muscles, motors) must also be able to act fast and provide high acceleration for the plant, as when hitting a ball or jumping. The accuracy of the fast components of the motion must be reasonably good but need not be as good as the accuracy of the slow mode action.

High-accuracy slow motion can be achieved with closed-loop control using eyes or other position sensors. The feedback configuration, however, doesn't suit the fast-action mode since the speed of the feedback action is bounded by the delays in the feedback loop. To make the actuators move the plant with their maximum speed and acceleration, while maintaining a reasonable accuracy, the actuators must be commanded directly.

The block diagram of a feedback/feedforward system that satisfies these conflicting requirements is shown in Figure 2.2. The *commander* transforms the general input command into commands for the individual actuators. The actuators are equipped with wide-band (fast) feedback loops using local sensors (*inner loops*) that make them accurate. The plant output variables (average speed, direction of motion, etc.) are controlled by the *outer feedback loops*.

At low frequencies (i.e., for slow motion), the compensators' gains are high and the system operates closed-loop. At higher frequencies (for fast motion), the compensators' gains roll down and the command feedforward paths become dominant. For each of the controlled output variables, the system input–output transfer function

$$\frac{1 + T_o^{-1}}{1 + T^{-1}}$$

approximates 1 over the entire frequency range of operation, although much more accurately at lower frequencies where the feedback is large. The transition in accuracy from the slow to the fast mode is gradual, since the gain of the compensator gradually decreases with frequency.

When the motion comprises both slow and fast components, the outer feedback loops control the slow components (as when adjusting the general direction of running), while for the fast components, only the inner loops are closed.

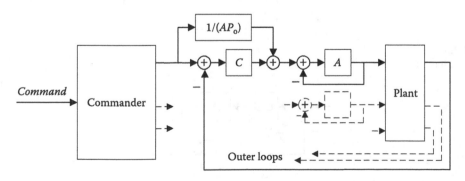

FIGURE 2.2
Feedforward system with fast actuator loops and slow outer loops.

2.2 Prefilter and the Feedback Path Equivalent

The block with transfer function R in the command path preceding the *command summer* (same as the feedback summer) in Figure 2.3 is called the *prefilter*. This system has return ratio T and closed-loop response $RT/(T+1)$. The system in Figure 2.4 has return ratio T and closed-loop response $B^{-1}T/(T+1)$. These two systems are equivalent if $B = 1/R$.

If $R = 1/B = 1/M_o$, the systems in Figures 2.3 and 2.4 have the closed-loop transfer function M/M_o, the same as that for the system in Figure 2.1. Therefore, all three systems are potentially equivalent. A system having all three links—feedforward path, prefilter, and feedback path—can always be equivalently transformed to any one of the systems in Figures 2.1, 2.3, and 2.4.

The three equivalent methods modify the input–output transfer function and can make it closer to the desired, compared with the system in Figure 1.1(b), but do not change the feedback or the sensitivity.

The system design is performed in two stages: (1) the compensator C is defined, and then (2) R, or B, or the feedforward path $(AP_o)^{-1}$, is defined.

The compensator transfer function must be chosen so as to maximize the feedback over the bandwidth of interest—as will be shown in Chapters 4 and 5. Once designed, the compensator should not be compromised during the next stage of the design, which is the implementation of a suitable nominal closed-loop response. This goal can be achieved by a proper choice of B, or R, or the feedforward path.

Since the feedback reduces the effects of the compensator parameter variations, these links need not be precise. The tolerances in the prefilter and the feedback link directly contribute to the output error, so these links do need to be precise. The required accuracy of the feedforward path implementation depends on the accuracy of the knowledge of the plant transfer function, and may be different at different frequencies.

It might seem attractive to integrate the prefilter, the compensator, and the feedback link into a generalized linear subsystem that can be designed using some universal performance index. However, this is not recommended since the sensitivities and the accuracies with which the blocks should be implemented are quite different; the design of these blocks is to a large extent independent, and it is much easier to design these blocks one at a time.

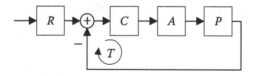

FIGURE 2.3
System with prefilter.

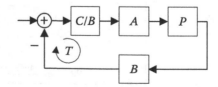

FIGURE 2.4
System with feedback path.

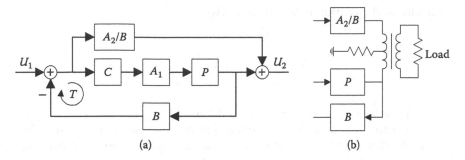

FIGURE 2.5
Error feedforward.

2.3 Error Feedforward

Figure 2.5(a) describes an entirely different scheme known as *error feedforward*.
The input–output transfer function is:

$$\frac{U_2}{U_1} = \frac{CA_1P}{1+CA_1PB} + \frac{A_2/B}{1+CA_1PB} \tag{2.1}$$

and if $A_2 = 1$, then $U_2 = U_1/B$. The sensitivity of the system input–output transfer function to the plant parameter variations (and to variations in C and A_1) can be calculated as $(1 - A_2)/F$. If A_2 is made close to 1, the sensitivity approaches zero.

Practical applications of this method of sensitivity reduction are restricted by the difficulties in the design of the output summer. For an electrical amplifier system with a known load, the output summer can be made using a bridge-type signal combiner as shown in Figure 2.5(b); the bridge prevents the output signal from the upper path from going into the feedback path. It is more difficult to implement a mechanical system with such properties.

2.4 Black's Feedforward

Finally, consider *Black's feedforward* method for sensitivity reduction, which was invented by Harold Black, around the same time he invented the feedback method. Figure 2.6(a) depicts the method. Note that no feedback appears in this system. The upper signal path is the main one, and the lower one is the error compensation path. The error signal is the difference between the command U_1 and the output of the main path, measured via the B-path. The error is amplified by the error path and added to the system's output, so as to compensate for the initial error in the main path.

The input–output transfer function is:

$$A_M + A_E - A_M A_E B$$

If either A_M or A_E, or both, equal $1/B$, the input–output transfer function is $1/B$.

The sensitivity of the output to variations in A_M is:

$$S = \frac{\dfrac{dU_2}{U_2}}{\dfrac{dA_M}{A_M}} = \frac{1 - A_E B}{1 + \dfrac{A_E}{A_M} - A_E B} \tag{2.2}$$

(A derivation of the formula is requested in Problem 11.)

If $B \approx 1/A_M$, the error signal is the difference between two nearly equal signals, and the error is small. In this case the actuator A_E can be low power. Such an actuator can be made very precise. When the gain coefficient of this actuator is $A_E = 1/B$, the sensitivity (Equation 2.2) becomes zero. In this case, the output effect of the disturbance D is also zero.

In order to make A_M and A_E each equal to $1/B$, and preserve these conditions in spite of variations in these actuators' parameters, both actuators are commonly stabilized by internal feedback or by some adaptive automatic gain adjustment. Also, in the case where both of the transfer functions are $1/B$, as in Figure 2.6(b), the feedforward scheme provides redundancy: if one of the actuators fails, the remaining one takes the full load and the input–output gain remains unchanged.

In some high-frequency physical systems, the links A_M, B, and A_E incorporate substantial delays: τ_M, τ_B, and τ_E, respectively, as shown in Figure 2.7. These delays do not prevent the use of feedforward if they are properly compensated by insertion of delay link $\tau_M + \tau_B$ in the signal path to the first summer, and delay link $\tau_B + \tau_E$ at the output of the main channel. Then, the phase difference between the signals reaching the summers remains the same, and the only difference in the resulting input–output transfer function of the system is the extra delay $\tau_M + \tau_B + \tau_E$.

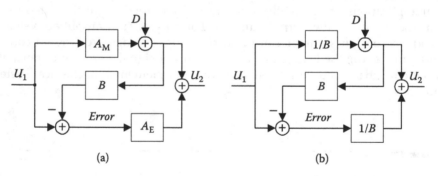

(a)   (b)

FIGURE 2.6
Black's feedforward system: (a) general and (b) ideal case.

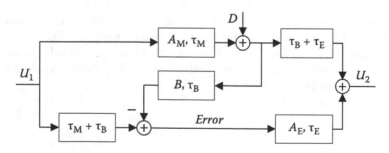

FIGURE 2.7
Black's feedforward with delay compensation.

This method is often employed in low-distortion amplifiers for telecommunication systems, for signals with frequencies from hundreds of MHz to tens of GHz, but it is not common in control systems. There may be applications to control systems when extreme accuracy is required.

2.5 Multi-Loop Feedback Systems

Linear systems can always be transformed to another configuration with a different number of loops, as illustrated in Figure 2.8.

In accordance with Bode's definition of physical multi-loop systems, only the loops that comprise *nonlinear* saturation-type elements are counted. For example, in Figure 2.9, system a is a single-loop system, and system b is a three-loop system. This definition is related to the problem of stability analysis of practical systems whose actuators are always nonlinear. Such systems will be studied in Chapters 4–13.

In this chapter, we will analyze feedback systems in only the linear state of operation, i.e., for small-amplitude signals, with saturation links equivalently replaced by unity links.

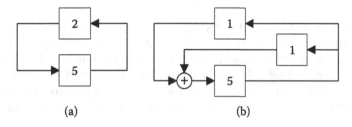

(a)   (b)

FIGURE 2.8
Modifications of a linear system.

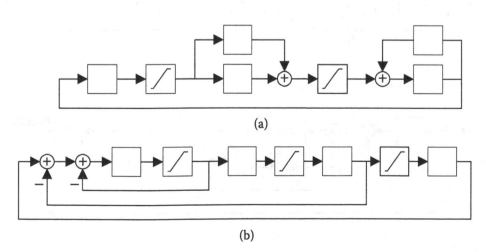

(a)

(b)

FIGURE 2.9
Single-loop (a) and three-loop (b) feedback systems.

2.6 Local, Common, and Nested Loops

Figure 2.10 depicts *local feedback loops*. The sensitivities to each link's parameter variations depend on the feedback in the local loop. The total gain of the chain of these links is reduced by the product of all these feedback values. This gain reduction effect is much larger in this arrangement than in the *common loop arrangement*, shown in Figure 2.10(b).

If each of the three links with nominal gain coefficient k has the same tolerances, then to provide the same degree of accuracy, we must *enloop* each stage by the same value of feedback. In this case the closed-loop gain $k^3/(T+1)$ of the single-loop system is much higher than the closed-loop gain $k^3/(T+1)^3$ of the system with local loops.

This is why common loop feedback is preferred in amplification techniques where the resulting gain is important. For control systems, this consideration is not important since the gain can be increased by adding inexpensive gain blocks, but this effect does need to be taken into account when designing analog compensators.

Example 2.2

Each stage of an amplifier has a gain coefficient of 50. The feedback about each stage needs to be at least 10 to make the gain coefficient stable in time. Then, when local feedback is used, the total amplifier gain coefficient will be 125. When the common loop configuration is used, the gain coefficient will be much higher, 12,500.

Local loops are often made about links with large parameter variations. These loops can be *nested* as shown in Figure 2.11. Here, the driver amplifier is enlooped by large feedback to make its gain accurate and stable in time, and also to manipulate the output impedance of the driver. The actuator loop makes the actuator-plus-driver subsystem accurate and stable in time. The outer loop improves the accuracy of all links in the forward path, including the plant. The nested loop arrangement is employed for several reasons, to be

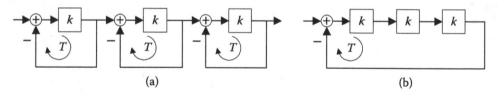

(a)   (b)

FIGURE 2.10
Local loops (a) and the common loop arrangement (b).

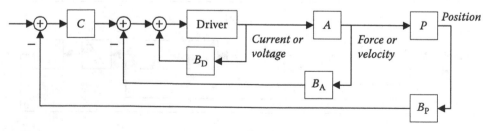

FIGURE 2.11
Nested feedback loops.

discussed in detail in Chapter 7, primarily because the feedback bandwidth in the outer loop cannot be made arbitrarily large. The wide-band inner loop is about the electrical amplifier (driver), the intermediate bandwidth loop is about the actuator (motor), and the rather narrow bandwidth outer loop is about the plant.

For example, consider the typical case of a driver implemented with an operational amplifier. The op-amps without feedback have very large gain uncertainty and variations due to power supply voltage and temperature changes. An easily implemented, large local feedback loop about the driver amplifiers will reduce the tolerances of the forward path to only those of the compensator, plant, and actuator. If op-amps are used in the compensator, they must also have large local feedback.

The variables that are fed back in the inner loops can be different: at the output of the driver, the variable can be the voltage or the current, and at the output of the actuator, the velocity (rate) or the force. The choice of these variables alters the plant transfer function P, which is the ratio of the output to the input variables of the plant. For example, in a position control system, force feedback about the actuator makes a rigid body plant a double integrator, while rate feedback makes the plant a single integrator. When the actuator is an electromagnetic motor, rate feedback about the motor is typically accompanied by voltage feedback about the driver. (These issues will be studied in more detail in Chapters 4–7.)

2.7 Crossed Loops and Main/Vernier Loops

Crossed feedback loops are shown in Figure 2.12. Such loops are often formed by parasitic coupling. Crossed dc feedback loops are frequently used in bias stabilization circuitry in amplifiers, as in the amplifier illustrated in Figure 2.12(b).

For high control accuracy over a large dynamic range, the actuator must be fast and powerful. If such an actuator is not available, and if large changes in the output variable need not be fast, then an arrangement of two complementary actuators can be employed: the *main actuator* and the *vernier actuator*, which is orders of magnitude faster but also orders of magnitude less powerful. Figure 2.13(a) and (b) show two equivalent block diagrams for the *main/vernier* loop arrangement.

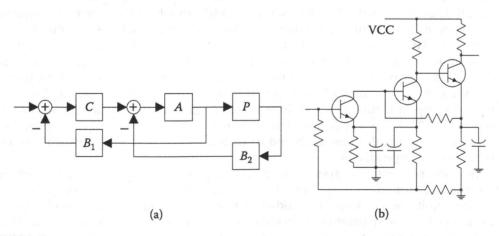

(a)   (b)

FIGURE 2.12
(a) Crossed feedback loops and (b) crossed dc loops in an amplifier.

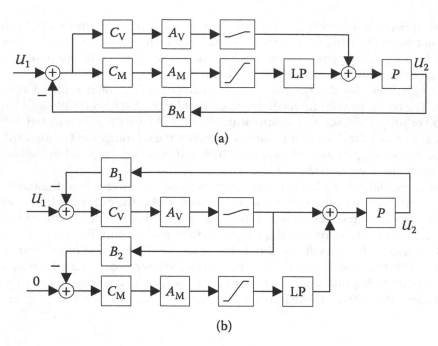

(a)

(b)

FIGURE 2.13
Feedback systems with the main and vernier loops.

The main actuator provides most of the action applied to the plant (force, voltage, etc.). However, due to its large inertia, which is represented in the block diagram by the low-pass link LP, the main actuator cannot render fast signal components. These components, smaller in amplitude but rapidly changing, are provided by the vernier actuator.

From the diagram in Figure 2.13(a), it is apparent what the actuators are doing, but it is rather difficult to figure out how to design the compensators in the main and vernier channels. For this purpose, the diagram is modified as shown in Figure 2.13(b).

Now, it is clear how the system operates. The command is given to the vernier loop summer, and the vernier actuator tries to reduce the error rapidly. However, when the error is large, the vernier actuator becomes saturated and cannot compensate high-frequency disturbances in the system. This situation is corrected—the vernier is de-saturated—by the main loop. The output signal of the vernier actuator is applied to the feedback summer of the main loop. The command for the main loop is zero since the desired value of the vernier actuator output for slowly varying signal components is zero. (Therefore, there is no physical command summer in the main loop in Figure 2.13(a); the command summer in the main loop in Figure 2.13(b) is shown only to simplify the explanation of how the system works.) The slow but powerful main actuator unloads the vernier actuator from slow but large-amplitude commands and disturbances. Two examples of such a system are described in Appendix 13.

By the same principle, the system can be extended to a three-loop configuration, etc. Each extra loop provides an economical way to improve the control accuracy by a few orders of magnitude. The feedback bandwidth of each subsequent loop increases. Due to the difference in the loop bandwidth, loop coupling is rather easy to account for during the system stability analysis, both in linear and nonlinear modes of operation.

Example 2.3

In the orbiting stellar interferometer (a high-resolution optical instrument to be placed in orbit about the Earth), the lengths of the optical paths from the two primary mirrors to the summing point must be kept equal to each other. The optical path lengths are measured with laser interferometers, and must be adjusted with nanometer accuracy.

For the purpose of this adjustment, in one of the paths a variable delay is introduced by bouncing the light between additional mirrors. The position of one of these mirrors is regulated by three means.

The mirror is mounted on a piezoelectric actuator. The piezoelectric actuator can be controlled with nanometer accuracy, but its maximum displacement (stroke) is only 50 μm. The small platform bearing the piezoactuator is moved by a voice coil. (A voice coil is an electromechanical actuator based on a coil placed in a field of a permanent magnet; voice coils are widely employed in loudspeakers and hard disk drives where they position the reading/writing heads.) The accuracy of the voice coil control loop is lower since its feedback is limited by some mechanical structural resonances, but the actuator maximum stroke is much longer, 1 cm. The voice coil is placed on a cart that can be moved on wheels along a set of rails.

The voice coil is the vernier for the cart, and the piezoelement is the vernier for the voice coil. The voice coil desaturates the piezoactuator, and the cart desaturates the voice coil. The entire control system is able to adjust the optical path length rapidly and very accurately. (The control system is described in detail in Appendix 13.14.)

2.8 Block Diagram Manipulations and Transfer Function Calculations

Equivalence block diagram transformations facilitate the conversions of various configurations to standard ones for the purpose of analysis. For example, the diagram in Figure 2.14a can be transformed into diagrams b and c by changing the node from which the signal is taken while preserving the signal value at the branch output. In this transformation, the forward path transfer function and the feedback loop return ratio are preserved.

For the transfer function calculation, the following evident rules apply:

1. Transmission along a forward path is reduced by the value of the feedback in the loop that includes links in the path.

2. When there are several parallel forward paths, the total transfer function can be found by superposition of the signals propagating along the paths, i.e., summing the path transfer functions.

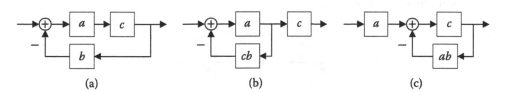

(a)   (b)   (c)

FIGURE 2.14
Feedback system equivalent transformations.

Example 2.4

The diagram in Figure 2.16 is obtained from the diagram in Figure 2.15 by equivalence transformations. The signals taken at different points are multiplied by additional blocks' coefficients so that the signals at the outputs of the branches remain the same.

The diagram in Figure 2.16 has *tangent loops*, i.e., loops with unity forward paths. According to Equation 1.3, each tangent loop reduces the signal transmission by the value of feedback in the loop.

There are two forward paths and two loops with return rations *bga* and *cdeh*, so that the transfer function is

$$\frac{af}{1+bga} + \frac{abcde}{(1+bga)(1+cdeh)} \tag{2.3}$$

Often, instead of the block diagram representations, systems are described by the signal flowchart exemplified in Figure 2.17.

Example 2.5

The gain coefficient of the graph shown in Figure 2.18 can be calculated as the sum of transmissions along two parallel paths, divided by the feedback in the tangent loop:

$$\frac{(3-2\times5)10}{1+5\times6} = -2.26$$

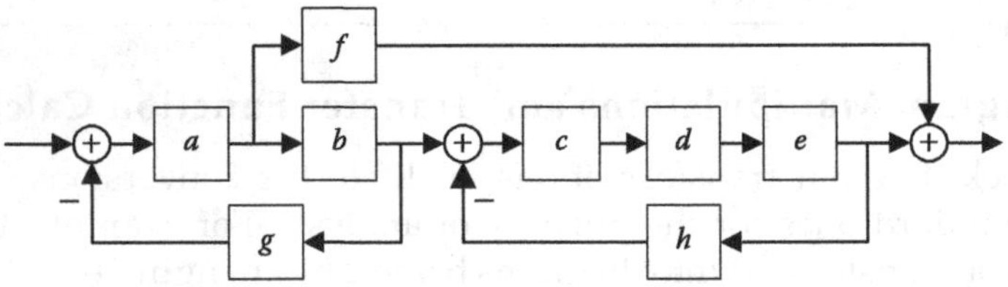

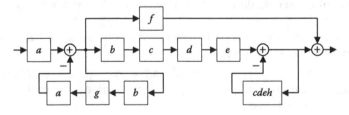

FIGURE 2.15
Block diagram of a feedback system.

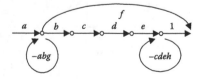

FIGURE 2.16
Feedback system with tangent loops.

FIGURE 2.17
System flowchart representation.

Next, consider nested loops. As shown in Figure 2.19, by converting all nested loops to the loops between the same nodes, we obtain several parallel loops. The equivalent single loop has the loop transfer function equal to the sum of all the nested loop transfer functions.

Thus, the third rule can be formulated:

3. When the loops are nested, the input–output transfer function is the forward path transmission divided by the sum of all loop return ratios and 1.

The three rules are sometimes referred to as *Mason's rules.*

Example 2.6

The transfer function for the system with nested loops shown in Figure 2.11 is:

$$\frac{CDAP}{DB_D + DAB_D + CDAPB_D + 1} \tag{2.4}$$

where D is the driver transfer function.

With block diagram manipulations, it is often possible to prove the equivalence of different control schemes. For example, the position command in a single-input single-output system is sometimes split into several paths to form position, velocity, and

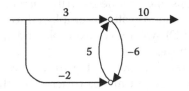

FIGURE 2.18
Flowchart.

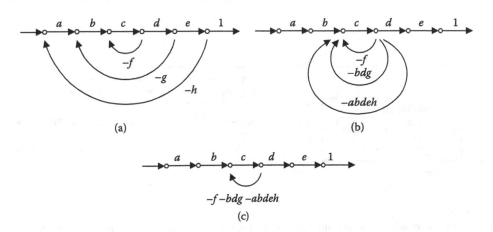

FIGURE 2.19
Transformation of (a) nested loops to (b) parallel loops and further to (c) a single equivalent loop.

acceleration commands, and these three signals are separately fed forward into three different summing points. The sensor output is often passed through a low-pass filter to attenuate the sensor noise (as well as some components of the signal), and then the filtered and unfiltered sensor signals are fed to different summing points. Linear filters are used to estimate the output position, velocity, and acceleration, and these signals are combined linearly to form the signal driving the actuator. Some block diagrams include linear time-invariant links named predictors, plant models, and estimators. If these block diagrams can be shown to be equivalent to the block diagrams in Figures 2.1, 2.3, and 2.4 (which is very often the case), then the achievable performance of these control schemes, whatever their name, is not better than that of the standard control system configurations. (In fact the performance could be inferior if block diagrams are chosen that inherently limit the order of the compensators, as in Example 8.1.) On the other hand, some of the potentially equivalent block diagrams may have certain advantages from an implementation point of view.

2.9 MIMO Feedback Systems

Multi-input multi-output systems have several command inputs, and several output variables are controlled simultaneously. For example, if the number of commands is two, and the number of outputs is three, this is a 2×3 system. The controlled variables could be, for example, angles of different bodies or angles in different dimensions of the same body.

The number of feedback *loops* does not necessarily correlate with the number of inputs and outputs. Very often, a multi-loop system is employed to improve the performance of a single-input single-output (SISO) system. For example, the systems shown in Figures 2.11 and 2.12 are multi-loop SISO systems.

An example of a MIMO system is shown in Figure 2.20, where different plant variables are regulated by separate loops. The transfer function of the plant for each loop is the plant transmission from the actuator output to the sensor input. The transfer function from the ith actuator output to the jth sensor input (i ≠ j) shown by the dashed line is called a *coupling transfer function*. If the coupling transfer functions are all zero, then the

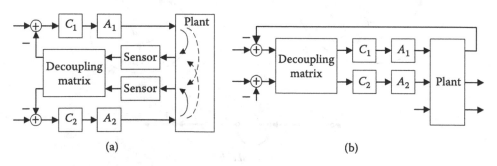

(a)   (b)

FIGURE 2.20
A 2×2 MIMO system with loops to control nearly independent variables: the decoupling matrix can be placed in (a) the feedback path or (b) in the forward path.

multi-loop system is just a set of individual single-loop systems. In many cases, coupling exists but is small.

Actuators are relatively expensive, so their number in engineering systems needs to be kept to a minimum. In general, only one actuator is assigned to do a specific job (Example 2.9 below offers exceptions): one actuator moves the plant in one direction, the second in another, etc. Or, in the case of an electrical signal generator, one actuator varies the signal frequency, the second one the signal amplitude, the third one the temperature of the quartz resonator, etc. Because of this, the actuator loops in the block diagram in Figure 2.20 are already to a large extent decoupled; i.e., the terms on the main diagonal of the plant matrix (from plant actuators to plant sensors) are substantially larger than the off-diagonal terms.

Coupling between loops can be compensated for by using a *decoupling matrix*, whose outputs only reflect the action of an appropriate actuator. The decoupling matrix makes the feedback loops independent of each other, simplifying the design and improving the system performance. The decoupling can be done in the feedback path by decoupling the sensor readings, or in the forward path by decoupling the signals going to actuators. Either method can make the loops independent of each other, but there is substantial difference between the methods: the matrix needs to be precise when placed in the feedback path, and can be less precise when placed in the forward path.

A decoupling matrix for linear plants can be found by inverting the matrix of known coupling transfer functions. If the coupling transfer functions do not contain pure delay, the decoupling matrix is causal and can be implemented with a digital or an analog computer. However, since the plant parameters are not known exactly, decoupling is never perfect.

The following types of multi-loop systems are most often encountered in practice: local actuator feedback, vernier-type control with actuators differing in speed and power, and nearly decoupled control where each of the actuators dominantly affects a specified output variable.

When fast action is of utmost importance, complex engineering and biological systems are typically arranged as an aggregation of several SISO mechanisms with large and relatively wide-band feedback in each loop and a complex precision *commander* producing commands to the mechanisms. When the action need not be very fast but the accuracy is of prime concern, additional slower common feedback loops are added to precisely control the output variables, as discussed in Example 2.1.

Example 2.7

The azimuth angle of the antenna, as well as the elevation, in Figure 1.1(b) might be regulated, and the result would be a two-input two-output system. The coupling between the elevation and the azimuth loops is typically small, and can be calculated and compensated, which results in practically decoupling the loops.

Example 2.8

Spacecraft attitude controllers are commonly arranged as three separate loops for rotating the spacecraft about the x-, y-, and z-axes. The spacecraft inertia matrix is not symmetrical about all the axes. Therefore, the transfer function about one axis depends on the rotation angle and velocities about the other axes, and the three controllers are coupled and cannot be considered as three separate SISO systems. Good decoupling can be achieved over most of the frequency bandwidth of interest where the spacecraft parameters are well known and the decoupling matrix transfer functions can be accurately calculated. However, over some frequency ranges, for example, at the slosh modes of the propellant in the fuel tanks, the spacecraft parameters have much larger uncertainty and the calculated decoupling matrix is not very accurate. The uncertain coupling necessitates a reduction in the feedback in the control loops, as will be discussed later, in Section 4.4.

Example 2.9

Multiple actuators of the same type can be used to achieve the appropriate power and balance. An example is the use of multiple power plants on jet transports, as shown in Figure 2.21. The output is one of the variables defining the airplane attitude and velocity (i.e., this block diagram shows only a part of the entire control system).

The arrangement provides redundancy, i.e., one-engine-out capability (OEOC). Special control modes (auto or manual) may be necessary to support this sort of operation. For this purpose, additional feedback loops using aerodynamic control surfaces are applied so that a single actuator can power the plant independently in the event that the other actuators fail. The system is a multi-loop MIMO system.

Example 2.10

In a TV set or VCR, there are several hundred feedback loops. More than 90% of the loops control electrical variables (currents, voltages), and some of the loops control image color and brightness, speed of the motors, and tension of the tape. The majority of the loops are analog, but some are digital, particularly those for tuning the receiver and for controlling the display. This, say, 300×300 MIMO feedback system is conventionally designed with frequency-domain methods, as if the loops were independent, i.e., as if the system were merely a combination of 300 SISO systems. The variables to be controlled are to a large extent independent; i.e., the diagonal terms are dominant in the 300×300 matrix. Only seldom is some primitive decoupling used in the forward path. The decoupling matrix is sometimes included in the feedback path to calculate the variables fed back from the sensors' readings.

The design of a MIMO controller as a combination of several independent loops has important advantages in terms of structural design. It simplifies the system testing and troubleshooting, improves reliability, and simplifies the work of modification and redesign. To meet these goals, most engineering devices are designed structurally, in spite of the mathematically attractive idea of combined optimization of the entire 300×300 multivariable system, which, ideally, must produce at least as good or better performance, but at the price of losing the advantages of the structural approach.

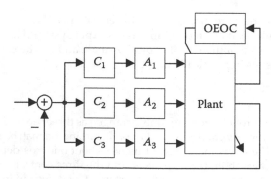

FIGURE 2.21
Several parallel power plants system.

PROBLEMS

1. For a tracking system ($B = 1$) with T equal to (a) 5, (b) 20, (c) −80, (d) 120, and (e) 2.72, find the value of the prefilter R that makes the closed-loop transfer function equal to 1.

2. Find the compensator and feedback path transfer coefficients for a system without a prefilter so that the system is equivalent to the system of Problem 1, with $AP = 10$ in both systems.

3. Find the compensator C and feedforward path gain coefficient FF for a system without a prefilter, with $B = 1$, and with $AP = 10$, so that the system is equivalent to the system of Problem 1.

4. Include link B in the feedback path in the block diagram depicted in Figure 2.1. Derive an expression for the input–output transfer function.

5. $C = 2$, $A = 1$. Plant gain coefficient P is uncertain within the 10 to 20 range, and nominal plant gain coefficient $P_o = 15$.

    Calculate the input–output gain coefficient ranges without and with a feedforward path. Does the feedforward affect the ratios of the maximum to the minimum input–output gain coefficient?

6. The loop gain coefficient is inversely proportional to the frequency. At what frequency ranges is the benefit of using feedforward most important?

7. Plant $P(s) = 10/(s + 10)$, the actuator model includes a linear gain block with gain coefficient $A = 10$, followed by a saturation link with threshold 1, and $C(s) = 0.3(s + 0.35)/(s + 3)$. The feedforward path transfer function is $0.1a(s + 10)/(s + a)$. The command is sinusoidal, with possible frequencies from 0 to 10 Hz.

    Plot the frequency responses with MATLAB. Choose coefficient a such that the signal amplitude at the input to the saturation block will not exceed the threshold. What is the bandwidth of the feedforward?

8. Same problem as the previous one, but the input signal is a step of 1 V. Make simulations with MATLAB, find a by trial and error.

9. Find the Bode sensitivity of transfer function $W_1 + W_2$ to W_1 if the following:

    a. $W_1 = 100$ and $W_2 = 2$
    b. $W_1 = 50$ and $W_2 = 50$
    c. $W_1 = -9$ and $W_2 = 10$
    d. $W_1 = 10/s$ and $W_2 = 100/s^2$

10. Find the Bode sensitivity to the transfer functions of the links P, A_V, and A_M in Figure 2.13. (*Hint*: Use the chain rule. First, employ Bode sensitivity for a single-loop system for the composite link, including the main and vernier channels; then, multiply the composite link Bode sensitivity by the sensitivity of the composite channel transfer function to variations in only one channel.) Give a numerical example.

11. Derive the expression for the sensitivity to variations in A_M of Black's feedforward system shown in Figure 2.6. Compare two cases:

    a. When $A_M = 10$ and the values of the rest of the links' gain coefficients are nominal, i.e., $A_E = 1$ and $B = 0.1$.
    b. When A_E deviates by 3 dB from the nominal value of 10.

12. In Figure 2.6, $A_E = 95$, $B = 0.01$. Find the sensitivity of the output to variations in A_M when A_M is the following:

   a. 100

   b. 105

   c. 150

13. For the conditions described in the previous problem, and considering the maximum output signal (saturation threshold) in the main amplifier to be $A_{MS} = 10$, find the maximum output signal in the error amplifier. What is the conclusion?

14. In Black's feedforward system shown in Figure 2.6, the error amplifier A_E has internal feedback F_E. Using the chain rule, find the sensitivity of the system's input–output transfer function to the error amplifier gain variations.

15. How many loops, according to Bode's definition, are in the systems diagrammed in Figure 2.22?

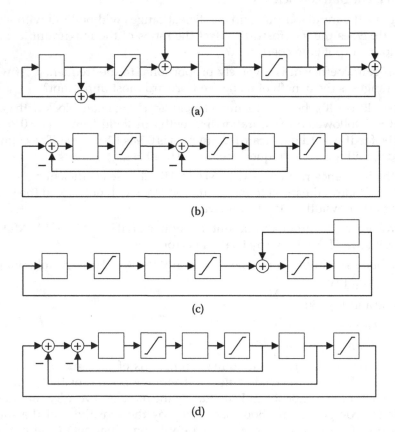

(a)

(b)

(c)

(d)

FIGURE 2.22
Feedback systems.

16. a. In Figure 2.11, the driver gain coefficient changes with temperature by ±30% from the nominal, the actuator changes by ±15% from the nominal, and the plant transfer function is uncertain within 2 dB. The loop gain in the driver

loop is 30 dB, and in the actuator loop (with the driver loop closed) is 10 dB. Find the total uncertainty in the plant loop gain.

b. Same problem, but the driver gain coefficient uncertainty is ±3 dB, the actuator uncertainty is ±2 dB, and the plant uncertainty is ±2 dB.

c. Same problem as in (b), but with a 40 dB loop gain in the driver loop and 20 dB in the actuator loop.

d. Same problem as in (b), but with a 20 dB loop gain in driver loop and 20 dB in the actuator loop.

17. Explain why a pair of actuators, one high power and sluggish, and one low power and very fast, should typically cost less than a single powerful and fast actuator.

18. Find the input–output transfer function for the system shown in Figure 2.10.

19. Find the input–output function of the system shown in Figure 2.15.

20. For the multi-loop feedback system described in Figure 2.11, find the input–output transfer function and the sensitivities to variations in driver, actuator, and plant.

21. Derive expressions for input–output transfer functions for the systems shown in Figure 2.23.

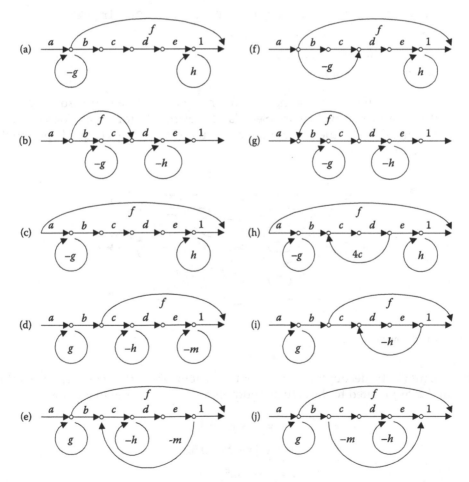

FIGURE 2.23
Flowcharts.

22. a. Calculate the decoupling matrix for the system where the sensor readings x', y', and z' are related to the actuator outputs x, y, and z by the following:

$$x' = 2x + 0.2y + 0.3z$$

$$y' = 0.1x + 2.1y + 0.1z$$

$$z' = 0.04x + 0.1y + 1.9z$$

Since the diagonal terms are dominant,

$$x \approx 0.5x', y \approx 0.5y', z \approx 0.5z'$$

A better approximation to x is found by substituting these first approximations to y and z into the first equation:

$$x \approx 0.5x' - 0.05y' - 0.075z'$$

Proceed with better approximations to y and z. Compare these expressions with the exact solution found with MATLAB script:

```
A = [2 0.2 0.3; 0.1 2.1 0.1; 0.04 0.1 1.9]; inv(A)
ans = 0.5038 -0.0443 -0.0772
-0.0235 0.4795 -0.0215
-0.0094 -0.0243 0.5291
```

Draw the flowchart for the solution in the form shown in Figure 2.24, and put the numerical values in. (See also Problem 6.10 for an analog computer implementation of the decoupling matrix using 6 op-amps (2 quad op-amp IC) and 18 resistors.)

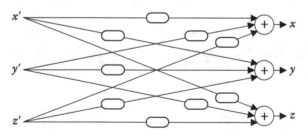

FIGURE 2.24
Decoupling matrix flowchart.

b. Calculate the decoupling matrix for the system where the sensor readings x', y', and z' are related to the actuator outputs x, y, and z by the following:

$$x' = 2x + y + 0.3z$$

$$y' = 0.1x + 2y + 0.5z$$

$$z' = 0.4x + 0.5y + 1.9z$$

by inverting the coefficient matrix with MATLAB.

c. Same as (b) for the following:

$$x' = 3x + 0.4y + 0.3z$$
$$y' = 0.3x + 2.1y + 0.2z$$
$$z' = 0.04x + 0.1y + 1.9z$$

d. Same as (b) for the following:

$$x' = 2x + 0.1y + 0.1z$$
$$y' = 0.1x + 3.1y + 0.1z$$
$$z' = 0.04x + 0.4y + 1.9z$$

e. Same as (b) for the following:

$$x' = x + y - z$$
$$y' = x - y + z$$
$$z' = -x + y + z$$

(This arrangement of three piezoactuators and three load cells has been used in the spacecraft vibration isolation system described in Section 6.4.2.)

23. The frequency of a quartz oscillator depends on the crystal temperature and on the power supply voltage (the voltage changes the capacitances of *pn*-junctions of the transistor that participate in the resonance contour). The temperature of the environment changes from 10°C to 70°C. The power supply voltage uncertainty range is from 5 V to 6 V.

The oscillator elements are placed in a small compartment (oven) equipped with an electrical heater and a temperature sensor. The temperature and the dc voltage are regulated by control loops. The thermal loop return ratio is 600. The dc voltage stabilizing loop return ratio is 200. The references are 70°C and 5 V, and the loops maintain the quartz temperature close to 70°C and the power supply voltage close to 5 V.

For the employed quartz crystal and the transistor, the dependencies of the frequency of oscillation on the crystal temperature and on the power supply voltage are well approximated in the neighborhood of the references by linear dependencies with coefficients of 10^{-4} Hz/°C and -10^{-3} Hz/V. The maximum disturbances in temperature and voltage are 60°C (when the environment temperature is 10°C) and 1 V (when the power supply voltage is 6 V). Figure 2.25 shows the flowchart for calculations of the effects of the disturbances.

The loops are coupled since the dc voltage also affects the power dissipated in the transistor and, consequently, the oven temperature with the rate 20°C/V.

The flowchart represents a two-input single-output system. No decoupling between the control loops is required since the coupling is small and one-directional, from the voltage to the temperature loop. (The effect of temperature on the voltage loop is negligibly small.)

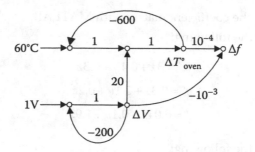

FIGURE 2.25
Flowcharts representing the effects of the voltage and temperature variations on variations of the oscillator frequency.

Calculate the total range of the frequency variations Δf due to the instability of the environment temperature and the power supply voltage.

24. Prove that, generally, when sensitivity is 0, redundancy is always provided. (*Hint*: Use the linear fractional transformation $W = (aw + b)/(cw + d)$ for the general dependence of a linear system transfer function W on a link transfer function w.)

3

Frequency Response Methods

Some requirements for control systems are typically expressed in the frequency domain (such as disturbance rejection), while some others are most often formulated in the time domain (such as rise time and overshoot). The latter need to be converted into frequency-domain specifications in order to use frequency-domain design methods. Formulations of the time-domain requirements are commonly very simple, and the equivalent frequency-domain formulations are also simple. The requirements can be translated between the domains with simple approximate relations.

Since most control systems are of the low-pass type, responses of standard low-pass filters are reviewed for future reference.

Typical closed-loop frequency responses for homing and tracking systems are considered.

The Nyquist stability criterion is derived and its applications reviewed. Stability margins are introduced and Nyquist stability and absolute stability are discussed.

The Nyquist–Bode criterion is developed for multi-loop systems' stability analysis with successive loop closure. Feedback systems with unstable plants are analyzed with the Nyquist criterion and the Nyquist–Bode criterion. The effect of saturation on the system stability is briefly discussed.

Static error reduction is considered for systems of different servomechanism types.

The notion of a minimum phase (m.p.) transfer function is introduced, and the Bode phase-gain relationship is presented. Its meaning and significance are clarified, and the procedure for calculating the phase from a given gain response is explained. Also, the problem of finding the Bode diagram from a given Nyquist diagram is considered.

Non-minimum phase lag is studied, and a criterion is derived for the parallel-path connection of two m.p. transfer functions to remain m.p.

The integral of the feedback in dB over the frequency axis is shown to be zero, and the implications for control system design are discussed. Also presented are the phase integral and the gain integral over finite bandwidth, both of which give specific insight into the evaluation of achievable performance.

Finally the resistance integral is applied to the evaluation of impedances.

When the book is used for a single-semester introductory control course, Section 3.14.4 can be bypassed.

3.1 Conversion of Time Domain Requirements to Frequency Domain

3.1.1 Approximate Relations

Signals can be represented by sums of their sinusoidal components, and in linear links, these signal components do not interfere (i.e., the superposition principle applies), so linear links are fully characterized by their frequency responses. The Laplace transform, with its complex argument $s = \sigma + j\omega$, is often used to make conversions between the time-domain and frequency-domain responses. For brevity, we will write $W(s)$ even when we only mean the frequency response $W(j\omega)$. We assume the reader is already familiar with using frequency responses. If not, Appendix 2 can be of some help.

Frequency responses are widely employed for characterizing links and design specifications. The feedback response required for disturbance rejection is commonly specified in the frequency domain since the disturbances are most often characterized by their spectral density, i.e., in frequency domain. High-order compensators and plants are also most often characterized by their frequency responses.

Time domain characterization, on the other hand, is commonly applied to systems that are required to transfer signals without distortions. A step function or a series of step functions is usually employed as the input test signal, and requirements are specified on the output.

Given a mathematical description of a linear system, conversion between the frequency and time responses and specifications is easily performed by computer. Analytical transformation between the time domain function and the Laplace transform expression can be obtained in MATLAB by functions `laplace` and `invlaplace`. The time and frequency responses can be plotted with standard MATLAB plotting commands (or with SPICE simulation). Yet, it is important to be able to make the approximate conversion mentally for the purposes of the creation and analysis of specifications to systems and subsystems, resolution of trade-offs, and comparison of available versions of conceptual design. This can be done using the simple rules described below.

The *3-dB bandwidth* is the bandwidth of a low-pass system up to the frequency where the gain coefficient decreases $\sqrt{2}$ times, i.e., by 3 dB. For the first-order low-pass transfer function $a/(s+a)$, the 3-dB bandwidth is the pole frequency

$$f_p = a / (2\pi)$$

as shown in Figure 3.1(a). The time response of such a link to a step function input is:

$$1 - \exp(-at)$$

(see Section A2.2). It is shown in Figure 3.1(b).

The line tangent to the time response at $t = 0$ is at. The time it takes the signal to rise to 0.9 is found from the equation:

$$1 - \exp(-at_r) = 0.9$$

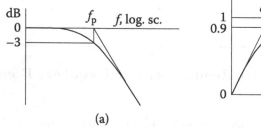
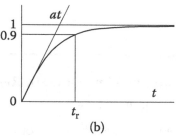

(a)   (b)

FIGURE 3.1

(a) Frequency response and (b) time response for first-order link $\dfrac{a}{s+a}$.

to be:

$$t_r \approx \frac{(1/f_p)}{3} \qquad (3.1)$$

In other words, rise time is approximately one-third of the period $1/f_p$ related to 3-dB bandwidth. This rule is employed for calculating the bandwidth required for the rise time not to be longer than prescribed.

Example 3.1

A telecommunication antenna, a 10-inch diameter dish, to be placed on a balloon flying in the Venus atmosphere, needs to be pointed to Earth with 0.5° accuracy. The rate of the attitude variations of the balloon can reach 5°/s. Therefore, the rise time of the antenna attitude control system must be smaller than 0.1 s, which translates into a 3 dB closed-loop bandwidth of at least 3 Hz or, approximately, the crossover frequency $f_b > 1.5$ Hz. These calculations of the required feedback bandwidth are sufficiently accurate for conceptual design, even though the closed-loop transfer function will not be first order, for which Equation 3.1 was derived, but higher order.

For higher-order low-pass transfer functions, the rise time is still roughly approximated by Equation 3.1, where f_p is understood to be the cutoff frequency of the frequency response. However, the transient response is more complicated, and the deviation of the output from the desired step function is commonly characterized by the five parameters shown in Figure 3.2: *delay time t_d, rise time t_r, settling time t_s* of settling within the dynamic error envelope, *overshoot*, and *steady-state (static) error*, all of them required to be small.

The Laplace transform initial and final value theorems relate the gain at lower frequencies to the time response at longer times, and the gain at higher frequencies to the time response at shorter times, as is indicated in Figure 3.3. More generally, for systems with relatively smooth gain responses, the gain over specific frequency intervals is related to the transient response at specific times. Numerically, according to Equation 3.1, the time response at 1 s is mostly affected by the gain at 0.3 Hz or, more realistically, by the gain over the 0.1 to 1 Hz frequency interval; the output at the time of 1 ms, by the gain over the 1–10 kHz interval, etc.

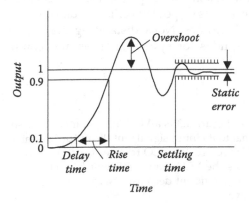

FIGURE 3.2
Time response to step function input.

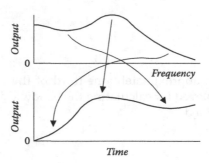

FIGURE 3.3
Relations between the time domain and frequency domain regions.

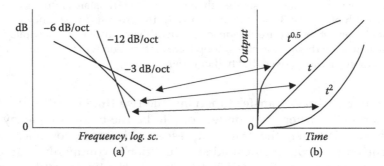

FIGURE 3.4
Correlation between the slope of gain frequency response and the curvature of the time domain step response.

Therefore, in Figure 3.2, the rise time corresponds to the operational bandwidth and the settling time corresponds to the lower-frequency gain. The static error corresponds to the dc gain. It is zero when the dc gain is 1.

An important correspondence also exists between the *slope* of the logarithmic gain frequency response (Bode diagram) and the *curvature* of the time response. For the gain responses with constant slopes shown in Figure 3.4(a), the Laplace transform gives the time responses shown in Figure 3.4(b). Particularly, the delay time increases with the high-frequency asymptotic slope of the gain Bode diagram. (At zero time, the first n time derivatives vanish for a system with a nth-order pole at high frequencies.)

From the gain response and its slope at specific frequencies, we can roughly reconstruct the time response at specific times. Notwithstanding the imprecision of these relations, they render very useful guides for system analysis and computer-aided iteration and tuning.

Example 3.2

The plant in the PLL in Figure 1.3 is a VCO. It is an integrator since the *frequency* ω of the VCO is proportional to its input signal, but the output variable applied to the phase detector is the *phase*. Therefore, the VCO transfer function is k/s, where k is some coefficient that characterizes the VCO gain coefficient k/ω. Thus, when ω increases twice (by an octave), the gain coefficient decreases twice (by 6 dB); i.e., the slope of the gain response is -6 dB/octave.

Plants that are double integrators k/s^2 and triple integrators k/s^3 have gain response slopes of -12 and -18 dB/octave, respectively.

Example 3.3

When the plant and the loop gain responses have a high-frequency asymptotic slope of −18 dB/octave, the closed-loop response also has this slope, since the loop gain vanishes there. Then, the closed-loop transient response at small times will be proportional to the third power of time.

Example 3.4

The frequency response for the transfer function

$$T(s) = \frac{9,000}{(s+30)(s+300)}$$

is plotted with MATLAB in Figure 3.5. The output time response to the step input for the same link is shown in Figure 3.5(b). We can trace the correspondences shown in Figures 3.3 and 3.4 on these responses.

At high frequencies, the transfer function behaves like $9,000/s^2$; i.e., it turns into a double integrator. The slope of the gain response becomes −12 dB/octave which is equivalent to −40 dB/decade. (Note that each decade contains $\log_2 10 \approx 2.3$ octaves. The octaves in Figure 3.5 are, for example, from 10 to 20, from 20 to 40, and from 30 to 60; each octave has the same width on the logarithmic frequency scale.)

We see some correlation between the slope of the gain response and the phase response: when the slope is zero, the phase is zero; when the slope of the gain response approaches −40 dB/decade, the phase approaches −180°.

3.1.2 Filters

Since most feedback control systems are of the low-pass type, their responses can be better understood from their similarity to the responses of the standard low-pass filters.

Low-pass filters are most often employed for attenuating high-frequency noise and disturbances outside of the filter passband. When the shape of the signals must be preserved, then:

1. The filter gain must be nearly the same for all important frequency components of the signal.

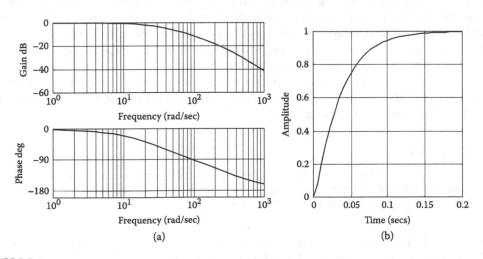

(a) (b)

FIGURE 3.5
(a) Frequency domain and (b) time domain responses for T(s) = 9,000/[(s + 30)(s + 300)].

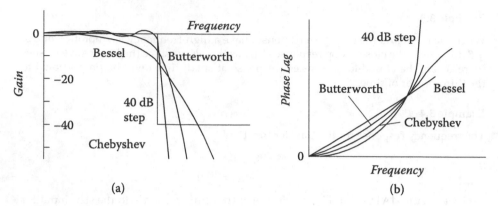

FIGURE 3.6
Frequency responses of the (a) gain and (b) phase for the step 40 dB low-pass filter and for Chebyshev, Butterworth, and Bessel filters.

2. The dependence of the filter phase shift on frequency must be close to linear; i.e., the slope of this dependence, which is the group time delay, must be the same for all these components.

The curvature of the phase response causes different delays for sinusoidal components of different frequencies and has a profound effect on the overshoot. At some moment, various signal components that are not in phase at the input nearly all come in phase at the output and cause the overshoot.

Figure 3.6(a) and (b) show the gain and phase responses of several low-pass filters. The phase response of an ideal filter with 40 dB selectivity is extensively curved (it follows the response in Figure 3.34, which is the weight function of the Bode integral, to be studied in Section 3.9).

The gain response of the *Chebyshev* (equiripple) filter bends sharply at the corner frequency, and its phase response is also significantly curved.

The *Butterworth* filter has a *maximum flat* gain frequency response; i.e., the first n derivatives of the gain response of the nth-order filter equal zero at zero frequency. The filter has less selectivity than the Chebyshev filter, and its phase response is less curved.

The higher the order of a Chebyshev or Butterworth filter, the sharper the gain response's selectivity is, and the phase shift response is more curved.

The curvature of the phase responses manifests itself in the overshoots, as shown in Figure 3.7. The higher the order of a Chebyshev or Butterworth filter, the higher the overshoot is and the longer the settling time.

Example 3.5

For the third-order low-pass Butterworth filter with normalized bandwidth 1 rad/s, the overshoot is 8% and the settling time to 10^{-5} accuracy is 25 s; for the eighth-order filter, the overshoot is 16% and the settling time is 60 s.

The phase shift of the *Bessel filter* (or *Thompson*, or *linear-phase filter*) is approximately proportional to frequency. The higher the order of a Bessel filter, the better the phase response

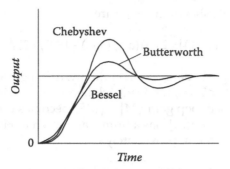

FIGURE 3.7
Time step responses for the filters.

linearity, and the smaller are the overshoot, the rise time, and the settling time. The filter transfer function is:

$$B(s) = \frac{b_o}{b_n s^n + b_{n-1} s^{n-1} + \dots b_k s^k + \dots + b_o}$$

where $b_k = (2n - k)! \ / \ [2^{n-k} (n - k)! \ k!)]$.

Example 3.6

For the settling error of 10^{-5}, the settling time for the first-order normalized Bessel filter is 11.5 s, for the third-order filter, 8 s, and for the eighth-order filter, 4 s. The overshoot of the eighth-order filter is 0.35%.

Random variations of ±1% of the denominator coefficients of the eighth-order Bessel filter transfer function increase the settling time up to 5 s (i.e., by 1 s).

Example 3.7

The transfer functions of the second- to fourth-order Bessel filters are the following:

$$\frac{3}{s^2 + 3s + 3}, \frac{15}{s^3 + 6s^2 + 15s + 15}, \frac{105}{s^4 + 10s^3 + 45s^2 + 105s + 105}$$

The gain and step responses for the three Bessel filters are shown in Figure 3.8. It is easy to recognize which of the three unmarked responses corresponds to the highest-order filter: the one with the steepest slope of gain at high-frequency asymptotes, the one with the largest negative phase at higher frequencies, and the one with the largest time delay at small times. The phase responses do not look linear since the phase shift is plotted against the frequency axis with logarithmic scale.

3.2 Closed-Loop Transient Response

The behavior of the closed-loop transfer function in the neighborhood of the critical point at −1 was discussed in Sec 1.4.3 in the context of the Nichols chart. The Nyquist diagram in the neighborhood of the critical point is shown in Figure 3.9(a). At the cross-over frequency f_b, $|T| = 1$.

From the isosceles triangle shown in the figure,

$$|F(f_b)| = 2\sin[(\arg T(f_b) - 180°)/2] \tag{3.2}$$

(Here, arg indicates the angle in degrees.) Commonly, the angle $\arg T(f_b) - 180°$ is less than 60°, and as a result, $|F(f_b)| < 1$; i.e., the feedback becomes positive. When a tracking system has no prefilter, its closed-loop gain $|M| = |T/F|$ becomes greater than 1 at f_b, and the closed-loop gain response $20\log|M|$ has a hump, as shown in Figure 3.9(b).

The value of this maximum is approximately:

$$\max|M| \approx |M(f_b)| = 1/[2\sin(180° - \arg T)/2] \tag{3.3}$$

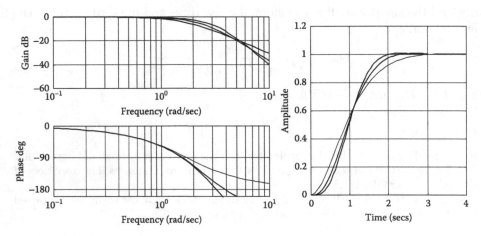

FIGURE 3.8
Frequency and step responses of Bessel filters of second to fourth orders.

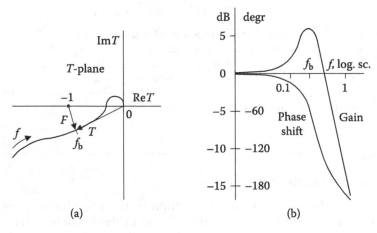

FIGURE 3.9
(a) Nyquist diagram and (b) closed-loop gain and phase responses.

The actual maximum of M is commonly at a frequency somewhat smaller than f_b. The hump is large when arg $T(f_b)$ is close to 180°, i.e., when the Nyquist diagram approaches the –1 point.

Example 3.8

When the angle of T approaches –180°, then $|F|$ decreases to 0 and $|M|$ grows infinitely. When the angle is –150°, then $|M| \approx 2$, and the hump is 6 dB high, as shown in Figure 3.9(b). The resulting overshoot is about 50%, as shown in Figure 3.10.

In a feedback system with a prefilter (or a non-unity feedback path, or a command feed-forward path), the open-loop and closed-loop responses can be optimized independently of each other as was shown in Sections 2.1 and 2.2, and there is no need to compromise the loop response in order to reduce the closed-loop overshoot.

For the overshoot in a closed-loop feedback system to be small, and the output to settle with high accuracy in a rather short time, the closed-loop gain response (together with the prefilter) must approximate a Bessel filter response. The prefilter must therefore equalize the hump of the closed-loop response from the summer to the output, i.e., it must incorporate a broad notch. An example of such a prefilter is given in Section 4.2.3.

If the prefilter (or the feedback path, or the command feedforward link) cannot be implemented to be exactly optimal, it should at least, on average, make the closed-loop phase response linear. This will prevent most of the harmonic components of the signal from reaching the output in phase at any time.

Homing systems do not have command summers, and therefore do not have prefilters. The homing system open-loop response must be made such that the closed-loop transient response is as desired.

Example 3.9

For a homing missile, the response of interest is that of the missile direction to the disturbance, which is the changing direction to the target caused by the target motion. Commonly, the disturbance is not measured or observed, and only the error is measured, i.e., the difference between the missile direction and the direction to the target. The closed-loop transfer function from the disturbance to the missile direction is $1/F$. If the phase stability margin at f_b is small, the frequency response of $-20 \log |F|$ has a large hump, the transient response of the output becomes too oscillatory, and there exists an effective maneuver for the target to avoid being hit. Typically, the *phase lag*, -arg $T(f_b)$, does not exceed 135° in such systems.

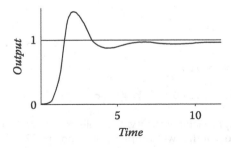

FIGURE 3.10
Closed-loop transient response to step input.

3.3 Root Locus

The transfer function $T(s)/F(s)$ of a closed-loop system is infinite for the signal components corresponding to the poles of the function. As long as the poles are in the left half-plane of the Laplace variable s, this doesn't create a problem. However, when one of the poles has a positive real part, i.e., the pole s_p is in the right half-plane, the transmission is infinite for the signal, which is a growing exponent. Then, some components of the random input noise will be continuously magnified, causing the system's output to grow exponentially. Such a linear system is unstable and cannot perform as a control system. Verification of system stability is one of the major tasks in control system design.

When the links in the loop are each inherently stable, i.e., all poles of their transfer functions are in the left half-plane of s, then certainly, all poles of the transfer function $T(s)$ are in the left half-plane of s, and the open-loop system is stable. When the loop is closed, some of the poles of the closed-loop transfer function $T(s)/F(s)$ can appear in the right half-plane of s, and in this case, the system is unstable. It is interesting to trace what happens when the loop is closed only gradually. To visualize this we place a link with the gain coefficient k in the loop, and we continuously increase k from 0 to 1. Correspondingly, the poles of the transfer function will move continuously from the poles of the open-loop system to the poles of the closed-loop system. Their trajectories on the s-plane are called *root loci*. Invasion of a root locus into the right half-plane of s indicates that the closed-loop system becomes unstable.

There exist several rules on drawing the root loci manually and for using the loci for feedback-system design. The rules provide for simple low-order system analysis, but the method becomes cumbersome when applied to high-performance systems that are high-order. The root locus design method will be further discussed in Section 8.2.

Example 3.10

$T(s) = 10(s+2)/(s^3+s^2+s)$. The open-loop and closed-loop poles can be calculated with MATLAB commands:

```
n= [0 0 10 20]; d= [1 1 1 0]; roots(d) %open-loop poles
ans = 0 -0.5000 ± 0.8660i
roots(n+d) %closed-loop poles
ans = -1.6551 0.3275 ± 3.4608i
```

The root loci in Figure 3.11 is plotted with:

```
n= [10 20]; d= [1 1 1 0];
rlocus(n,d)
hold on
k= [0.05 0.1 0.2 0.5 1];
rlocus(n, d, k)
title('k= [0.05 0.1 0.2 0.5 1]'); hold off
```

As k increases and approaches infinity, the pole at the origin moves to the left and approaches 1.655. The loci of the two complex poles end at $0.3275 \pm j3.4608$. The system is stable with the coefficient k up to 0.1, when the complex poles become purely imaginary $\pm j1.42$. After that, the complex poles migrate to the right half-plane of s and the system becomes unstable.

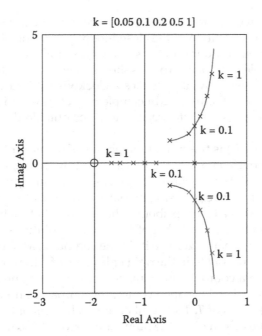

FIGURE 3.11
Root loci for $T(s) = 10(s+2)/(s^3+s^2+s)$.

When a pole of $T(s)/F(s) = T(s) / [T(s)+1]$ crosses the $j\omega$-axis at a certain frequency, the transfer function at this frequency becomes ∞, which can only happen if $T(j\omega)$ at this frequency becomes -1, i.e., $|T| = 1$ and arg $T = \pm n\pi$, where n is an odd integer.

From here, a simple stability criterion follows: if an open-loop system is stable, the closed-loop system is stable if $|T| < 1$ at all frequencies at which arg $T = \pm n\pi$, i.e., in practice, when arg$T = \pm\pi$. This criterion is convenient but not necessary. A necessary and sufficient stability criterion based on the open-loop frequency response will be derived in the next section.

3.4 Nyquist Stability Criterion

Stability criteria allow verification of system stability without pole position calculations or experiments. The most convenient among them is the *Nyquist criterion*. It allows passing a judgment on whether the system is stable by observing the plot of the open-loop frequency response, measured or calculated.

We will consider a single-loop feedback system that consists of linear time-invariant links whose transfer functions are rational with real coefficients. We assume that the system is stable when the loop is disconnected; i.e., the transfer function $T(s)$ does not have poles in the right half-plane of s.

Therefore, the closed-loop transfer function T/F can only have poles in the right half-plane of s if some of the $F(s)$ zeros are in the right half-plane of s.

The zeros of the function:

$$F(s) = \frac{(s-s_i)(s-s_j)(s-s_k)\ldots}{(s-s_p)(s-s_q)(s-s_r)\ldots} \tag{3.4}$$

are s_i, s_j, s_k, etc. Let us derive a condition for one or several of the zeros to appear in the right half-plane of s. This is the condition of the closed-loop system instability.

Consider a simple closed contour c_i in the s-plane crossing neither poles nor zeros of $F(s)$ and encompassing no poles and one zero s_i, as shown in Figure 3.12(b). The rest of the zeros of $F(s)$ are outside the contour. While s makes a clockwise round-trip about the contour c_i, the vector $s - s_i$ shown in Figure 3.12(b) completes a clockwise revolution. It is easy to notice that the vectors $s - s_j$, $s - s_k$, etc., related to zeros outside the contour c_i exercise no such revolutions.

The phase of the vector $F(s)$ is the sum of the phases of its multipliers. Therefore, a revolution of a multiplier about s_i changes the phase of $F(s)$ by 2π; i.e., it makes the vector $F(s)$ complete a clockwise revolution about the origin, as shown in Figure 3.12(a).

Consider next a contour c encompassing no poles and several zeros s_i, s_j, s_k, as shown in Figure 3.13(b). When s makes a full trip about c, the argument of each multiplier of the kind $s - s_i$ changes by 2π. Therefore, the number of clockwise revolutions of the locus $F(j\omega)$ about the origin indicates the number of zeros within the contour c, as shown in Figure 3.13(a).

To find the number of zeros of $F(s)$ in the right half-plane of s, the contour c should envelop the right half-plane. Such a contour can be made of the $j\omega$-axis and an infinite radius arc, as shown in Figure 3.14(b). The D-shaped contour encompasses no poles because $F = T + 1$, so the poles of F are the poles of T. The number of revolutions of the locus of $F(j\omega)$ shown in Figure 3.14(a) about the origin gives the number of zeros of $F(s)$ in the right half-plane of s. For this particular diagram, the number is 2, reflecting two complex conjugate zeros s_i and s_j in the right half-plane of s. Therefore, this particular closed-loop system is unstable.

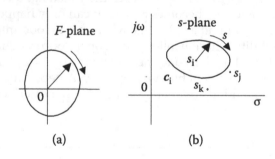

(a)   (b)

FIGURE 3.12
(a) Revolution of a rational function F(s) caused by (b) the trip of s about a closed contour ci on the s-plane.

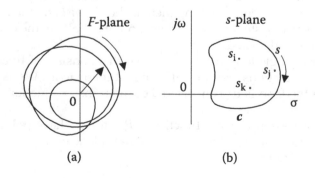

(a)   (b)

FIGURE 3.13
(a) Revolutions of a rational function F(s) caused by (b) s moved about a closed contour c.

As mentioned before, the return ratio disappears in physical systems when s is large. Therefore, $F(s)$ becomes 1 and remains 1 as s moves along the infinite radius arc. Thus, the locus in Figure 3.14(a) is the mapping of the $j\omega$-axis, and the interior of the F-plane contour is the mapping of the entire right half-plane of s.

The function $F(s)$ is rational with real coefficients, so it satisfies the *real symmetry* condition:

$$F(\tilde{s}) = \tilde{F}(s)$$

Therefore, the locus of F consists of two image-symmetrical halves relating to positive and negative frequencies, respectively.

The part of the locus of F drawn for positive ω is shown by the thicker curve and is called the *Nyquist diagram*. The diagram makes half the number of revolutions of the whole locus, which reflects the existence of the zero s_i in the first quadrant of the s-plane.

The *Nyquist criterion* follows: if a linear system is stable with the feedback loop open, it is stable with the loop closed if and only if the Nyquist diagram for F does not encircle the origin of the F-plane.

Although the Nyquist criterion has been proven here for only rational transfer functions, it is also valid for transcendental transfer functions which can be closely approximated by rational functions, which includes all practical systems. For instance, the transcendental transfer function for transport delay in a medium with distributed parameters can be approximated by a rational transfer function corresponding to a system with many small-value lumped elements. Therefore, the Nyquist criterion can be applied to all practical systems described by calculated or measured gain and phase frequency responses.

The Nyquist diagram is commonly drawn for $T = F - 1$, as shown in Figure 3.15 (compare this plot with the locus in Figure 3.14(a)). On the T-plane, the locus goes to the origin as s goes to infinity, and it is the *critical point* at -1, rather than the origin, whose encirclements must be analyzed.

Most importantly, the Nyquist diagram indicates not only whether the system is stable, but also how to stabilize the system by reshaping the loop response with the compensator. The loop gain might be reduced over a specific frequency range, or the loop phase lag could be reduced at certain frequencies. *Loop shaping* will be described in detail in Chapter 4.

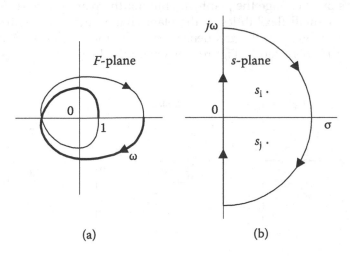

(a) (b)

FIGURE 3.14
(a) Nyquist diagram for F and (b) the contour surrounding the right half-plane of s.

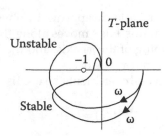

FIGURE 3.15
Nyquist diagrams on T-plane.

As mentioned in Chapter 1, Nyquist diagrams can also be plotted on the *L*-plane, where the critical points are (180° ± *n* 360°, 0 dB). Notice that when a Nyquist diagram on the *T*-plane encircles the critical point (−1,0) in the clockwise direction, the diagram on the *L*-plane encircles the critical point (−180°,0) in the counterclockwise direction.

3.5 Robustness and Stability Margins

Practical systems are required to be not only stable but also *robust*, i.e., remain stable when the plant parameters, and consequently the return ratio, deviate from the nominal values. The *stability margins* guard the critical point. They are often formed as shown in Figure 3.16(a) by a segment on the *T*-plane, or equivalently, by the rectangles on the *L*-plane as shown in Figure 3.16(b). If the Nyquist diagram for the nominal plant does not penetrate the boundary of the stability margin, the system will remain stable for a certain range of variations of the plant parameters.

The shape of stability margins shown in Figure 3.16 assumes that plant parameter variations in gain and in phase are not correlated. This is typical for many practical plants. Consider, for example, a force actuator driving a rigid body plant. Variations in the force and in the plant's mass change the plant gain but not the plant phase shift. On the other hand, variations of small flexibilities in the plant change the phase shift but don't significantly change the gain in the frequency region of the crossover. Another example is the volume control depicted in Figure 1.19, where the loop gain changes up to 10,000

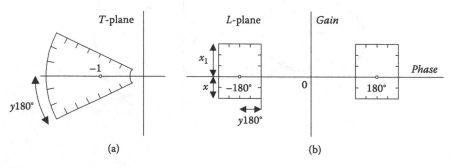

(a)   (b)

FIGURE 3.16
Amplitude and phase stability margins on the (a) T- and (b) L-planes.

times without any change in the loop phase shift. The disk stability margin, as shown in Figure 3.20(c), is less suitable for practice.

The values of the lower and upper *amplitude stability margins* x,x_1 shown in Figure 3.16, are typically 6 to 10 dB, and the values of the *phase stability margin* 30° to 45° or $y = 1/6$ to 1/4. (These values are also sufficient to guard from some nonlinear phenomena in the control loop, to be studied in Chapter 12.)

As illustrated in Figure 3.17, over the frequency range where the loop gain is within the interval $[-x, x_1]$, the system is *phase stabilized*, and over the frequency range where the angle of *T* is within the interval $[-180°(1+y), -180°(1-y)]$, the system is *gain stabilized*. Figure 3.17(b) shows the Nyquist diagram of a system that is both phase and gain stabilized at frequencies below f_1 and above f_2, and either gain or phase stabilized between these frequencies.

Nyquist stability refers to a stable system whose Nyquist diagram crosses the negative axis to the left from the point −1. Examples are shown in Figures 3.17(a) and 3.18(a). Such systems are only gain stabilized, not phase stabilized, at some frequencies ($f_1 < f < f_2$ in Figure 3.18(a)) where the loop gain is larger than 1.

Practical systems all include nonlinear links, at least the saturation of the actuator, and are required to be *globally stable*, i.e., to remain stable after any set of initial conditions. No *limit cycle* conditions, i.e., conditions of periodic oscillation, are typically allowed. (These issues will be discussed in more detail in Chapters 9–11.)

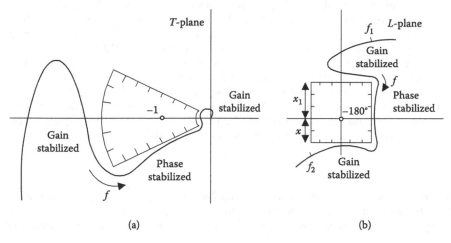

(a)　　　　　　　　　　　　**(b)**

FIGURE 3.17
Gain and phase stabilizing in (a) the T-plane and (b) the L-plane.

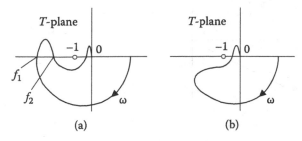

(a)　　　　　　**(b)**

FIGURE 3.18
Nyquist diagrams for (a) Nyquist-stable and (b) absolutely stable systems.

When the only nonlinear link in the loop is the actuator, then, loosely speaking, its saturation may reduce the equivalent loop gain while retaining the loop phase shift. This causes the equivalent Nyquist diagram to "shrink." While shrinking, the diagram of a Nyquist-stable system crosses the point −1. At this specific frequency and signal level, the equivalent return ratio becomes −1, which is the condition of self-oscillation. Therefore, using Nyquist stability should be avoided in those feedback systems that have no nonlinear links other than the actuator saturation.

Absolute stability refers to systems with a saturation link and whose Nyquist diagram is like that in Figure 3.18(b), i.e., not crossing the critical point while shrinking. For absolute stability, margins are typically chosen as shown in Figure 3.19.

Absolute stability will be studied in Chapter 10 in a more precise manner.

Example 3.11

According to the Nyquist criterion, the systems with Nyquist diagrams shown in Figure 3.20(a)–(e) are stable, while system (f) is unstable. The feedback is positive (i.e., $|T+1| < 1$) if and only if T is on the unit radius disk centered at the point −1.

In Figure 3.20(c), T is real and less than −1 at two frequencies, f_1 and f_2. This system is Nyquist stable. In other words, the return signal at these frequencies is in phase and of a larger amplitude than the signal applied to the loop input. That a feedback system with this sort of T is stable was first found experimentally during the development of feedback amplifiers at Bell Laboratories, and was later proven theoretically by H. Nyquist.

(Nyquist-stable designs are acceptable and even beneficial as long as special nonlinear dynamic links are introduced in the loop to exclude the possibility of self-oscillation. Designing such nonlinear links is described in Chapters 10 and 11.)

One of the Nyquist criterion's advantages is the simplicity of estimating the effects of multiplicative variations of the loop gain coefficient. For example, it is seen from the diagram in Figure 3.20(a) that reducing the loop gain without changing the loop phase shift will not make the system oscillate, that increasing the gain by a factor of 2 will cause oscillation, and that increasing the loop phase delay by 60° at the frequency where $|T| = 1$ (the crossover frequency f_b) will also cause oscillation.

Example 3.12

The linear scale is inconvenient for drawing Nyquist diagrams for practical systems, where the loop gain changes by several orders of magnitude between the lowest frequency of interest and the crossover frequency. The inventor of feedback amplifiers,

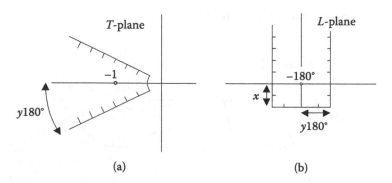

(a)   (b)

FIGURE 3.19
Stability margins for a single-loop system with saturation on the (a) T- and (b) L-planes.

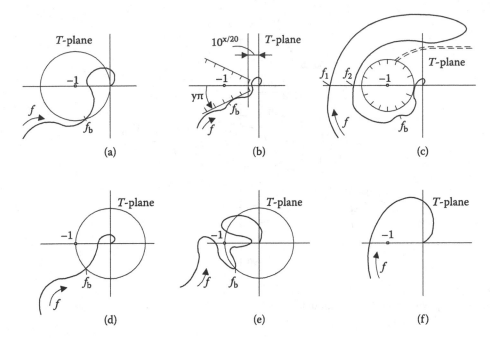

FIGURE 3.20

Nyquist diagrams: (a) feedback is positive at frequencies at which the Nyquist diagram is on the disk, (b) Bode stability margins, (c) disk stability margin, (d) robust system, (e) non-robust system, and (f) unstable system.

Harold Black, employed a circular coordinate system with gain in dB and phase in degrees, as illustrated in Figure 3.21. Another example of using these coordinates is given in Figure A13.26.

Example 3.13

The L-plane diagrams equivalent to the corresponding T-plane diagrams of Figure 3.20 are shown in Figure 3.22. The vertical axis corresponds to $-180°$.

Example 3.14

In some of the contemporary literature, stability margins are defined as if they need only apply at discrete points instead of a solid boundary, and this concept is referred to as guard-point stability margins. The *guard-point phase margin* is the phase margin at the frequency where the gain margin is zero, i.e., at f_b. The *guard-point gain margin* is the gain margin at the frequency where the phase margin is zero.

This interpretation is acceptable and convenient when the order of $T(s)$ is low, and even when the order of $T(s)$ is high but the response is smooth, but it does not suffice in general. Because of the inescapable trade-off between stability margins and the available feedback, if a high-order compensator is optimized for closed-loop performance while only the guard-point stability margins are enforced, the Nyquist diagram might end up looking like Figures 3.20(e) and 3.22(e)—these loop responses approach the critical point much too closely. This sort of misuse of the guard-point stability margins may be responsible for the misconception that high-order compensators generally lead to non-robust systems.

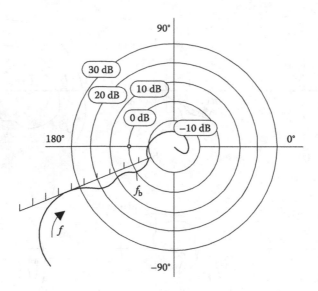

FIGURE 3.21
Nyquist diagram with logarithmic magnitude scale.

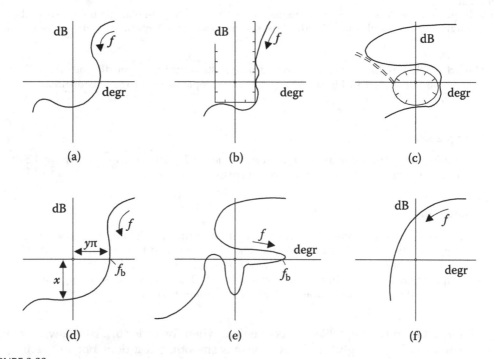

FIGURE 3.22
Nyquist diagrams on the L-plane: (a) negative and positive feedback areas, (b) Bode stability margins, (c) disk stability margin, (d) robust system, (e) non-robust system, and (f) unstable system.

3.6 Nyquist Criterion for Unstable Plants

The systems discussed until now were considered stable with the feedback loop disconnected. However, some physical systems are unstable without the feedback; i.e., $T(s)$ has poles in the right half-plane of s. We would like to know what feedback will make such systems stable.

Consider the function $F(s) = T(s) + 1$, which may now have poles and zeros within the contour of interest in Figure 3.14. Having a *pole* within the contour causes the function to rotate in the direction opposite of the revolution produced by a *zero* within the contour. Consequently, when s completes a trip about the contour, the number of revolutions of the function locus equals the difference between the numbers of zeros and poles of $F(s)$ within the contour. The corresponding Nyquist diagram encircles the critical point a number of times equal to the difference between the number of poles and the number of zeros in the first quadrant of the s-plane. The rule follows: in order for the system to become stable when the feedback is applied, the Nyquist diagram must encircle the critical point a number of times equal to the number of open-loop poles in the first quadrant of the s-plane, and these encirclements must be in the counterclockwise direction.

It is common to describe the plant's instability as the result of an internal feedback loop, where the links characterize physical processes and relations between the variables in the plant. This method is routinely applied to the analysis of various unstable plants: plants with aerodynamic instability, wind flutter, thermal flutter, gas turbulence, and the inverted pendulum. Consider two examples:

Example 3.15

In the system diagrammed in Figure 3.23(a), the plant is unstable since the internal feedback path transmits the signal back to the input of the gain block in phase, at the frequency of the feedback path resonator. The Nyquist diagram for the internal loop encloses the critical point clockwise, as shown in Figure 3.23(b).

Because of the internal feedback, the plant transfer function

$$P(s) = \frac{800\left(s^2 + 0.2s + 4\right)}{\left(s^2 - 0.8 + 4\right)s^2}$$

possesses a pair of complex conjugate poles with positive real parts, as can be judged by the negative sign of the damping coefficient in the denominator polynomial. With the compensator transfer function shown in Figure 3.23(a), the main loop transfer function at cross section 1 is:

$$T_1(s) = \frac{800(s + 10)\left(s^2 + 0.2s + 4\right)}{(s + 40)\left(s^2 - 0.8 + 4\right)s^2}$$

The locus of $T_1(j\omega)$ shown in Figure 3.23(c) encircles the critical point –1 in the counterclockwise direction. This indicates that the closed-loop system is stable. The locus on the L-plane shown in Figure 3.23(d) correspondingly encircles the critical point (0 dB,180°) in the clockwise direction.

Another example of this system stability analysis is in cross section 2, where the loop transfer function is found to be:

$$T_2(s) = \frac{-0.1s^4 + 76s^3 + 960s^2 + 1920s + 3,200}{s^5 + 42s^4 + 84s^3 + 160s^2}$$

The Nyquist diagram for this cross section is shown in Figure 3.23(e). The system is stable.

Example 3.16

The high-gain amplifier shown in Figure 3.24(a) would be stable by itself, but a small parasitic input–output capacitance C causes self-oscillation in the absence of the standard feedback loop via R_2. The transfer function for the parasitic feedback path is shown in Figure 3.24(b). The Nyquist diagram for the parasitic loop is shown in Figure 3.24(c). When the amplifier is operated with the conventional feedback path $B = -R_1/(R_1 + R_2)$

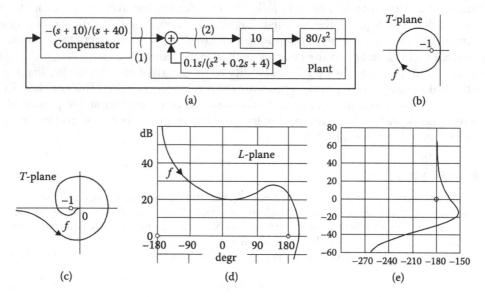

(a)   (b)

(c)   (d)   (e)

FIGURE 3.23
(a) Feedback system with unstable plant, (b) Nyquist diagram for the internal plant loop, (c) Nyquist diagram for T1, not to scale, (d) Nyquist diagram for T1 on the L-plane, and (e) L-plane Nyquist diagram for T2.

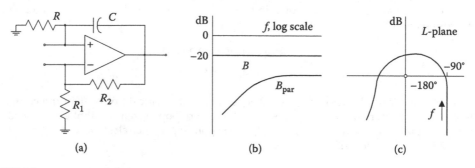

(a)   (b)   (c)

FIGURE 3.24
System with parasitic feedback via capacitor C: (a) schematic diagram, (b) Bode diagrams for the feedback paths, and (c) Nyquist diagram for parasitic feedback loop.

closed, the amplifier gain is much smaller (say, only 10 when $B = 0.1$, as indicated in Figure 3.24(b)), the parasitic feedback loop gain is small, and the system is stable. Another point of view on the problem is that the parasitic feedback path transfer function $B_{par} = s/[s + 1/(RC)]$ is negligible when compared with the normal feedback path transfer function B, with which the amplifier is stable. Hence, the system can be analyzed without taking into account the parasitic feedback.

3.7 Successive Loop Closure Stability Criterion (Bode–Nyquist)

Example 3.17

Consider the multi-loop system shown in Figure 3.25.

Assume the system is stable when all loops are disconnected, and start closing the feedback loops successively, which will eventually lead to the system with all the loops closed. A series of five such Nyquist diagrams is exemplified in Figure 3.26.

It is seen in Figure 3.26 that, after the first loop is closed, the system remains stable; after the second loop is closed, the system becomes unstable and the system transfer function possesses one pole in the first quadrant of the s-plane; after the third loop is closed, the system remains unstable since the Nyquist diagram indicates no change in the difference between the numbers of poles and zeros in the right half-plane of s; after the fourth loop is closed, the system becomes stable since the diagram encircles the critical point once in the counterclockwise direction; and the system remains stable after the fifth loop is closed since the fifth Nyquist diagram does not encircle the critical point.

By generalizing the procedure given in the Example 3.17, Bode formulated the *Nyquist criterion for multi-loop systems (Bode–Nyquist)* as follows: when a linear system is stable with certain loops disconnected, it is stable with these loops closed if and only if the total numbers of clockwise and counterclockwise encirclements of the point (–1,0) are equal to each other in a series of Nyquist diagrams drawn for each loop and obtained by beginning

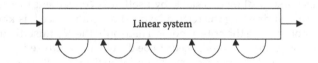

FIGURE 3.25
Block diagram of a multi-loop system.

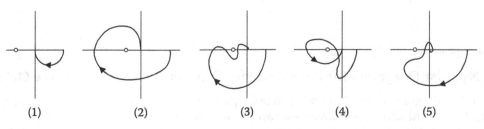

(1)   (2)   (3)   (4)   (5)

FIGURE 3.26
Nyquist diagrams for the successive loop closure stability analysis.

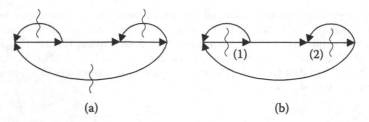

FIGURE 3.27
Stability analysis with (a) three and (b) two cross sections of the feedback loops.

with all loops open and closing the loops successively in any order, leading to the normal system configuration.

The order in which the loops are closed can be chosen at the convenience of the designer.

Example 3.18

The system diagrammed in Figure 3.27 contains two local loops and a common loop. It is analyzed as three-loop system in Figure 3.27(a). In this case, drawing three Nyquist diagrams for the indicated cross sections is required for the stability analysis. The order of closing the loops and making the Nyquist diagrams can be the following: (1) drawing the Nyquist diagram for the common loop (with local loops open), (2) closing the common loop and drawing the first local loop diagram, and (3) drawing the Nyquist diagram for the remaining local loop. Thus, we would have to draw and analyze all three diagrams since it is not evident that any of the three diagrams avoids encircling the critical point.

It is often evident, however, that the local loops are inherently stable because the phase lag in the loops is small. When this is the case, it is worth starting with these loops and not drawing the related diagrams since it is already known that the diagrams will not enclose the critical point. It remains, therefore, only to draw the diagram for the common loop, with the local loops closed.

Alternatively, the system can be analyzed using the two cross sections in the forward paths, as shown Figure 3.27(b). Breaking the signal flow at these cross sections eliminates all feedback, so the analysis can be performed using only two Nyquist diagrams. Further, when the first local loop is stable by itself, it is convenient to start by closing cross section 1 without drawing the related Nyquist diagram since it is known that the diagram should not encircle the critical point. Then, only the Nyquist diagram for cross section 2, with cross section 1 closed, needs to be drawn and analyzed.

Nevertheless, the simplest approach might be inconsistent with the convenience of verifying global stability. For this purpose, the order of closing the loops can be chosen such that it reflects the order in which the loops desaturate when the signal level gradually decreases [33].

3.8 Nyquist Diagrams for Loop Transfer Functions with Poles at the Origin

Figure 3.28 diagrams a mechanical rigid body plant driven by a force actuator. Depending on the type of the sensor employed—an accelerometer, a rate sensor, or a position sensor—the plant transfer function is correspondingly a constant $1/M$, a single integrator $1/(Ms)$, or a double integrator $1/(Ms^2)$.

The frequency responses for these plant functions are shown in Figure 3.29(a). The gain for the single and double integrators is infinite at zero frequency. The poles (the singularities) for the single- and double-integrator plants are at the origin of the s-plane, as shown in Figure 3.29(b).

The Nyquist diagram is the mapping of the contour encompassing the right half-plane of s. The contour should be chosen such that the function on the contour does not go to infinity. Therefore, if the transfer function $T(s)$ possesses poles on the $j\omega$-axis, the contour should avoid the poles. Particularly, when n poles are at the origin, i.e.,

$$T(s) = \frac{(s - s_{z1})(s - s_{z2})(s - s_{z3})(s - s_{z4})\ldots}{s^n (s - s_{p1})(s - s_{p2})(s - s_{p3})\ldots}$$

the origin can be avoided by an infinitesimal radius arc, as shown in Figure 3.29(b). On this arc, in close vicinity of the pole, $|T|$ is infinitely large, and its phase changes by $n\pi$ as the phase of s changes by π along the arc. Therefore, the small arc in the s-plane maps onto the T-pane as an $n\pi$ arc of infinite radius. Half of this arc becomes a part of the Nyquist diagram.

The Nyquist diagram for a single-integrator plant is shown in Figure 3.30(a). For a double-integrator plant, i.e., for a loop transfer function with a double pole at the origin, the arc is twice as long as shown in Figure 3.30(b).

Feedback control loops are often classified as servomechanisms of *type 0, type 1*, and *type 2*. This is the number of poles of $T(s)$ at the origin. The Nyquist diagram shown in Figure 3.30(a) is of type 1, and the diagram shown in Figure 3.29(b) is of type 2. A Nyquist diagram of "servo" type 0 was shown in Figure 3.18. The shape of the Nyquist diagram and the related loop frequency response at lower frequencies define the steady-state response to commands and the steady-state error reduction.

The properties of these three types of servomechanisms can be understood by considering the feedback system shown in Figure 3.31. The plant output variable is the angle θ. Two

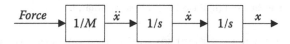

FIGURE 3.28
Mechanical rigid body plant.

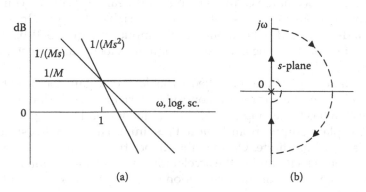

FIGURE 3.29
(a) Bode diagrams for the plant as a constant, a single integrator, and a double integrator; also (b) the related contour on the s-plane.

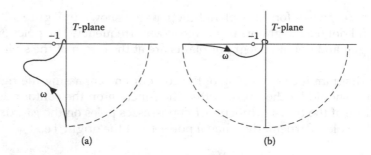

FIGURE 3.30
Nyquist diagram for a stable feedback system with (a) a single integrator in the loop and (b) a double integrator in the loop.

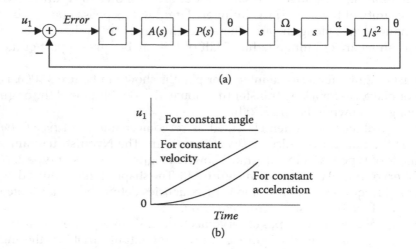

FIGURE 3.31
(a) Feedback-system block diagram and (b) the profiles of the command.

differentiators are added after the plant for the sake of analysis, to calculate the angular velocity Ω as well as the acceleration α. A double integrator is added to keep the return ratio unchanged.

Assuming the error is small, i.e., the return signal nearly equals the command, let us consider the problem of keeping constant one of the variables: θ, Ω, or α. In order to do this, specific commands $u_1(t)$ need to be applied to the input. In Figure 3.31(b), the time functions are shown for constant angle, constant angular velocity, and constant acceleration commands, respectively.

To reduce the static error of the controlled variable, the dc gain coefficient (i.e., the gain coefficient at $s \to 0$) from the error to the controlled variable must be large.

The type 0 system has finite loop gain at zero frequency, and the dc gain from the summer output to the plant output (from *Error* to θ) is finite. Therefore, the steady-state error of the angle θ is small but finite. On the other hand, the forward path gain coefficients from the *Error* to the velocity and to the acceleration at $s \to 0$ are infinitely small. If in the commanded $u_1(t)$ the velocity or the acceleration is constant, the output will not track the command.

The type 1 system has an infinite loop gain coefficient at zero frequency. The gain from the error to θ is also infinite, but the gain to velocity Ω is finite, and the gain to acceleration

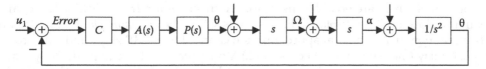

FIGURE 3.32
Disturbances in angle, velocity, and acceleration in a feedback system.

α is zero. Therefore, when constant angle θ is commanded, the error is the angle error and it is zero; when constant velocity is commanded, the velocity error is finite; and when constant acceleration is commanded, the acceleration error is not corrected at all.

Next, consider the effects of disturbances entering the feedback system at different points: disturbance in position, disturbance in velocity, and disturbance in acceleration, as shown in Figure 3.32. Disturbance in position is commonly caused by misalignment of mechanical parts; disturbance in velocity, by an extra velocity component of moving parts of the plant or by drift in time of the values of some of the plant's parameters; and the disturbance in acceleration, by disturbance torque due to wind, magnetic forces, etc.

In a type 1 system, a disturbance entering at the point of velocity is reduced infinitely since the feedback is infinite, but this disturbance causes a finite change in the angle since the gain coefficient at dc from the angle to the velocity is infinitely small. Thus, the constant velocity disturbance causes a "hang-up" error in position (i.e., an error that does not decay in time). To eliminate this error, a type 2 system should be employed. The type 2 servomechanism is also referred to as a zero-velocity error system. In this system, the steady-state errors in position and velocity are zero, and there is finite reduction in the steady-state error in acceleration.

In some systems, the return ratio has a triple pole at zero frequency. These systems have larger loop gain and better accuracy at low frequencies, but the low-frequency phase lag in such systems approaches 270° and the system is not absolutely stable. For such a system to remain stable after the actuator becomes overloaded, the compensator must be made nonlinear, as will be discussed in Chapters 9–13.

3.9 Bode Phase-Gain Relation

Synthesis (design) of a stable feedback system using the Nyquist criterion is not quite straightforward. For example, if, to correct the shape of the Nyquist diagram, one decided to reduce the gain at some frequencies, he might find out that this gain change affected the phase shift at other frequencies, and the system is still unstable, although with a quite different shape of the Nyquist diagram.

The Nyquist criterion uses three variables: frequency, loop gain, and phase shift. These variables are interdependent. Bode showed that in most practical cases, using only two of them (the frequency and the gain) suffices for feedback-system design. This greatly simplifies the search for the optimal design solution.

3.9.1 Minimum Phase Systems

In order to analyze this interdependence we will utilize the D-shaped contour that was used to derive the Nyquist stability criterion, but on the *T*-plane. In a system that is open-loop stable there are no poles inside (or on) this contour. If, in addition, there are no zeros in the closed right half-plane, the system is said to be *minimum phase (m.p.)*. If part of *T* is transcendental, the rational approximation cannot have RHP zeros and this prohibits transport delay (more on this in Section 3.12.) More precisely, to be m.p., the loop transfer function and its inverse need to be causal and stable—refer to Appendix 3. Compared to a LHP zero, a RHP zero has the same effect on the gain frequency response but adds 90° of phase lag instead of 90° of phase lead. Similarly, transport delay introduces phase lag with no effect on the gain. This "extra" phase lag is referred to as *non-minimum phase (n.p.)* lag.

As will be shown further, the phase delay in the feedback loop limits the available feedback. Therefore, it is desirable for the transfer functions of the feedback loop links to be m.p. Designers of control loops use m.p. functions in compensators and, if possible, employ actuators and plants with m.p. transfer functions. The phase-gain relations for m.p. functions are of special interest for feedback-system designers.

Example 3.19

A passive two-pole is stable in the conditions of being open or shorted. Therefore, its impedance and admittance have no zeros or poles in the right half-plane of *s*, so these functions are m.p.

Example 3.20

Figure 3.33 shows a ladder passive electrical network. The output-to-input ratios of the network can be expressed as voltage and current transfer functions, the transimpedance (ratio of the output voltage to input current), and the transadmittance (ratio of the output current to the input voltage). All of the logarithm transfer functions are minimum phase functions.

It is evident that a signal applied to the input of a ladder network arrives at the output unless at least one of the series branches is open, or one of the parallel branches is shorted. Therefore, the output-to-input ratio zeros are produced by poles of the impedances of the series two-poles and by zeros of the impedances of the shunting two-poles. When the two-poles are passive, their poles and zeros are not in the right half-plane of *s*, so the network transfer function does not have zeros in the right half-plane and it is minimum-phase.

3.9.2 Phase-Gain Relation

The gain and phase responses are the real and imaginary parts of the (complex) natural logarithm of the loop transfer function:

$$\ln T(s) = \ln |T(s)| + j \arg T(s)$$

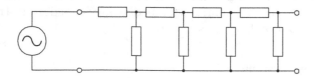

FIGURE 3.33
Ladder network.

which is analytic in the right half-plane when $T(s)$ is m.p. To align with Bode [6], we adopt the following notation for the gain and phase of T as real functions of the frequency:

$$A(\omega) = \ln|T(j\omega)|, \text{ and } B(\omega) = \arg T(j\omega)$$

By integrating $\ln T(s)$ with a frequency-dependent weight function along the usual D-shaped contour (with cut-outs for the singularities), and using Cauchy's integral theorem:

$$\oint_D \frac{\ln|T(s)|}{s^2 + \omega_c^2} ds = 0$$

The phase response at frequency ω_c can be determined by integrating the gain response, using a version of the residue theorem to integrate around the singularities:

$$B(\omega_c) = \frac{2\omega_c}{\pi} \int_0^\infty \frac{A(\omega) - A(\omega_c)}{\omega^2 - \omega_c^2} d\omega$$

The details are in Appendix 4, where this is shown to be equivalent to:

$$B(\omega_c) = \frac{1}{\pi} \int_{-\infty}^\infty \frac{dA(u)}{du} \ln \coth \frac{|u|}{2} du \qquad (3.5)$$

where $u = \ln(\omega/\omega_c)$. It can be seen that the phase shift is proportional to the slope of the gain response vs. the frequency with a logarithmic scale.

It is always possible to add a constant to the gain without affecting the phase, and to add additional (non-minimum) phase lag without affecting the gain.

Equation 3.5 is in terms of the natural logarithm with the phase measured in radians, which we need to convert to degrees and decibels. For example, for a single integrator with transfer function $1/s$, the slope is −6 dB/octave and the phase shift is −90°. For a double integrator, the slope is −12 dB/octave and the phase shift is −180°. For the gain response having constant slope −10 dB/octave, the phase shift, proportionally, is −150°. The convenience of relating the phase response to the slope of the gain response with this formula is the reason why the gain responses drawn with logarithmic frequency scale are called Bode diagrams.

Since the integral in Equation 3.5 is taken from −∞ to ∞ along the u-axis, i.e., over the frequencies from 0 to ∞, the phase shift at any specified frequency depends on the gain response slope at all frequencies. However the extent of this dependence is determined by the weight function $\ln \coth |u/2|$, charted in Figure 3.34 with the script

```
u=linspace(-3,3,200);
b=log(coth(abs(u/2))); plot(u,b,w);
grid
```

Due to the selectiveness of the weight function, the slope of the gain in the neighborhood of $u = 0$, i.e., the neighborhood of the frequency ω_c at which the phase shift is being calculated, contributes much more to the phase than does the slope at remote parts of the Bode diagram.

The Bode phase-gain relation alleviates the difficulty of design using the Nyquist stability analysis. The gain vs. frequency, expressed in the form of the Bode diagram, also contains the required information about the phase shift. Therefore, the system design can be based on the gain responses with the implied phase responses easily anticipated. Bode diagram methods are widely applied to practical systems that are m.p., or whose n.p. components are small or can be calculated and accounted for separately.

The sequence of frequency responses in Figure 3.35 illustrates the structure of the Bode formula: the Bode diagram is shown (a), the slope of the Bode diagram is plotted (b), the weight function is centered at ω_c (c), and the slope response is multiplied by the weight function (d).

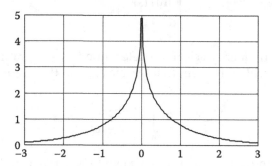

FIGURE 3.34
Weight function $\ln \coth |u/2|$.

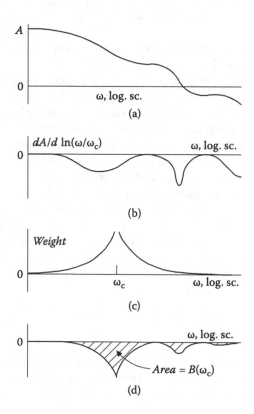

FIGURE 3.35
Phase shift calculation at frequency ω_c: (a) Bode diagram, (b) Bode diagram slope, (c) weight function centered at ω_c, and (d) product of the slope and the weight.

The area under response (d) gives the phase shift at ω_c. Notice that Figure 3.41 is only illustrating the structure of the phase-gain relation. A practical method for the phase calculation will be described in the next section.

Example 3.21

Calculate the phase at frequency $\omega_c < 1$, which relates to the following low-pass response: the gain is 0 dB up to the corner frequency $\omega = 1$, then decreasing with unit slope, $dA/du = -1$.

From Equation 3.14, the phase is:

$$B(\omega_c) = -\frac{1}{\pi} \int\limits_{-\ln\omega_c}^{+\infty} \ln\coth\frac{|u|}{2}\,du$$

where the lower integration limit corresponds to the corner frequency $\omega = 1$, which is the frequency where the gain slope changes from zero to -1, i.e., $u = \ln(1/\omega_c)$. Since

$$\ln\coth\frac{|u|}{2} = \ln\left|\frac{1 + \omega_c/\omega}{1 - \omega_c/\omega}\right|$$

and $du = d\ln(\omega/\omega_c) = \omega^{-1}d\omega$, we can rewrite the expression for $B(\omega_c)$ as:

$$B(\omega_c) = -\frac{1}{\pi}\int\limits_{1}^{\infty}\ln\left|\frac{1 + \omega_c/\omega}{1 - \omega_c/\omega}\right|\frac{d\omega}{\omega}$$

At small ω_c, the logarithm equals $|2\omega_c/\omega|$. Then,

$$B(\omega_c) \approx -\frac{2}{\pi}\omega_c\int\limits_{1}^{\infty}\frac{d\omega}{\omega^2}$$

and finally,

$$B(\omega_c) \approx -\frac{2}{\pi}\omega_c \tag{3.6}$$

The phase shift is negative and proportional to the frequency.

Corollary 3.1

Since the gain response of any low-pass filter can be approximated in a piece-linear manner, the phase of a low-pass filter at low frequencies is a sum of functions, each proportional to the frequency, so that this sum as well is proportional to the frequency.

3.10 Phase Calculations

Accurate calculation of the phase lag from the gain response is rarely needed in engineering practice, and computer programs developed for calculation of the integral in Equation 3.5 are used rather infrequently. However, approximate calculation of the phase is quite often required during the conceptual stage of the design, and for small re-adjustments of the loop frequency responses. For these purposes, a modified version of a graphical procedure suggested by Bode is described below.

In Figure 3.37, the phase responses are plotted for the gain ray that originates at f_c with the slope of $-6n$ dB/octave (dashed line), and for the segments (ramps) of the gain response with the slope of $-6n$ dB/octave over w octaves centered at f_c, i.e., at $u = 1$.

In general, if the segment's slope is a dB/octave, then the left scale of the phase should be multiplied by $a/6$, or the right scale multiplied by $a/10$.

Bode diagrams can be approximated piecewise-linearly by segments and rays, and the phase responses related to these can be added. Bode proved that even a crude approximation of A renders a fairly accurate phase frequency response. For the responses typical in automatic control, the number of segments need not be large.

Example 3.22

The gain response is approximated by two segments with nonzero slope, three segments with zero slope (no phase is related to these segments), and a ray as illustrated in Figure 3.36. The phase response is then obtained as the sum of the elementary phase responses, each related to a single segment or a ray of the gain frequency response. The total phase response is the sum of the three-phase responses.

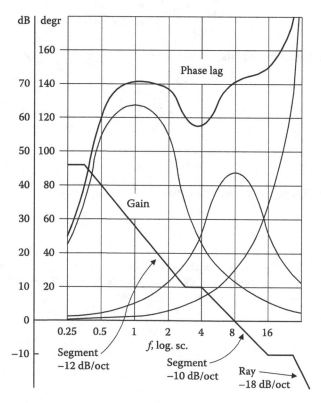

FIGURE 3.36
Phase calculation for piece-linear approximation of a Bode diagram.

Example 3.23

A loop response crosses the 0 dB line at the frequency 800 Hz. In an attempt to increase the feedback at lower frequencies, it is contemplated to make the Bode diagram steeper by 6 dB/octave over an octave centered at 200 Hz. What will be the effect of this change on the guard-point phase margin? The effect can be calculated with the help of the chart in Figure 3.37. From the curve marked 1, at the distance of two octaves from the center, the phase is 13°. That is, the guard-point phase stability margin will be reduced by 13°.

Piecewise-linear approximation of $A(\omega)$ is particularly useful for trial-and-error procedures of finding a physically realizable response for $\ln T(j\omega)$ that maximizes a certain norm while complying with a set of heterogeneous constraints (such as weighted maximization of the real component over a given frequency range under the limitation in the form of a prescribed boundary for the frequency hodograph of the function).

Starting with some initial response for A, say, A', we could calculate the related response B' and get a physically realizable $\ln T' = A' + jB'$. Next, changing the gain response as seems reasonable, we would find the related phase response, etc. As a rule, the process converges

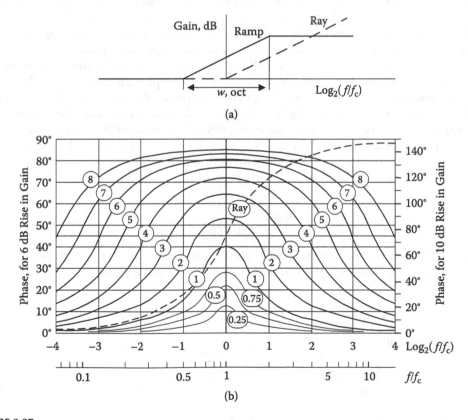

FIGURE 3.37
(a) Ramp gain response with constant slope over w octaves centered at f_c, and the gain ray starting at f_c, as well as (b) phase responses corresponding to this gain response, for different w, and (dashed line) phase response corresponding to 6 dB/octave ray.

rapidly, and the accuracy of the graphical procedure is sufficient. (The appropriate scales are 10 dB/cm and 1 octave/cm for sketches, and 5 dB/cm and 0.5 octave/cm for more accurate calculations.)

Example 3.24

A piecewise-linear gain response can be viewed as a sum of several ray responses. A ray starting at ω_o with the slope $-12n$ dB/octave can be approximated by $(s^2 + \omega_o s + \omega_o^2)^{-n}$; here n is not necessarily an integer. MATLAB function BONYQAS, described in Appendix 14, is based on this approximation. It calculates and plots the m.p. phase response and the Nyquist diagram related to a piecewise-linear gain response specified by the vector of corner frequencies, the vector of the gains at these frequencies, and the low-frequency and high-frequency asymptotic slopes.

3.11 From the Nyquist Diagram to the Bode Diagram

From a known Bode diagram, the phase can be calculated and a Nyquist diagram can be plotted. The inverse problem is, given the function $B(A)$, i.e., the shape of the Nyquist diagram for an m.p. function, find the Bode diagram. Although no analytical solution exists to this problem, the solution can be found numerically by approximating the Nyquist diagram with a high-order rational function. Alternatively, the responses important in practice can be found rather easily with an iterative procedure utilizing the Bode method of finding $B(\omega)$ from $A(\omega)$.

The iterative process consists of the following steps: (1) plotting a first-guess Bode diagram composed of segments and rays, (2) calculating the related phase lag, (3) plotting the Nyquist diagram, and (4) correcting the first-guess Bode diagram, etc. The process converges rapidly for smooth-shaped Nyquist diagrams. However, for Nyquist diagrams with sharp angles (which are optimal for many systems), the convergence of this procedure is slow. The convergence can be improved by including a sharp-cornered response in the set of basis functions. For this purpose we can use the function:

$$b(jf) = \frac{1}{\sqrt{1 - f^2} + jf} \tag{3.7}$$

which is plotted in Figure 3.38(b). This response is low pass. It has the peculiar property of having a gain of 0 dB over the frequency band $f \leq 1$, and a phase lag of $\pi/2$ for $f \geq 1$.

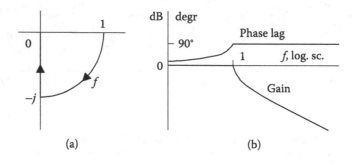

(a)   (b)

FIGURE 3.38
(a) Locus of the ratio in Equation 3.7 and (b) frequency responses for the gain and phase of Equation 3.7.

The frequency locus of the ratio in Equation 3.7 is plotted in Figure 3.38(a). The asymptotic slope is −6 dB per octave.

The high-pass response shown in Figure 3.39(a) can be calculated by substituting −1/f for f in Equation 3.7, and the band-pass response shown in Figure 3.39(b) can be obtained by substituting $(f - 1/f)$ for f [6].

Raising b to the power n preserves the function's m.p. property. The asymptotic phase lag will then be −$n\pi$/2. For example, for the phase lag of −150°, the power coefficient is $n = 5/3$, and the asymptotic slope is −10 dB/octave, as shown in Figure 3.40(a), (b).

Combining these frequency responses with piecewise-linear responses, one can compose Nyquist diagrams with sharp angles. For example, the sum of the four responses, (1) low-pass, (2) high-pass, (3) constant-slope, and (4) a constant (not shown in the picture), produces the response shown in Figure 3.41. This response can be used as a part of the

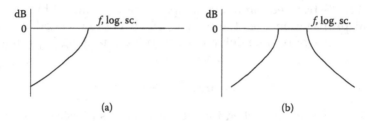

FIGURE 3.39
Bode diagrams for high-pass and band-pass transforms of Equation 3.7.

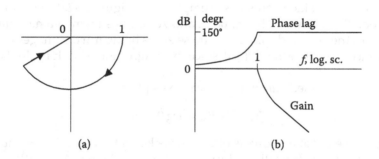

FIGURE 3.40
(a) Nyquist diagram and (b) frequency responses for gain and phase of Equation 3.7 raised to the power 5/3.

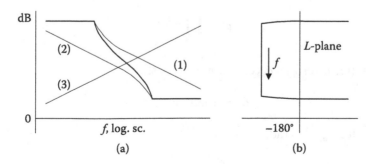

FIGURE 3.41
(a) Example of a gain response composed of several elementary responses and (b) the response on the *L*-plane.

Nyquist diagram for a Nyquist-stable system. This part can be connected to the rest of the diagram by the same procedure as that used in Figure 3.41 to "glue" together the low-pass and high-pass responses.

3.12 More on Non-Minimum Phase Lag

Non-minimum phase lag is a link's lag in excess of that given by the Bode phase-gain formula. Non-minimum phase lag can appear in lumped-parameter systems whose transfer functions have zeros in the right half-plane, and also in systems with distributed parameters that are described by transcendental functions.

In systems with distributed parameters, n.p. lag can be caused by the time τ that it takes for signal propagation over the media, which is often called *transport lag*. The Laplace transform of a pure time delay τ is $e^{-s\tau}$ so the phase lag is proportional to the frequency:

$$B_n = \arg e^{-j\omega\tau} = -\omega\tau$$

The transport lag is substantial when the feedback loop is physically long or the speed of the signal propagation in the media is low, as might happen in thermal, pneumatic, or acoustical systems. The transport lag of electrical signals in feedback amplifiers can be significant when the feedback bandwidth reaches hundreds of MHz.

The effect of the right-sided zeros s_i is exemplified in Figure 3.42. In (a), poles and zeros of a m.p. function θ_m are shown. In (b), poles and zeros are shown of a n.p. function θ that has some zeros in the right half-plane of s, these zeros being mirror images of some of the zeros of θ_m. Notice that $|\theta_m(j\omega)| = |\theta(j\omega)|$ since the magnitude of each multiplier $(j\omega - s_i)$ is preserved.

The ratio $\theta_n = \theta/\theta_m$ is called *pure n.p. lag* because its phase

$$B_n = B - B_m = \arg \theta_n(j\omega)$$

represents the phase lag of θ in excess of the phase lag of θ_m. Since $|\theta_n| = 1$ at all frequencies, the function $\theta_n(s)$ is also called *all-pass*. As shown in Figure 3.42(c), the zeros s_i of θ_n are in the right half-plane, and they are the mirror images of the poles that are in the left half-plane. Since the zeros are either real or come in complex conjugate pairs, θ_n can be expressed as:

$$\theta_n = \prod_i \frac{s + s_i}{s - s_i}$$

In particular, each *real* zero $s_i = \sigma_i$ contributes a n.p. shift:

$$B_{ni} = 2\arctan(\omega/\sigma_i) \tag{3.8}$$

For $\omega < 0.4\sigma_i$, i.e., for $B_{ni} < 0.8$ rad,

$$B_{ni} \approx 2\omega/\sigma_i$$

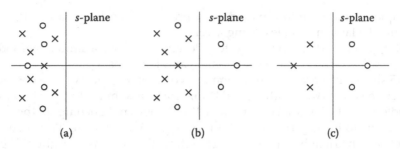

FIGURE 3.42
Poles and zeros of (a) m.p. function θm, (b) n.p. function θ, and (c) the all-pass function $\theta/\theta m$.

A sum of linear functions is a linear function. Therefore, generally, if the n.p. lag B_n is less than 0.8 rad, its frequency dependence can be approximated by the linear function:

$$B_n \approx \left| B_n(f_c) \right| \frac{f}{f_c} \qquad (3.9)$$

for some frequency f_c.

Rational approximations (Padé) to the pure time delay are all-pass functions, for example:

$$e^{-j\omega\tau} \approx \frac{1+(j\omega/2)}{1-(j\omega/2)}$$

3.13 Ladder Networks and Parallel Connections of m.p. Links

The ladder network diagrammed in Figure 3.43 consists of parallel and series two-poles.

The transfer function for translational motion propagation in the *x*-direction via the mechanical system depicted in Figure 3.44 is equivalent to that of the electrical ladder network shown in Figure 3.43(b), if we use the force-to-current, velocity-to-voltage electromechanical analogy (described in detail in Section 7.1.1).

In ladder networks, transfer function zeros can only result either from infinite impedances in the series branches or from zero impedances in the shunting branches. Hence, as long as the branch impedances do not have right half-plane poles and zeros, the transfer function cannot possess zeros in the right half-plane. More generally, the branch impedances need to be *positive real*, which implies passivity—refer to Appendix 3. Therefore, the transfer function of a passive ladder network is always m.p.

A general network transfer function can be presented as a sum of several m.p. transfer functions connected in parallel. Therefore, the right half-plane zeros of a general network transfer function can only result from mutual cancellation of the output signals of two or several parallel paths, which is feasible even for *s* in the right half-plane. In other words, an n.p. transfer function may result from the parallel connection of several ladder networks.

Similarly, a mechanical system transfer function can become n.p. when the signal propagates from the input to the output along different paths or modes. The incident signal may excite rotational and translational modes, which both contribute to the motion of the target. Examples include pendulum models for propellant slosh in spacecraft, and longitudinal bending of a launch vehicle during ascent.

It is therefore of interest for control system designers to have a simple method of detecting when a parallel connection of several m.p. paths becomes n.p.

Figure 3.45 shows the parallel connection of two links, W_1 and W_2. The composite link's transfer function is $W_1 + W_2$. If both W_1 and W_2 are stable and m.p., then the function $W_1 + W_2$ is m.p. if and only if the Nyquist diagram for W_2/W_1 does not encompass the point -1.

The proof is the following: the sum $W_1 + W_2 = W_1(1 + W_2/W_1)$. Here, W_1 possesses neither zeros nor poles in the right half-plane of s. Therefore, $W_1 + W_2$ has no right half-plane zeros if $(1 + W_2/W_1)$ has no such zeros, i.e., if $1/(1 + W_2/W_1)$ has no such poles. The latter expression is the transfer function of system (b), the stability of which can be verified with the Nyquist diagram for W_2/W_1.

When the Nyquist criterion is used, the index 1 should be assigned to the path with larger gain at higher frequencies in order for the ratio $|W_2/W_1|$ to roll off at higher frequencies. Applications for this criterion will be exemplified in Sections 5.9 and 6.3. The minimum phase property of a system including more than two parallel paths can be verified with the analog of the Bode–Nyquist criterion for successive loop closure.

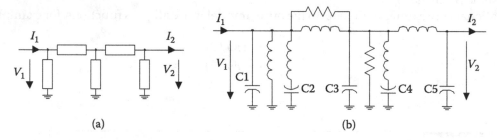

(a)   (b)

FIGURE 3.43
Electrical ladder network: (a) general and (b) example.

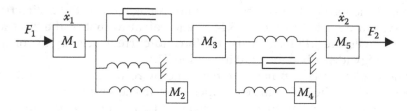

FIGURE 3.44
Mechanical ladder network.

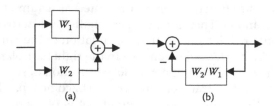

(a)   (b)

FIGURE 3.45
(a) A parallel connection of two links and (b) closed-loop system.

3.14 Other Bode Definite Integrals

Besides the Bode phase-gain relationship, several results can be derived using integrations along the D-shaped contour that give insights into control system design as well as practical computations regarding achievable performance. Appendix 4 organizes the derivations in the manner presented by Bode [6].

3.14.1 Integral of the Feedback

Consider integrating the log of the feedback $\ln|F(s)|$ along the D-shaped contour. If $T(s)$ is open-loop stable, then there are no RHP poles in T or F. If the closed-loop system is stable, then there are also no RHP zeros in F so $\ln|F(s)|$ is analytic in the right half-plane, and the integral around the contour will be zero. As s goes to infinity, $T(s)$ goes to zero and $\ln|1+T(s)|$ goes like $|T(s)|$. For the contribution on the infinite arc to vanish, it is necessary for $s\,|T(s)|$ to go to zero and this is true if T is rational with a relative degree ≥ 2. This is satisfied by almost all practical systems. Then the remaining integral along the imaginary axis, which is the *integral of the feedback*, is zero:

$$\int_{-\infty}^{\infty} \ln|F|\,d\omega = 0 \tag{3.10}$$

When the feedback is negative, $|F| > 1$ and therefore $\ln|F| > 0$, and vice versa. Hence, as illustrated in Figure 3.46, the feedback integral over the frequency region where the feedback is negative is equal to the negative of the integral over the range of positive feedback. The larger the negative feedback and its frequency range, the larger must be the area of positive feedback. Typically, as illustrated in back in Figure 1.8, positive feedback concentrates near the crossover frequency f_b. The positive feedback area in Figure 1.8 looks much smaller because this picture was drawn with a logarithmic frequency scale.

Therefore, if negative feedback reduces the effect of disturbances in certain frequency regions, there must exist a frequency region of positive feedback where these effects are increased. In practice, the feedback decreases the output mean square error since the feedback becomes positive only at higher frequencies where the error components are already reduced by the plant, which is typically a kind of low-pass filter.

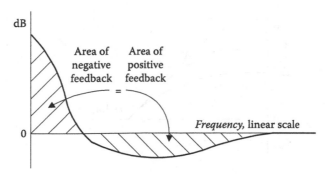

FIGURE 3.46
Negative and positive feedback areas.

Corollary 3.2

If the area of negative feedback over the functional frequency bandwidth needs to be maximized, the area of positive feedback must be maximized as well. The value of the positive feedback depends on the distance from the Nyquist diagram to the critical point. Therefore, this distance should be kept minimal over the bandwidth of positive feedback, or in other words, the Nyquist diagram should follow the stability margins' boundary as closely as possible, and the chosen stability margins should not be excessive.

Corollary 3.3

Since the positive feedback is concentrated within a few octaves near the crossover frequency, the accuracy of loop shaping in the crossover area is of extreme importance for achieving maximum negative feedback over the functional frequency bandwidth.

Equation 3.8 is sometimes expressed in terms of the sensitivity $S = 1/F$, and then referred to as the *Bode sensitivity integral*. Equation 3.10 can be modified in the case that there are RHP poles in T(s), i.e., the open-loop system is unstable.

3.14.2 Integral of the Imaginary Part

If we integrate $|\ln(T(s))|/s$ around the contour (with cut-out at $s = 0$), the relation known as the *phase integral* is derived:

$$\frac{2}{\pi} \int_{-\infty}^{\infty} B(u)\,du = A_\infty - A_0 \tag{3.11}$$

where B is the phase shift of a m.p. function, A_0 and A_∞ are the values of the gain at zero and infinite frequency, respectively, $u = \ln(\omega/\omega_c)$, and ω_c can be arbitrary. In other words, the integral is taken along the frequency axis with logarithmic scale.

The integral can be conveniently applied to the difference between the two gain frequency responses A'' and A' joining at higher frequencies as shown in Figure 3.47, and therefore having the same value of A_∞. By Equation 3.11, this difference is:

$$\Delta A_0 = \left(A_0'' - A_0' \right) = -\frac{2}{\pi} \int_{-\infty}^{\infty} \left(B'' - B' \right) du$$

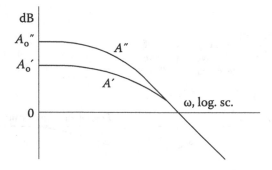

FIGURE 3.47
Two gain responses having a common high-frequency asymptote.

In Equation 3.11, the phase, the gain, and the frequency units are related to the natural logarithm. When the units are converted to degrees, dB, and decades, the low-frequency gain difference is:

$$\Delta A_o, dB = 0.56a \, (\text{decades} \times \text{degrees}) \tag{3.12}$$

where a is the difference in the phase integrals, i.e., the difference in the areas under the phase responses. (The use of this formula will be illustrated in Section 5.5.)

It follows that an increase in the loop gain in the band of operation is accompanied by an increase of the area under the frequency response of the loop phase lag. Hence, the larger the feedback, the larger must be the area of the phase lag. In particular, the available feedback is larger in Nyquist-stable systems because of their larger loop phase lag.

3.14.3 Gain Integral over Finite Bandwidth

Still another important relation is:

$$\int_{\omega=0}^{\omega=1} (A - A_\infty) \, d\arcsin(\omega) = -\int_1^\infty \frac{B}{\sqrt{\omega^2 - 1}} \, d\omega \tag{3.13}$$

With application to feedback systems, Equation 3.13 means that if the phase lag at frequencies $\omega \geq 1$ needs to be preserved, then the loop gain response can be reshaped in the functional band $\omega \leq 1$ as long as the area of the gain plotted against arcsinω is not changed. Applications of this rule will be studied in Chapter 4.

3.14.4 Integral of the Resistance

Next, let $Z(j\omega) = R(j\omega) + jX(j\omega)$ be the impedance of a parallel connection of a capacitance C and a two-poles with impedance Z', as shown in Figure 3.48, where Z' is assumed not to reduce to zero at infinite frequency and to be limited at all frequencies. Then, at higher frequencies, $Z \approx 1/(j\omega C)$. Integrating the impedance along the usual contour results in the following equation called the *resistance integral*:

$$\int_0^\infty R \, d\omega = \frac{\pi}{2C} \tag{3.14}$$

It is seen that the area under the frequency response of the resistance R is exclusively determined by the parallel capacitance C. The frequency responses of R in Figure 3.49(d) relate to the two-poles of Figure 3.49(a)–(c) with different Z' but the same C. The area under the curves is the same. It is also seen that the maximum value of R over the desired frequency band can be achieved if R equals zero outside the operational band, which can be achieved by using Z' as a filter loaded at a matched resistor.

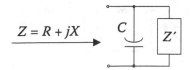

$$Z = R + jX$$

FIGURE 3.48
Two-pole Z' shunted by capacitance C.

The similarly derived integral of the real part of the admittance Y,

$$\int_0^\infty \text{Re}\, Y(\omega)\, d\omega = \frac{\pi}{2L} \tag{3.15}$$

is valid for the dual circuit shown in Figure 3.50, where the admittance Y' of the remaining part of the circuit does not turn to 0 at infinite frequency, i.e., does not contain a series inductance. Equations 3.8 and 3.9 are widely applied in radio frequency and microwave engineering for the evaluation of the available bandwidth performance product in systems where the stray reactive element, C or L, becomes critical (in particular, in the input and output circuits of wide-band high-frequency amplifiers, or in the parallel or series feedback paths of such amplifiers).

The resistance integral is also useful for estimating the available performance of the control and active damping of mechanical flexible structures. Important classes of flexible plants include active suspension systems, micromachined mechanical systems, and large, relatively lightweight actively controlled and damped structures in zero-gravity environments.

In mechanical flexible structures where some flexible modes need to be damped, sometimes the damper can be connected only to the port where a mass or a spring limits the bandwidth of a disturbance isolation system. In this case, diagrammed in Figure 3.51, to achieve maximum performance over a specified bandwidth, Z' can be implemented electrically and connected to the mechanical structure via an electromechanical transducer.

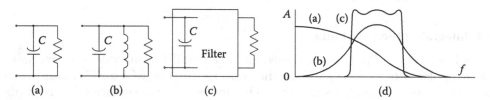

FIGURE 3.49
Two-poles made of reactive two-ports loaded at resistors: (a) low pass, (b) resonance, (c) Chebyshev band-pass filter, and (d) their resistive components.

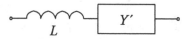

FIGURE 3.50
Stray inductance limiting the real part of admittance.

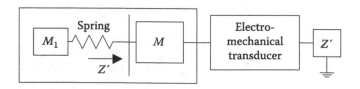

FIGURE 3.51
Active damping of a mechanical structural mode by connecting an active damper with impedance Z'.

PROBLEMS

Many problems on Laplace transform and frequency responses can be found at the end of Appendix 2.

1. Derive Equation 3.1.

2. What frequency bandwidth is required for the rise time to be less than the following:
   a. 5 s
   b. 0.5 s
   c. 2 ms
   d. 1 ns
   e. 2.72 ps

3. Calculate the rise time for the system with the highest-order frequency response from those shown in Figure 3.8.

4. The plant is a rigid body, the input signal is a force, the output is an (a) acceleration, (b) velocity, or (c) position. What is the slope of the plant gain response, in dB/octave and dB/decade?

5. The plant is a capacitor, the input signal is a current, the output is a (a) voltage or (b) charge. What is the slope of the plant gain response, in dB/octave and dB/decade?

6. Plot with MATLAB the Bessel filter responses using transfer functions given in Section 3.1.2, with linear and logarithmic frequency scales (use MATLAB commands linspace, logspace, plot).

7. Using some filter design software, print the plots for the gain and phase of a third-order Butterworth filter.

8. The phase stability margin in a homing system is (a) 25°, (b) 35°, (c) 45°, or (d) 55°. What is the hump in dB on the closed-loop response (calculate or use the Nichols plot)?

9. Use the relationship shown in Figure 3.8 to explain why the Bessel filter has nearly no overshoot.

10. Why are Nyquist-stable systems stable? (The gain about the loop in such systems exceeds 1, the phase is 0, and hence the return signal comes back in phase and with increased amplitude. It is counterintuitive to suggest that such a system is stable, but it is.)

11. Prove the Nyquist criterion using the notion of continuous dependence of the roots of a polynomial on its coefficients. *Hint*: Modify the system under analysis to be definitely stable, and further change the coefficients gradually and continuously until they reach their true values, while observing the topology of the locus of $F(s)$.

12. If the $j\omega$-axis maps onto the locus of $F(j\omega)$, onto what area is the right half-plane of s mapped? What is the mapping of the zeros of $F(s)$?

13. The system is stable open loop. The Nyquist diagram makes two revolutions about the critical point. Is the closed-loop system stable? How many poles in the right half-plane does the closed-loop transfer function have?

14. Plot the Bode diagram and the *L*-plane Nyquist diagram for the system with an unstable plant, shown in Figure 3.23. Change the value of damping in the local feedback path of the unstable plant, and check the system stability. The following SPICE program analyzes the equivalent electrical schematic diagram in Figure 3.52.

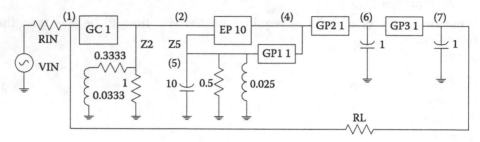

FIGURE 3.52
SPICE model (de-floating 1 MEG resistors not shown) of the system with an unstable plant shown in Figure 3.23.

```
** unst_pl.cir for feedback loop with unstable plant
* compensator, inverting, zero s=10, pole s=40,
* crossover s=20
GC 2 0 1 0 1
RC1 1 0 1MEG ; for SPICE, to make node (1) non-floating
RC2 2 0 1
RC3 2 3 0.33333
LC 3 0 0.033333 ; Z2=(s+10)/(s+4)
* plant loop, non inverting, closed loop 0.1s/(s^2+0.2s+4)
EP 5 0 2 4 10
GP1 4 0 5 0 1
RP 4 0 0.5
CP1 4 0 10
LP 4 0 0.025 ; Z5=0.1s/(s^2+0.05s+4)
* plant integrators, (4) to (6) to (7)
GP2 6 0 0 5 80
GP3 7 0 0 6 1
RP2 5 0 1MEG ; for SPICE
RP3 6 0 1MEG ; for SPICE
RP4 7 0 1MEG ; for SPICE
CP2 6 0 1
CP3 7 0 1
* loop closing resistor
RL 7 1 1MEG ; to close the loop, place semicolon
 ; after M to reduce RL
*
```

```
VIN 8 0 AC 1
RIN 8 1 1
.AC DEC 100 0.1 20
.PROBE ; (if using PSPICE)
.END
```

When using the PROBE postprocessor to plot open-loop responses with convenient scales, plot vdb(7) and 0.2*vp(7). To plot the Nyquist diagram, make the x-axis scale linear, the x-variable being vp(7), and the range −50, 400; plot vdb(7). To plot the plant loop Bode diagram, plot vdb(4)-vdb(5)+20. Alternatively, use MATLAB, Simulink, or some other analysis program.

15. In the plant of the system shown in Figure 3.23, change the gain block coefficient to 5. Is the system stable?

16. Depending on the angle of attack of the horizontal stabilizer, the airplane in Figure 3.53(a) can be statically unstable. The block diagram for the pitch autopilot feedback loop is shown in Figure 3.53(b). Here, τ is the torque and J is the moment of inertia about the center of gravity CG. Consider $J = 2{,}000$, the plant aerodynamics transfer function $-1{,}000s + 2{,}000$, the control surface gain coefficient of 1,000, and the compensator transfer function $C = 400(s+5)/(s+20)$. Use SPICE, MATLAB, or Simulink. Is this airplane stable with the autopilot?

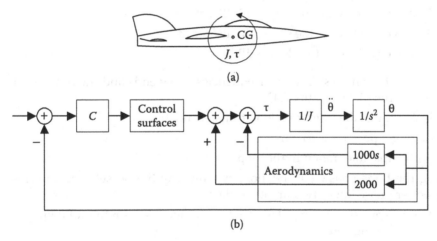

FIGURE 3.53
System with aerodynamic instability in the plant.

17. Show that Equation 3.16 is p.r.

18. Which of the following functions can be used as a model of a passive physical plant, and for plotting a Nyquist diagram for stability analysis:

a. $T = 250(s+5)/[(s+5)(s+5)(s+5)]$

b. $T = 250(s+5)/[(s+5)(s+5)]$

c. $T = 250(s+5)(s+50)(s+50)/(s+500)$

d. $T = 250(s+5)(s+5)(s+50)(s+500)$

19. In literature, you can often read the sentence: "The feedback in this system is negative." Does this statement need qualification? Of what should you be aware?

20. What is the slope of the Bode diagram and the phase shift of the following functions:

    a. $s^{-0.6}$

    b. $s^{-1.6}$

    c. $s^{-0.2}$

    d. $s^{-0.8}$

    e. $s^{-2.7}$

21. What are the guard-point gain and phase stability margins if $T = 6{,}200(s+2.5)/[(s+5)(s+5)(s+50)]$? Is the feedback negative at the crossover frequency?

22. If the slope of a Bode diagram at all frequencies is (a) –9 dB/octave, (b) –40 dB/decade, (c) –8 dB/octave, or (d) –30 dB/decade, what is the phase shift?

23. If the phase shift at all frequencies is (a) 90°, (b) –40°, (c) –150°, or (d) –210°, what is the gain response?

24. The gain response can be approximated by a segment:

    a. 3 octaves wide with the slope –10 dB/octave

    b. 1.5 octaves wide with the slope –6 dB/octave

    c. 2 octaves wide with the slope –12 dB/octave

    d. 3 octaves wide with the slope –20 dB/decade

    e. 1 octave wide with the slope 5 dB/octave

    What is the phase shift at the frequencies at the ends and the center of the segment (use the plot in Figure 3.42)?

25. Draw a Nyquist diagram for a loop transfer function:

    a. having a pole at the $j\omega$-axis

    b. with triple integration in the loop

26. Using the Bode phase-gain relation, explain why the Bessel-type gain response has better phase linearity than other filters.

27. Find the expression for n.p. lag at lower frequencies for the following transfer functions having zeros:

    a. $(3, -5, 0.1)$

    b. $(-1, -3, -10)$

    c. $(10, 5, -3)$

    d. $(5 \pm j12)$

    e. $(4, -15, 0.1)$

    f. $(-1, -4, 10)$

    g. $(10, -5, -3)$

    h. $(6 \pm j1.2)$

28. Sketch a Nyquist diagram on the L-plane related to the Bode diagram in Figure 3.18(a).

29. Plot phase responses for piecewise-linear approximations of certain Bode diagrams shown by responses 1–4 in Figure 3.54. For convenience, copy the pages with Figures 3.42 and 3.54 using the same magnification (either 1:1 or 1:1.5). To copy the plots from one sheet of paper to another, superimpose the two sheets on a light table (or against a window glass), or make and use a transparent paper copy of the page with the problem. For $w = 1.5$ or 2.4, interpolate.

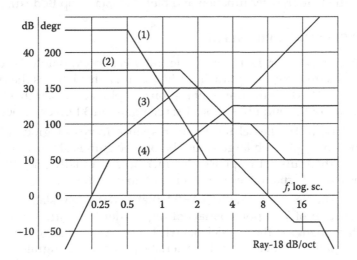

FIGURE 3.54
Piece-linear Bode diagrams for phase calculation exercises.

Which of the four responses is suitable for the loop transfer function? Is this system absolutely or Nyquist stable?

30. Figure 3.55 shows the loop gain responses of the thruster attitude control about the x- and z-axes of a spacecraft. Approximate the responses by segments and rays. Calculate the phase shift.

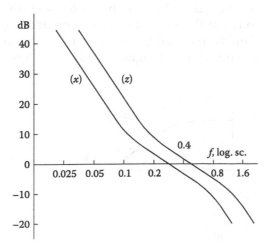

FIGURE 3.55
Bode diagrams for attitude control loops of a spacecraft.

31. Determine whether $W_1 + W_2$ is m.p. for the following:

    a. $W_1 = 1,000/[s^2(s+10)]$, $W_2 = (s+2)(s+5)/[(s+200)(s+500)]$

    b. $W_1 = 10,000/[s^2(s+20)]$, $W_2 = (s+4)(s+10)/[(s+400)(s+1,000)]$

    c. $|W_1| < |W_2|$ at all frequencies, and W_1 and W_2 are m.p.

32. In a homing system, the closed-loop response is that of the fourth-order Bessel filter. Find the loop transfer function and plot the open-loop Bode diagram.

ANSWERS TO SELECTED PROBLEMS

3. The cutoff frequency for the highest-order filter is approximately 2 rad/s, and the period for this frequency is approximately 3 s, and a third of it is the rise time ≈ 1 s. From Figure 3.8 where the transient response is plotted, we see that the rise time (from 0.1 to 0.9 of the output) is about 1 s; i.e., Equation 3.1 gives a good result.

9. Since the slope of the Bessel filter gain response increases gradually with frequency and does not reach large values before the gain is already very small, the curvature of the transient response also increases gradually and does not reach large values, which guarantees the absence of an overshoot.

11. Start with a simple system with $T(s) = 0.1$ that is definitely stable. Augment the transfer function of T to polynomials of high order but with infinitesimal coefficients. Start changing the coefficients while observing changes in the resulting Nyquist diagram. If a root of the characteristic polynomial migrates into the right half-plane of s, it must cross the $j\omega$-axis. At the crossing frequency the closed-loop gain must become infinite, and F, therefore, must become 0. If this never happens, i.e., the locus of F during the modifications never crosses the origin, i.e., the locus of T never crosses the point -1, then the system must be stable.

24. (a) The phase shift is 110° at the center of the segment, 63° at its ends.

26. Bessel filter gain response can be approximated by a piecewise-linear response, as shown in Figure 3.56 (not to scale). It is seen that the phase response is rather close to linear because of the contribution to the phase lag at lower frequencies made by the gain response slope at lower frequencies. For higher selectivity filters (Butterworth, Chebyshev, Cauer), the gain response slope at lower frequencies is negligible, but the slope becomes very steep right after the passband ends, and within the passband, correspondingly, the phase lag is smaller at lower frequencies and higher at higher frequencies, thus making the phase response curved.

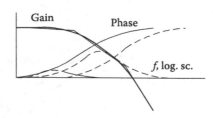

FIGURE 3.56
Bessel filter gain response and phase response.

Another explanation is that to make the phase lag proportional to frequency, the phase lag increase over an octave should double such an increase over the preceding octave. Therefore, roughly speaking, the increase in the slope of the gain response should double each octave.

27. (a) $B_n = 2[\arctan(10\omega) + \arctan(\omega/3)]$.

32. $105/(s^4 + 10s^3 + 45s^2 + 105s)$. The response is plotted in Figure 3.57.

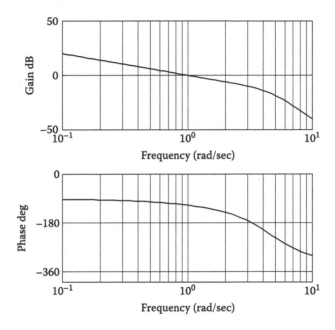

FIGURE 3.57
Bode diagram.

4

Shaping the Loop Frequency Response

The problem of optimal loop shaping encompasses two fairly independent parts that can be solved sequentially (thus making the design structural.) The first part is *feedback bandwidth maximization,* which is solved by appropriately shaping the feedback loop response at higher frequencies (in the region of crossover frequency and higher). The second part is the *distribution of the available feedback* over the functional feedback band.

The feedback bandwidth is limited by the sensor noise effect at the system output, the sensor noise effect at the actuator input, plant tolerances (including structural modes), and non-minimum phase lag (analog and digital) in the feedback loop. The optimal shape of the loop gain response at higher frequencies, subject to all these limitations except the first, includes a Bode step.

The Bode step is presented in detail as a loop shaping tool for maximizing the feedback bandwidth. The problem of optimal loop shaping is further described, and formulas are presented for the calculation of the maximum available feedback over the specified bandwidth.

The above solution is then generalized by the application of a Bode integral to reshaping the loop gain response over the functional bandwidth (thus solving the second part of the loop shaping problem). It is shown that the feedback is larger and the disturbance rejection improved in Nyquist-stable systems.

Loop shaping is described for plants with flexible modes for collocated and non-collocated control, and for loops where the plant is unstable. The effect of resonance mode coupling on loop shaping in MIMO systems is considered.

It is described how to shape the responses of parallel feedback channels to avoid non-minimum phase lag, while providing good frequency selection between the channels.

When the book is used for an introductory control course, Sections 4.2.5–4.2.7, 4.3.3, 4.3.6, 4.3.7, 4.4, and 4.5 can be omitted.

4.1 Optimality of the Compensator Design

Webster's Collegiate Dictionary defines the word *optimal* as "most favorable or desirable," and in practical engineering the optimal design is that which maximizes customer satisfaction. The best controller is defined by some combination of the accuracy of implementation, repeatability, reliability, disturbance rejection, and the quality of design vs. price. These trade-offs are easily resolved with Bode integrals.

Improving the performance of compensators is many times less expensive than improving the precision of the plants they are controlling. Hence, it is worthwhile to make them close to optimal, even at the price of making them complex. To be reasonably close to optimal, the order of the compensator is, typically, 8 to 25.

Example 4.1

Increasing the compensator order from 4 to 12 and including in the compensator several nonlinear links will only add 20 to 30 lines of code to the controller software or several extra resistors, capacitors, and operational amplifiers, if the compensator is analog. For example, if the settling time of some expensive manufacturing machinery with a short repetition cycle of operation can be reduced by 20% while retaining the same accuracy, the resulting time per operation might be reduced by, say, 5%, and the number of pieces of the equipment at the factory can be correspondingly reduced by 5%, with additional savings on maintenance. Or, a fighter's maneuverability can be noticeably improved. Or, the yield of a chemical process can be raised by 2%, etc. This is why the compensators should be designed to provide close to optimal performance, and not just the performance specified by a customer representative who doesn't know in advance what kind of performance might be available.

Even when the accuracy of the system with a simple controller suffices, it still pays to improve the controller, since with larger margins in accuracy, the system will remain operational when some of the system parameters degrade to the point that, without the better controller, the system fails.

Linear compensators are fully defined by their frequency responses. Therefore, the problem of optimal linear compensator design is the problem of optimal loop response shaping. The theory of optimal feedback loop shaping should be able to provide the answers to the questions: (1) What performance is feasible? (2) What loop response achieves this performance limit?

Commonly, control systems are initially designed as linear with some idealized plant model. Still, the design must result in a sound solution when the idealized plant is replaced by high-fidelity models of the plant and actuator with nonlinearities, as well as when used with the actual hardware.

Physical system models must reflect the uncertainty in the system parameters, the asymptotic behavior of the transfer functions at higher frequencies, the sensor noise, and the significant nonlinearities.

Example 4.2

In the paper "When Is a Linear Control System Optimal?" [27], considered by many to be a cornerstone of "modern" control, the following definition is given: "A feedback system is optimal if and only if the absolute value of the return difference is at least one at all frequencies." However, this theory cannot be applied to physical systems because at higher frequencies the loop gain drops faster than an integrator, and according to Equation 3.10 and Figure 3.46, the integral from 0 to ∞ of the return difference is 0, not ∞.

Optimal control law design must provide timely performance information to the system engineers, so that requirements to the system hardware can be evaluated, possibly allowing the use of simpler and/or cheaper components.

Example 4.3

In Chapter 1, we analyzed the design of an antenna elevation angle controller—without proving that this design is the best possible. (It is not. The real-life controller is multi-loop, and includes signal feedforward and high-order nonlinear compensators.) Reasonable questions for the customer to ask are: Is this design the best? If not, by how much can it be improved?

Normally, we should not even ask: at what cost? As we already mentioned, the cost of the compensator is small compared to the cost of the antenna dish (although better or additional sensors, bearings, and servomotors can add to the system cost).

Also, the loop responses in this example do not include the plant structural modes. Given the modes, the loop gain at higher frequency should roll off fast to avoid instability.

The system engineer asked the control designer to find the shape of the control loop response that is optimal for some specific task, and to estimate the available performance without completing the lengthy compensator design. Can this be done?

Example 4.4

A flexible mode of a feedback structure may be anywhere between 30 and 40 Hz. This resonance accentuates the vibration noise, thus decreasing the accuracy of the device. The feedback system must therefore be able to reject the noise over the entire bandwidth 30 to 40 Hz. How do we implement the maximum feedback, and what is the value of this feedback? Can it be quickly calculated without designing the controller?

4.2 Feedback Maximization

4.2.1 Structural Design

We already mentioned the advantages of structural design in Section 2.9 (and will further discuss them in Section 7.2.1). The structural design flowchart for the basic feedback control system is shown in Figure 4.1.

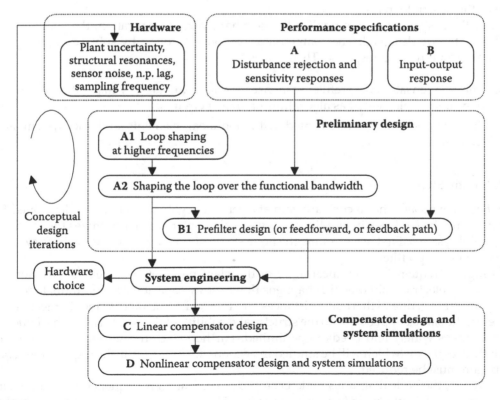

FIGURE 4.1
Feedback-system design flowchart.

The performance specifications comprise the desired responses of:

A: The disturbance rejection and the sensitivity, i.e., the loop gain.

B: The nominal command-to-output transfer function.

The preliminary design addresses specification **A**, first, by *loop response shaping*. This can be subdivided into:

A1: Achieving maximum feedback bandwidth by appropriate loop shaping at higher frequencies.

A2: Distributing the achieved feedback over the functional frequency range so as to exceed the most stringent specifications.

Since the first subproblem is to a large extent independent from the second one, it makes sense to solve them sequentially. At this stage, the responses need not be expressed by rational functions. They can be expressed by rational or transcendental functions, by plots, or by tables.

After the loop is shaped, the command-to-output response is modified (if required) by adding an appropriate prefilter or command feedforward to meet the specifications **B**.

During the conceptual design of complex engineering systems, only steps **A1** and **A2** (and sometimes **B1**) need to be performed to provide the system engineers with accurate data on the *available* control performance. The system engineers evaluate different versions of the hardware/software configurations iteratively, using the preliminary design results on the available control performance. Inclusion of only stages **A1**, **A2**, and **B** in the iteration makes the design loop fast.

After the system engineers finalize the best hardware configuration, the compensator and prefilter are designed. The compensator response is found by subtracting the plant response from the loop response. Then:

C: The compensator and prefilter responses are approximated by m.p. rational functions and implemented as algorithms or analog circuits.

D: The compensator is augmented with nonlinear elements and the system is simulated.

4.2.2 Bode Step

The *Bode step response* has a constant gain at a certain negative level. It was developed for maximizing the feedback in amplifiers. As we explain later, it also optimizes the feedback in general systems. First, we examine the Bode step response and show why these systems work well with prefilters.

At higher frequencies, the uncertainty of the plant parameters increases variations of the gain. In electrical circuits, this happens due to stray capacitances and inductances; in thermal systems, due to thermal resistances and thermal capacitances; and, in mechanical systems, the same happens due to the structural flexibility of the plant. The high frequency gain changes widely and needs to be limited. Further, since the sensor noise increases with the frequency (which will be studied in Sections 4.3.2 and 4.3.3), the higher-frequency loop gain must be reduced.

Therefore, the Bode diagram should be approximated at high frequencies by a line with the *asymptotic slope* of $-6n$ dB/octave, where $n \geq 2$. The loop gain $|T(j\omega)|$ decreases at higher frequencies at least as ω^{-2}, and the integral of feedback (Equation 3.10) is therefore zero.

The crossover is the region of transition between the frequency of positive feedback and the higher band where the feedback becomes negligible. As was stated in the corollaries in Section 3.14, proper shaping of the Bode diagram in the crossover region (step **A1** from Section 4.2.1) is critical to achieving the maximum area of positive feedback near the crossover, and therefore achieving the maximum area of negative feedback in the functional feedback band.

When the system contains saturation, its absolute stability is required with the stability margin shown in Figure 4.2(a).

From the Bode integral of phase it follows that the Nyquist diagram should follow the boundary curve as closely as possible.

The required Nyquist diagram is produced by the Bode diagram, as shown in Figure 4.2(b). The loop response is piecewise linear. The corner frequencies are f_d and f_c. The loop phase lag is less than $(1-y)180°$ until the loop gain becomes smaller than $-x$ dB.

The system is phase stabilized with the margin $y180°$ up to the frequency f_d, where the loop gain has dropped to $-x$. As can be seen from the Bode phase-gain relation, the slope of the Bode diagram at these frequencies approximates $-12(1-y)$ dB/octave.

The high-frequency asymptotic loop response is defined by:

1. The asymptotic slope $-6n$ dB/octave.

2. The point on this asymptote with coordinates $(f_c, -x)$, as shown in Figure 4.2(b).

3. The non-minimum phase lag at this frequency $B_n(f_c)$.

The transition between the slope $-12(1-y)$ dB/octave and the high-frequency asymptotic slope must be short to maximize the gain in the functional frequency range while reducing the loop gain at higher frequencies. The Bode step is made at the gain level $-x$ dB, as shown in Figure 4.2(b).

Without the step, the phase lag in the crossover area would be too large due to the steep high-frequency asymptote and the non-minimum phase lag. The step reduces the phase lag at the crossover frequency, but also reduces the loop selectivity (i.e., given the high-frequency asymptote, reduces the feedback bandwidth). Therefore, the length of the step must not be excessive.

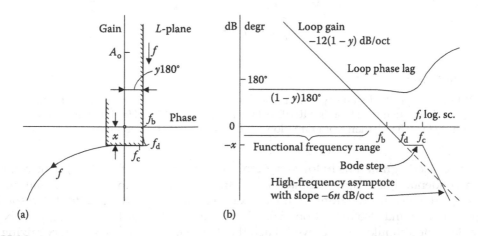

FIGURE 4.2
Bode step: (a) absolute stability boundary on the *L*-plane, and (b) piece-linear gain response with related phase lag response that produces the Nyquist diagram shown in (a), which approximates the boundary.

The non-minimum phase lag $B_n(f_c)$ is assumed to be less than 1 radian, which is true in well-designed systems. With the linear approximation (Equation 3.9), the non-minimum phase lag at frequencies lower than f_c is:

$$B_n \approx |B_n(f_c)| \frac{f}{f_c}$$

The phase lag due to the slope of the asymptotic ray that starts at f_c can be approximately expressed with Equation 3.9 as:

$$\frac{2}{\pi} n \frac{f}{f_c}$$

Consider next the "discarded" dashed-line ray in Figure 4.2(b), which is the extension of the main slope line. It starts at the frequency f_d. The phase lag related to this ray is approximately expressed with Equation 3.9 as:

$$\frac{2}{\pi} 2(1-y) \frac{f}{f_d}$$

To make the loop lag at frequencies $f < f_d$ equal to $(1-y)180°$, the sum of the phase contribution of the asymptotic slope and the non-minimum phase lag should equal the phase contribution of the discarded dashed-line ray. This consideration is seen as:

$$\frac{2}{\pi} n \frac{f}{f_c} + |B_n(f_c)| \frac{f}{f_c} \approx \frac{2}{\pi} 2(1-y) \frac{f}{f_d}$$

From this equation, the Bode step frequency ratio is

$$\frac{f_c}{f_d} \approx \frac{n + \frac{\pi}{2} |B_n(f_c)|}{2(1-y)} \tag{4.1}$$

For the typical phase stability margin of 30°, i.e., $y = 1/6$,

$$\frac{f_c}{f_d} \approx 0.6n + B_n(f_c) \tag{4.2}$$

Example 4.5

When the specified stability margins are 30° and 10 dB, and the crossover frequency $f_b = 6.4$ kHz, then the slope is -10 dB/octave and $f_d = 2f_b = 12.8$ kHz. Further, when $n = 3$ and $B_n(f_c) = 0.5$ rad, then from Equation 4.2, $f_c/f_d = 2.3$, so that $f_c \approx 30$ kHz.

The Nyquist diagram thus follows the stability boundary in Figure 4.2(a). This response is transcendental but is approximated by a rational transfer function, thus making the corners on the Nyquist diagram rounded. Examples of designs with Bode step responses are given below, and also in Sections 5.6, 5.7, and 5.10, in Chapter 13, and in Appendix 13.

The Bode step should be employed when the dominant requirement is maximizing the disturbance rejection, i.e., maximizing the feedback in the functional frequency range. This case is common but not ubiquitous. Noise reduction and certain implementation issues may require Bode diagrams to be differently shaped in the crossover area. The Bode diagram must be made shallow over some range in the crossover frequency region to ensure the desired phase stability margin. There are several options for where to do this: to the right of f_d by the Bode step, to the left of the crossover frequency, as in systems where the sensor noise is critical, and over a frequency range nearly symmetrically situated about the crossover, as in the so-called *PID controller*, which will be discussed in Chapter 6. Loop responses without any Bode step typically provide 4 to 20 dB less feedback in the functional frequency range.

The output transient response to a step disturbance in a homing system with a Bode step and 30° to 40° phase stability margin has substantial overshoot. If this overshoot exceeds the specifications, the loop gain response in the neighborhood of f_b should be made shallower. However, this will reduce the available feedback and the disturbance rejection.

The prefilter or one of its equivalents may be introduced to ensure good step responses without reducing the available feedback.

4.2.3 Example of a System Having a Loop Response with a Bode Step

In this section an example of the implementation of a loop response with a Bode step is considered. This response has been used as a prototype for several practical control systems used for the Chemistry spacecraft.

Example 4.6

The plant

$$P(s) = \frac{1}{s^2} \frac{10 - s}{10 + s}$$

is a double integrator with n.p. lag. The compensator function $C(s)$ is a ratio of a third-order to a fourth-order polynomial (the compensator design methods are described in the next chapter). The loop transfer function

$$T(s) = C(s)\frac{1}{s^2}\frac{10 - s}{10 + s} = \frac{11s^3 + 55s^2 + 110s + 36}{s^4 + 7.7s^3 + 34s^2 + 97s + 83}\frac{1}{s^2}\frac{-s + 10}{s + 10} = \frac{n(s)}{d(s)}$$

is plotted in Figure 4.3 with:

```
n = conv([11 55 110 36], [-1 10]);
d = conv([1 7.7 34 97 83 0 0], [1 10]);
w = logspace(-1,1,200);
bode(n,d,w)
```

The crossover frequency $\omega_b = 1$ rad/s. The *L*-plane Nyquist diagram is plotted in Figure 4.4 with:

```
[mag, phase] = bode(n,d,w);
plot(phase,20*log10(mag),'w', -180,0,'wo'); grid
```

The slope of the loop gain at lower frequencies approaches −12 dB/octave, and the loop phase shift approaches −180°. This response provides larger feedback at lower frequencies than the response in Figure 4.2(b), following the stability margin boundary in Figure 4.2(a) but violating the stability margin at these frequencies. To ensure stability

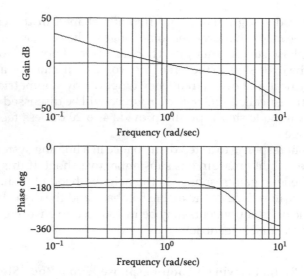

FIGURE 4.3
Open-loop frequency response.

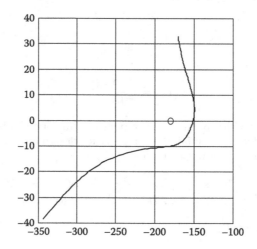

FIGURE 4.4
L-plane Nyquist diagram.

and good transient responses to large commands and disturbances (which saturate the actuator), the loop compensation must be nonlinear. Such compensation is described in the end of this example and explained at length in Chapters 10, 11, and 13.

The closed-loop transfer function $M = n/g$, where $g = n + d$, is plotted in Figure 4.5 with:

```
n = conv([0 0 0 1],n);
g = n + d;
w = logspace(-1,1,200);
bode(n,g,w)
```

(The zeros are added to n to make the dimensions of vectors n and d equal so that MATLAB® can calculate g.)

As should be expected, the closed-loop step response plotted by step(n,g) and shown in Figure 4.6 has a large (55%) overshoot. The overshoot can be reduced by introduction of a prefilter $R(s)$ without any sacrifice in feedback.

The command-to-output response of the system with the prefilter should be close to the response of a linear phase (Bessel or Gaussian) filter. This can be achieved by using a notch prefilter:

$$R(s) = \frac{s^2 + \omega_b s + 0.81\omega_b^2}{s^2 + 2\omega_b s + 0.81\omega_b^2}$$

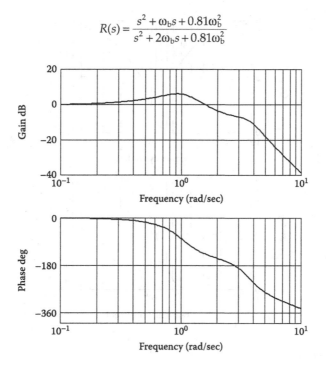

FIGURE 4.5
Closed-loop frequency response.

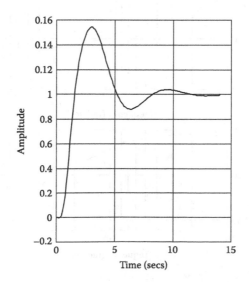

FIGURE 4.6
Closed-loop step response.

The notch is tuned to $0.9\omega_b$ with the gain at this frequency equal to $20 \log(1/2) = -6$ dB. In the case considered, $\omega_b = 1$. The prefilter response is plotted in Figure 4.7 with:

```
nr = [1 1 0.81]; dr = [1 2 0.81]; bode(nr, dr, w)
```

The closed-loop response including the prefilter is plotted in Figure 4.8 with:

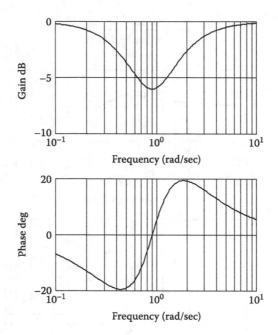

FIGURE 4.7
Prefilter frequency response.

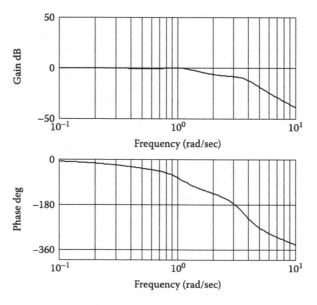

FIGURE 4.8
Closed-loop response with a notch prefilter.

```
nc = conv(nr, n); gc = conv(dr, g);
bode(nc, gc, w)
```

The phase response in Figure 4.8 has less curvature than that in Figure 4.5. We might therefore expect a better transient response. Indeed, the overshoot is only 7% on the step response. The step response with the prefilter is shown in Figure 4.9. The rise time and settling time remain approximately the same.

The gain response in Figure 4.8 still has a hump at $\omega=1$, which contributes to the phase nonlinearity and the overshoot. By compensating for the hump with a small (1.6 dB) additional notch with:

```
nr2 = [1 0.5 1]; dr2 = [1 0.6 1];
nc2 = conv(nr2, nc); gc2 = conv(dr2, gc);
bode(nc2, gc2, w)
step(nc2, gc2)
```

one can further reduce the overshoot, as shown in Figure 4.10. A still better equalization can be devised and implemented.

The equalization becomes less accurate when the plant parameters deviate from the nominal. The plant gain variations can be specified by a multiplier k. The effects of variations in k can be calculated as follows:

```
n = conv([11 55 110 36], [-1,10]);
d = conv([1 7.7 34 97 83 0], [1
 10]);
k = 1; n = conv(n, k);              % specify k
w = logspace(-1,1,200);
bode(n, d, w)                       % open-loop response
[mag, phase] = bode(n, d, w);
plot(phase, 20*log10(mag))          % Nyquist diagram
grid
n = conv([0 0 1], n);
```

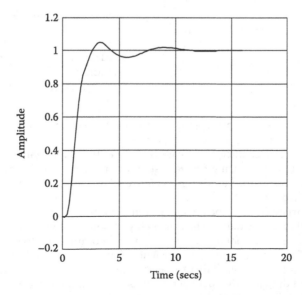

FIGURE 4.9
Closed-loop step response with a prefilter.

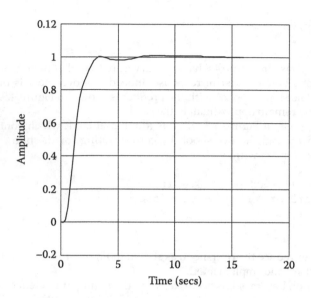

FIGURE 4.10
Closed-loop response of the system with a prefilter composed of two notches.

```
g = n + d;
bode(n, g, w)                    % closed-loop response
nr = [1 1 0.81]; dr = [1 2 0.81];
bode(nr, dr, w)                  % prefilter notch response
nc = conv(nr, n); gc = conv(dr,
 g);
bode(nc, gc, w)                  % closed-loop response with
                                   notch
nr2 = [1 0.5 1]; dr2 = [1 0.6 1];
bode(nr2,dr2, w)                 % second notch response
nc2 = conv(nr2, nc); gc2 = conv
 (dr2, gc);
bode(nc2, gc2, w)                % closed-loop response with
                                   notches
step(nc2, gc2)                   % step response with notches
```

For ±20% variations in the plant gain coefficient, the step responses are shown in Figure 4.11. The overshoot remains less than 10%.

When the plant's gain coefficient increases to 2.5, i.e., the plant's gain increases by 8 dB from the nominal, thus reducing the gain stability margin to only 2 dB, the transient response becomes oscillatory, as shown in Figure 4.11(b); still, this is not a catastrophic failure of the controller.

With ± 40% variations in k, the overshoot remains under 20%, and with $k \geq 3.2$ (i.e., with $20 \log k = 10.1 > x$), the output starts growing exponentially; the proof (simulation) is left as a recommended student exercise. It is also recommended to make simulations with small (say, 5%) variations in the coefficients of the polynomials in the compensator transfer function and to observe that these changes do not critically affect the system performance, and therefore the coefficients can be appropriately rounded.

To ensure that large-amplitude commands that overload the actuator will not trigger self-oscillation, and to improve the transient responses for large-amplitude commands,

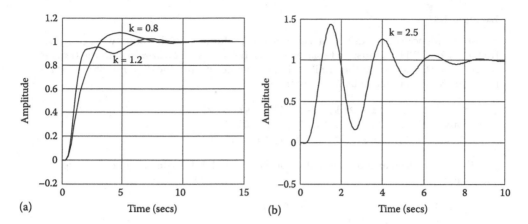

FIGURE 4.11
Closed-loop step responses of the system with notches: (a) for $k=1.2$ and $k=0.8$ and (b) for $k=2.5$.

the compensation can be made nonlinear. This is done by splitting the compensator transfer function $C(s)$ into the sum of two parallel paths with transfer functions, $C_1(s)$ and $C_2(s)$, the first path being dominant at lower frequencies. A saturation link with an appropriate threshold is placed in front of the first path's linear link. The related theory and design methods are described in Chapters 10, 11, and 13. The transfer functions of the paths can be found as:

$$C_1 = a_0 / (s + p_1) \text{ and } C_2 = [C(s) - C_1(s)]$$

where p_1 is the lowest pole of $C(s)$ (p_1 can be 0), and a_0 is its residue. For finding p_1 and a_0, the MATLAB function `residue` or the function `bointegr` from the Bode step toolbox in Appendix 14 can be used.

Example 4.7

A dc motor rotates a spacecraft radiometer antenna whose moment of inertia is $J = 0.027$ Nm². θ is the antenna angle, $s\theta$ is the antenna angular velocity, and U is the voltage applied to the motor. The motor winding resistance is $R = 2\,\Omega$. The motor constant (the torque-to-current ratio) is $k = 0.7$ Nm/A. The torque is $k(U - E_B)/R$, where the back electromotive force $E_B = ks\theta$. The angle $\theta = [k(U - E_B)/R] / (s^2 J)$.

From the latter two equations, the voltage-to-angle transfer function is

$$\frac{\theta}{U} = \frac{k}{RJ} \frac{1}{s\left[s + k^2 / (RJ)\right]} = \frac{13}{s(s + 9.07)}$$

If the loop n.p. lag equals that in Example 4.6, and if the loop response needs to be the same as in Example 4.6, the compensator and driver transfer function should be $C(s)$ from Example 4.6 multiplied by:

$$C_1 = \frac{U}{\theta} \frac{1}{s^2} = \frac{RJ}{k} \frac{s + k^2 / (RJ)}{s} = \frac{0.077(s + 9.07)}{s}$$

The MATLAB code for this transfer function's numerator and denominator is:

```
k = 0.7; res = 2; ja = 0.027; a = k*k/(res*ja);
nc1 = conv(res*ja/k,[1 a]); dc1 = [1 0];
```

Example 4.8

To use the response in Example 4.6 as a prototype for a control system with crossover frequency $f_b = 12$ Hz, s must be replaced by s/ω_b (where $\omega_b = 12 \times 2\pi \approx 75.4$) in the transfer functions for the loop and the prefilter. The numerator ns and the denominator ds of the loop transfer function can be found with:

```
n = conv([11 55 110 36], [-1,10]); wb = 75.4;
d = conv([1 7.7 34 97 83 0 0], [1 10]);
format short e; [ns, ds] = lp2lp(n, d, wb)
```

and the numerators and denominators of the two notch transfer functions of the prefilter with:

```
nr = [1 1 0.81]; dr = [1 2 0.81];
[nrs, drs] = lp2lp(nr, dr, wb)
nr2 = [1 0.5 1]; dr2 = [1 0.6 1];
[nr2s, dr2s] = lp2lp(nr2, dr2, wb)
```

The results will be displayed with single precision (`format short e`); although, the calculations will be performed and the numbers stored with double precision.

(Notice that if we apply the `lp2lp` transform not to the entire loop but only to the compensator and retain $1/s^2$ as the plant, the new compensator transfer function must be multiplied by ω_b^2.)

Example 4.9

We can now write an example of a design specification for a time-invariant controller for a plant with a prescribed nominal transfer function P_n, with high-frequency flexible modes, and with hard saturation in the actuator:

At $\omega > 8$, the nominal loop gain coefficient should not exceed $0.1/\omega^3$ in order to gain stabilize the uncertain flexible modes.

The output effect of disturbances must be minimized.

With the plant $P = kP_n$ where $0.8 < k < 1.2$, in the linear state of operation (for small commands), the rise time must be less than 2s.

The overshoot/undershoot for small-amplitude step commands must be:

For the nominal plant, under 5%
For $0.8 < k < 1.2$, under 10%
For $0.6 < k < 1.4$, under 20%
For $0.4 < k < 2.5$, under 50%

These norms on the overshoot are also applied to large-amplitude step commands (up to overloading the actuator 10 times).

No step command may trigger self-oscillation for $0.4 < k < 2.5$.

The specifications on the overshoot guarantee minimum performance for the worst case of maximum plant parameter variations, and also ensure that most systems in production (with the plant parameters close to the nominal) have much better performance than the worst case.

From Example 4.6 it is seen that these specifications can be met when the stability margins are 30° and 10 dB. If smaller stability margins are chosen, the sensitivity in the frequency neighborhood of f_b increases and the variations in k cause larger deviations of the transient response from the nominal.

Example 4.10

The linear prefilters described in Figures 4.5–4.10 helped the transient response of a linear feedback system within the working range of the signals. However, the prefilter response changes when the input working range largely extends the feedback system output working range, and especially when the nonlinear elements become more complicated than simple saturation. In this case we might employ nonlinear elements in the prefilter.

For example, the prefilter in Figure 4.12 uses two parallel paths, one with unity gain and another the cascade combination of a dead zone and a linear filter. It was employed successfully in a space system giving two times reduction in the transient response and three times less overshoot. The bandwidth of the system is several hertz; the parameters of the system should be experimentally modified.

4.2.4 Reshaping the Feedback Response

We proceed next with step **A2** from Section 4.2.1, shaping the loop response over the functional feedback bandwidth.

In many practical cases the amplitude of the disturbances decreases with frequency, and the loop gain should also decrease with frequency. If, further, the system is phase stabilized with a constant stability margin, the loop gain response is similar to that shown in Figure 4.2 and to response 3 in Figure 4.13(a).

To match a specified disturbance rejection response, the constant slope gain response must be modified. Figure 4.13(a) gives several examples of feasibly reshaped loop gain responses. The responses redrawn in Figure 4.13(b) on the arcsin f scale have the same area under them over the frequency interval [0, 1]. Therefore, the phase response at frequencies $f > 1$ and the stability margins can be the same for all shown gain responses (recall Equation 3.13).

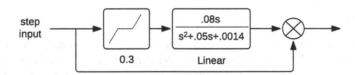

FIGURE 4.12
A prefilter including a nonlinear parallel branch.

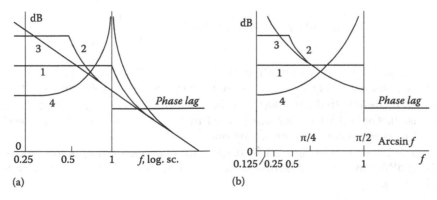

FIGURE 4.13
Reshaped loop gain responses on (a) the logarithmic frequency scale and (b) the arcsin f scale.

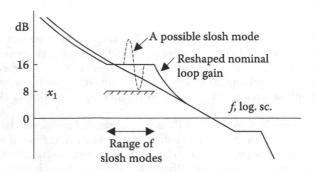

FIGURE 4.14
Reshaping the loop gain response.

Any frequency can be chosen to be the normalized frequency $f=1$ (or $\omega=1$). For control system design it is commonly convenient to use the frequency at which the constant slope response gain is approximately 10 dB.

Example 4.11

Liquid sloshing in the propellant tanks of a spacecraft appears as structural modes with gain magnitude up to 8 dB. To provide gain stabilization with an 8 dB upper stability margin over the range where the modes can be, the nominal loop gain needs to be at least 16 dB. Figure 4.14 shows the loop gain response to suit the problem, obtained by reshaping the constant slope response. Numerically, the loss in feedback at lower frequencies can be estimated by application of the rule in Equation 3.13 (preservation of the area of the loop gain) when the plot is redrawn on the arcsin f scale with the normalized frequency $\omega=1$ at the upper end of the slosh mode range.

4.2.5 Bode Cutoff

When the frequency components of the expected disturbances within the functional band $f<1$ have the same amplitudes, the loop gain within the functional band should be constant, as shown in Figure 4.15. The value $A_o=20 \log |T|$ must be maximized.

To find this response, Bode made use of the function $b(jf)$ defined by Equation 3.7, scaling by A_0 and raising to the power $2(1-y)$:

$$\log \left|10^{A_0} b(jf)^{2(1-y)}\right| = A_0 + 2(1-y)\log \left|\frac{1}{\sqrt{1-f^2}+jf}\right| \tag{4.3}$$

This has the high-frequency asymptote with the slope $-12(1-y)$ dB/octave. It replaces the constant slope response in Figure 4.2, as shown in Figure 4.15(a). It is seen from the picture (and from the formulas) that this loop gain at $f=1$ equals the value A_o that the constant slope response has at $f=0.5$. In other words, the functional bandwidth of A_o dB feedback in the *Bode optimal cutoff* becomes extended by one octave.

From the triangle shown in Figure 4.15(a), the available loop gain A_o is the product of the slope $12(1-y)$ dB/octave and the feedback bandwidth in octaves plus 1 (the extra octave is that from 0.5 to 1):

$$A_o \approx 12(1-y)(\log_2 f_b +1) \tag{4.4}$$

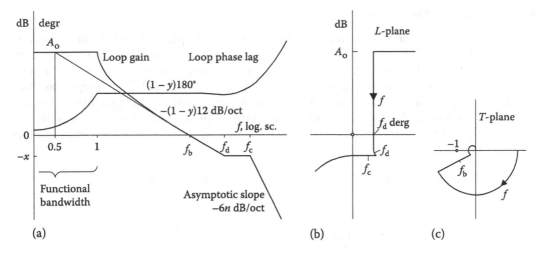

FIGURE 4.15
Bode optimal cutoff: (a) Bode diagram, (b) Nyquist diagram on *L*-plane, and (c) Nyquist diagram on *T*-plane.

In the common case of 30° stability margin, i.e., $y = 1/6$,

$$A_o \approx 10(\log_2 f_b + 1) \tag{4.5}$$

This simple formula is quite useful for rapid estimation of the available feedback.

Example 4.12

The prescribed stability margins are 30° and 10 dB, the feedback is required to be constant over the bandwidth of [0, 200] Hz, and the crossover frequency f_b is limited by the system dynamics to 6.4 kHz. From Equation 4.5, the available feedback is 60 dB.

4.2.6 Band-Pass Systems

In some systems, such as vibration suppression systems, or in a frequency band receiver, the frequency of functional feedback does not include dc. The band can be viewed as centered at some finite frequency f_{center}. Generally, the physically realizable band-pass transfer function can be found by substituting

$$s + (2\pi f_{center})^2 / s \tag{4.6}$$

for s in a low-pass prototype transfer function [6]. The loop response obtained with Equation 4.6 from the low-pass Bode optimal cutoff is shown in Figure 4.16. Notice that in Figure 4.16(b), two critical points for the Nyquist diagram to avoid are shown: −180° and 180°, each of the points being a mapping of the *T*-plane point −1 onto the *L*-plane.

The absolute bandwidth Δf of the available feedback is an invariant of the transform in Equation 4.6, as is illustrated in Figure 4.17, and it equals the bandwidth of the low-frequency prototype f_o. (The bandwidths of the three responses in the picture do not look equal because the frequency scale is logarithmic.) It is seen that a higher f_{center} corresponds to a smaller relative bandwidth and steeper slopes of the band-pass cutoff.

When the relative functional bandwidth is fairly wide, more than two octaves, the steepness of the low-frequency slope has only a small effect on the available feedback since the absolute bandwidth of the entire low-frequency roll-off is rather small. This case is shown in Figure 4.18.

4.2.7 Nyquist-Stable Systems

As mentioned in Section 4.2.2, and as will be detailed in Chapters 10 and 11, the responses shown in Figures 4.2, 4.13, 4.17, and 4.18 are tailored to guarantee stability of a system with a saturation link in the loop. If, however, the system is furnished with an extra dynamic nonlinear link of special design (described in Chapters 10, 11, and 13), the loop phase lag, the slope of the Bode diagram, and the available feedback can be increased by using the Nyquist-stable system loop response shown in Figure 4.19 instead of the phase-stabilizing response shown by the thin line. Here, x_1 and x represent the upper and lower amplitude

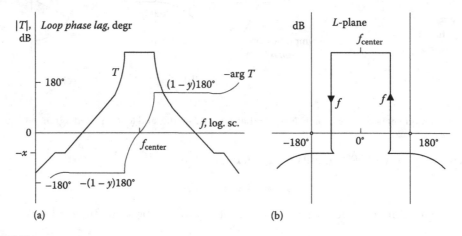

(a)   (b)

FIGURE 4.16
Band-pass optimal Bode cutoff: Bode diagram (a) and Nyquist diagram (b).

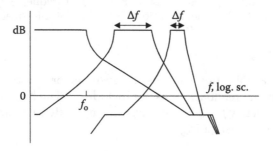

FIGURE 4.17
Preservation of operational bandwidth of the band-pass transform.

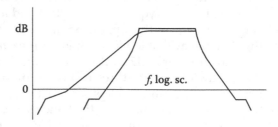

FIGURE 4.18
Bode diagrams for a wide-band band-pass system.

stability margins. At frequencies where $A > x_1$, the system is only gain stabilized. The integral of the phase lag in this system is larger than in the absolutely stable system, and as a result, the feedback over the functional bandwidth is larger. This response can be generated by pasting together several elementary responses [33].

Figure 4.20(a) shows a simplified response that is easier to implement (although it provides somewhat less feedback). The essential features of the response are the steep $-6n_1$ dB/octave slope before the upper Bode step and the presence of two Bode steps, the width of the lower step calculated with Equation 4.1, and of the upper step from f_g to f_h with the similarly derived formula:

$$\frac{f_h}{f_g} = 0.6n_1 \tag{4.7}$$

The larger the integral of phase, the larger is the available feedback. The phase lag, however, cannot be arbitrarily large. A certain boundary curve $A(B)$ exemplified in Figure 4.21 is specified by the features of nonlinear links in the loop (nonlinear compensators will be discussed in Chapters 10–13).

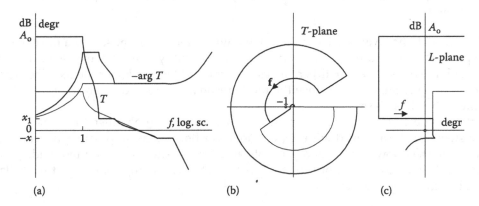

FIGURE 4.19
Comparison of a Nyquist-stable system with an absolutely stable system (thin lines): (a) Bode diagrams, (b) Nyquist diagrams (not to scale), and (c) Nyquist diagrams on the L-plane.

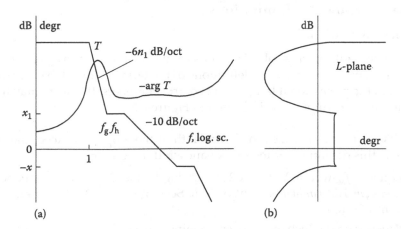

FIGURE 4.20
Simplified Nyquist-stable loop response: (a) Bode diagram and (b) Nyquist diagram on the L-plane.

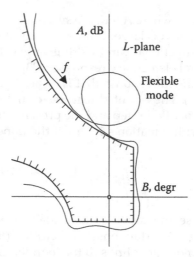

FIGURE 4.21
The Nyquist diagram should not penetrate the boundary curve specified by the properties of the nonlinear dynamic compensator.

In Figure 4.21, the Nyquist diagram is shown with a loop on it caused by a flexible mode of the plant. At frequencies of this mode, the phase stability margin is excessive. In accordance with the phase integral, this reduces the achieved feedback, but the feedback deficit due to the loop is rather small since the mode resonance is narrow and the excess in the integral of the phase is small.

Practical type 1 and type 2 systems (recall Section 3.8) are Nyquist stable. Stability in such systems can be achieved with upper and lower Bode steps. In practice, the transition between the steep low-frequency asymptote and the crossover area is often made gradually to simplify the compensator transfer functions, thus reducing somewhat the available feedback at lower frequencies.

4.3 Feedback Bandwidth Limitations

4.3.1 Feedback Bandwidth

In Section 4.3 we will discuss physical constraints on the high-frequency loop gain. However, first we need to clarify the definitions of the term "feedback bandwidth." In the literature and in the professional language of control engineers, this term may have any of the following three interpretations indicated in Figure 4.22:

1. The *crossover frequency* f_b, i.e., the bandwidth of the loop gain exceeding 0 dB. In this book, this definition for feedback bandwidth is accepted.

2. The frequency f_M, where $|M| = 1 / \sqrt{2}$, i.e., $20 \log |M| \approx -3$ dB. This frequency is the *tracking system 3 dB bandwidth* mentioned in Section 3.1. The 3 dB frequency is typically from $1.3f_b$ to $1.7f_b$.

3. The frequency up to which the loop gain retains a specified value (e.g., the bandwidth of 30 dB feedback). This bandwidth is also called the *bandwidth of functional feedback*.

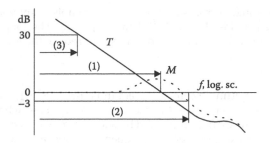

FIGURE 4.22
Feedback bandwidth definitions.

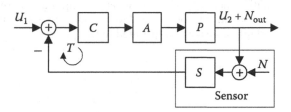

FIGURE 4.23
Sensor noise effect at the system's output.

4.3.2 Sensor Noise at the System Output

Consider the system shown in Figure 4.23. The source N represents the sensor noise. (Here, N is understood to be the root mean square of the noise.) From the noise input to the output, the system can be viewed as a tracking system with unity feedback, so the root mean square of the noise at the system's output is:

$$N_{out} = NT / F$$

In shaping the loop gain response, the trade-off is between output noise reduction and disturbance rejection. Larger feedback bandwidth leads to larger output noise but smaller disturbances. The output root mean square error caused by the noise can be found by simulation of the output time responses, or by numerical (or even graphical) integration of the frequency-domain noise responses.

Example 4.13

Consider the Bode diagram shown in Figure 4.24. This loop gain response has a rather steep cutoff after f_b to reduce the output noise effect, but shallower gain response and smaller feedback at lower frequencies. The phase stability margin is large. The hump on the response of $|M|$ is small; therefore, the overshoot in the transient response is also small.

This response is employed when the plant is already fairly accurate and there is no need for large feedback at lower frequencies, and positive feedback near the crossover frequency should be reduced to lessen the output effect of the sensor noise. In such a system, command feedforward is commonly used to improve the closed-loop input–output response.

When the loop response is steep, as shown in Figure 4.25, the output noise increases because of the positive feedback at the crossover frequency and beyond. This causes a substantial increase in the output noise since the contribution of the noise spectral density to the mean square error is proportional to the noise bandwidth. On the other hand, this response provides better disturbance rejection. The loop response should therefore be shaped in each specific case differently to reduce the total error.

Example 4.14

Consider the spacecraft attitude control system in Figure 4.26, which uses a gyro as a sensor. The system is accurate except at the lowest frequencies, where the gyro drift causes attitude error. The drift is eliminated by a low-frequency feedback employing a second sensor, a star tracker. The optimal frequency response for the star tracker loop is that which reduces the total noise from the two sensors, i.e., which reduces the mean square error of the system output variable. The calculations can be performed in the frequency domain or by using the LQG method described in Chapter 8. Since the star tracker noise varies with time, depending on whether bright stars are available in its field of view, the feedback path responses need to be varied to maintain the minimum of the error. Such an adaptive system is illustrated in Chapter 9.

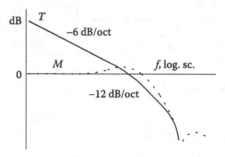

FIGURE 4.24
Shallow slope response.

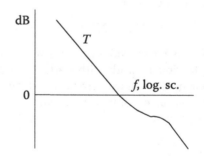

FIGURE 4.25
Steeper slope response.

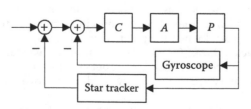

FIGURE 4.26
Spacecraft attitude control system using two sensors.

4.3.3 Sensor Noise at the Actuator Input

In the control system diagram in Figure 4.27, the noise source N represents the noise from the sensor and the noise from the preamplifier in the compensator. When the signal at the input of the saturation link is below the saturation threshold, the noise effect at the input of the saturation link is:

$$N_A = \frac{NCA}{CAP+1}$$

With the typical responses of P and T shown in Figure 4.28, $|CAP| \gg 1$ at lower frequencies. At these frequencies the noise $N_A \approx N/P$ does not depend on C. On the other hand, at frequencies higher than f_b, $N_A \approx NCA$, and reducing $|C|$ decreases the noise. With proper loop shaping, the noise N_A is most prominent at the frequencies within two to four octaves above f_b.

It is seen from the figure that the increase of the feedback from $|T|$ to $|T'|$ is attained at the price of increasing $|CA|$ to $|C'A|$, i.e., at the price of a bigger noise effect at the input to the nonlinear link in Figure 4.27. The noise amplitude and power increase not only because $C' > |C|$, but also because the noise power is proportional to the frequency bandwidth.

When the noise overloads the actuator, the actuator cannot transfer the signal. As a result, the effective gain of the actuator drops, the distortions of the signal increase, and the control system accuracy decreases. Hence, the noise effect at the input to the actuator must be bounded. This restricts the available feedback in the operational band by constraining $|C|$.

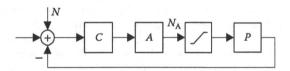

FIGURE 4.27
Noise source in a feedback system.

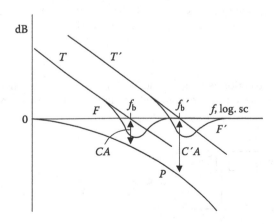

FIGURE 4.28
Noise level at saturation link input.

The optimal shape of the Bode diagram that provides maximum feedback bandwidth while limiting the noise effect can be found by experimenting with computer simulation. Typically, the responses that are best in this sense contain Bode steps.

Example 4.15

In an existing system, the bandwidth is limited by the noise at the input to the actuator. If a better amplifier and better sensors become available with half the noise mean square amplitude, the feedback bandwidth can be increased. Maintaining the same mean square noise amplitude at the actuator input, the feedback bandwidth can be increased 1.4 times (since the mean square amplitude of the white noise is proportional to the square root of the noise bandwidth).

4.3.4 Non-Minimum Phase Shift

As mentioned, the n.p. lag in the feedback loop typically should be less than 1 rad at f_c, or else to compensate for it, the Bode step would have to be very long. This limitation on the feedback can be critical for loops that include a substantial delay. In some systems, transport delay, which, as we have seen in Section 3.12, causes a phase lag proportional to frequency, can be particularly large.

Let us consider two examples of audio systems with large transport delay.

Example 4.16

Because speaker systems are expensive and their frequency responses are difficult to equalize, and since the sound wave reflections from the walls of the room change the frequency responses on the way to the listener's ears, and since good quality, inexpensive microphones are easily available, it would be commercially advantageous to make acoustical feedback from a microphone placed in the vicinity of one's ears. Is it possible to maintain good sound quality over the entire range of audio signals up to 15,000 Hz using a feedback system like that shown in Figure 4.29? Probably not, since nobody does this. There must be a good reason. We might suspect that the reason is the excessive time of the signal propagation about the feedback loop. Let us check it out.

The speed of sound being 330 m/s, and the distance between the speakers and the microphone being 2 m, the transport delay is 6.6 ms. For the frequency $f_c = 15 \times 4 = 60$ kHz, the phase lag $B_n(f_c) = 2\pi \times 60,000 \times 0.0066 \approx 2500$ rad, which is 2,500 times the allowable limit. Thus, real-time feedback in this system is not possible. (The response *can* be equalized by an adaptive system using plant identification.)

Example 4.17

A system for noise rejection is diagrammed in Figure 4.30. The microphone together with the speaker is placed at the point where noise is to be minimized. When the

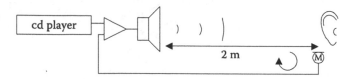

FIGURE 4.29
This type of real-time acoustical feedback system is not feasible.

feedback is big, the noise is canceled. The feedback must be taken at the output voltage of an amplifier since the load for the current must depend on the load impedance.

The acoustic signal propagation between the speaker and the microphone introduce non-minimum phase lag into the feedback loop. The assembly of the microphone and the speaker is commonly mounted in a helmet.

When 30 dB of feedback is required up to 5 kHz, then the frequency $f_d = 20$ kHz. At this frequency, B_n should not exceed 1 rad. Therefore, the pure delay is $(1/40,000)/2\pi \approx 0.000008$ s and the distance between the speaker and the microphone should be shorter than 2.6 mm.

Another source of non-minimum phase lag is the delay in analog-to-digital control.

4.3.5 Plant Tolerances

The plant gain typically decreases with frequency, and the tolerances of the plant transmission function increase with frequency, as illustrated by the limiting curves in Figure 4.31. In plants whose transfer functions have only real poles and zeros, the plant responses and their variations are commonly smooth and monotonic. Such responses are typical for temperature control and rigid-body position control. Since the minimum necessary stability margins must be satisfied for the worst case, which is typically the case of the largest plant gain, the feedback will be smaller in the case of the minimum plant gain. In this way the plant response tolerances reduce the minimum guaranteed feedback.

For plants with monotonic responses, it is convenient to consider some *nominal* plant response, as shown in Figure 4.31. The feedback loop design is then performed for the nominal plant. Larger plant tolerances—larger deviations from the nominal—require increased stability margins for the nominal response, and thereby limit the nominal available feedback.

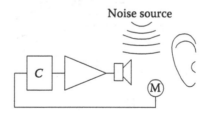

FIGURE 4.30
Sound suppression feedback system.

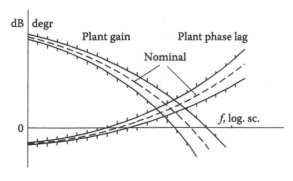

FIGURE 4.31
Boundaries of monotonic plant transfer functions.

For computer design and simulation, the plant uncertainty is most often modeled as *multiplicative uncertainty*, that is, multiplication of the loop transfer function by some error response (i.e., addition of some uncertainty to the gain and phase responses). The multiplicative uncertainty is typically either a constant, as in Example 4.6 (see Figure 4.11) or a function of frequency, as those shown in Figure 4.31.

In general, the dependence of the plant transfer function on its varying parameters can be complicated. For some plants parameter uncertainty causes deviations from the nominal plant responses which are neither symmetrical nor monotonic, like those shown in Figure 4.32. Uncertainties are also sometimes modeled by vector addition of some error response to the transfer function (*additive uncertainty*).

Flexible plants have structural resonances corresponding to specific modes of vibration. They may be composed of rigid bodies connected with springs and dampers, or they may be continuous such as a beam undergoing longitudinal bending. Stiffness and mass variations in flexible plants change the pole and zero frequencies, as shown in Figure 4.33. Similar responses are obtained in low-loss electrical systems such as transformers and filters. The resonances typically produce loops nearly 180° wide on the *L*-plane Nyquist diagram, as shown in Figure 4.34. Neither multiplicative nor additive uncertainty conveniently characterizes these effects. It is better to describe such plants by specified uncertainties in the transfer function poles and zeros.

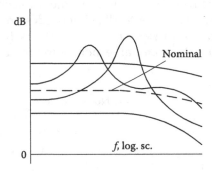

FIGURE 4.32
Plant gain frequency responses.

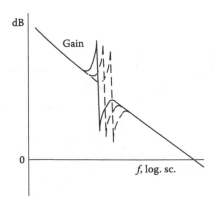

FIGURE 4.33
Plant structural resonance on a Bode diagram.

4.3.6 Lightly Damped Flexible Plants: Collocated and Non-Collocated Control

Some of the poles and zeros of flexible plants' transfer functions are only lightly damped, with the damping coefficients as small as 1% and even 0.1%. The loop gain responses with such plants exhibit sharp peaks and notches, as in Figures 4.33 and 4.35.

For the closed-loop system to be stable, the modes should be gain or phase stabilized. The modes that need attention are those that are not already gain stabilized, i.e., those resulting in the loop gain being within the interval from $-x$ to x_1, as modes 2 and 3 shown in Figure 4.35(a). Increasing the modal damping can reduce the size of the modal peak/notch and gain stabilize the mode. Otherwise, the mode needs to be phase-stabilized as shown in Figure 4.35(b).

To phase stabilize mode 3, a phase lag might be added to the loop to center the mode at the phase lag of $-360°$, so the resonance loop will be kept away from the critical points $-180°$ and $-540°$, as illustrated in Figure 4.35(b). The required phase lag can be obtained by introducing a low-pass filter in the loop (this is a better solution than adding a n.p.

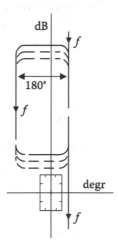

FIGURE 4.34
L-plane Nyquist diagram.

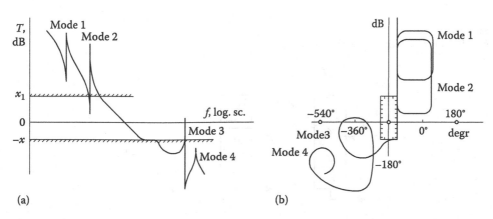

FIGURE 4.35
Modes (a) on the loop gain response and (b) on the *L*-plane Nyquist diagram.

lag since the filter will provide the additional benefit of attenuating modes of higher frequencies).

If the plant's phase uncertainty at the frequency of the flexible mode is large, phase stabilization is not feasible and the mode must be gain stabilized. A typical case of the gain stabilization of a structural mode is shown in Figure 4.36. The Bode step allows a steep roll-off at frequencies beyond the step. Gain stabilization of the mode reduces the feedback in the functional band. The average loop gain at the frequency of the mode must be no higher than $(20 \log Q + x)$ dB. Increasing the damping of the high-frequency mode would allow increased feedback bandwidth. This situation will be addressed with nonlinear multi-loop control in Chapter 14 (see Figure 14.8).

Next, consider the flexible plant model for translational motion shown in Figure 4.37, consisting of rigid bodies with masses M_1, M_2, etc., connected with springs. The actuator applies a force to the first body. The motion sensor S_1 is located with the actuator and senses the motion of the first body, so the control is called *collocated*. For distributed systems like the bending beam, this definition of collocated is imprecise but still meaningful in a qualitative way.

If the sensors are velocity sensors, the transfer function from the actuator force to the sensor S_1 is, in fact, the plant driving point impedance (or mobility). The driving point

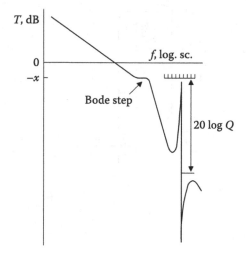

FIGURE 4.36
Gain stabilization of a high-frequency mode.

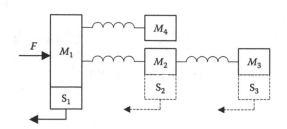

FIGURE 4.37
Mechanical plant with flexible appendages and the sensor collocated with the actuator.

impedance of a passive system is *positive real* (see Appendix 3), and its phase belongs to the interval [–90°, 90°]. The impedance function of a lossless plant has purely imaginary poles and zeros that alternate along the frequency axis with the phase alternating between 90° and –90° (Foster's theorem). Flexible plants are discussed in more detail in Chapter 7.

Example 4.18

The plant of the control system having the loop response shown in Figure 4.36 is *non-collocated* since the plant has a pole at zero frequency, and then the mode's pole and zero follow, i.e., a pole follows a pole. If the mode pole-zero order were reversed, the control would be collocated.

Example 4.19

A collocated force-to-velocity translational transfer function of a body with two flexible appendages resonating respectively at 3.32 and 7.35 rad/s and negligible damping is:

$$P(s) = \frac{1}{s} \frac{s^2 + 10}{s^2 + 11} \frac{s^2 + 50}{s^2 + 54}$$

The masses of the appendages in this example are approximately 10 times smaller than the mass of the main body, which explains why the poles are rather close to the zeros. The gain and phase responses are plotted in Figure 4.38 with:

```
n = conv([1 0 10],[1 0 50]);
d = conv([1 0],[1 0 11]); d = conv(d,[1 0 54]);
w = logspace(0, 1, 1000); bode(n,d,w)
```

When the sensor is a position sensor or an accelerometer, an extra integrator or differentiator should be added to this transfer function that changes the slope by –6 or 6 dB/ octave, respectively.

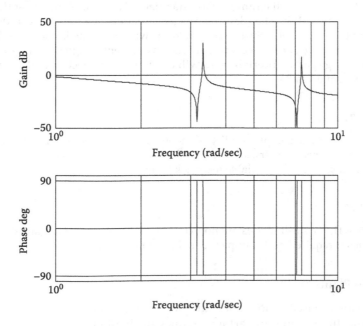

FIGURE 4.38
Frequency response for a collocated mechanical plant with two flexible appendages.

When the control is collocated, a flexible appendage adds a zero-pole pair to the loop response, as shown in Figures 4.35 (two lower-frequency modes) and 4.38. The modes do not destabilize the system since the phase lag only *decreases* by 180° between the added zero-pole pair. (However, the mode reduces the integral of phase, and therefore the average gain slope and the available feedback somewhat decrease.)

Placing the sensor on any other body makes the control *non-collocated*. In this case, the spring connecting the bodies introduces an extra unwelcome phase lag into the loop. Thus, the sensor location defines whether the control is collocated or non-collocated. The trade-off associated with where to place the sensor, on M_1, M_2, or M_3 in Figure 4.36, is very often encountered in practice. The sensor must be placed within the power train someplace from the actuator to the tip of the tool or other object of control. When the sensor is placed closer to the actuator, i.e., on M_1, the feedback bandwidth can be widened, but it is the position of the first body that is controlled. The flexibility between the bodies will introduce the error in the position of the tip. However, when the sensor is placed on the tool, i.e., on M_2, or on the tip of the tool, on M_3, then we are controlling exactly the variable that needs to be controlled, but the feedback bandwidth must be reduced, as shown in Figure 4.36. The best results can be obtained by combining these sensors. Collocated and non-collocated control will be further discussed in Sections 7.8.3 and 7.8.4.

Example 4.20

A Nyquist diagram for a flexible plant (Saturn V controller) is given in Appendix 13, Figure A13.26. Several slosh modes and structural flexibility modes are seen in this diagram. While the controller was being designed, alternative locations were being considered for the rate gyros in order to make the multiple bending modes easier to stabilize. It was eventually decided to keep the rate gyros in the instrument ring with the other control electronics, which necessitated a higher-order compensator.

Most of the modes are gain stabilized; only the large first-bending mode seen in the upper-right sector of the diagram is phase stabilized. The loop phase lag at the mode's central frequency is 315°. The plant parameter variation must not reduce this phase lag by more than 45°, or else the stability margins will be violated. The control at these frequencies is analog. A digital controller with insufficiently high sampling frequency would cause large phase uncertainty (as will be discussed in Section 5.10.7) and would make the system unstable.

Example 4.21

In pneumatic systems, compressibility of the air in the cylinder of an actuator creates a series "spring" between the actuator and the plant. This makes the control non-collocated and reduces the available feedback bandwidth.

Example 4.22

In Example 4.11, the slosh modes are non-collocated. Because of this, gain stabilization of the modes is required and is implemented as shown in Figure 4.14.

4.3.7 Unstable Plants

Unstable plants are quite common. For example, many launch vehicles and some airplanes are aerodynamically unstable in certain regimes of flight; a slug formed in the combustion chamber (or even turbulence in the chamber) can make a spin-stabilized upper-stage

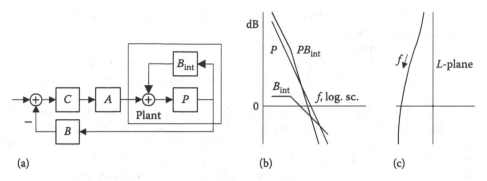

FIGURE 4.39
(a) Plant with internal feedback, and the diagrams of (b) Bode and (c) Nyquist for the internal feedback.

unstable; the rotation of a prolate spacecraft is unstable; a large-gain electronic amplifier without external feedback circuitry is often unstable. For the purposes of analysis and design, an unstable plant can be equivalently presented as a combination of a stable forward path link P with internal feedback path B_{int} that makes the plant unstable, as shown in Figure 4.39(a).

Example 4.23

Consider the system diagrammed in Figure 4.39(a). Assume that the plant is a double integrator with an internal feedback path having a low-pass transfer function $B_{int} = b/(s+a)$. The Bode diagrams are shown in Figure 4.39(b). The internal loop phase lag exceeds 180° at all frequencies, and the plant becomes unstable as seen from the Nyquist diagram in Figure 4.39(c).

There are two convenient ways of analyzing and designing such systems.

When the loop is disconnected at the input to the link P, the loop transfer function is $T = (B_{int} + BCA)P$. After the desired frequency response for T is specified, the required transfer function for the compensator is $C = (T/P - B_{int})/(AB)$. The method is especially convenient when B_{int} is small compared with BCA.

The compensator can be directly designed for the unstable plant. In this case, the main-loop Nyquist diagram must encompass the critical point in the counterclockwise direction, as required by the Bode–Nyquist stability criterion.

4.4 Coupling in MIMO Systems

As explained in Section 2.9, coupling is typically negligible in well-designed control systems where the number of the actuators is kept small. However, at some frequencies the coupling can be large, uncertain, and create stability problems.

In mechanical structures with multidimensional control, the actuators are typically applied in mutually orthogonal directions so that the coupling between the corresponding feedback loops is relatively small. However, the plant might include some flexible attachment, like an antenna, a solar panel, or a magnetometer boom on a spacecraft, as shown in Figure 4.40. The attachment's flexible mode can be excited by any of the actuators (reaction wheels, thrusters), and will provide signals to all the sensors. This coupling may occur at any frequency within a certain frequency range defined by the uncertainties in the mass and stiffness of the appendage.

Because of the coupling, the block diagram for the coupled loops looks like that shown in Figure 4.41. Here, $K(s)$ is the coupling transfer function. The return ratios for the controllers in x and y calculated without taking the coupling into account are $T_x = C_x A_x P_x B_x$ and $T_y = C_y A_y P_y B_y$.

The system can be designed with the Bode–Nyquist stability criterion as follows. First, the y-actuator is disabled and T_x is shaped so that the x-loop is stable and robust. Then, A_y is switched on (while the x-actuator is kept on). The transfer function in the y-channel between A_y and P_y is $1 + (T_x/F_x)K^2$. The compensator C_y is then shaped properly to make the system stable (sometimes, this is not possible). Gain stabilization in the range of the flexible modes is the best choice—phase stabilization is difficult because the transfer function $K^2 = a^2 M/(s^2 + 2\omega + \omega^2)^2$ contains complex poles with large associated phase uncertainty; here, a is some coefficient. If gain stabilization cannot be used and the system needs to be phase stabilized, C_x should be modified to make the response in the x-loop shallower over the frequency range of coupling. The associated reduction of available feedback is unavoidable. Bode diagrams for stability analysis with the successive loop closure criterion, for the x- and y-loops, may look like those shown in Figure 4.42.

Coupling between x- and y-controllers can also be caused by the effects of rotation about the z-axis (even in a spacecraft without a flexible appendage). In this case, x-actuators

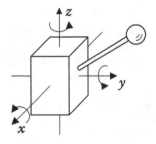

FIGURE 4.40
Mechanical plant with a flexible appendage.

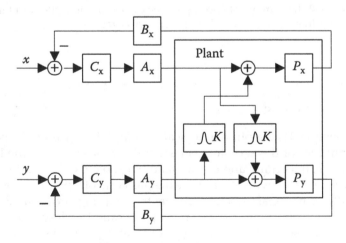

FIGURE 4.41
Block diagram for coupled attitude control loops.

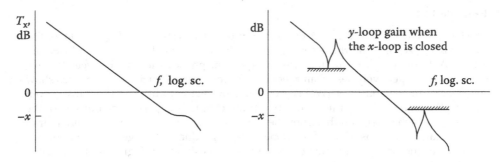

FIGURE 4.42
Gain responses of the attitude control loops: the x-loop with the y-loop open (left), and the y-loop with the x-loop closed (right).

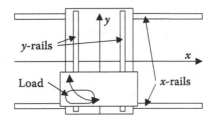

FIGURE 4.43
x–y positioner.

produce rotation about the y-axis, and y-actuators produce rotation about the x-axis. The effect is only profound at frequencies close to the frequency of rotation about the z-axis. The system analysis and design are similar to those in the case of flexible mode coupling.

An x–y positioning table is shown in Figure 4.43. The translational motions may become coupled via rotational motion due to the load asymmetry and structural flexibility, especially at the frequency of the structural mode of the rotation.

When the number of the actuators is large, each control loop should use position (or velocity) and force sensors. This is common in biological systems; for example, there are many thousand separate muscles in the trunk of an elephant. This compound feedback makes the loop transfer function less sensitive to plant parameter variations, and makes the output mobility of the actuator dissipative, damping the plant. This also reduces the variations of the loop coupling that are caused by variations of the load and the plant parameters. Loop decoupling algorithms can then be used effectively. Design of a loop with prescribed actuator mobility will be discussed in Chapter 7.

4.5 Shaping Parallel Channel Responses

In MIMO and even in SISO control systems, several paths are often connected in parallel, especially when several actuators or sensors are employed. As was demonstrated in Chapter 3, if two stable m.p. links W_1 and W_2 are connected in parallel, as shown in Figure 4.44(a), then the total transfer function $W_1 + W_2$ can become n.p. The frequency responses of the parallel channels should be shaped properly for the combined channel to be m.p.

Example 4.24

The low-pass links W_1 and W_2 are connected in parallel as shown in Figure 4.44(a). The steep roll-off W_1 and the three versions of shallower roll-off W_2 are shown in Figure 4.44(b). At frequency f_1 where the gains are equal, the phase difference between the two channels is, respectively, less than π, equal to π, and more than π. The thin lines show the logarithmic responses of $|W_1 + W_2|$ obtained by vector addition of the links' output signals. When the phase difference between the channels at f_1 is π, the outputs of the links cancel each other, and therefore, the composite link transfer function has a pair of purely imaginary zeros $\pm j2\pi f_1$. If the slope of W_2 is gradually changed, the root loci for the zeros of the transfer function cross the $j\omega$-axis as shown in Figure 4.44(c), and the total transfer function becomes n.p.

As has been proven, the sum $W_1 + W_2$ is m.p. if and only if the Nyquist diagram for W_1/W_2 does not encompass the point -1. Since the ratio W_1/W_2 is also stable and m.p., one can determine whether the Nyquist diagram encloses the critical point by examining the Bode diagram for W_1/W_2.

When the tolerances of the parallel channel transfer functions are not negligible, they can produce large variations in $W_1 + W_2$. The sensitivities of the sum to the components:

$$\frac{\dfrac{d(W_1 + W_2)}{W_1 + W_2}}{\dfrac{dW_1}{W_1}} = \frac{W_1}{W_1 + W_2}\frac{d(W_1 + W_2)}{dW_1} = \frac{1}{1 + W_2/W_1}$$

and

$$\frac{\dfrac{d(W_1 + W_2)}{W_1 + W_2}}{\dfrac{dW_2}{W_2}} = \frac{1}{1 + W_1/W_2}$$

become unlimited as the ratio W_1/W_2 approaches -1. To constrain the sensitivities, the hodograph of W_1/W_2 should be required not to penetrate the safety margin around the point -1.

Analogous to the stability margins, the *phase safety margin* is defined as $y\pi$, and the *amplitude safety margins* as x and x_1.

A common practical reason for using two parallel links is that one of the links (actuators or sensors) works better at lower frequencies, and the second link works better at higher frequencies. Combining them with frequency selection filters generates a link

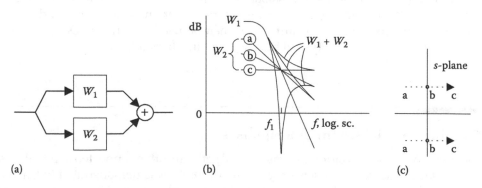

(a)   (b)   (c)

FIGURE 4.44

(a) Parallel channels, (b) Bode diagrams for W_1, W_2, and $W_2 + W_1$, and (c) s-plane root loci for $W_1 + W_2$.

(actuator or sensor) that is good over a wide frequency range. The composed link transfer function must be m.p. so that it can be included in the feedback loop. However, excessive selectivity of the filters can make the composite link n.p.

Figure 4.45(a) shows the responses of the low-pass link W_1 and the high-pass link W_2, with different selectivities. Figure 4.45(b) shows the Bode diagrams for the ratio W_1/W_2. It is seen that when the difference in the slope between W_1 and W_2 responses increases, the Bode diagram for the ratio steepens, the related phase lag increases, and the critical point becomes enclosed by the related Nyquist diagram.

In order to preserve sufficient safety margins while keeping the slope of W_1/W_2 steep, the Bode diagram for W_1/W_2 can be shaped as in a Nyquist-stable system (Figure 4.46(b)). Then, W_1 and W_2 can be as illustrated as in Figure 4.46(a). Responses of this kind are particularly useful for systems with the main/vernier actuator arrangement described in Section 2.7.

An alternative to shaping the responses and then approximating them with rational functions is direct calculation of the channel transfer functions. Given the transfer function of the first link, the transfer function of the second link can be found directly as $W_2 = 1 - W_1$. This method works well if the links are precise (as when sensors' readings are combined). If the links are imprecise (like actuators and different signal paths through the plant), and the selectivity is high, then the link parameter variations should be accounted for and sufficient safety margins introduced.

Example 4.25

If W_1 is low pass,

$$W_1 = \frac{a}{s+a}$$

then the second channel transfer function

$$W_2 = 1 - W_1 = \frac{s}{s+a}$$

is high pass, and the ratio is:

$$\frac{W_1}{W_2} = \frac{a}{s}$$

The phase safety margin is 90° at all frequencies. The margin is large, but the selectivity between the channels is not high.

Example 4.26

To improve the selectivity, the first link is chosen to be a second-order low-pass Butterworth filter:

$$W_1 = \frac{1}{s^2 + s\sqrt{2} + 1} \tag{4.8}$$

The second link transfer function is then:

$$W_2 = 1 - W_1 = \frac{s\left(s + \sqrt{2}\right)}{s^2 + s\sqrt{2} + 1} \tag{4.9}$$

and the ratio:

$$\frac{W_1}{W_2} = \frac{1}{s(s+\sqrt{2})}$$

The plots shown in Figure 4.47 indicate that the safety margins are less than in the previous example but still quite wide. With higher-order filters, safety margins become smaller.

These diagrams are for single-loop systems. For multi-loop systems, the feedback may be much larger as will be shown in Chapter 14 (see Figure 14.9.)

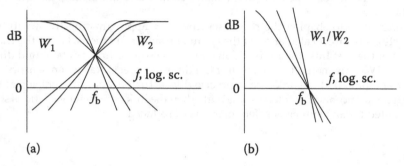

(a)   (b)

FIGURE 4.45
Bode diagrams for (a) W_2, W_1 and (b) W_1/W_2.

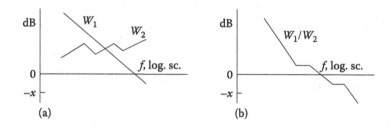

(a)   (b)

FIGURE 4.46
(a) Frequency-selective responses for W_1 and W_2 and (b) Nyquist-stable shape of Bode diagram for W_1/W_2, which preserves the m.p. character of $W_1 + W_2$.

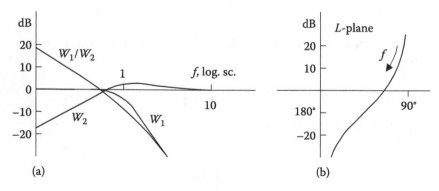

(a)   (b)

FIGURE 4.47
Bode (a) and Nyquist (b) diagrams for W_1, $W_2 - W_1$, W_1/W_2.

PROBLEMS

1. The slope at the crossover frequency 40 Hz is (a) −6 dB/octave, (b) −9 dB/octave, (c) −10 dB/octave, or (d) −12 dB/octave. What is the feedback in dB at 1.25 Hz if the slope remains the same down to this frequency? What would be the loop phase lag if the slope is constant at all frequencies?

2. What is the length of the Bode step and f_b for the following:

   a. The asymptotic slope is −18 dB/octave, $x = 10$ dB, $y = 1/6$, the asymptote passes the −10 dB level at 1 kHz, and the n.p. lag at 1 kHz is 0.5 rad

   b. The asymptotic slope is −12 dB/octave, the asymptote passes the 0 dB level at 5 kHz, and the n.p. lag at 5 kHz is 1 rad

   c. The asymptotic slope is −12 dB/octave, the asymptote passes the −10 dB level at 100 Hz, and the n.p. lag at 100 Hz is 0.5 rad

   d. The asymptotic slope is −24 dB/octave, the asymptote passes the −10 dB level at 50 kHz, and the n.p. lag at 100 kHz is 0.6 rad

3. Sketch the phase lag response and the Nyquist diagram for the optimal Bode cut-off in which the Bode step is omitted, with low-frequency slope −10 dB/octave and asymptotic slope −18 dB/octave. Is the system stable?

4. At the frequency of structural resonance $f_{st} = 120$ Hz or higher, there is a narrow resonance peak in the plant gain response and the plant phase at this frequency is completely uncertain. (The phase uncertainty is the result of using a digital controller with sampling frequency 100 Hz. Digital controllers will be studied in detail in Chapter 5.) The system must therefore be gain stabilized at the frequency of the structural resonance f_{st}.

   Using the asymptotic slope −18 dB/octave, 30°, and 10 dB stability margins, and assuming that $f_d = 2f_b$, and $f_c = 4f_b$ (which are typical numbers and should be used for initial estimates), express f_b as a function of f_{st} and Q. Make a sketch of the loop gain response similar to that shown in Figure 4.48, but with numbers on it, for the peak value 20 log Q equal to the following:

   a. 20 dB

   b. 25 dB

   c. 30 dB

   d. 40 dB

   e. 50 dB

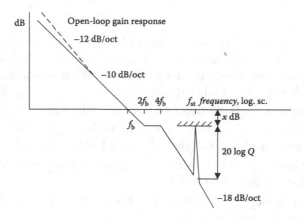

FIGURE 4.48
Feedback bandwidth limitation due to a structural resonance.

5. The same as in Problem 4, but the resonance uncertainty range starts at 170 Hz.

6. The same as in Problem 4, but the resonance uncertainty range starts at 85 Hz.

7. The crossover frequency f_b is 1 kHz, $y = 1/6$. The system is phase stabilized at all frequencies up to f_b, with Bode optimal loop response. What is the available feed-back over the following bandwidths?

   a. [0, 50]

   b. [0, 30]

   c. [30, 60]

8.

   a. What is the feedback in dB at 1.5 Hz if the crossover frequency is 300 Hz and the main slope is −10 dB/octave?

   b. What is the maximum available feedback in dB at 1.6 Hz when the feedback is kept constant at frequencies below 1.6 Hz, the system is phase stabilized with 30° stability margins, and the crossover frequency is 300 Hz?

9. The required feedback at 10 Hz is 40 dB, and the feedback should increase at lower frequencies with the slope −10 dB/octave. What is the crossover frequency? What are the frequencies at the beginning and the end of the Bode step if the step's length is 0.8 octave and $x = 10$ dB?

10. In a system with feedback bandwidth 100 Hz, amplitude stability margin 10 dB, phase stability margin 30°, and no n.p. lag, the attenuation in the feedback loop is required to be large, over 1.5 kHz, where there might be flexible modes in the plant. Calculate what attenuation at 1.5 kHz is available in the loop response with a Bode step if the asymptotic slope is chosen to be (a) −12 dB/octave or (b) −18 dB/octave. What is the conclusion? Explain the result by referring to the shape of the weight function in the Bode phase-gain relation.

11. In a GaAs microwave feedback amplifier, there are two gain stages, and the length of the feedback loop is 1 mm. The speed of signal propagation is 150,000 km/s. At what frequency is $B_n = 1$ rad? Considering this frequency as f_c, what is the length of the Bode step? What is the available feedback over the following bandwidths:

a. 0 to 3 GHz

b. 1.5 to 3 GHz

c. 0 to 2 GHz

d. 2 to 3 GHz

12. Find the loop transfer function and the prefilter for the systems with the following crossover frequencies. Use the 1 rad/s crossover prototype described in Section 4.2.3.

a. 0.2 Hz

b. 6 Hz

c. 2 kHz

d. 6 kHz

e. 2 kHz

f. 6 MHz

g. 2 MHz

h. 4 rad/s

i. 100 rad/s

13. The actuator and plant transfer function AP are the following:

a. $4(10-s)(s+2)/[s^2(10+s)(s+7)]$

b. $2(10-s)(s+2)/[s^2(10+s)(s+8)]$

c. $3(10-s)(s+4)/[s^2(10+s)(s+9)]$

d. $0.6(10-s)(s+5)/[s^2(10+s)(s+10)]$

e. $2.72(10-s)(s+6)/[s^2(10+s)(s+11)]$

Find the compensator that makes the loop transfer function the same as in the example studied in Section 4.2.3, where:

$$T(s) = C(s)\frac{1}{s^2}\frac{10-s}{10+s} = \frac{11s^3+55s^2+110s+36}{s^4+7.7s^3+34s^2+97s+83}\frac{1}{s^2}\frac{-s+10}{s+10}$$

14. Determine the band-pass transform from the low-pass optimal cutoff with the following frequency ranges:

a. [0, 1] rad/s to the bandwidth [30, 70] Hz

b. [0, 2] rad/s to the bandwidth [50, 70] Hz

c. [0, 3] rad/s to the bandwidth [60, 70] Hz

d. [0, 4] rad/s to the bandwidth [40, 70] Hz

e. [0, 5] rad/s to the bandwidth [30, 120] Hz

f. [0, 1] rad/s to the bandwidth [30, 100] Hz

15. Draw a Nyquist diagram on the *T*-plane (not necessarily to scale, only to show the shape) for the Nyquist-stable system in Figure 4.19.

16. Initial analysis with a low-order compensator has shown that in the plant hardware configuration A, larger feedback is available than in configuration B. The system

engineer assumed (wrongly) that feedback will be larger in configuration A even when, later, a better controller will be developed. Therefore, s/he decided that configuration A should be chosen. Devise a counterexample to prove that optimal shaping for the Bode diagram must be used for initial analysis as well. (*Hint*: use a plant with a flexible mode.)

17. Using the phase-gain chart in Figure 3.42 (or the program from Appendix 5), calculate the phase response for a Nyquist-stable system whose loop gain response is: from 0 to 10 Hz, 60 dB; from 10 to 20 Hz, −50 dB/octave; from 20 to 80 Hz, 10 dB; from 80 to 320 Hz, −10 dB/octave; from 320 to 640 Hz, −10 dB; and from 640 Hz, −18 dB/octave.

18. An extra management level was added to a four-level management system. How will it affect the speed of accessing the market and adjusting the product quantities and features (make a rough estimate)?

19. In a Nyquist-stable system with a response like that shown in Figure 4.19, $f_b = 100$ Hz, the phase stability margin is 30°, and the upper and lower gain stability margins are 10 dB. Calculate the frequencies at the ends of the upper and lower Bode steps if the slope at lower frequencies is (a) −12 dB/octave, (b) −18 dB/octave, and the asymptotic slope is (a) −18 dB/octave, (b) −24 dB/octave.

20. The loop gain plot crosses the 0 dB level at 200 kHz, and the rest is as in Problem 19, versions (a) and (b).

21. The unstable plant can be equivalently represented as a stable plant with a feedback path B_1. The path from the plant output to the plant input via the regular feedback loop links is 40 dB larger than via path B_1 at most frequencies, but only 20 dB larger over some narrow frequency range. How should you approach the design of the feedback system?

22. Two parallel links have the following transfer functions, respectively:

a. $W_1(s) = \dfrac{(s+5)(s+8)}{(s+40)(s+100)}$ and $W_2(s) = \dfrac{1000}{(s+3)(s+10)(s+20)}$

b. $W_1(s) = \dfrac{(s+6)(s+10)}{(s+50)(s+125)}$ and $W_2(s) = \dfrac{1800}{(s+4)(s+13)(s+27)}$

c. $W_1(s) = \dfrac{(s+2.5)(s+4)}{(s+20)(s+50)}$ and $W_2(s) = \dfrac{125}{(s+1.5)(s+5)(s+10)}$

d. $W_1(s) = \dfrac{(s+10)(s+16)}{(s+80)(s+200)}$ and $W_2(s) = \dfrac{8000}{(s+6)(s+20)(s+40)}$

Is the composite link $W_1(s) + W_2(s)$ m.p.?

23. Two minimum phase links in parallel made a non-minimum phase link. A third link with constant gain coefficient k has been added in parallel to the two links. How would you find the minimum k for the total transfer function to be m.p.?

24. Find transfer function $G(s)$ of a high-pass in parallel to a low-pass $12/[(s+3)(s+4)]$ so that the total transfer function is 1. Is this $G(s)$ realizable?

25. Prove that if a transfer function of linear passive two-ports is m.p. with some passive impedances of the signal source and the load, then the transfer function is m.p. with any other passive source and load impedances.

ANSWERS TO SELECTED PROBLEMS

1a. 1.25 Hz is five octaves below 40 Hz. Therefore, the feedback is $6 \times 5 = 30$ dB.

2a. The frequency $f_c = 1$ kHz; then, from Equation 4.2, $f_c/f_d = 0.6 \times 3 + 0.5 = 2.3$ (i.e., the step length is 3.32 log $2.3 = 1.2$ oct), $f_d = 0.435$, and $f_b = f_d/2 = 0.22$ kHz.

3a. Since $y = 1/6$, the slope is -10 dB/octave. There are 3.32 log$(1,000/50) = 4.3$ octave down to 50 Hz from the crossover. Then, according to Equation 4.5, the available feedback is $(4.3 + 1) \times 10 = 53$ dB.

4a. 18 $\log_2(f_c/f_{st}) = 20 \log_{10}Q$, and $f_b = f_c/4$, then $f_b = f_{st}/2^{(20 \log Q)/18+2}$, i.e.,

$$f_{st} = 120; Q = 10; f_b = f_{st}/2^{[20^* \log_{10}(Q)/18 + 2]}$$
$$f_b = 13.8881$$

The diagram is shown in Figure 4.49.

Given the slope of the Bode diagram (-10 dB/octave), disturbance rejection in dB at frequency $f < f_b/2$ depends on the feedback bandwidth f_b approximately as 10 $\log_2(f_b/f)$. At lower frequencies $f < f_b/6$, the slope can be increased to -12 dB/octave as shown by the dashed line. (In this case, a nonlinear dynamic compensator must be employed to ensure global stability as described in Chapters 10 and 11.)

When discussing the trade-off between the resonance frequency and the resonance quality (Q) of the object of control and the available disturbance rejection at a meeting with mechanical designers, it is helpful for control engineers to have prepared plots like those shown in Figure 4.50, exemplifying the available disturbance rejection for two structural resonance frequencies, $f_{st} = 50$ Hz and 100 Hz.

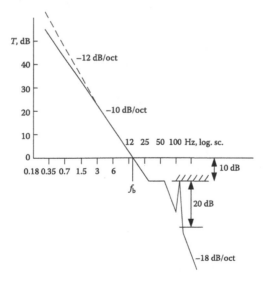

FIGURE 4.49
Loop response Bode diagram.

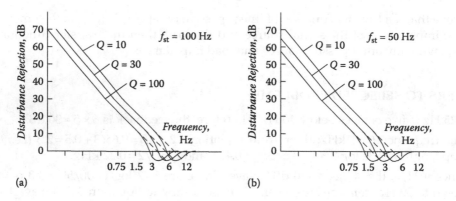

FIGURE 4.50
Dependence of disturbance rejection on the structural mode frequency and damping.

5

Compensator Design

Compensators, in software or hardware, are commonly built with rational transfer functions.

An asymptotic Bode diagram is piecewise linear. The slope of the segments of the diagram is $6n$ dB/octave, where n is an integer. The asymptotic Bode diagram is used to approximate the conceptual Bode diagram for the compensator (which may include segments of any slope). From the asymptotic Bode diagram, the poles and zeros of the compensator transfer function are immediately evident.

The approximation of an arbitrary constant slope gain response based on asymptotic Bode diagrams is described.

Lead and lag links are defined, their asymptotic responses shown, and it is demonstrated how to use them to increase or reduce the slope of the compensator Bode diagram.

A set of normalized plots is presented for second-order low-pass functions having complex poles with different damping coefficients.

The allocation of transfer function poles and zeros to cascaded links affects the compensator dynamic range. This is demonstrated and recommendations are formulated about how to construct a compensator by cascading links of first and second order.

Compensators consisting of parallel connections of links can be made in such a way that each of the links dominates over a certain frequency range. This is a convenient way to make steps on the loop response.

A digital compensator can be viewed as a modification of an analog compensator in which the analog integrators are replaced by discrete trapezoidal integrators. The trapezoidal integrator is analyzed, the z-transform introduced, and the conversions of the polynomial coefficients tabulated.

The Laplace and Tustin transforms are compared, and it is shown that the Tustin transform is sufficiently accurate in practice.

A design sequence is recommended for digital controllers. It is explained how to generate block diagrams, equations, and computer code for the first- and second-order digital compensator links. A complete compensator design example is presented with the derivation of a prototype analog compensator, including linear and nonlinear links, Tustin transform z-functions, equations of compensator links, and computer code.

The effect of aliasing is described, and the loop response is considered for the reduction of the aliasing errors with the necessary reduction in the available feedback.

5.1 Loop Shaping Accuracy

The desired frequency response of the compensator has been presented in previous chapters in terms of curves of logarithmic magnitude and phase vs. logarithmic frequency. In general, these curves imply transcendental functions of s. For implementation, a rational approximation to the ideal transcendental response is almost always required.

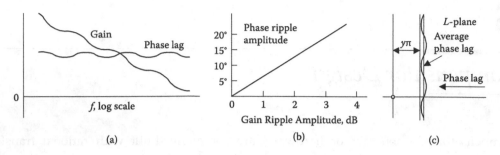

FIGURE 5.1
(a) Gain and phase lag diagrams, (b) relations between the ripple amplitudes, and (c) the ripples on the Nyquist diagram.

The accuracy of the constant slope implementation directly affects the value of the available feedback. Figure 5.1(a) shows the ripples on the gain and phase responses caused by the Chebyshev (equiripple) approximation of a desired constant slope response. The higher the order of the rational function, the better the approximation can be and the smaller the ripples. Figure 5.1(b) shows the relation between the ripple amplitudes of the gain and phase responses. In order for the *maximum* phase lag not to exceed $(1-y)\pi$, the *average* phase lag must be reduced by the value of the phase ripple, as seen in Figure 5.1(c), and the average slope of the gain response must be reduced correspondingly.

Example 5.1

A 5° phase ripple amplitude, i.e., 10° peak-to-peak phase ripples, will force an increase in the average phase margin by 5°. The average slope of the gain response will therefore be reduced by

$$(5° / 180°) \times 12 dB / octave = 1/3 dB / octave$$

For the typical four octaves length of the cutoff, the loss in the feedback will be 1.2 dB, which is marginally acceptable. The corresponding gain ripple amplitude is 0.75 dB, as seen in Figure 5.1(b), and the peak-to-peak ripples in the gain response are 1.5 dB.

Thus, the required accuracy in the loop gain response is typically not better than ±0.5 dB and the compensator need not be precise.

The higher the order of the compensator, the smaller the ripples and the better the accuracy of the loop response. Typically, the designer can achieve feedback within 2–5 dB of the theoretically available using a compensator of order 8–15.

5.2 Asymptotic Bode Diagram

The gain for transfer function s^{-n} is expressed as $-20n \log \omega$, and the Bode plot is a straight line with the slope $-20n$ dB/dec, or $-6.021n$ dB/octave $\approx -6n$ dB/octave, as shown in Figure 5.2. The use of octaves is preferred since decades do not provide the necessary resolution. The corresponding phase shift is constant at $-n90°$.

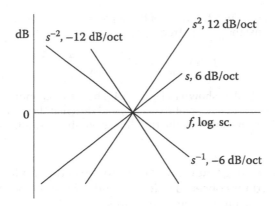

FIGURE 5.2
Constant slope Bode diagrams.

Generally, a rational transfer function of the Laplace variable s can be expressed as:

$$L(s) = k \prod \frac{(s - s_{zi})}{(s - s_{pj})}$$

where k is a real coefficient and the zeros s_{zi} and the poles s_{pj} can be real or come in complex conjugate pairs. The frequency response is calculated by replacing s by $j\omega$. The gain frequency response is:

$$20 \log |L(j\omega)| = 20 \log k + \Sigma 20 \log |j\omega - s_{zi}| - \Sigma 20 \log |j\omega - s_{pj}|$$

The gain plots (Bode diagrams) for transfer functions with the single real negative zero 20 log $|(j\omega + 2\pi f_z)/2\pi f_z|$ and the single real negative pole 20 log $|2\pi f_p/(j\omega + 2\pi f_p)|$, respectively, are shown in Figure 5.3(a) and (b) by thin lines.

The responses can be approximated by their high- and low-frequency asymptotes, shown by the thicker lines. These are two straight lines that have slopes 0 and ±6 dB/octave, and which intersect at the *corner frequency* equal to the frequency of the zero f_z or of the pole f_p.

For the calculation of the asymptotic gain, the expression $(j\omega + 2\pi f_z)$ is replaced by its larger component, whether real or imaginary. The error of such approximation is shown in Figure 5.3(c). The error is 3 dB at the corner, 1 dB one octave away from the corner, and less than 0.1 dB at two octaves.

The diagrams are drawn against the f or ω frequency axis. A convenient scale for drawing the diagram is 10 dB/cm, 1 octave/cm.

For a function having several real poles and zeros, the piecewise-linear *asymptotic Bode diagram* turns upward at each zero corner frequency by 6 dB/octave, and turns downward by −6 dB/octave, at each pole frequency.

Example 5.2

The asymptotic Bode diagram for the function

$$L(j\omega) = \frac{(j\omega + 0.5)(j\omega + 2)(j\omega + 20)}{(j\omega + 0.2)(j\omega + 0.3)(j\omega + 10)}$$

can be drawn as follows. First, the value of the asymptotic Bode diagram is calculated at some frequency, for instance, at the frequency $\omega = 1$. It is:

$$L(j\omega) = \frac{j\omega \times 2 \times 20}{j\omega \times j\omega \times 10} = \frac{4}{j\omega} = \frac{4}{j}$$

That is, the gain is 12 dB as shown in Figure 5.4. Then, the asymptotic diagram can be drawn by turning the response up at the zeros, and down at the poles, by 6 dB/octave. The actual Bode diagram shown by the thin line is obtained by adding the asymptotic error responses.

As seen, the asymptotic diagram is fairly close to the actual Bode diagram. Asymptotic diagrams are widely used for conceptual design, and for the purpose of finding a rational function that approximates a gain response defined by a plot.

The shape of the actual Bode diagrams is commonly verified numerically. For the above example, the polynomial coefficients of the numerator and denominator of the transfer function can be found from the known roots of the numerator and denominator with:

```
rn = [-0.5 -2 -20];
rd = [-0.2 -0.3 -10];
num = poly(rn);
denum = poly(rd);
```

and the Bode plot can be obtained with:

```
bode(num, denum)
```

or by using the command `freqs`.

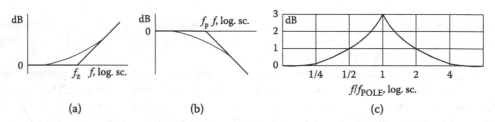

FIGURE 5.3
Bode diagrams for a (a) single zero, (b) single pole, and (c) the error of the asymptotic gain approximation.

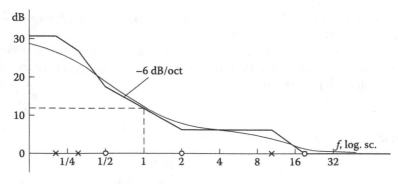

FIGURE 5.4
Asymptotic and actual Bode diagrams of a rational transfer function.

5.3 Approximation of Constant Slope Gain Response

As mentioned in Chapter 4, constant slope segments are important components in shaping the loop frequency response.

Any constant slope function can be decomposed into a product of a rational and an irrational function of s: $s^{-p} = s^{-m} s^{-q}$, where p and q are real, m is an integer, and $0 < q < 1$. The irrational function s^{-q} can be approximated by a rational function whose poles and zeros are real and alternate along the frequency axis as illustrated in Figure 5.5(a). The Bode plot of a rational approximation to s^{-q} is shown in Figure 5.5(b). The pole-zero spacings a and b result in an average slope of $-6b/(a+b)$ dB/octave, which should equal $-q$ dB/octave.

Example 5.3

Figure 5.6 shows the gain and phase responses for a simple rational approximation to $s^{-1/2}$ using three zero-pole pairs spaced evenly, an octave apart. The function is:

$$F(s) = \frac{(32s+1)(8s+1)(2s+1)}{(16s+1)(4s+1)(s+1)} \tag{5.1}$$

The gain slope is nearly constant at -3 dB/octave over a wide frequency range. At the corner frequencies, as the gain transitions to the flat asymptotes, the phase lag is nearly $45°/2$, in accordance with the Bode phase-gain relation. The $-38°$ phase at the center differs from the $-45°$ phase of the half-integrator due to the effects of the gain asymptotes.

Example 5.4

A rather extreme example is the following approximation to $s^{-1/3}$, which was generated by a curve-fitting program:

$$C(s) =$$

$$\frac{0.4415s^6 + 2.234s^5 + 1.861s^4 + 0.4276s^3 + 0.02954s^2 + 0.0005682s + 0.000002178}{s^6 + 2.462s^5 + 1.3037s^4 + 0.2007s^3 + 0.009201s^2 + 0.00010989s + 0.0000001979} \tag{5.2}$$

Over the frequency range 0.01–10 Hz, the gain response deviates from the ideal -2.007 dB/octave by less than 0.05 dB, while the phase remains within 0.05° of the ideal $-30°$. This high accuracy is rarely necessary since the required implementation accuracy for compensators in control systems is typically only 0.5 dB or so.

Curve-fitting programs can be used for approximating the desired compensator responses. This can work well; however, the programs may produce transfer functions that are too sensitive to the polynomial coefficient variations. Therefore, the transfer

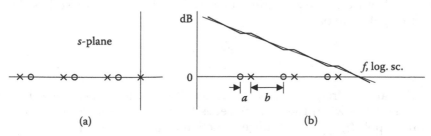

(a)   (b)

FIGURE 5.5
Alternating pole-zero approximation to s^{-q}.

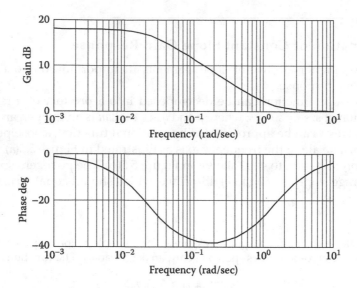

FIGURE 5.6
Gain and phase responses of a function with alternating real poles and zeros.

functions must be checked over by perturbing the coefficients by increments that reflect the round-off errors and the tolerances of the hardware implementation. On the other hand, approximating the desired curve by the careful combination of the poles and zeros implemented by cascading low-order links always results in a robust design. Refer to Section 5.6.

5.4 Lead and Lag Links

The ideal compensator response should be determined by subtracting the known actuator/plant frequency response from the desired loop response, as discussed in Chapter 4. Computer generation of frequency responses makes the iterative design of the compensator quick and effective. When linear models of the actuator and plant are available, the iteration can be carried out using a trial compensator in series with the actuator and plant, until the desired loop frequency response is achieved. With some experience, convergence will require no more than 5–10 iterations.

For the purpose of iterative structural design, it is best to regard the compensator as being composed of elementary building blocks. The simplest of these has the transfer function with a pole-zero pair:

$$\frac{s+z}{s+p} \tag{5.3}$$

where f_z and f_p are the frequencies of the zero and the pole, respectively. This transfer function is called *lead* when the zero precedes the pole; i.e., the zero is at a lower frequency than the pole ($f_z < f_p$). The transfer function is called *lag* when the pole comes first

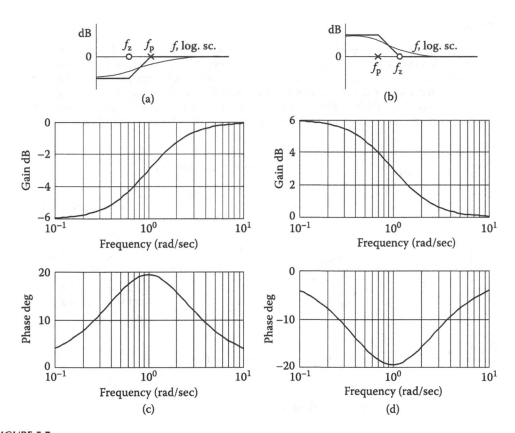

FIGURE 5.7
(a) Lead and (b) lag Bode asymptotic diagrams and Bode diagrams (thin lines). Frequency responses for (c) lead $(s+0.7)/(s+1.4)$ and (d) lag $(s+1.4)/(s+0.7)$.

$(f_p < f_z)$. Figure 5.7 shows asymptotic Bode diagrams and Bode diagrams for lead and lag transfer functions.

The larger the pole-zero separation, the larger the gain change is from the lower to the higher frequencies, and the larger the phase lead (or phase lag) is.

Adding a lead link makes the loop gain Bode diagram locally shallower, thus reducing the gain at low frequencies and locally reducing the phase lag.

Figure 5.8 shows the use of a compensator to change the –12 dB/octave slope of a double integrator to the desired slope of –10 dB/octave with a single lead and with two leads. The use of two lead links provides a closer approximation to the desired Bode diagram than could be achieved with just one lead link with a larger pole-zero separation.

Lag compensation locally steepens the Bode diagram, thus increasing the loop gain at lower frequencies, as shown in Figure 5.9. As in the case of lead compensation, several lag links are sometimes needed for better approximation accuracy.

Before deciding to introduce further compensation to improve the approximation to the desired response, one can use Bode's phase integral (Equations 3.11 and 3.12) to estimate the available improvement in the feedback.

Example 5.5

In a system, the desired loop gain response is a straight line with slope −10 dB/octave, which corresponds to a constant 150° phase lag, as shown in Figure 5.10. The current gain response is somewhat shallower, and there is an excessive phase stability margin with the area a expressed in decades × degrees. The phase integral indicates that elimination of this excess will yield an additional $0.56a$ dB of feedback at lower frequencies. The trade of the excessive phase margin for larger low-frequency loop gain can be made by introducing lag links.

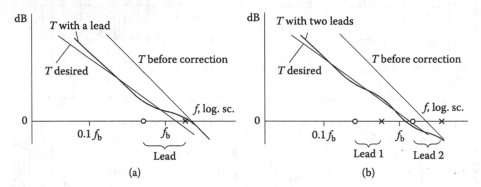

FIGURE 5.8
Compensation of a double integrator with (a) one and (b) two lead links.

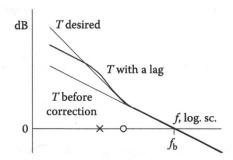

FIGURE 5.9
Lag compensation.

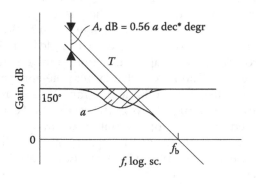

FIGURE 5.10
Use of the phase integral.

5.5 Complex Poles

Upper and lower Bode steps need to be reasonably sharp. To implement these sharp angles on the Bode diagram, complex poles and zeros are required. Complex poles can also be required to compensate for the plant response, and to shape the loop gain over the functional frequency band.

The normalized gain and phase lag frequency responses for the complex pole pair function

$$\frac{\omega_0^2}{s^2 + 2\zeta\omega_0 s + \omega_0^2} \tag{5.4}$$

are presented in Figure 5.11. The magnitude of Equation 5.4 at the resonance where $s = j\omega_0$ and the first and last terms cancel each other is the *quality factor* $Q = 1/(2\zeta)$, where ζ is the *damping coefficient*.

The asymptotic Bode diagram is calculated by retaining only the last term in the denominator of Equation 5.4 at $\omega < \omega_0$, and only the first term at $\omega > \omega_0$.

Example 5.6

A simple low-pass filter can be obtained by using the transfer function in Equation 5.4 with $\zeta = 0.5$.

Example 5.7

A third-order low-pass filter can be obtained by cascading a single-pole link and a link with the transfer function in Equation 5.4. The frequency of the complex pole must be higher than the frequency of the real pole, and ζ should be chosen such that the peak of the complex poles compensates the roll-off of the real pole. This method will be used to form the Bode step in Figure 5.14.

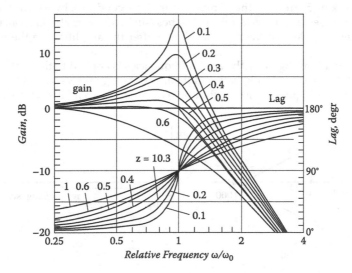

FIGURE 5.11
Gain and phase lag responses for a complex pole pair.

Example 5.8

The difference between the two logarithmic responses in Figure 5.11 is a notch or a peak response. Its transfer function is the ratio of functions (Equation 5.4):

$$\frac{s^2 + 2\zeta_n \omega_0 s + \omega_0^2}{s^2 + 2\zeta_d \omega_0 s + \omega_0^2}$$

This function equals 1 at zero and infinite frequencies, and ζ_n/ζ_d at the frequency of the resonance where $s = j\omega_0$. When $\zeta_n < \zeta_d$, a notch response results. When $\zeta_n > \zeta_d$, a peak response follows. The width of the notch or the peak depends on the chosen damping. Such notches have been used in the prefilter described in Section 4.2.3.

5.6 Cascaded Links

When the elementary links of the compensator are cascaded, attention should be paid to the signal level at the link junctions so as not to impair the compensator's *dynamic range*. This is the amplitude range of the signals, the largest of which is not yet distorted by saturation in the nonlinear links, and the smallest is still substantially larger than the disturbance and noise mean square amplitudes.

Example 5.9

Consider the implementation of the transfer function

$$W(s) = W_1(s)W_2(s) = \frac{s+2}{s+1000}\frac{s+500}{s+10} \tag{5.5}$$

as a cascade connection of the two links. The asymptotic gain responses for W_1 and W_2 are shown in Figure 5.12(a). It is seen that the signals at lower frequencies are attenuated in the first link by 54 dB, and then amplified in the second link by 34 dB. This way of making the compensator is certainly not the best since, after the attenuation, the signal drops dangerously close to the noise level, and after the amplification the noise floor will be raised.

Assume that a 1 mV signal with various frequencies is applied to the input as shown in Figure 5.13(a). At the junction between the links, the signal levels at 1 Hz and 1 kHz differ, as shown in Figure 5.13(a). Assume also that there is a 5 μV disturbance source at

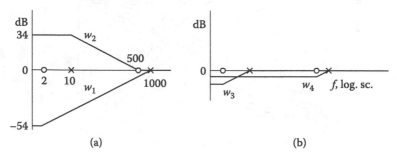

(a)   (b)

FIGURE 5.12
Gain responses of two different implementations of the same transfer function.

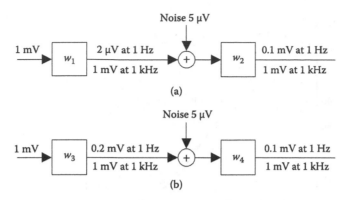

FIGURE 5.13
Signal levels at link junctions for the responses in Figure 5.12.

the junction of the links. Such disturbances may be caused by noise or interference in analog systems, and by round-off errors in digital systems. At 1 Hz, the signal amplitude is only 2 μV, so that the signal will be heavily corrupted with the noise.

It is better to implement the same transfer function by cascading the following links:

$$W(s) = W_3(s)W_4(s) = \frac{s+2}{s+10}\frac{s+500}{s+1000} \tag{5.6}$$

The frequency responses for W_3 and W_4 are shown in Figure 5.13(b). The signal levels have a much smaller dynamic range, as indicated in Figure 5.13(b), and even the smallest signal amplitude (0.2 mV at 1 Hz) remains much larger than the 5 μV noise.

The general rule is to avoid creating links with excessive attenuation or gain at any frequency, i.e., to keep in the same link the poles and zeros that are close to each other. When this rule is followed, the link affects the slope of the total Bode diagram over a relatively small frequency range, which also simplifies iterative adjustments of the frequency responses.

Example 5.10

The plant is a single-integrator $1/s$. The loop transfer function must behave as a single integrator at zero frequency. The gain and phase stability margins must be not less than 10 dB and 30°. The crossover frequency must be at least 0.9 rad/s, but the loop gain at frequency 10 rad/s and higher must not exceed −35 dB. For this, the roll-off at higher frequencies must be −18 dB/octave or steeper, and the loop response must include a Bode step.

Let us design the compensator as a cascade connection of several links.

When $\omega_b = 1$ and $x = 10$ dB, then $\omega_d = 2$. The desired width of the Bode step (Equation 4.2) is $0.6 \times 3 = 1.8$, i.e., $\omega_c \approx 3.6$.

The asymptotic Bode diagram shown in Figure 5.14 is composed of pieces with slopes −6, −12, 0, and −18 dB/octave. At lower frequencies, the compensator makes the slope −6 dB/octave. The average slope is −10 dB/octave in the crossover region at frequencies 0.15 to 2 rad/s.

A pair of complex zeros at ω_c make the corner at the beginning of the Bode step.

A real pole at $\omega = 2.8$ and a pair of complex poles at a frequency somewhat smaller than ω_c form a third-order low-pass filter (as explained in Example 5.7) at the end of the Bode step and effect the desired asymptotic slope, −18 dB/octave.

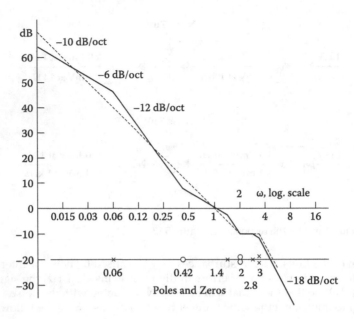

FIGURE 5.14
Asymptotic Bode diagram.

A damping coefficient of 0.5 is chosen for the complex zeros in order for the gain response to pass through the corner point, as seen in Figure 5.11. A damping coefficient of 0.4 is chosen for the complex poles to compensate for the rounding effect of the real pole at $\omega = 2.8$. The resulting loop transfer function is:

$$T(s) = k \frac{1}{s+0.06} \frac{s+0.42}{s+1.4} \frac{2.8}{s+2.8} \frac{s^2+2s+4}{s^2+2.4s+9} \frac{1}{s}$$

The compensator is composed of four first- or second-order links. This example shows that even when the order of the plant transfer function is low, the order of the compensator transfer function must be reasonably high in high-performance controllers.

From the condition that at $\omega = \omega_b = 1$, the asymptotic loop gain coefficient:

$$k(1 \times 1 \times 1 \times 2.8 \times 4 \times 1)/(1 \times 1.4 \times 2.8 \times 9 \times 1) = 1$$

when the coefficient $k \approx 3.15$. By using this value of k initially, it was found that for the actual Bode diagram to pass close to the 0 dB level at $\omega = 1$, k must be increased to 4.

After multiplication of the polynomials in the numerator and denominator:

```
n = 4 * 2.8 * conv([1 0.42],[1 2 4])
n = 11.2000 27.1040 54.2080 18.8160
d = conv(conv([1 0.06],[1 1.4]), conv([1 2.8],[1 2.4 9 0]))
d = 1.0000 6.6600 23.3960 48.5880 38.1125 2.1168 0
```

the loop transfer function becomes:

$$T(s) = \frac{11.2s^3 + 27.1s^2 + 54.2s + 18.8}{s^6 + 6.66s^5 + 23.4s^4 + 48.6s^3 + 38.1s^2 + 2.12s}$$

The loop responses are shown in Figure 5.15(a). The Nyquist diagram on the *L*-plane is plotted by:

```
w = logspace(-1, 1);        [mag, phase] = bode(n, d, w);
plot(phase,20*log10(mag),'r', -180, 0,'wo')
title('L-plane Nyquist diagram')
set(gca,'XTick',[-270 -240 -210 -180 -150 -120 -90])
grid
```

and is shown in Figure 5.15(b).

It is recommended for students to play with this response, to modify it by changing the poles and zeros (or the coefficients of the polynomials) in order to get a feeling for the sensitivity of the response to the poles, the zeros, and to the polynomial coefficients. For example, by increasing the complex pole damping coefficient, we can make the Nyquist diagram more rounded; an n.p. lag can be added to the plant and the Bode step made, correspondingly, wider; the high-frequency asymptotic slope can be made steeper by adding complex poles or a notch, which will also require lengthening the step; the gain response at lower frequencies can be made steeper. The response can be shifted along the frequency axis by replacing *s* by *as*, and along the gain axis, by multiplying the function by a constant.

Since the loop is certain only to the accuracy of the plant, the loop should be accurate only within the physical plant accuracy. With analog devices in the loop, very high order compensators might be excessive. But when the devices in the compensator are digital (Section 5.10), they can be easily implemented with an order of 20 or 30, making the compensator function closer to ideal.

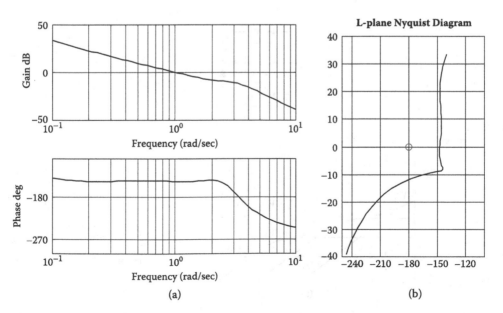

FIGURE 5.15

(a) Loop frequency response and (b) the Nyquist diagram.

5.7 Parallel Connection of Links

The compensator may be implemented also by connecting several links in parallel, as shown in Figure 5.16. The compensator transfer function is equal to the sum of the transfer functions of the elementary links $W_1 + W_2 + W_3$. The poles of the compensator are just the poles of the elementary links. The zeros of the compensator result from the interactions between the elementary links: the output is 0 at that value of s at which the sum of all the links' outputs is 0.

It is convenient to design the compensator such that each one of the parallel links dominates the response over a certain frequency band, as shown in Figure 5.17(a). This way the links can be designed and adjusted one at a time. (This configuration provides also for an option of placing separate nonlinear links in the parallel paths, if required, as described in Chapters 11 and 13.) At the frequency at which the link responses cross (i.e., at f_{12}, f_{23}, etc.), the output can be found by vector addition of the links' transfer functions. Depending on the phase difference between the output signals of frequency-adjacent channels, the summed signal amplitude may be larger or smaller than the amplitude of the components.

Example 5.11

The plant is an integrator $1/s$ in the system of Figure 5.18(a). The crossover frequency must be 1 rad/s.

The compensator is the parallel connection of the link $C_1 = 4/(s^2 + 4s)$ and the low-pass filter $C_2 = 5/(s^2 + 2.4s + 16)$ with $\zeta = 0.3$. In the filter, the resonance is four times the crossover frequency, and the low-frequency gain is −10.1 dB. The loop response:

$$T(s) = \frac{C_1(s) + C_2(s)}{s} = \frac{9s^2 + 29.6s + 64}{s^5 + 6.4s^4 + 25.6s^3 + 64s^2}$$

is shown in Figure 5.18(b). It is seen that the Bode step is well implemented, but the phase lag at lower frequencies is too large. This can be remedied with lead links placed in C_1 or in the common path.

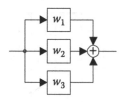

FIGURE 5.16
Parallel connection.

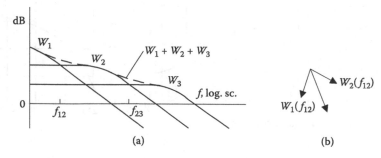

(a)   (b)

FIGURE 5.17
(a) Parallel links' gain responses and (b) output signals' addition at f_{12}.

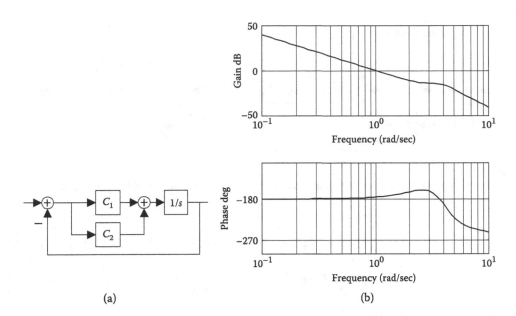

FIGURE 5.18
(a) Bode step implementation and (b) Bode diagram.

5.8 Simulation of a PID Controller

The *PID controller* consists of three parallel branches: I/s, P, and Ds. The coefficients I, P, and D define the transfer function of the controller. A saturation link is commonly placed in front of the I-channel to improve the controller performance in the nonlinear mode of operation for large-level signals (this issue will be considered in Chapters 10–13). This controller does not implement a Bode step and is not optimal in most applications, but it is simple and, as such, quite popular.

We will consider an example with a double-integrator plant with a flexible mode. The system block diagram is shown in Figure 5.19. The plant includes two masses, $CP1$ and $CP2$, connected by a spring and a dashpot (a device providing viscous friction).

The simulation can be performed in MATLAB, Simulink, or SPICE. When MATLAB is used, the transfer function for the plant must be determined analytically. With Simulink and SPICE, the plant equations can often be implemented using the available links. (Using Simulink block diagrams for ladder network analysis is described in Section 7.6.1.)

Example 5.12

In this example, we will use SPICE. In spite of a somewhat longer input file and the necessity to draw an equivalent schematic diagram, using SPICE has certain advantages: there is no need to generate a mathematical description of the plant if we use the common electromechanical analogies (force to current, velocity to voltage, mass to capacitor, spring to an inductor with the inductance equal to the inverse of the spring stiffness coefficient, and the dashpot to a resistor with the resistance equal to the inverse of the damping coefficient; the analogy is explained in detail in Section 7.1.1). The schematic diagram for the simulation is shown in Figure 5.20. The current to node 13 represents

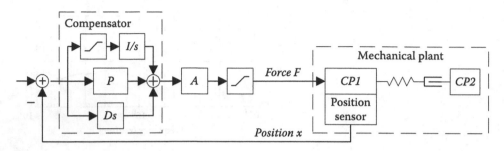

FIGURE 5.19
Block diagram for a control loop, a *PID* compensator, and a double-integrator plant having a flexible mode.

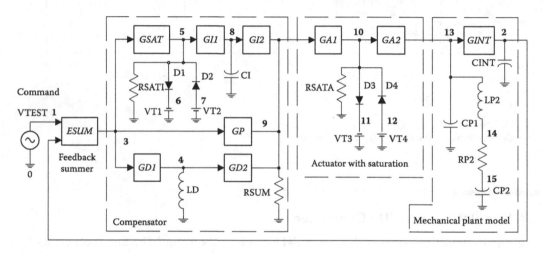

FIGURE 5.20
SPICE model for the system shown in Figure 5.19.

the force *F*. The position *x* (voltage at node 2) is calculated from the plant velocity (voltage at node 13) by integration.

The output currents of the three parallel paths in the compensator pass through the 1Ω summing resistor RSUM, and the voltage on the resistor represents the sum of the outputs of the parallel paths. The integrators are imitated by ideal controlled current sources loaded into capacitors. Saturation links are implemented using opposite-biased diodes shunting a resistive load.

This diagram represents a simple way of simulating the performance of the block diagram in Figure 5.19 in SPICE, and it does *not* describe the implementation of an analog compensator (compensator implementation will be considered in Chapter 6).

The SPICE input file is shown below. Included in the file but not shown on the picture are the high-resistance leakage resistors connecting nodes 1, 2, 8, and 13 to the ground, as required by SPICE.

```
*** PID example Figs. 5.19, 5.20 ***
ES 3 0 1 2 1            ; input signal summer
RSR1 1 0 1MEG           ; leakage resistor
RSP2 2 0 1MEG           ; leakage resistor
***
GSAT 5 0 0 3 0.001      ; saturation in I-path
RSAT 5 0 1K             ; threshold=(0.7+VT1)*GI1/GSAT
```

```
D1 5 6 DIODE
D2 7 5 DIODE
.MODEL DIODE D
VT1 6 0 1V
VT2 0 7 1V
***
GI1 8 0 0 5 1          ; I-path
CI2 9 0 0 8 10         ; integral coefficient
RSP8 8 0 1MEG          ; leakage resistor
***
GP 9 0 0 3 2           ; proportional coefficient
***
GD1 4 0 0 3 3          ; differential coefficient
LD 4 0 1
GD2 9 0 0 4 1
***
RS 9 0 1               ; summing resistor
***
GA1 10 0 9 0 1         ; actuator gain coefficient
***
RSATA 10 0 1K          ; saturation in actuator
D3 10 11 DIODE
D4 12 10 DIODE
VT3 11 0 1V
VT4 0 12 1V
GA2 13 0 0 10 1        ; force source
***
CP1 13 0 5             ; mass of the main body
RSP13 13 0 1MEG        ; leakage resistor
LP2 13 14 0.1          ; spring of flexible mode
CP2 12 0 0.5           ; mass of second body
RP 14 15 0.02          ; losses in the flexible mode
GINT 2 0 0 13 1
CINT 2 0 1             ; integrator to generate position
*** V10 is force, V13 is velocity, V2 is position
***
VTEST 1 0 AC 1         ; use only when frequency responses
                      ; are tested
.AC DEC 20 0.01 10     ; use only when frequency responses tested
** Pulse              (Vmin Vmax delay rise fall width period)
* VPULSE 10 PULSE (0V 10V 0S 0S 0S 500 500)
                      ; when transient responses tested
*.TRAN 0.1 10          ; when transient responses tested
.PROBE                ; or other graphical postprocessor
.END
```

Since the number of the nodes is only 15, this system can be simulated using a student version of SPICE, which presently allows up to 25 nodes, available free of charge from Intusoft, or PSPICE® from Microsim. (In SPICE, the summers and nonlinearities can also be specified by algebraic expressions; in most versions of SPICE, the transfer functions can also be specified by their poles and zeros.)

To plot the closed-loop response, plot vdb(2), vp(2).

To plot the loop response, connect the ESUM second input to 0 and then plot vdb(2), vp(2).

To plot transient response, disable with an asterisk lines VTEST and .AC and enable VPULSE and .TRAN.

To plot the Nyquist diagram with logarithmic scales, change the abscissa scale to linear and make it vp(2).

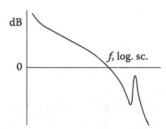

FIGURE 5.21
Loop gain response.

Running the program is recommended as a student exercise. The plots should be made for the following:

1. Frequency response of loop gain and phase so that stability margins can be checked.
2. Closed-loop frequency response; responses of the linear links (as the difference between the input and output signal levels and phases).
3. Transient closed-loop response for different coefficients *P*, *I*, *D* and saturation thresholds.

The loop gain response should be similar to that shown in Figure 5.21.

The loop response for *PID* controllers can be augmented by additional low-pass filters and notches to reduce the loop gain at the frequencies of the plant structural resonance and provide gain stabilization over the frequency range of the resonance.

Example 5.13

A triple-pole low-pass filter (two complex poles and one real) with a cutoff frequency approximately four times larger than the crossover frequency is placed in front of a single-integrator plant. The coefficient *D* is chosen such that with only this path of the compensator connected, the loop gain at lower frequencies is –*x* dB. The *P* coefficient is chosen such that with only this path on, the crossover frequency is where it must be. The integral term can be chosen such that with only this path on, the loop gain at the crossover frequency is –*x* dB. The resulting response has the high-frequency asymptotic slope –18 dB/octave, and the lower-frequency loop gain is lower than that of the response with the crisp Bode step by about 5 dB.

Drawing the asymptotic Bode diagrams and making the MATLAB simulation is left as a recommended student exercise.

5.9 Analog and Digital Controllers

While many factors may affect the choice between an analog and a digital system, three important considerations are accuracy, bandwidth, and price.

When considering accuracy, it should be remembered that only the accuracies of the prefilter, summer, and feedback path link directly affect the output accuracy. The accuracy required of the compensator and also of command feedforward links is much lower, so generally these can be implemented with less accurate elements.

Analog circuitry can be made very accurate and stable in time. For example, the analog reference voltage source, resistors, and chopper-stabilized amplifiers inside bench-type digital voltmeters are accurate to seven or eight digits. The dynamic range of a common op-amp is 1 μV to 10 V. This dynamic range is equivalent to 23 bits.

In digital systems, the accuracy and the dynamic range are frequently limited by the dynamic range of the employed A/D converter, which is typically 12–23 bits. An additional principal drawback of digital controllers is that the sample rate (typically, up to 1 MHz) and computational delay limit the bandwidth and the available feedback. Because of this, feedback loops with $f_b > 100$ kHz should be analog.

Other considerations are the type of sensors used, prices of a microcontroller and of D/A and A/D converters, power consumption, and ease of troubleshooting. Another important issue is the compensator's flexibility in being adjusted for different plants.

For the great majority of applications, either type of controller, analog or digital, can be designed such that it will perform well, and the choice between the two is determined mostly by the price of development and fabrication.

5.10 Digital Compensator Design

5.10.1 Discrete Trapezoidal Integrator

As is explained in Section 5.6, for the purpose of reducing the required dynamic range, it is advantageous to break the compensator into several cascaded links, each link related to different time constants, i.e., different frequency regions. For a similar reason, to reduce the rounding errors, it is common in digital signal processing (DSP) to implement a high-order transfer function as a cascade connection of first-order and second-order functions, each such function having poles reasonably close to zeros.

The second-order transfer function

$$\frac{p_2 s^2 + p_1 s + p_o}{s^2 + q_1 s + q_o} \tag{5.7}$$

or *biquad*, can be implemented using analog integrators $1/s$ with the feedback block diagram shown in Figure 5.22 (verification of the correspondence of the equation to the block diagram is left as a student exercise).

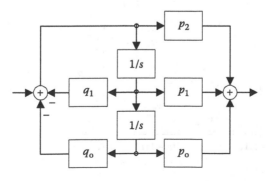

FIGURE 5.22
Feedback implementation of a biquad transfer function.

A similar block diagram can be used for digital implementation of the biquad operator, only the integration must be implemented in discrete steps performed at sampling instants.

Discrete trapezoidal integration is shown in Figure 5.23. The input analog signal $u(t)$ is sampled at time intervals T_S, $2T_S$, $3T_S$, etc. The output $v(t)$ represents the shadowed area, which approximates the integral of the input. The increment in $v(t)$ for the time interval from the sampling # $n - 1$ to the sampling # n is:

$$v_n - v_{n-1} = \frac{u_{n-1} + u_n}{2} T_S \tag{5.8}$$

Next, let multiplication by z signify an increase in time by one sample period so that:

$$zv_{n-1} = v_n$$
$$zu_{n-1} = u_n \tag{5.9}$$

With this symbolism, Equation 5.8 can be rewritten as:

$$v_n - z^{-1}v_n = \frac{z^{-1}u_n + u_n}{2} T_S$$

and the transfer function of the discrete trapezoidal integrator is:

$$\frac{v_n}{u_n} = \frac{1 + z^{-1}}{1 - z^{-1}} \frac{T_S}{2} \tag{5.10}$$

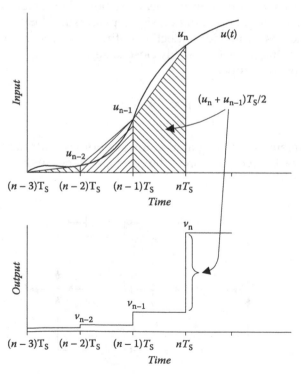

FIGURE 5.23
Trapezoidal digital integration.

This formula is presented as a flowchart in Figure 5.24. The operator z^{-1} can be implemented by storing a sampled value and recalling it after the sampling time T_S.

The biquad (Equation 5.7) can be implemented digitally by replacing integrators $1/s$ in the flowchart in Figure 5.22 by digital integrators. An equivalent but simpler flowchart can be obtained by substituting Equation 5.10 into Equation 5.7, and simplifying the resulting expression to obtain a function of z as a ratio of two second-order polynomials. The resulting flowchart looks like the block diagram in Figure 5.22, but with different coefficients and with $1/z$ replacing $1/s$.

Next, three important comments need to be made:

1. As seen in Figure 5.23, the input signal value averaged over the interval of sampling is calculated by discrete integration only at the end of the sampling interval, while analog integrator output reaches the average value in the middle of the sampling interval. In other words, compared with analog integration, digital integration delays the signal by $T_S/2$.

2. While a digital feedback system is being designed, the delay $T_S/2$ can be treated as non-minimum phase lag. As mentioned in Sections 4.2.2 and 4.3.4, this lag should not exceed 1 rad at the crossover frequency. Therefore, the *sampling frequency* $f_S = 1/T_S$ should be at least $2\pi/1 \approx 6$ times larger than f_b. With such a high sampling frequency, trapezoidal integration is quite accurate.

3. Since it needs at least two samples per period to identify a sinusoid, digital signal processing is only feasible for sinusoidal signals with frequencies less than the *Nyquist frequency* $f_S/2$.

5.10.2 Laplace and Tustin Transforms

To get an additional insight into the problem, we might re-analyze the results of the previous section using the Laplace transform. Since $1/z$ signifies a delay by T_S, we can use the Laplace transform of a pure time delay to imply:

$$z = \exp \frac{s}{f_S} \tag{5.11}$$

From here,

$$s = f_S \ln z \tag{5.12}$$

Equation 5.11 maps the strip of width $2\pi f_S$ of the left-hand plane of s, bounded by the dash-dotted lines in Figure 5.25(b), onto the unit radius disk of the z-plane shown in

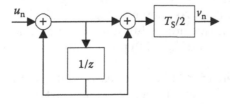

FIGURE 5.24
Digital trapezoidal integration flowchart.

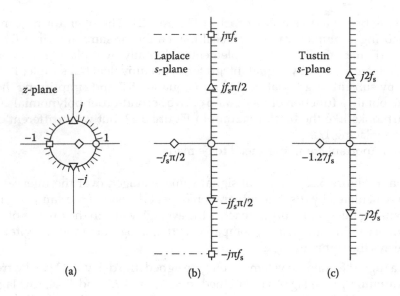

FIGURE 5.25
Mapping of (a) the z-plane onto the s-plane with (b) function $s = f_s \ln z$ and with (c) Tustin transform
$s = 2f_s(z-1)/(z+1)$.

Figure 5.25(a). The origin of the s-plane maps onto the point (1,0) of the z-plane, and the
points $-j\pi f_s$ and $j\pi f_s$, onto the point (–1,0) of the z-plane.

The Nyquist frequencies $\pm j\pi f_s/2$ of the s-plane map onto points $\pm j$ in the z-plane.

Near the origin, i.e. when x is small, the exponent is well approximated by $\exp(2x) \approx (1+x)/(1-x)$, which is the Padé approximation from Section 3.12. In DSP applications, this approximation is known as the *Tustin transform*, and sometimes as the *bilinear transform*:

$$z = \frac{2f_s + s}{2f_s - s} \tag{5.13}$$

i.e.,

$$s = 2f_s \frac{z-1}{z+1} \tag{5.14}$$

From Equation 5.14, the expression for the integrator $1/s$ is:

$$\frac{1}{s} = \frac{1+z^{-1}}{1-z^{-1}} \frac{T_s}{2}$$

This expression is the same as Equation 5.10, signifying that the Tustin transform is equivalent to trapezoidal integration. As shown in Figure 5.25, the Tustin transform maps the entire left half-plane of s onto the unit radius disk in the z-plane. The frequencies $\pm j2f_s$ map onto points $\pm j$ in the z-plane, and the (–1,0) point is only reached as the frequency goes to infinity.

In practical control systems $f_b < 0.1f_s$ for the reasons cited at the end of Sections 5.10.1 and later in 5.10.6. Therefore, the design uses only a small part of the first quadrant of the mapping in Figure 5.25(a). Within this part of the quadrant, the Tustin transform is quite accurate. At frequencies of the Bode step, from $2f_b$ to $4f_b$, the accuracy of the Tustin transform, although decreased, still remains adequate since the required accuracy of the loop gain response implementation over this frequency range also decreases.

In fact, in application to compensator design there is not much difference between the exact equivalent Laplace transform and the Tustin transform, so either one can be used. The (rational) Tustin transform is used most frequently, despite the slight distortion at practical frequencies of interest. The transform can be adjusted to precisely match the Laplace transform at a specified frequency by *prewarping*. The advantages of using prewarping, however, are typically insignificant for control system design, and the method will not be presented here.

It is seen from Equation 5.13 that z remains the same when s and f_S are similarly scaled up or down; for example, when f_S is expressed in kHz, s must be expressed in krad/s.

The following evident properties of the z-transform can be useful for verifying properties of discrete-time transfer functions:

A transfer function of s for $s = 0$ is equal to the transfer function of z for $z = 1$.

A transfer function of s for $s = \infty$ is equal to the Tustin transform of z for $z = -1$.

Multiplying a function of s by a constant is equivalent to multiplying the function z by the same constant.

Another approximating z-transform can be found by mapping the poles and zeros of the function of s onto the z-plane, and then appropriately scaling the gain coefficient.

Example 5.14

The Tustin transform of the lead

$$\frac{s+5}{s+10}$$

with sampling frequency 100 Hz can be found by substituting for s with Equation 5.14. Alternatively, Table 5.1 below in Section 5.10.4 gives algebraic expressions for the coefficients of the discrete transfer function. We obtain $m = 210$, $a_1 = 0.9762$, $a_o = -0.9286$, and $b_o = -0.9048$, so the z transfer function is:

$$\frac{0.9762 - 0.9286 / z}{1.0 - 0.9048 / z}$$

Example 5.15

The same problem is solved with MATLAB by:

```
n = [1  5];
d = [1  10];
fs = 100; % sampling frequency in Hz
[nd, dd] = bilinear (n, d, fs)
nd = 0.9762  -0.9286
dd = 1.0000  -0.9048
```

Alternatively, the function c2d can be used.

Example 5.16

The Tustin transform properties for the numeric values of the previous example are verified in the following.

For $s = 0$, the function of s is 0.5; for $z = 1$, the Tustin transform is also 0.5.

For $s = \infty$, the function of s is 1; for $z = -1$, the Tustin transform is also 1.

The zero of the function of s is -5, and the pole is -10. Using Equation 5.14, the zero and the pole of the function of z are found to be, respectively,

$$z_z = \frac{2f_S + s}{2f_S - s} = \frac{200 - 5}{200 + 5} = 0.95122$$

and

$$z_p = \frac{2f_S + s}{2f_S - s} = \frac{200 - 10}{200 + 10} = 0.904762$$

Mapping the zero and pole using 5.11 gives $z_z = 0.95123$ and $z_p = 0.90484$. So for typical sampling rates the errors caused by the fact that the Tustin transform is only an approximation to the Laplace transform do not matter much for the digital functions in the forward path. However, in some cases the error might not be acceptable for the links in the feedback path or in the prefilter, and in this case, a good option is to use command feedforward instead.

5.10.3 Design Sequence

Digital control systems can be designed as follows:

Given the required feedback bandwidth f_b, the sampling frequency f_S is chosen: commonly, $f_S \geq 10f_b$, and for better performance, $f_S \geq 50f_b$.

By approximation of the optimal response, a rational transfer function of the analog compensator is determined and broken appropriately into second-order rational functions (as described in Section 5.5). The sampling delay $T_S/2$ is approximated by introducing in the loop an all-pass with pole-zero pair $(s-f_S)/(s+f_S)$, as mentioned in Section 3.12, or more simply by placing a pole at frequency $f_S/3$. For simulation, MATLAB, Simulink, or SPICE can be used.

The Tustin transform is used to find functions of z that correspond to the second-order functions of s.

The system is simulated using a digital control software package (for example, Simulink), or directly in C or in another implementation language.

The functions of z are coded in, and the system is either tested or simulated (including the nonlinearities).

The stability margins are verified by increasing the loop gain and/or the loop phase lag until self-oscillation begins.

5.10.4 Block Diagrams, Equations, and Computer Code

The functions of z can be specified by block diagrams, by equations, or by computer code. The block diagrams for the first-order and second-order functions are shown in Figure 5.26. These block diagrams have forward and feedback paths.

Using Mason's rules, the transfer function for block diagram a is found to be:

$$\frac{a_1 + a_0 z^{-1}}{1 + b_0 z^{-1}} \quad \text{or} \quad \frac{a_1 z + a_0}{z + b_0} \tag{5.15}$$

TABLE 5.1

Coefficients for Linear Fractional Function of z

a_1	a_0	b_0	m
$(p_1 \times 2f_S + p_0)/m$	$(-p_1 \times 2f_S + p_0)/m$	$(q_0 - 2f_S)/m$	$2f_S + q_0$

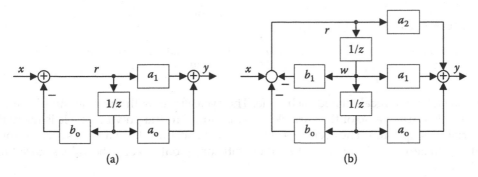

FIGURE 5.26
Block diagrams for (a) the first-order link and (b) the second-order link.

and for block diagram b:

$$\frac{a_2 + a_1 z^{-1} + a_0 z^{-2}}{1 + b_1 z^{-1} + b_0 z^{-2}} \quad \text{or} \quad \frac{a_2 z^2 + a_1 z + a_0}{z^2 + b_1 z + b_0} \tag{5.16}$$

The coefficients a_i, b_i can be obtained by substituting Equation 5.14 into the expressions for the transfer functions of s. For the function of s:

$$\frac{p_1 s + p_0}{s + q_0}$$

which is a linear fractional function; the coefficients are given in Table 5.1.
   For the biquad of s:

$$\frac{p_2 s^2 + p_1 s + p_0}{s^2 + q_1 s + q_0}$$

the coefficients for the biquad of z are given in Table 5.2.
   As already mentioned, the coefficients of a rational function of z can be calculated from a rational function of s with given numerator num and denominator den by the MATLAB function:

```
[numd, dend] = bilinear(num, den, fs)
```

where fs is the sampling frequency, or by using the function c2d. To verify the transform, frequency responses of digital filters can be found with the MATLAB command freqz.
   For the first-order link in Figure 5.26(a), the following C code equations can be used, with a_0, a_1, b_0 renamed A0, A1, B0:

TABLE 5.2

Coefficients for Biquad of z

a_2	a_1	a_0
$(4f_s 2 p_2 + 2f_s\, p_1 + p_0)/n$	$(-8f_s^2 p_2 + 2p_0)/n$	$(4f_s^2\, p_2 - 2f_s\, p_1 + p_0)/n$

b_1	b_0	n
$(-8f_s^2 + 2q_0)/n$	$(4f_s^2 - 2f_s\, q_1 + q_0)/n$	$4f_s^2 + q_1 2f_s + q_0$

```
y = A0 * r;                      /* using previous value of r to find **
** first component of the output */
r = x - B0 * r;                  /* updating r */
y += A1 * r;                     /* adding second component to **
                                 ** the output using updated r */
```
$$(5.17)$$

The variables are recalculated each cycle. The cycle starts with the new sample value of the input. First, the value of the variable r that is stored in the previous cycle is used; then this variable is updated. The cycle repeats f_s times per second. The variables are commonly initialized to zeros, and they must be either static or global to keep the values stored to be used in the next cycle.

For the second-order link, similarly, the following code can be used:

```
y = A0 * w + A1 * r;             /* using previous values of r,w */
r1 = x - B0 * w - B1 * r;        /* updating r */
w = r;                           /* updating w */
r = r1;                          /* updating r */
y += A2 * r;                     /* adding component to the output */
```
$$(5.18)$$

5.10.5 Compensator Design Example

A small parabolic telecommunication link antenna tracking the Earth has been placed on the Mars Pathfinder lander. Two identical brushless motors with internal analog rate feedback loops articulate the antenna in two orthogonal directions. The motors are controlled by two independent identical SISO controllers.

The sampling frequency is 8 Hz. Were the delay caused only by the sampling, the crossover frequency would be $f_s/5 \approx 1.6$ Hz. However, since the computer must handle not only the motor control loops but also other, higher-priority tasks, there is an additional 500 ms delay caused by four real-time interrupt (RTI) delays, 125 ms each. Also, due to limited bandwidth of the analog rate controllers for the motors (already designed), the motors have a 50 ms delay. Since the total delay is not only 62.5 ms (of sampling) but $62.5 + 500 + 50 \approx 600$ ms, the realizable crossover frequency is lower in proportion to this delay, i.e.,

$$f_b < 1.6 \times 62.5 / 600 \approx 0.17 \, \text{Hz}$$

The design was done initially in the s-domain. The controller is nonlinear, and includes two cascaded linear links, $C_1 + 1$ and C_2. A saturation link placed in front of C_1 makes the transfer function of the compensator dependent on the signal level. When the signal level is below the saturation threshold, the compensator transfer function is $(C_1 + 1)C_2$. When the signal is high, the compensator transfer function is reduced to C_2. The operation of such nonlinear dynamic compensators (NDCs) will be further described in Chapters 10 and 13.

For small-signal amplitudes, the compensator function is:

$$C = (C_1 + 1)C_2$$

where C_1 is a single-pole low-pass filter,

$$C_1 = 2.5 / (0.0833 + s)$$

and C_2 is a lead link,

$$C_2 = (0.106 + s) / (2.23 + s)$$

The asymptotic gain frequency responses of the compensators are shown in Figure 5.27.

The lead C_2 provides phase advance and partially compensates the following lags: the phase lag of up to seven RTI (for extra robustness), i.e., up to 1.875 s delay, and the delay of 0.05 s of the closed analog rate loop.

The path C_1 is parallel to the path with unity gain. At zero frequency, C_1 becomes 30.

The asymptotic loop gain frequency responses are shown in Figure 5.28 for the case of both C_1 and C_2 operational, and for the case of $C_1 = 0$ (lower curve). The Bode step is very long because of the necessity to compensate for a large time delay of up to seven RTI, and to reduce or eliminate the overshoot.

The system with these analog compensators was simulated in SPICE and MATLAB. The phase delay of the sampling was imitated by an extra pole placed at frequency $f_S/3$. After several small adjustments to the initial response were made to obtain the desired stability margins, the compensator design was converted into digital.

The following digital compensator equations were obtained from the analog controller functions with the Tustin transform:

$$C_1 = (0.15 + 0.15\,/\,z)\,/\,(1 - 0.99\,/\,z) \tag{5.19}$$

$$C_2 = (0.9 - 0.8883\,/\,z)\,/\,(1 - 0.75\,/\,z) \tag{5.20}$$

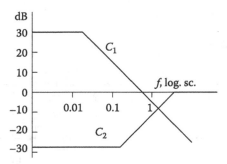

FIGURE 5.27
Asymptotic Bode diagrams of compensators.

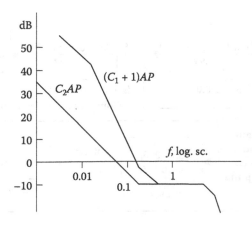

FIGURE 5.28
Open-loop asymptotic Bode diagrams for small error (upper curve) and large error (lower curve).

The coefficients in the equations have been rounded to the required accuracy. The second expression can be rewritten as:

$$C_2 = 0.9(1 - 0.987) / (1 - 0.75 / z)$$

The accuracy of the coefficient 0.987 should be rather high since this value is subtracted from 1 at lower frequencies where z approaches 1. Thus, for the accuracy of the low-frequency gain coefficient to be better than 6% (i.e., 0.5 dB), the difference $1 - 0.987 = 0.013$ must be accurate to 6%, i.e., to 0.0008, so that the number 0.8883 in Equation 5.20 should not be further rounded.

Equations 5.19 and 5.20 correspond to the flowchart shown in Figure 5.29(a) and (b).

The simplified feedback loop block diagram is shown in Figure 5.30. The block diagram includes a saturation link in the higher-gain, low-frequency path, linear links C_1 and C_2, a scaling block that has saturation and a dead zone, a delay block, and a model of the plant (of the motor with its analog control electronics).

The C code for the compensator follows:

```
#define PAR1  .15
#define PAR2  .99
#define PAR3  .9
#define PAR4  .75
#define PAR5  .987
#define THRESHP 1000
#define THRESHN -THRESHP
```

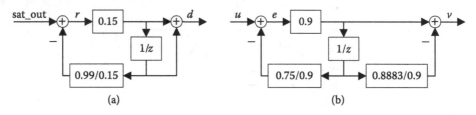

(a)   (b)

FIGURE 5.29
Flowcharts corresponding to Equations 5.19 and 5.20.

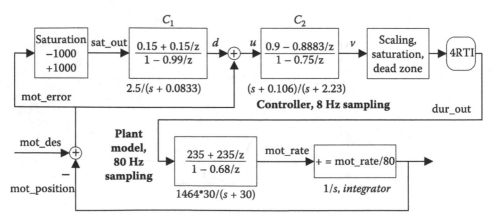

FIGURE 5.30
Motor controller flowchart.

```
global double r=0.0;
global double d=0.0;
global double e=0.0;
global double u=0.0;
global double v=0.0;
global double sat_out=0.0;
sat_out=mot_error; /* saturation */
if (mot_error > THRESHP)
sat_out=THRESHP;
if (mot_error < THRESHN)
sat_out=THRESHN;
d=PAR1 * r; /* compensator C1 */
r=sat_out + PAR2 * r;
d += PAR1 * r;
u=d + mot_error;
v=-PAR5 * e; /* lead C2 */
e=u + PAR4 * e;
v += PAR3 * e;
```

The variable dur_out is the duration of time that the motor is on during the sampling period of 125 ms. The motor is rate-stabilized by an analog loop with 30 ms rise time. The motor transfer function is therefore that of an integrator (the angle of rotation is proportional to the time the motor is on), with an extra pole caused by the limited bandwidth of the analog rate loop. The motor (plant) transfer function of s is shown in Figure 5.30, under the block.

For computer simulations in C, a digital motor model was employed. The transfer function of z is shown in Figure 5.30 in the block. The rectangular sample-and-hold integrator was used for simplicity, and for better accuracy the sampling frequency for the model was set to 80 Hz, 10 times higher than that of the compensator. The data in the motor model are updated 10 times after each update in the controller. This system, shown in Figure 5.30, is an example of a multirate system (although the controller itself is single rate). In multivariable controllers, different rates are frequently used: faster rates for processing rapidly changing variables, and lower rates for slowly varying variables.

5.10.6 Aliasing and Noise

An A/D converter contains a *sample-and-hold* (S/H) link, i.e., a device that samples the signal and keeps this value at its output until the time of the next sampling. An example of such a link is shown in Figure 5.31. The switch samples the input signal by closing for a short duration at the sampling times. The capacitor charges and holds the sampled value of the signal until the next sampling. The output of the S/H is processed digitally and then

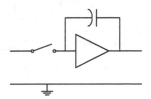

FIGURE 5.31
Sample-and-hold link circuit diagram.

returned to the analog form by a D/A converter at the input to the actuator, or directly to the plant (when the actuator is also digital).

The S/H link is a linear time-variable circuit and, as such, works as an amplitude modulator. Modulation of the high-frequency noise in the S/H link by the sampling frequency and its harmonics produces frequency difference products that fall within the signal bandwidth. This effect, called *aliasing*, is illustrated in Figure 5.32. It is seen that on the basis of the information sampled at discrete points, it is impossible to distinguish between the low-frequency signal with frequency f and the high-frequency signal with frequency nf. From here, two important implications for the control system design follow.

First, the effects of the high-frequency noise are added to the baseband signal at the output of the A/D converter.

Aliasing might introduce substantial error in the A/D conversion. To reduce this error by rejecting the high-frequency input noise, a high-order antialiasing low-pass filter is commonly installed at the input to the sample-and-hold link (or A/D converter) of DSP systems, as shown in Figure 5.33.

In closed-loop feedback systems like that shown in Figure 5.34, the high-frequency sensor noise N causes the output noise N_{out} in the functional frequency band. The noise is reduced by the antialiasing filter.

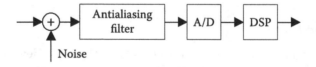

FIGURE 5.32
Aliasing.

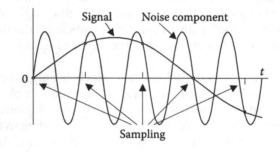

FIGURE 5.33
Antialiasing filter.

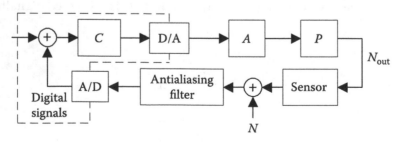

FIGURE 5.34
Control system with antialiasing filter.

The filter selectivity is limited by its the effect of the loop gain and the input–output closed-loop response. To a certain extent, the attenuation of the antialiasing filter at higher frequencies can be equalized by an increase of the gain of the digital compensator (thus making the loop gain as desired) and the introduction of a prefilter in the command path to reduce the input–output closed-loop gain.

The sensor noise is transformed into the baseband largely from the frequencies close to f_S and its harmonics. Therefore, the attenuation of the antialiasing filter can be smaller between these frequency bands. The optimal response of the feedback antialiasing filter is therefore not monotonic, and the loop gain response, which is close to the optimum, looks like that shown in Figure 5.35. Its feedback bandwidth is wider by 0.3 to 0.5 octaves than the monotonic response shown by the dotted line.

Second, due to aliasing, the gain of the digital filter for a sinusoidal signal with Nyquist frequency is the same as the dc gain. Therefore, a combination of a large low-frequency gain of the digital compensator and a small attenuation in the analog plant near the Nyquist frequency may result in large loop gain and oscillation. In these situations, the sampling frequency must be substantially increased.

5.10.7 Transfer Function for the Fundamental

In this section, we present yet another view on the effect of digital compensation on the feedback loop. A linear digital compensator is a linear time-variable (LTV) link. As will be shown in Section 7.11, time dependencies of linear systems can reduce the stability margins and result in oscillation.

Consider the case when the signal is sinusoidal. The input and output signals of the S/H link with 12 samples per period are plotted in Figure 5.36. Let's define the gain coefficient in fundamentals as the ratio of the amplitude of the output signal fundamental to the amplitude of the input signal. It is seen that the magnitude of this gain coefficient is approximately 1, and the phase lag is approximately 15°. It is clearly seen that the lag, as long as the sampling frequency is relatively high, is inversely proportional to the sampling frequency.

When the number of samples per signal period is only two (the Nyquist frequency case), the output of the S/H circuit is Π shaped, as shown in Figure 5.37, and the output amplitude is sin φ, where φ is the phase shift between the sampling and the input signal.

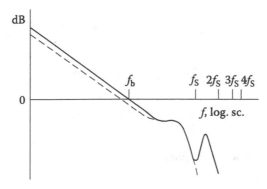

FIGURE 5.35
Open-loop Bode diagrams for rejection of aliasing noise.

As follows from Fourier analysis, the amplitude of the fundamental of the output is $(4/\pi)\sin\varphi$.

When, in particular, $\varphi = 90°$, then the phase lag is 90°, the output amplitude is 1, and the equivalent gain coefficient is $4/\pi$; i.e., the gain is 2.1 dB. When φ approaches 0 or 180°, the gain coefficient approaches 0. The uncertainty in the gain and phase (due to the uncertainty of φ) gradually increases with the decrease of the sampling frequency, as can be seen by comparing Figure 5.36 and Figure 5.37. The system must be made stable with sufficient margins for all possible φ.

Figure 5.38 shows the stability margin boundary and an example of an L-plane Nyquist plot of a well-designed LTI system. Consider next the effect of introducing an S/H link in this loop at the Nyquist frequency.

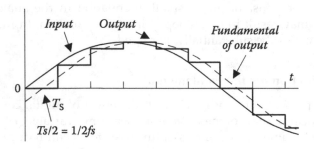

FIGURE 5.36
Signal at the output of a sample-and-hold link.

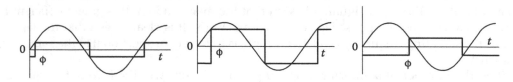

FIGURE 5.37
Effect of the phase difference between the signal and the sampling.

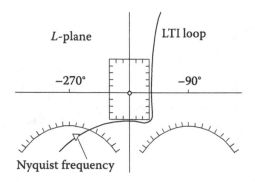

FIGURE 5.38
Stability margin boundary and the loop transfer function uncertainty at Nyquist frequency.

Oscillation in control systems, if it happens, is usually periodical, and the shape of the oscillation at the plant output is close to sinusoidal due to the plant low-pass filter properties (this will be discussed further in Chapter 11). Because of the relations between the phase and amplitude illustrated in Figure 5.37, and because of the extra 2.1 dB of gain on the fundamental, the gain of the LTI loop at the Nyquist frequency must be below the boundary curves with maximum $-x-2.1$ dB centered at -270 and $-90°$, to rule out an oscillation at this frequency with any possible φ, with x dB margin. The resulting penalty in the available feedback is typically not large. On the other hand, if the Nyquist frequency should fall on the part of the Nyquist diagram that is phase stabilized, the penalty would be up to $90°$, which would require reducing the slope of the Bode diagram and the feedback. This is why the sampling frequency must be kept sufficiently high.

PROBLEMS

1. The crossover frequency is 100 Hz. The system must be phase stabilized at all frequencies below the crossover with margin $30°$. The loop gain approximates a constant slope Bode diagram in the Chebyshev sense. By how much (approximately) can the feedback be increased at (a) 10 Hz, (b) 1 Hz, (c) 0.5 Hz, (d) 20 Hz, and (e) 2.72 Hz if, by using a higher-order compensator, the phase peak-to-peak ripples are reduced from $15°$ to $2°$?

2. By addition of a real pole and a real zero, an asymptotic Bode diagram was made steeper by 6 dB/octave over the frequency interval from 10 to 30 Hz. What are the pole and the zero frequencies? Will the new Bode diagram be more concave or more convex? What happens to the diagram if the pole and zero are interchanged?

3. Draw an asymptotic Bode diagram for the functions having the following:
    a. Gain coefficient 10 at $\omega = 0$, zeros (in ω, i.e., in rad/s) -1, -3, -6, and poles (in ω) -0.5, -4, -8
    b. Gain 10 dB at $\omega = 2$, zeros (in ω) -2, -5, -5, and poles (in ω) -1, -3, -20
    c. Gain coefficient 10 at $f = \infty$, zeros (in Hz) -15, -30, -400, and poles (in Hz) -60, -100, -200
    d. Gain 20 dB at $f = 200$, zeros (in Hz) -100, -200, $-1,000$, and poles (in Hz) 0, -10, $-1,600$

    Use scales 10 dB/1 cm and 1 octave/1 cm. Find the Bode diagrams from asymptotic responses using the rule for the error: 3 dB at pole, 1 dB one octave from the pole, $0.1 \approx 0$ dB two octaves from the pole.

4. Use MATLAB to make Bode plots for the following functions:
    a. $T(s) = 100/[s(s+15)(s+100)]$
    b. $T(s) = 1,000/[s(s+100)(s+500)]$
    c. $T(s) = 5,000/[s(s+200)(s+6,000)]$
    d. $T(s) = 200/[s^2(s+100)(s+1,000)]$

5. Find a rational function approximation of the following constant slope functions:
    a. Slope -6 dB/octave, frequency range 1–10 Hz

   b. Slope −9 dB/octave, frequency range 1–10 Hz

   c. Slope −12 dB/octave, frequency range 1–10 Hz

   d. Slope −15 dB/octave, frequency range 1–10 Hz

   e. Slope −27 dB/octave, frequency range 1–10 rad/s

   f. Slope −9 dB/octave, frequency range 1–10 rad/s

   g. Slope −12 dB/octave, frequency range 1–10 rad/s

   h. Slope −12 dB/octave, frequency range 1–10 rad/s

   i. Slope −18 dB/octave, frequency range 1–10 rad/s

   j. Slope −6 dB/octave, frequency range 1–10 rad/s

6. Draw asymptotic Bode diagrams and make the plots with MATLAB for the following leads:

   a. $(s+2)/(s+15)$

   b. $(s+0.1)/(s+0.2)$

   c. $(s+0.5)/(s+2.5)$

   d. $(s+2)/(s+4)$

   e. $(s+2.72)/(s+21)$

   f. $(s+1)/(s+16)$

7. Draw asymptotic Bode diagrams and make the plots with MATLAB for the following lags:

   a. $(s+15)/(s+2)$

   b. $(s+1)/(s+0.2)$

   c. $(s+5)/(s+2.5)$

   d. $(s+8)/(s+4)$

   e. $(s+7)/(s+2.72)$

   f. $(s+16)/(s+2)$

   g. $(s+8)/(s+4)$

8. The phase stability margin is excessive by 10° over one decade. Find the lost feedback at lower frequencies.

9. If the peaking must be 8 dB, what is the damping coefficient ζ (use the plots in Figure 5.10)? Find the polynomial corresponding to the peaking frequency 300 Hz.

10. Plot with MATLAB the normalized low-pass frequency response with a pair of complex poles, with the following damping coefficients:

   a. 0.0125

   b. 0.125

   c. 0.25

   d. 0.5

   e. 0.99

    Use the MATLAB function lp2lp to convert the transfer function to that having the resonance frequency 5 Hz.

11. Plot with MATLAB the normalized band-pass frequency response with a pair of complex poles, with the following damping coefficients (obtain the response by multiplying the low-pass transfer function by s):

    a. 0.01
    b. 0.1
    c. 0.2
    d. 0.4
    e. 0.99

    Use the MATLAB function lp2lp to convert the transfer function to that having the resonance frequency 50 Hz.

12. Plot with MATLAB the normalized high-pass frequency response with a pair of complex poles, with the following damping coefficients (obtain the response by dividing the low-pass transfer function by s^2):

    a. 0.02
    b. 0.2
    c. 0.3
    d. 0.5
    e. 0.99

    Use the MATLAB function lp2lp to convert the obtained response to that having the resonance frequency 15 Hz.

13. Plot a series of five notches with the notch amplitude 6 dB and various widths, centered at:

    a. 1 rad/s
    b. 10 rad/s
    c. 10 Hz
    d. 1 kHz
    e. 2.72 kHz

14. Break the compensator function into cascaded links:

    a. $5{,}000(s+1)(s+2)(s+1{,}000)/[s(s+20)(s+6{,}000)]$
    b. $100(s+0.1)(s+8)(s+200)/[s(s+20)(s+600)]$
    c. $5{,}000(s+1)(s+2)(s+1{,}000)/[s(s+20)(s+6{,}000)]$
    d. $100(s+0.1)(s+8)(s+200)/[s(s+20)(s+600)]$
    e. $5{,}000(s+1)(s+2)(s+1{,}000)/[s(s+20)(s+6{,}000)]$
    f. $100(s+0.1)(s+8)(s+200)/[s(s+20)(s+600)]$

15. The m.p. component of the plant is 1/s, and the n.p. lag of the plant is 1 rad at 2 kHz. The amplitude stability margin must be 10 dB. The asymptotic slope must be −18 dB/octave, the asymptote crossing the −10 dB level at 2 kHz. The loop must have a Bode step and −10 dB/octave constant slope down to 100 Hz. Design an analog compensator composed of cascaded links.

16. Same plant and requirements as in Problem 15; design an analog compensator composed of parallel links.

17. The feedback bandwidth is limited by the effect of the sensor noise. The loop gain response must be steep right after f_b, and to provide the stability margin, the slope of the loop gain must be only -6 dB/octave for two octaves below f_b. $B_n(f_b) = 1$ rad. Design the compensator for the following plant and feedback bandwidth:

    a. $1/[s\ (s+300)(s+1{,}000)]$; $f_b = 3$ kHz

    b. $10^{-3}/[s\ (s+30)(s+100)]$; $f_b = 300$ Hz

    c. $10^{-6}/[s\ (s+3)(s+10)]$; $f_b = 30$ Hz

18. Verify that Equation 5.7 follows from the diagram in Figure 5.22.

19. A high-order digital compensator was implemented in C without breaking the transfer function into second-order links. It was found that a single-precision simulation on different processors or using different compilers gave slightly different results, while double-precision simulation showed nearly identical results. After the DSP was modified by properly breaking the function of z into second-order multipliers, single precision became sufficient to obtain the same results on all computers. Explain why this might have happened.

20. The poles of an analog compensator, in s, are:

    a. $-3, -6, -8$

    b. $-12, -60, -80$

    c. $-13, -16, -85$

    d. $-10, -600, -1{,}500$

    With sampling frequency $f_S = 50$ Hz, find the poles of the function of z using Equation 5.14 or MATLAB command bilinear. (*Hint:* each pole can be found by applying the function bilinear to the function $1/(s - s_{pole})$.)

21. Find the Tustin transforms from:

    a. $C(s) = (s+3)/(s+2)$

    b. $C(s) = 5(s+2)/(s+3)$

    c. $C(s) = 10(s+5)/[s(s+4)]$

    d. $C(s) = 3(s+7)/(s+20)$

    e. $C(s) = 15(s+8)/[(s+100)s]$

    f. $C(s) = 2(s+3)/s$

22. For sampling frequency $f_S = 10$ Hz, convert to $C(s)$ from the following:

    a. $f(z) = (0.2174 + 2174/z)/(1 - 0.7391/z)$

    b. $f(z) = (0.1200 + 0.1200/z)/(1 - 0.600/z)$

    c. $f(z) = (1.33 - 0.4444/z)/(1 - 0.1111/z)$

    For sampling frequency $f_S = 100$ Hz, convert to $C(s)$ from the following:

    d. $f(z) = (0.22 + 22/z)/(1 - 0.74/z)$

    e. $f(z) = 0.272 + 0.272/z)/(1 - 0.600/z)$

    f. $f(z) = (1.1 - 0.4/z)/(1 - 0.1/z)$

23. Design a digital compensator for the analog plant $P(s) = 50{,}000(s+200)/(s+300)$, with the slope of the loop Bode diagram at frequencies below f_b approximately -10

dB/octave. Assume $f_s = 10$ kHz, the aliasing noise is of critical importance, and the gain and phase stability margins are, respectively, 10 dB and 30°. Consider:

a. A version with $f_b = 1$ kHz, a Bode step, monotonic response, and asymptotic slope −12 dB/octave

b. A version with $f_b = 1.4$ kHz, a Bode step, and a notch at f_s, as in Figure 5.35

24. Write a program in C for $f(z)$ in (a)–(c) of Problem 22.

25. Consider Example 5.10. Remove the Bode step. In the function $T(s)$, remove the step-forming complex poles and zeros, and move the two real poles from $\omega = 2$ to the right until the guard-point phase stability margin becomes 30°. Where will these poles be? What will be the loop gain at $\omega = 10$? Are the technical specifications satisfied?

26. Make simulations of the system with the *PID* controller shown in Figure 5.19 with (a) MATLAB and (b) Simulink.

27. In a spacecraft scanning interferometer, a carriage with retroreflectors is being moved by a motor via a cable, as shown in Figure 5.39, to change the lengths of the optical paths. The carriage position range is 20 cm, the position must be accurate within 0.1 mm, and the velocity, within 3%. The lowest structural mode with the frequency in the 100–150 Hz range results from the cable flexibility.

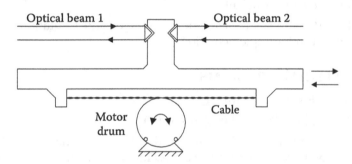

FIGURE 5.39
Retroreflector carriage.

In the block diagram in Figure 5.40(a), the prefilter, the feedback summer, and the compensator are digital.

The sampling frequency is limited to 100 Hz since the calculations are performed by the flight computer on a time-sharing basis with several other tasks. The position sensor (16-bit encoder) is connected directly to the motor shaft, so the control is collocated. (A more accurate position sensor, a laser interferometer, is used to measure the exact position of the carriage for taking the science data. This sensor is not used for closed-loop position control and is not shown in the pictures.) The D/A converter is placed at the input of the motor driver. The sensor output is digital. As discussed in Problem 4 in Chapter 4, the control bandwidth is limited to 6 Hz.

Estimate and compare the available feedback bandwidth and the control accuracy (a) in this case and (b) when the sensor data are read with a rather high sampling rate; D/A converters are placed in the command and sensor paths; and the

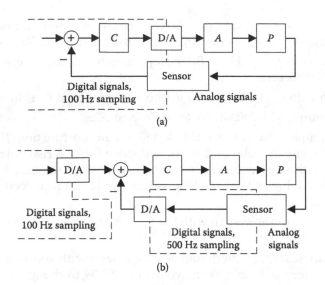

FIGURE 5.40
Block diagrams of the carriage control options.

prefilter, the command summer, and the compensator are analog. Is the accuracy of the analog circuitry sufficient?

Draw block diagrams. Consider the advantages and limitations of these two modes of the controller implementation.

ANSWERS TO SELECTED PROBLEMS

6a. The diagram is shown in Figure 5.41.

14a. $5,000(s+1)/s$; $(s+2)/(s+20)$; $(s+1,000)/(s+6,000)$.

15. The frequency f_c must be 2 kHz since at this frequency n.p. lag is 1 rad. From Equation 5.2, the Bode step ratio is 2.8 (i.e., 1.5 octave). Thus, $f_d = f_c/2.8 = 0.7$ kHz, and $f_b = f_d/2 = 0.35$ kHz, or $\omega_b = 2.2$ krad/s. The required ideal loop response shown in Figure 5.42 is similar to that shown in Figure 5.15a and Figure 5.16 (Example 5.9), but with a wider Bode step. The asymptotic diagram shown by the solid line approximates the general shape of the ideal Bode diagram.

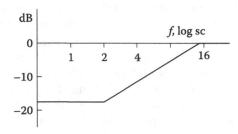

FIGURE 5.41
Asymptotic Bode diagrams for the lead $(s+2)/(s+15)$.

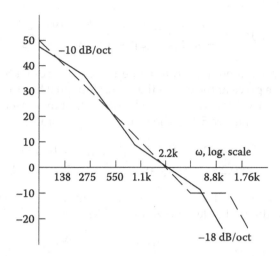

FIGURE 5.42
Ideal and asymptotic Bode diagrams.

The expression in Example 5.9 for the return ratio,

$$T(s) = 10 \frac{s+0.4}{s+0.1} \frac{1}{(s+2)^2} \frac{s^2+1.6s+4}{s^2+2.4s+9} \frac{1}{s}$$

needs to be modified:

a. It must be scaled to change ω_b from 1 to 2,200. To avoid large numbers, we express ω in krad/s (Figure 5.43) and ω_b becomes 2.2. For the scaling, s should be replaced by $s/2.2$. The return ratio becomes:

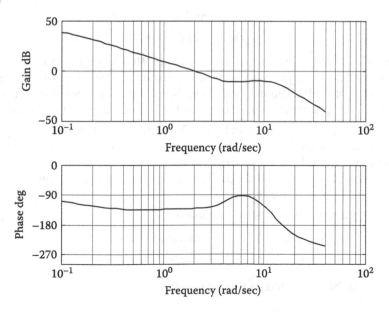

FIGURE 5.43
Bode diagrams, ω in krad/s.

$$T_1(s) = 10\frac{s+0.88}{s+0.22}\frac{2.2^2}{(s+4.4)^2}\frac{s^2+3.52s+19.4}{s^2+5.28s+43.6}\frac{2.2}{s}$$

b. The Bode step frequency ratio must be increased from 1.5 to 2.8, and correspondingly, the poles at the end of the Bode step must be shifted up $2.8/1.5 = 1.9$ times. Also, the two poles at 4.4 must be somewhat increased, say, to 5.5 (this is already shown in Figure 5.40). The return ratio becomes:

$$T_2(s) = 10\frac{s+0.88}{s+0.22}\frac{2.2^2\times(5.5/4.4)^2}{(s+5.5)^2}\frac{s^2+3.5s+19.4}{s^2+10.7s+157}\frac{1.9^2}{1}\frac{2.2}{s}$$

(It can be seen that the corners in the asymptotic diagram in Figure 5.40 correspond to the real poles and zeros of $T_2(s)$.) Or,

$$T_2(s) = 601\frac{(s+0.88)(s^2+3.5s+19.4)}{s(s+0.22)(s+5.5)^2(s^2+10.7s+157)} \tag{5.21}$$

MATLAB function `conv` is used to multiply the polynomials in the numerator:

```
a= [601]; b= [1 0.88]; c= [1 3.5 19.4];
ab=conv(a,b); num=conv(ab,c)
```

and in the denominator:

```
d= [1 0]; e= [1 0.22]; f= [1 5.5]; g= [1 10.7 157];
de=conv(d,e); def=conv(de,f); deff=conv(def,f);
den=conv(deff,g)
```

The resulting return ratio is:

$$T_2(s) = \frac{601s^3 + 2632s^2 + 13510s + 10260}{s^6 + 21.9s^5 + 309.7s^4 + 2117.8s^3 + 5200.4s^2 + 1044.8s}$$

The Bode diagram for this function is plotted in Figure 5.43 with `w= logspace(-1,1.6)`, to properly scale the gain axis. The diagram is close to the desired. Notice that the frequency axis is erroneously labeled in rad/s since we used, for simplicity, the `bode` command. The axis must be labeled in krad/s.

The loop phase response in Figure 5.43 does not yet include the n.p. lag. The lag can be modeled as described in Section 4.11, or instead, we can just add this phase lag (which is linearly proportional to the frequency) to the phase response in Figure 5.43. If we do this, we will see that the system is stable with the desired stability margins.

The compensator transfer function is:

$$\frac{T_2(s)}{P(s)} = 601\frac{(s+0.88)(s^2+3.5s+19.4)}{(s+0.22)(s+5.5)^2(s^2+10.7s+157)} \tag{5.22}$$

and can be presented as three cascaded links:

$$C_1(s) = 60.1\frac{(s+0.88)}{(s+0.22)(s+5.5)^2}, C_2(s) = \frac{5}{(s+5.5)^2}, C_3(s) = 2\frac{s^2+3.5s+19.4}{s^2+10.7s+1570}$$

In these expressions s is in krad/s. To convert the compensator functions to functions of s in rad/s (if desired), s should be replaced by $s/1,000$.

16. Three solutions (among many possible) are given below.

(1) We might start with the compensator from Section 5.7 having two parallel paths:

$$C_1 = 4/(s^2 + 4s)$$

and

$$C_2 = 5/(s^2 + 2.4s + 16)$$

for the plant $1/s$, with the loop response:

$$T(s) = \frac{C_1(s) + C_2(s)}{s} = \frac{9s^2 + 29.6s + 64}{s^5 + 6.4s^4 + 25.6s^3 + 64s^2}$$

shown in Figure 5.19(b). This response must be modified to widen the Bode step. This can be done by reducing C_1 approximately 1.2 times, and increasing C_2 1.5 times. The return ratio becomes:

$$T_1(s) = \left(\frac{3.3}{s^2 + 4s} + \frac{7.5}{s^2 + 2.4s + 16} \right) \frac{1}{s} = \frac{10.8s^2 + 37.9s + 52.8}{s^5 + 6.4s^4 + 25.6s^3 + 64s^2}$$

This response plotted with MATLAB commands:

```
n= [10.8 37.9 52.8]; d= [1 6.4 25.6 64 0 0];
w= logspace(-1,1);
bode(n,d,w)
```

is shown in Figure 5.44. The step length looks about right.

Now, a lead must be introduced into C_1 to reduce the slope at lower frequencies. With the lead,

$$T_2(s) = \left(\frac{3.3(s+0.3)}{(s^2+4s)(s+1)} + \frac{7.5}{s^2+2.4s+16} \right) \frac{1}{s}$$

With:

```
n1= conv(3.3,[ 1 0.3]);
% during iterations, adjust the zero
% (try also zero 0.75, pole 1.5)
d1= conv([1 4 0], [1 1]);
% during iterations, adjust the pole
d2= [1 2.4 16];
d1d2= conv(d1,d2);
d= conv(d1d2,[1 0])
```

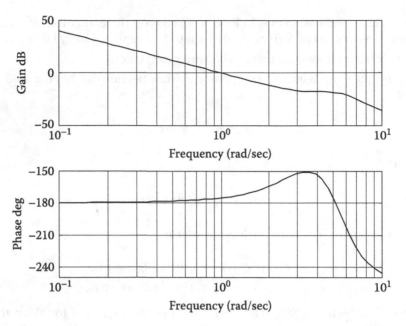

FIGURE 5.44
Bode step adjustments.

```
n=conv(n1,d2)  +  conv(7.5,d1)
% vectors have equal length since
% both polynomials are cubic
w=logspace(-1,1);
bode(n,d,w)
```

$T_2(s)$ is converted to the ratio of polynomials:

$$T_2(s) = \frac{10.8s^3 + 46.41s^2 + 85.18s + 15.84}{s^6 + 7.4s^5 + 32s^4 + 86.9s^3 + 64s^2} \tag{5.23}$$

and plotted. The plot is shown in Figure 5.45. Its shape is acceptable. The crossover frequency on the plot is 0.95 rad/s. It remains now only to scale the response for the crossover frequency to be at 2,200 by changing s to $s(0.95/2,200) = s/2,316$.

(2) We will use the already obtained solution to Problem 9. The compensator function:

$$\frac{T_2(s)}{P(s)} = 601\frac{(s+0.88)(s^2+3.5s+19.4)}{(s+0.22)(s+5.5)^2(s^2+10.7s+157)} \tag{5.24}$$

or

$$\frac{T_2(s)}{P(s)} = \frac{601s^3 + 2632s^2 + 13510s + 10260}{s^5 + 21.9s^4 + 309.7s^3 + 2117.8s^2 + 5200.4s + 1044.8} \tag{5.25}$$

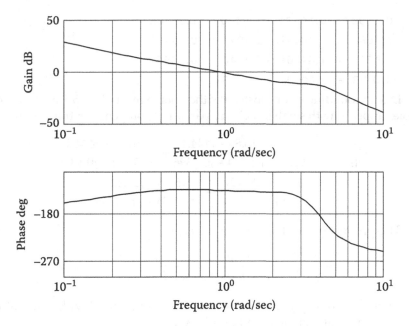

FIGURE 5.45
Loop Bode diagrams.

should be decomposed into a sum of transfer functions of lower order. Each such transfer function represents a link, and all the links are connected in parallel. There might be several options for such decomposition.

The function can be decomposed into the sum of partial fractions:

$$\frac{T_2(s)}{P(s)} = \frac{r_1}{s+0.22} + \frac{r_{21}s+r_{22}}{(s+5.5)^2} + \frac{r_{31}s+r_{32}}{s^2+10.7s+157} \tag{5.26}$$

The numerator of the fraction with a single pole can be found by assigning s the value of the pole and solving the resulting equation for the residue. During this exercise, all other fractions in the right-hand side can be neglected. Generally, the coefficients in the numerators of the fractions (including those with multiple poles) can be found by adding the fractions in Equation 5.26, which results in a ratio of polynomials in s, and solving a system of linear equations that result from comparing the numerator coefficients of s at specific powers to those in Equation 5.25.

With MATLAB, Equation 5.26 can be decomposed into a sum of partial fractions as follows (the calculation method is ill-conditioned and requires high accuracy in the initial data):

```
num= [601 2632 13510 10260];
den= [1 21.9 309.7 2117.8 5200.4 1044.8];
[Res,Pol,K] =residue(num, den)
 Res = Pol = K= []
 1.0e+002 *
 0.1598 - 0.2042i  -5.3460 +11.3430i
```

```
 0.1598 + 0.2042i -5.3460 -11.3430i
-0.1684 - 4.4410i -5.4940 + 0.1396i
-0.1684 + 4.4410i -5.4940 - 0.1396i
 0.0172 - 0.0000i -0.2200
```

Due to rounding errors, instead of the double real poles −5.5, a pair of complex poles appears, with small imaginary parts. The compensator transfer function is:

$$\frac{T_2(s)}{P(s)} = \frac{1.72}{s+0.22} + \frac{-16.84 - j444.1}{s+5.494 - j0.1396} + \frac{-16.84 + j444.1}{s+5.494 + j0.1396} +$$
$$\frac{15.98 - j20.42}{s^2 + 5.346 - j11.343} + \frac{15.98 + j20.42}{s^2 + 5.346 + j11.343}$$

The product of two fractions of the type:

$$\frac{a + jb}{s - (c + jd)} \frac{a - jb}{s - (c - jd)}$$

is a ratio of two polynomials (first order to second order) with real coefficients. The polynomials can be found as follows:

```
num _ prod= [2*a  (-2*(a*c + b*d))]
den _ prod= [1  (-2*c)  (c*c + d*d)]
```

When d is the result of calculation inaccuracy (for the double real poles) and can be neglected, then, approximately,

```
num _ prod= [2*a  (-2*a*c)]
den _ prod= [1  (-2*c)  (c*c)]
```

After making this conversion, and neglecting the small imaginary parts of the double poles that we know must be real, we obtain the compensator transfer function:

$$\frac{T_2(s)}{P(s)} = \frac{1.72}{s+0.22} + \frac{2.029s - 11.48}{(s+5.5)^2} + \frac{32s + 634}{s^2 + 10.7s + 157}$$

(3) In the solution to Problem 9, the compensator function:

$$\frac{T_2(s)}{P(s)} = 601 \frac{(s+0.88)(s^2 + 3.5s + 19.4)}{(s+0.22)(s+5.5)^2(s^2 + 10.7s + 157)} \tag{5.27}$$

can be presented as the product of two fractions:

$$\frac{T_2(s)}{P(s)} = \frac{601}{s+5.5} \frac{(s+0.88)(s^2 + 3.5s + 19.4)}{(s+0.22)(s+5.5)(s^2 + 10.7s + 157)} \tag{5.28}$$

With:

```
num=conv([1 0.88],[ 1 3.5 19.4])
d1=conv([1 0.22], [1 5.5]); den=conv(d1, [1 10.7 157])
```

the second fraction is converted to the ratio of two polynomials:

$$\frac{s^3 + 4.38s^2 + 22.48s + 17.072}{s^4 + 16.42s^3 + 219.414s^2 + 910.987s + 189.97} \tag{5.29}$$

With MATLAB, Equation 5.29 can be decomposed into a sum of partial fractions as follows (more digits are used here since the calculation method is ill-conditioned and sensitive to rounding errors):

```
num= [1  4.38  22.48  17.072];
den= [1  16.42  219.414  910.987  189.97];
[Res,Pol,K] =residue(num,  den)
 Res = Pol = K= []
0.3889 + 0.2973i  -5.3500 +11.3304i
0.3889 - 0.2973i  -5.3500 -11.3304i
0.2072 -5.5000
0.0151 -0.2200
```

The sum of the two complex pole fractions of the type:

$$\frac{a + jb}{s - (c + jd)} \frac{a - jb}{s - (c - jd)}$$

is a ratio of two polynomials with real coefficients that can be found as follows:

```
a=0.3889;  b=0.2973;  c=-5.35;  d=11.3304;
prod _ num= [2*a  (-2*(a*c + b*d))]
prod _ den= [1  (-2*c)  (c*c + d*d)]
```

Finally, the compensator transfer function is:

$$\frac{T_2(s)}{P(s)} = \frac{601}{s + 5.5}\left(\frac{0.0151}{s + 0.22} + \frac{0.2072}{s + 5.5} + \frac{0.7778s - 2.5758}{s^2 + 10.7s + 157} \right)$$

which is the function of the parallel connection of three links preceded or followed by the link $601/(s + 5.5)$.

There are many options of the compensator's implementation. Some of them might be better suited for implementing multiwindow nonlinear controllers, described in Chapter 13.

24a. Using Equations 5.15 and 5.17, $f(z) = (16.67z + 10)(z - 1.667)$. The C code is:

```
y=10 * r;  r=x + 1.667 * r;  y += 16.67 * r;
```

6

Analog Controller Implementation

This chapter walks through a variety of issues concerning the design and implementation of analog electrical compensators. Since the sensors' outputs and the actuators' inputs are most often analog electrical signals, it is convenient and economical to make the command summers and the compensators analog.

Operational amplifier circuits are considered: a summer, an integrator and a differentiator, leads and lags in inversion and noninversion configurations, a constant slope compensator, and compensators with complex poles, including computer-controlled analog compensators.

Basic types of *RC* active filters that are employed in feedback-system compensators are examined. *RC* circuit design in the element value domain is described, and the use of the *RC*-impedance chart is explained. Switched-capacitor circuits are reviewed, with an example of a band-pass tunable compensator design.

Implementations of dead-zone, saturation, and amplitude window circuits are briefly discussed.

The most important issues of analog compensator breadboarding are surveyed.

Tunable compensators, *PID* and *TID*, are introduced, and also, tunable compensators with one-variable parameters.

Methods of loop gain and phase measurements are outlined.

This is the last chapter in the introductory control course.

6.1 Active RC Circuits

6.1.1 Operational Amplifier

Sensors of electrical, mechanical, hydraulic, thermal, etc., variables produce, as a rule, electrical output signals. Most frequently, the input signals for the actuators are also electrical. Further, the commands are commonly generated as electrical signals. In many cases, these signals are analog, and the feedback summer subtracts the analog fed-back signal from the analog command. In such cases, the compensator, prefilter, command feedforward, and feedback path links can be economically and easily implemented with active *RC* circuits. The main building block for these circuits is the operational amplifier.

The dc gain of an op-amp is typically 100–120 dB. Following the flat response range, the amplifier gain drops linearly with a gain coefficient very close to f_T/f up until the *unity gain bandwidth* f_T, as shown in Figure 6.1. Depending on the type of op-amp, the unity gain bandwidth can be from 100 kHz to 1 GHz. The amplifier phase lag is close to $\pi/2$ up to the frequency f_T. With a feedback circuit added, the system is stable if the phase lag of the feedback path is less than $\pi/2$ at all frequencies where the loop gain is more than unity.

Op-amps are used with large feedback to make the gain stable in time. The available gain, therefore, must be high, in anticipation of the feedback-induced gain reduction. For example, if the gain coefficient of an op-amp with feedback of 10 is required to be 50 at 10 kHz, then the op-amp gain coefficient with no feedback needs to be at least 500 at 10 kHz, which puts f_T at 5 MHz.

The inverting configuration of an op-amp with feedback circuitry is shown in Figure 6.2. Two-pole Z_1 connects the signal source to the amplifier input, and Z_2 is the feedback path impedance.

Over the bandwidth where the feedback is large, the error voltage across the op-amp input is very small compared with the input voltage U_1 and the output voltage U_2. Therefore, $U_1 = I_1 Z_1$ and $U_2 = I_2 Z_2$, where I_1 and I_2 are the input and output currents. Next, since the input current of the op-amp itself is negligible, $I_2 = -I_1$. It follows that the transfer function of the inverting amplifier is:

$$K = -\frac{Z_2}{Z_1} \tag{6.1}$$

The input impedance of the inverting amplifier is Z_1 since at the op-amp input the voltage is the feedback loop error; i.e., it is very small, so that this node potential is very close to that of the ground. The amplifier can be used as a unity gain inverter (when $Z_1 = Z_2$) and as

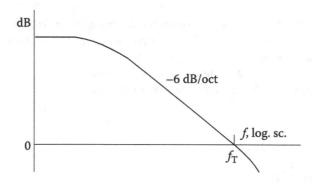

FIGURE 6.1
Op-amp gain frequency response.

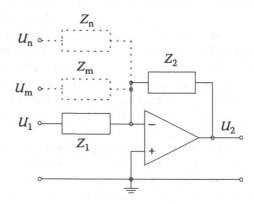

FIGURE 6.2
Inverting amplifier and inverting summer (dotted lines).

a summer amplifier combining signals from different sources, as is shown by the dotted lines, so that the output signal is $- (U_1/Z_1 + U_m/Z_m + U_n/Z_n)Z_2$.

Typically, $|Z_1|$ and $|Z_2|$ are chosen from 5 kΩ to 2 MΩ. The impedance Z_2 should not be too small or else the current in the impedance and the consumed power will be too big. The impedance Z_1, on the other hand, should not be too large since it reduces the signal at the amplifier input and increases the thermal noise whose mean square voltage at room temperature, according to the Johnson-Nyquist formula, is:

$$\bar{E} = 0.4 \cdot 10^{-8} \sqrt{\Delta f R}$$

Here, R is the resistance faced by the input port of the amplifier; it is the parallel connection of the input and feedback resistances.

6.1.2 Integrator and Differentiator

An inverting integrator with transfer function $-1/(R_1 C_2 s)$ results when the impedances in the schematic diagram in Figure 6.2 are chosen to be $Z_1 = R_1$, $Z_2 = 1/(sC_2)$. The Bode diagrams for the integrator are shown in Figure 6.3. The slope of the closed-loop gain is -6 dB/octave over a wide frequency range.

When the open-loop gain is being calculated, the left end of the two-pole Z_1 must be connected to the ground. Therefore, the open-loop gain is the product of the op-amp gain coefficient and the coefficient of the voltage divider $R_1/(R_1 + Z_2)$.

The loop gain is small at very low frequencies where the impedance of the feedback capacitor is very large. Therefore, the integrator does not perform well at these frequencies; i.e., the integrator is not accurate when the integration time is very long.

At medium and higher frequencies, the gain coefficient about the feedback loop is:

$$\frac{f_T}{f} \frac{R_1}{R_1 + Z_2}$$

It is large and nearly constant over the frequency range where $R_1 < |Z_2| = 1/(\omega C_2)$. At higher frequencies, R_1 becomes dominant in the denominator and the loop gain decreases as a single integrator, with 90° phase stability margin.

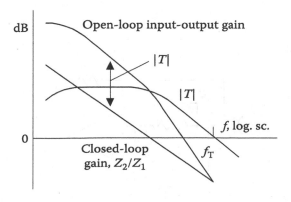

FIGURE 6.3
Integrator gain responses.

An inverting differentiator can be implemented with $Z_1 = 1/(\omega C_1)$, $Z_2 = R_2$. The open- and closed-loop responses for the differentiator are shown in Figure 6.4. The external feedback circuit for the differentiator is a low-pass filter that, together with the op-amp itself, produces a double integrator in the feedback loop. To provide some phase stability margin, a lead compensator C can be introduced in the feedback loop in front of the op-amp to reduce the gain at lower frequencies, as shown in Figure 6.5.

It is seen in Figure 6.4 that the effective bandwidth of the differentiator is much smaller than that of the integrator. This is one of the reasons why integrators, but not differentiators, are usually employed in analog computers.

6.1.3 Noninverting Configuration

The unity gain amplifier shown in Figure 6.6(a), or the *voltage follower* described briefly in Section 1.3, Figure 1.5, has the feedback path transmission coefficient $B = 1$. The schematic diagram of a more general noninverting amplifier configuration is shown in Figure 6.6(b). The feedback path transfer function is $B = Z_1/(Z_1 + Z_2)$; the feedback amplifier transfer function is $1/B$, i.e.,

$$K = \frac{Z_2}{Z_1} + 1 \tag{6.2}$$

The fed-back signal in both the inverting and noninverting amplifier configurations is proportional to the output voltage. Therefore, this feedback loop stabilizes the output voltage,

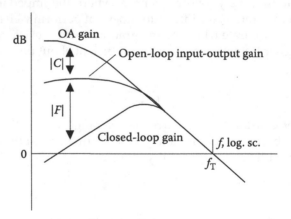

FIGURE 6.4
Differentiator gain responses.

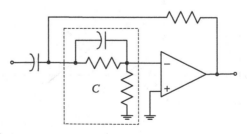

FIGURE 6.5
Differentiator schematic diagram.

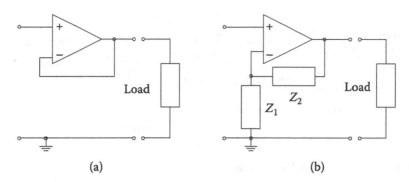

FIGURE 6.6
(a) Follower and (b) noninverting amplifier.

i.e., makes the output voltage nearly independent of various disturbances and of the load impedance variations. Hence, the output impedance of the circuits in Figure 6.6 is low over the range where the feedback is large.

6.1.4 Op-Amp Dynamic Range, Noise, and Packaging

The dynamic range of the op-amp is limited from above by the output voltage swing, and from below by the input noise and drift. The input noise is typically comparable with the thermal noise of a resistor of several hundred ohms to several kilo-ohms connected in series to the input. Therefore, external resistances in series with the op-amp input, such as R_1 in Figures 6.2 and 6.6(b), do increase the device noise. It is desirable to keep these resistances small, especially in the op-amp immediately following the feedback summer.

The input voltage dc offset (the internal parasitic dc bias) is typically in the 10 μV to 1 mV range, with the thermal drift in the 100 nV/°C to 10 μV/°C range.

While op-amp circuits are tested, neither of the input pins should be left open or else the voltage on the open pin will depend on the initial charge. Due to the extremely high impedance of the op-amp input, this charge will remain for an unpredictable time, thus producing confusion for the experimenter by altering the readings in an often irreproducible fashion.

The choice of f_T depends on the desired gain and the power consumption. As a rule, the higher f_T is, the larger must be the current in the transistors and, consequently, the dc current consumed by the op-amp from the power supply.

Op-amps come packaged as single, double, or quad in one case. The standard pin-outs for the double and the quad are shown in Figure 6.7. The pin-out is the same for the dual in-line package (DIP) and surface mount (SM) package (there also exist much smaller SM packages).

One quad op-amp is sufficient for all the needs of a typical analog compensator. Typically, it consumes an order of magnitude less power, occupies an order of magnitude smaller space, and costs an order of magnitude less than a digital microcontroller. Breadboarding and testing such an analog controller takes, typically, an order of magnitude less time than doing so with a digital controller.

The values of the RC constants correspond to the poles and the zeros of the transfer functions. For low-speed processes, these time constants can be large and the capacitors can become bulky.

The higher the resistances are, the smaller the capacitors can be. The resistances are, however, limited. The resistance in series with the op-amp input is limited by the requirements

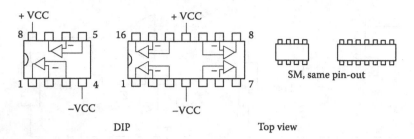

FIGURE 6.7
Typical double and quad OA, DIP, and SM packages.

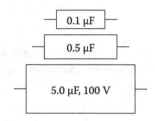

FIGURE 6.8
Typical size of mylar capacitors.

of keeping the thermal noise below a certain level, and the feedback resistance is limited by stray capacitances. Typically, resistor values should not exceed several megohms. Thus, for example, $R = 2$ MΩ and $C = 0.5$ μF can produce a pole or a zero at a frequency of $f = 1/(2\pi RC) \approx 0.16$ Hz.

The capacitance must be stable in time, like that of mylar capacitors. Figure 6.8 depicts typical sizes of various capacitors, but they can also be somewhat smaller.

6.1.5 Transfer Functions with Multiple Poles and Zeros

With the feedback amplifiers shown in Figure 6.6 where Z_1 and Z_2 are RC two-poles, transfer functions can be realized with multiple real poles and zeros. Noninverting lag and lead compensators are shown in Figure 6.9a and b, respectively.

Inverting lead compensators are shown in Figure 6.10(a) and (b). An inverting lag compensator is shown in Figure 6.10(c). (Design of these compensators will be discussed further in Sections 6.2.1 and 6.2.2, and in solutions to Problems 7–9 at the end of the chapter.)

Figure 6.11 shows the implementation of transfer function $s^{-1/3}$, which uses Equation 5.2 shifted to cover the band from 1 to 100 Hz.

Figure 6.12(a) shows the *bridged T-circuit* employed in front of an op-amp to implement complex zeros. The mutual compensation of the phase-advanced output signal of the upper path B_1 with the phase-delayed output signal of the path B_2 produces a broad notch on the Bode diagram.

The bridged T-circuit in the feedback path shown in Figure 6.12(b) allows implementation of a complex pole pair. The gain responses for the amplifier using the feedback path B_1, feedback path B_2, or both, are shown in Figure 6.12(c).

Implementation of a complex pole pair using parallel feedback paths is also shown in Figure 6.13. (This circuit has been used as a part of the compensator for the 100 kV, 1.6 MW precision power supply for a klystron transmitter.) The lower feedback path dominates

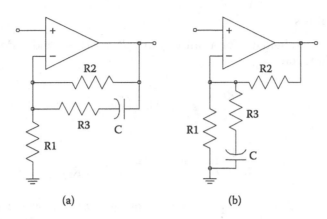

FIGURE 6.9
(a) Lag and (b) lead implementation, noninverting.

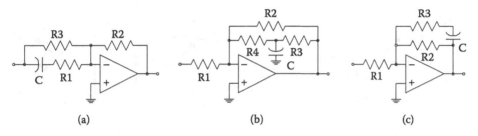

FIGURE 6.10
(a, b) Lead and (c) lag implementation, inverting.

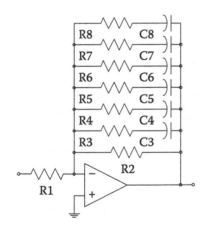

R8	555.2	C8	500
R7	563.9	C7	123.7
R6	423.3	C6	51.44
R5	308.13	C5	23.7
R4	185.0	C4	13.1
R3	39.43	C3	9.87
R2	599.0	R in kΩ	
R1	54.53	C in nF	

FIGURE 6.11
Implementation of transfer function $s^{-1/3}$ over the band from 1 to 100 Hz.

at lower frequencies and the upper path dominates at higher frequencies. At the crossing frequency, the output signals of the feedback paths have nearly opposite phase so that the total feedback path has a pair of complex conjugate zeros. As a result, the closed-loop transfer function possesses a complex pole pair. The quality factor Q depends on the difference in phase shift between the two paths. The difference can be adjusted by adding series or parallel resistors to the capacitors.

6.1.6 Active RC Filters

Figure 6.14 shows the schematic for a unity gain *Sallen–Key* second-order low-pass filter with the transfer function:

$$K(s) = \frac{\omega_o^2}{s^2 + Q^{-1}\omega_o s + \omega_o^2} \tag{6.3}$$

where

$$\omega_o = \frac{1}{\sqrt{R_1 R_2 C_1 C_2}} \quad \text{and} \quad Q = \frac{1}{\omega_o (R_1 + R_2) C_2} \tag{6.4}$$

This filter is well suited to the implementation of low Q complex poles such as those required to make Bode steps in the examples in Sections 5.6 and 5.7. The pole frequency ω_o and the damping coefficient $\zeta = 1/(2Q)$ are prescribed. Two of the circuit elements can be

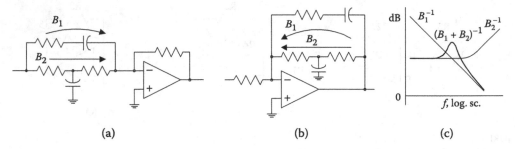

| (a) | (b) | (c) |

FIGURE 6.12
Implementation of (a) complex zeros and (b, c) complex poles.

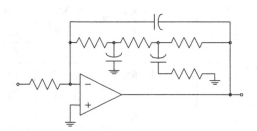

FIGURE 6.13
Compensator with parallel feedback paths.

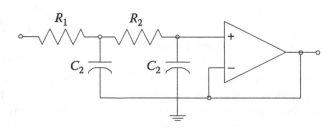

FIGURE 6.14
Sallen–Key low-pass filter.

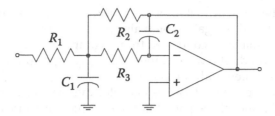

FIGURE 6.15
Multiple feedback low-pass filter.

chosen, and values for the remaining two can be found from the equations. For example, when the resistor values are initially chosen, $C_2 = 1/[\omega_0(R_1 + R_2)Q]$ and $C_1 = 1/(\omega_0^2 C_2 R_1 R_2)$.

A *multiple-feedback* second-order low-pass filter, shown in Figure 6.15, has reduced sensitivity to component parameter variations. It implements the transfer function:

$$K(s) = \frac{-H_o \omega_0^2}{s^2 + Q^{-1}\omega_0 s + \omega_0^2} = \frac{-(R_1 R_3 C_1 C_2)^{-1}}{s^2 + s(R_1^{-1} + R_2^{-1} + R_3^{-1})C_1^{-1} + (R_2 R_3 C_1 C_2)^{-1}} \tag{6.5}$$

The low-frequency gain coefficient H_o is limited by the prescribed Q. Particularly, H_o must be chosen less than 100 when $Q \geq 1$, and less than 10 when Q approaches 10. (The element values of this filter become inconvenient when Q exceeds 10.) For this filter, two of the element values can be chosen and the remaining three calculated from the following equations:

$$|H_o| = \frac{R_2}{R_1} \tag{6.6}$$

$$\omega_0^2 = \frac{1}{R_2 R_3 C_1 C_2} \tag{6.7}$$

$$Q = \frac{\sqrt{\dfrac{C_1}{C_2}}}{\sqrt{\dfrac{R_3}{R_2}} + \sqrt{\dfrac{R_2}{R_3}} + \dfrac{\sqrt{R_2 R_3}}{R_1}} \tag{6.8}$$

Table 6.1 gives the elements' values for the Chebyshev second-order low-pass filter with $H_o = -1$ and 1 kHz cutoff frequency. The filter response remains the same when all resistances are increased and all capacitances reduced by the same factor. The corner frequency can be changed by changing the product of the capacitances, and Q, by changing the ratio of the capacitances. The gain coefficient can be changed by changing R_1 and then adjusting the ratio of the capacitances to preserve the desired Q and changing both capacitances by some coefficient to preserve ω_0.

The circuit shown in Figure 6.16(a) is often called a *state-variable filter*. This is an analog computer consisting of a summer with gain adjustment resistors and two integrators. The block diagram is similar to that shown in Figure 5.22. The circuit can mimic second-order linear differential equations describing low-pass, high-pass, band-pass, and band rejection filters. This circuit enables reliable implementation of poles and zeros with Q up to 100. (An example of using band-pass compensators with state-variable filters will be given in Section 6.4.2.)

The state-variable filters are available as ICs, with only four resistors to be added to set the cutoff frequency, Q, and the gain. There also exist IC filters in which all resistors are built in and are programmable or controlled from a computer port. These circuits combine the best of the analog and digital worlds: they are easily reprogrammed, and they do not introduce the delay associated with sampling.

Programs for calculating element values and plotting frequency responses for Sallen–Key, multiple-feedback, and state-space filters are available from many manufacturers (e.g., Burr-Brown, Harris, MAXIM, National Semiconductor).

A notch $(s^2+\omega_0^2)/(s^2+2\zeta\omega_0+\omega_0^2)$ can be implemented with the twin-T bridge shown in Figure 6.16(b). The resonance frequency $\omega_0=1/(RC)$. The variable resistors allow the adjustment of ω_0 and the damping of the numerator (which must be 0). The potentiometer allows the adjustment of the denominator damping coefficient ζ over the range 0.01–0.16.

Example 6.1

Two cascaded notches with $\zeta=0.1$ and slightly different resonance frequencies, 1% under and 1% over the resonance, can be used to reject a plant structural resonance by at least 40 dB over the range ±1.4% of the nominal resonance frequency, while introducing only 5° lag at the frequency two octaves below the resonance.

6.1.7 Nonlinear Links

A saturation link can be implemented with an op-amp, as shown in Figure 6.17(a). This arrangement uses Zener diodes in the feedback path that have small differential resistance when the voltage across the diode exceeds the threshold. The 5 V saturation threshold

TABLE 6.1

Chebyshev 1 kHz Low-Pass Filter

p-p Ripple, dB	1	2	3	4
$R_1=R_2$, kΩ	8.13	7.55	6.17	3.01
R_3, kΩ	12.8	12.8	12.3	12.6
C_1, nF	47	68	100	220
C_2, nF	4.7	4.7	4.7	4.7
Q		0.956	1.129	1.305	1.493
f_0		1.05	0.907	0.841	0.803

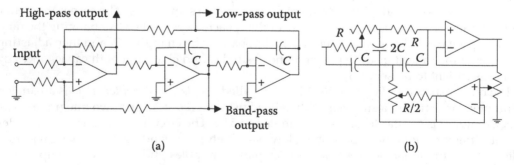

(a)   (b)

FIGURE 6.16

(a) State-variable filter and (b) twin-T notch.

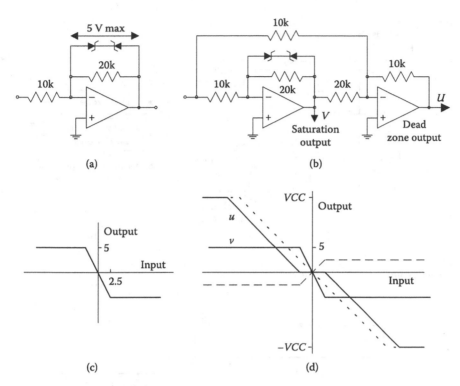

FIGURE 6.17
(a) Saturation link and (b) its characteristic, as well as (c) dead-zone link and (d) its characteristic.

shown in the figure results from the sum of the voltage drops across the open diodes: 4.3 V on the Zener with inverse polarity, and around 0.7 V on the open diode of the other Zener. The resulting saturation response, with the gain coefficient of −2, is shown in Figure 6.17(c).

The dead-zone link can be formed by summing the input of a saturation link with its inverted output. Figure 6.17(b) shows an implementation of such a circuit, and the resulting response is shown in Figure 6.17(d). The dotted line indicates signal transmission via the upper path, and the dashed line, via the lower path. The resulting characteristic is not a pure dead zone—it includes saturation at the VCC level. By using different resistors and Zeners, characteristics with different dead zones and saturation thresholds can be obtained.

Sometimes it is desirable to direct signals that have amplitudes within specified windows to separate outputs for further processing. To an extent, the circuits shown in Figure 6.17(b) do just that, for two windows. Figure 6.18(a) exemplifies a circuit with three windows that directs the input signal into three outputs for further processing, depending on the signal amplitude. The circuit to combine signals via different amplitude windows, whose block diagram is shown in Figure 6.18(b), can be designed in a similar way.

Nonlinear dynamic links can be designed by combining nonlinear and linear links. For example, rate can be limited by placing a saturation link in front of an integrator and closing a tracking feedback loop with sufficient gain coefficient k, as shown in Figure 6.19. Rate limiters are often included in the command path to prevent the plant from being damaged by excessive velocity, and also to reduce the overshoot in the control system response to large commands. Further examples of nonlinear dynamic links will be given in Chapters 11 and 13.

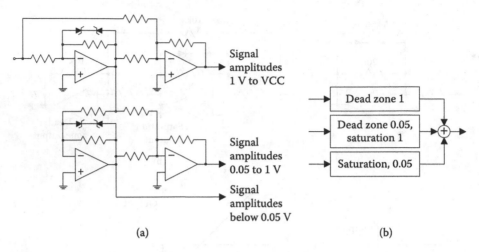

(a)   (b)

FIGURE 6.18
Three-window (a) splitter and (b) combiner.

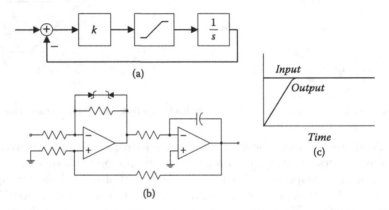

FIGURE 6.19
Rate-limiting follower: (a) block diagram, (b) schematic diagram, and (c) ramp output following step input.

6.2 Design and Iterations in the Element Value Domain

6.2.1 Cauer and Foster *RC* Two-Poles

RC two-poles are widely used as components of analog compensators. Not every function can be implemented as an impedance of an *RC* two-pole but only one whose poles and zeros are real, these poles and zeros alternating along the real axis of the *s*-plane, and the closest to the origin being a pole (this constitutes a part of Foster's theorem that generally considers *RC*, *RL*, and *LC* two-poles). Any *RC*-impedance function can be implemented in either of the Foster canonical forms shown in Figure 6.20. The parallel branch form renders smaller total capacitance.

(Foster's theorem also states that the impedance of a lossless (*LC*) system has alternating purely imaginary poles and zeros; this relates to collocated control discussed in Chapters 4 and 7. Also refer to Appendix 3.)

Also, any *RC*-impedance function can be implemented in any of the ladder (Cauer's) forms shown in Figure 6.21. The element values are the coefficients of a chain fraction expansion of the impedance function.

The Foster and Cauer two-poles are employed in analog compensators in Figures 6.9–6.11, and many others.

Example 6.2

Two parallel branches, the first a series connection of R_1 and C_1, and the second a series connection of R_2 and C_2, are often placed in the feedback path of an op-amp to form a compensator making the loop Bode diagram steeper. The frequencies of the two zeros ω_{z1} and ω_{z2}, and the pole ω_{p1}, and the high-frequency asymptotic value of the impedance are expressed by the following equations:

$$\omega_{z1} = 1/(R_1 C_1), \quad \omega_{z2} = 1/(R_2 C_2)$$

$$\omega_{p1} = (R_1 + R_2)/(1/C_1 + 1/C_2)$$

$$Z(s)|_{s\to\infty} = R_1 R_2 /(R_1 + R_2)$$

from which the element values can be easily found.

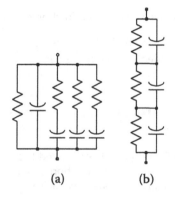

(a)   (b)

FIGURE 6.20
Foster canonical forms of an *RC* two-pole.

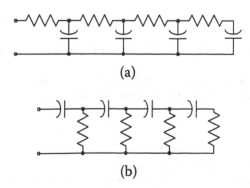

(a)

(b)

FIGURE 6.21
Cauer (ladder) canonical forms of an *RC* two-pole.

In general, the capacitances and resistances for the Foster form in Figure 6.20(a) can be calculated by expanding $1/[sZ(s)]$ into a sum of elementary first-order functions. The impedance frequency response of this two-pole with known elements in each branch R_iC_i can easily be found with SPICE. In MATLAB, the response can be found by first calculating the residues $r_i = 1/R_i$ and poles $p_i = -1/(R_iC_i)$, and then using:

```
r = [r1 r2 .. ri ..]; p = [p1 p2 .. pi ..];
[num, den] = residue(r,p); bode(num,den)
```

Example 6.3

Impedance

$$Z(s) = \frac{1,000,000(s+100)(s+2000)}{s(s+500)}$$

can be implemented as an *RC* circuit since the poles and the zeros are real, they alternate, and the lowest among them is a pole. We find the elements' values for the Foster form in Figure 6.21(a) by expanding the function:

$$\frac{Y(s)}{s} = \frac{1}{sZ(s)} = \frac{num(s)}{den(s)} = \frac{0.000001s + 0.0005}{s^2 + 2100s + 200,000}$$

into the sum of two components of the partial fraction expansion:

$$\frac{num(s)}{den(s)} = \frac{r_1}{s-p_1} + \frac{r_2}{s-p_2} + \cdots$$

The MATLAB code

```
num = [0.000001 0.0005]; den = [1 2100 200000];
[r, p] = residue(num, den)
```

calculates the residues $(0.7895 \times 10^{-6}, 0.2105 \times 10^{-6})$ and the poles $(-2,000, -100)$.

Next, by comparing a term in the partial fraction expansion of $Y(s)/s$ with the admittance of a series connection of a resistor and a capacitor:

$$\frac{s/R}{s+1/(RC)}$$

we identify $R = 1/r$ and $C = -r/p$. Therefore, the resistors and the capacitors are:

$$R_1 = 1.266\,M\Omega, C_1 = 390\,pF, R_2 = 4.75\,M\Omega, C_2 = 2.1\,nF$$

While working in the laboratory, it is often more convenient to think in terms of the element values R and C, rather than in terms of poles and zeros. In this case, Cauer forms can be of use.

Example 6.4

Consider the circuit shown in Figure 6.22(a).

We use in this example $C_2 \ll C_1$. The capacitance C_1 is dominant at lower frequencies where its impedance $X_1 = 1/(2\pi fC_1)$ is much higher than R_1. Starting at frequency f_1 where X_1 reduces below R_1, the resistor R_1 becomes dominant. At even higher frequencies,

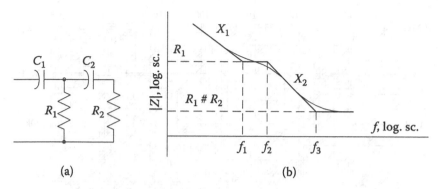

FIGURE 6.22
Example of (a) an *RC* two-pole and (b) its impedance modulus response.

starting with f_2, where $X_2 = 1/(2\pi f_2 C_2)$ equals $R_1 + R_2$, the impedance of the small capacitor C_2 becomes small enough to shunt R_1 and becomes dominant. Beyond the frequency f_3 where the impedance $X_2 = 1/(2\pi f_3 C_2)$ approximately equals R_2, the capacitances' impedances can be viewed as negligibly small, and the total impedance becomes R_1 # $R_2 = R_1 R_2/(R_1 + R_2)$.

From this analysis it is seen that it is easy to identify the elements that need to be adjusted.

Example 6.5

If the total impedance of the circuit in Figure 6.22(a) needs to be increased at specific frequencies, it is easy to see from the plot in Figure 6.22(b) what needs to be done: at lower frequencies, C_1 needs to be reduced; over the interval $[f_1, f_2]$, R_1 needs to be increased; over the interval $[f_2, f_3]$, C_2 needs to be reduced; at higher frequencies, R_2 needs to be increased. In this way the loop response can be adjusted in the laboratory. The same method is convenient to use for loop adjustments with SPICE simulation.

When the final version of the design is implemented, it might be desirable to reduce the values of the capacitors and convert the Cauer form into the parallel Foster form. With some experience, it is also possible to start with and make the iterations using the Foster form.

6.2.2 *RC*-Impedance Chart

A chart for calculating the corner frequencies and determining the values of resistors and capacitors is shown in Figure 6.23. Using the chart is especially convenient in the laboratory environment where the accuracy of calculation need not be high, the number of iterations is substantial, and the computer might be on a far-away desk. The chart accuracy is approximately 10%, i.e., 1 dB, which is sufficient for compensator design (at least, for preliminary design). With this chart, many design questions can be instantly answered.

Example 6.6

If the source impedance is 100 kΩ and the load resistance is 400 kΩ, what is the series capacitance that will place the pole at 20 Hz?

Since the total contour resistance is 500 kΩ, the capacitance must have 500 kΩ reactive impedance at 20 Hz; i.e.; it is 15 nF.

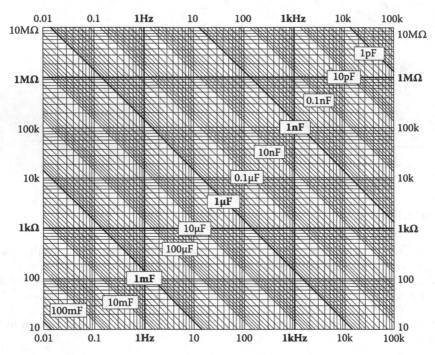

FIGURE 6.23
RC-impedance frequency responses.

Example 6.7

If the source impedance is 20 kΩ, the load impedance is 20 kΩ, and the shunting capacitance is 1 μF, where is the pole frequency?

The reactive impedance of the capacitance at the pole frequency equals the parallel resistance of the source and the load, which is 10 kΩ. Therefore, the pole frequency is at the crossing of the 1 μF line and the 10 kΩ line; i.e., it is 16 Hz.

Consider next the design of the inverting compensators depicted in Figure 6.10.

Example 6.8

For the lead link in Figure 6.10(a), what is the capacitance of the capacitor that is shunting the 100 kΩ resistor R_3, for the compensator zero to be at 200 Hz?

The answer is 7 nF.

What resistance needs to be placed in series with this capacitor to make a pole at 1 kHz?

The answer is 20 kΩ.

Example 6.9

For the lag link in Figure 6.10(c), what is the capacitance of the capacitor that is shunting the 100 kΩ feedback resistor R_2, for the compensator pole to be at 200 Hz?

The answer is 7 nF.

What is the resistance R_3 that makes a zero at 1 kHz?

The answer is 20 kΩ.

Example 6.10

For the lead link in Figure 6.10(b), the resistance at the connection point of the shunting capacitor is the parallel connection of R_3 and R_4. When $R_3 = R_4 = 200$ kΩ, it is 100 kΩ. What is the capacitance of the capacitor for the compensator zero to be at 200 Hz?

The answer is 7 nF.

What is the resistance R_2 for the pole to be at 1 kHz? This resistor must provide the same transconductance of the feedback path at this frequency as the T-branch at the pole frequency, which is five times smaller than the transconductance at dc (since the ratio of the pole to the zero is 5).

Therefore, $R_2 \approx 1$ MΩ.

In this way, design of the compensators is done in no time. Design of the compensators in Figure 6.10 is left as an exercise.

6.3 Analog Compensator, Analog or Digitally Controlled

A multiplying D/A converter is an attenuator with a digitally controlled attenuation coefficient. Using the converter, it is possible to change the parameters of an analog compensator via a computer command. This method combines the best of the digital and analog worlds: the compensator is analog, without digital delay, yet its frequency response can be modified by software. Figure 6.24 shows the application of a multiplying D/A converter (i.e., controllable attenuator) for changing the coefficients P and I in a compensator with transfer function $P + I/s$. There exist ICs for implementation of second-order transfer functions with built-in multiplying D/A converters that are controlled from an external parallel bus.

Analog multipliers can be used to change the gain coefficient of a link under the control of an analog signal, as shown in Figures 1.19 and 6.25.

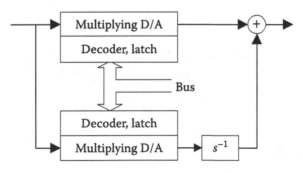

FIGURE 6.24
Analog compensator, digitally controlled.

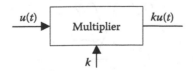

FIGURE 6.25
Gain regulation with a multiplier.

6.4 Switched-Capacitor Filters

6.4.1 Switched-Capacitor Circuits

Since the size of capacitors in analog circuitry is limited by economy considerations, to achieve the required large time constants, the resistors must be big. However, when active *RC* circuits are implemented as silicon ICs, large resistors tend to occupy substantial real estate. As an alternative, the resistor can be replaced by a circuit that transfers the charge in small, discrete steps. The resistor is imitated by charging a small capacitor C, f_s times per second, and then discharging it into the load using an electronic switch. This method of charge transfer (charge pump) is equivalent to a resistance $R = 1/(f_s C)$. Figure 6.26 shows a switched-capacitor integrator with transfer function $f_s C_1/(s C_2)$.

Switched-capacitor active *RC* circuits use switched-capacitor integrators, summers, and amplifiers. When the switching frequency is varied, all poles and zeros of the transfer function will change by the same factor. Therefore, the frequency response of the switched-capacitor circuit will shift on the logarithmic frequency scale without changing its shape, as shown in Figure 6.27. This makes the switched-capacitor filter easy to tune.

Switched-capacitor Chebyshev, Butterworth, and Bessel filters, and switched-capacitor biquads programmable from an external parallel bus or by connection to ground or to the VCC of some of the IC pins are available from several manufacturers.

6.4.2 Example of Compensator Design

As an example of the application of switched-capacitor circuits in control compensators, we consider a three-input three-output control system for rejecting the vibrations of a spacecraft camera's focal plane sensors. The sensors are cooled by a cold finger connected thermally and mechanically to a cryogenic cooler, as shown in Figure 6.28. The cooler's

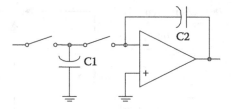

FIGURE 6.26
Switched-capacitor integrator.

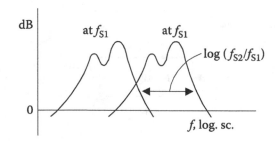

FIGURE 6.27
Change in the frequency response with changing sampling frequency.

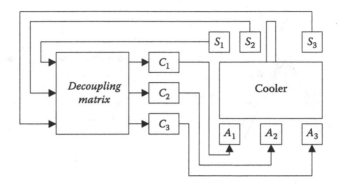

FIGURE 6.28
Control loops for vibration suppression of a cryogenic cooler for a camera focal plane.

vibration is counteracted by three piezoelement actuators placed orthogonally between the cooler and its armature. This disturbance rejection feedback system is a kind of homing system, with zero command.

The feedback bandwidth is limited by the flexible modes of the cooler armature. The modes' frequencies are kept over 300 Hz, although their exact values remain unknown. The modes' damping can be quite low so that the peaks on the Bode diagram can be up to 40 dB high. As can be calculated, if the feedback system is implemented as a low-pass system, the feedback of 100 cannot be implemented over a bandwidth wider than 5 Hz.

The frequency of the cooler operation is between 50 and 60 Hz. The frequency is controlled by a spacecraft computer in accordance with the temperature of the cooler. Therefore, in order to effectively reject the disturbances with frequencies in the 50–60 Hz range, the system should be the band-pass type.

Due to the system parameter variations, the reliable bandwidth of the feedback is only 3 Hz, while the frequency of the cooler motor varies by 10 Hz. Therefore, the system needs to be designed as adaptive, with the feedback frequency response adjustable. Thus, the Bode diagram should be slightly shifted along the frequency axis without substantial variations of its shape, which can be easily implemented by changing the sampling frequency for the switched-capacitor compensator.

The three-axis sensor arrangement is shifted by 90° from the arrangement for the actuators. A decoupling matrix with constant coefficients was calculated and implemented as an analog circuit. With the matrix, experiments show that coupling from the ith piezoelement to the jth sensor is less, by an order of magnitude, than the coupling from the ith piezoelement to the ith sensor. Therefore, the three loops can be considered fairly independent.

The loop Bode diagrams for each of the three channels are identical, and shown in Figure 6.29(a). The response is a band-pass version of a *PID* controller, with shallow slopes at the crossover frequencies.

The block diagram for the compensator is shown in Figure 6.29(b). Here, W_1 is a band-pass filter with quality factor 30, and W_2 is a band-pass filter with quality factor 3. Two cascaded high Q filters have steep roll-off with phase shifts of ±180°. These filters dominate at frequencies in the functional band (approximately 2 Hz wide) as seen in Figure 6.29(a). At the higher and lower frequencies, the low Q channel is dominant with shallower slopes of the Bode diagram and smaller phase shift, which provides the desired stability margin.

Each of the second-order band-pass filters was implemented using a band-pass second-order switched-capacitor filter IC. The sampling frequency, nominally of 10 kHz, was controlled from the flight computer to vary the central frequency, as required, within the range of 50–60 Hz.

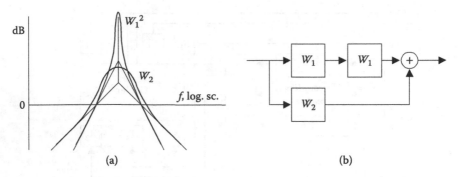

FIGURE 6.29
Compensator (a) frequency response and (b) block diagram.

6.5 Miscellaneous Hardware Issues

6.5.1 Ground

Practical implementation of analog compensators is simple and forgiving, and can be performed by people without a special background in electrical engineering. Still, some assorted hints and warnings are worth noting, as in the following.

Signal distortion, noise, and signal interference may result from improper grounding. The resistance of a ground wire or a ground plane (a grounded layer of a multilayer printed circuit board) is commonly very small, and for simplicity is commonly considered to be zero. This assumption is acceptable in most cases, but not in all. Consider, for example, a power supply for the circuitry with the schematic diagram shown in Figure 6.30. The diagram includes a power down-transformer, a two-way diode rectifier, a Π-type *RC* low-pass filter to smooth the ripple of the output voltage, and a voltage stabilizer IC.

The ac current through the filter capacitors is relatively large, especially through the first capacitor. The currents flowing via the ground plane to the central tab of the transformer winding create voltage drops on the ground plane that are typically of several μV and can even reach several mV. To reduce the voltage variations on the analog ground, first, the transformer tap wire should be placed close to the capacitors, and second, the rectifier ground and the analog ground should be connected in a single point, as shown in Figure 6.30. Any extra connection between the grounds would create paths for the ac currents to flow via the analog ground and produce voltage drops.

In systems employing digital controllers, commonly, the actuator and the plant of a control loop are analog, and the sensor in most cases is also analog. The digital part of the loop starts with an A/D converter following the sensor. The currents in the digital part of an A/D converter are typically several orders of magnitude larger than the lowest analog signal to be measured. Therefore, provisions should be made for the digital currents not to interfere with the analog input signal.

In particular, attention needs to be paid to maintaining proper ground configuration. The digital currents should be prevented from flowing over the analog ground and there producing voltage drops; otherwise, different points of the analog ground would have different potentials. This is achieved by connecting the analog and digital grounds at one point only. Since commonly the analog and digital grounds are already connected within an A/D converter, this point must be the only one connecting the grounds, as shown in Figure 6.31.

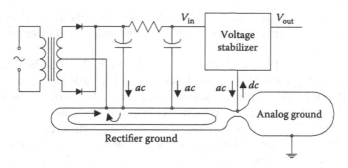

FIGURE 6.30
Ground for voltage regulator.

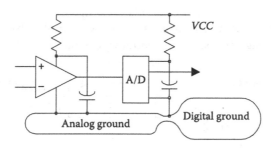

FIGURE 6.31
Grounding for A/D converter.

The ac current also flows via the power supply VCC line, and care should be taken to avoid parasitic coupling between the grounds. The low impedance ac connection between the grounds via power supply lines must be eliminated. To do this, the blocking capacitors from the power lines should be connected to the proper ground points, as shown in Figure 6.31.

In the digital part of the circuit, it is commonly sufficient to connect the VCC line to the ground by capacitors placed close to the IC that consume large pulsed current from the power supply line. In the analog part of the circuit, low-pass filters are usually placed in the power supply line. The capacitors of the filters should be placed close to the IC, or else at higher frequencies, the impedance of the wires might introduce parasitic feedback and coupling due to imperfect ground.

6.5.2 Signal Transmission

In some control systems, the sensors are far away from the controller and need to be connected with long cables. The resulting communication link is subject to noise and electromagnetic interference. Figure 6.32(a) shows signal transmission of the asymmetrical transmitter output voltage V_t to an asymmetrical receiver using common ground as the return wire. The ground potential at the input to the receiver differs by V_g from the transmitter ground potential because of various ac and dc currents from different sources flowing through the imperfect (nonzero impedance) ground wire. As a result, the received signal $V_r = V_t - V_g$ is corrupted with the ground noise V_g.

To avoid errors caused by imperfect ground, differential amplifiers are used with a separate return wire that is only connected to the ground at the transmitter output, as shown in Figure 6.32(b). Precision amplifiers with differential inputs are called *instrumentation amplifiers*. They are commonly composed of three op-amps or, better yet, purchased as complete ICs.

A simple differential amplifier with the gain coefficient of 2 can be built according to the schematic diagram in Figure 6.33. The amplifier is an acceptable replacement for an instrumentation amplifier when the common signal component does not exceed half of the VCC (so it will not saturate the amplifiers), which is commonly the case.

Electromagnetic interference from various sources in an industrial environment into the signal wires could contaminate relatively small analog signals in the control loop, especially when the links are connected with rather long wires. Twisting the pair of wires reduces the difference in the voltages induced by external magnetic fields in each of the wires, and these voltages, due to their opposite polarity, cancel each other at the input to the differential receiver.

As shown in Figure 6.34(a), bifilar series coils are employed for suppression of the interference at higher frequencies, where it is rather easy to make the coil with high inductive impedance. For the signal, the coil impedance is small since the signal current flows

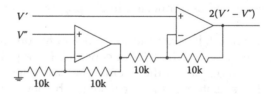

FIGURE 6.32
Parasitic feedback due to common ground.

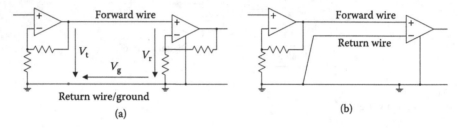

FIGURE 6.33
Differential amplifier.

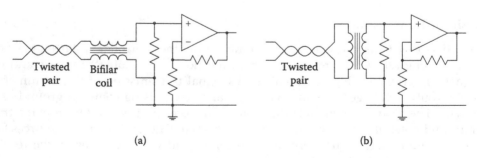

FIGURE 6.34
Using (a) bifilar coil to reject high-frequency interference and (b) balun transformers to reject low-frequency and medium-frequency interference.

through the two coil windings in opposite directions, producing no magnetic field. For the interfering signal affecting only one wire, the coil impedance is high, thus suppressing the interference.

To reduce interference at lower and medium frequencies while using asymmetrical receivers, balanced-to-unbalanced transformers (baluns) can be used as shown in Figure 6.34(b). Since transformers do not convey dc, this method is only suitable for signals not containing dc components (such as sensor signals in feedback systems for vibration suppression).

6.5.3 Stability and Testing Issues

The internal feedback in an op-amp is maximum when the op-amp is used as a follower. Some op-amps in this configuration are only marginally stable, especially at higher frequencies and when the wires are long, thus introducing stray inductances and capacitances in the feedback loop. Placing a resistor in the feedback path as shown in Figure 6.35 commonly provides sufficient stability margins.

Amplifiers are often connected with the rest of the equipment with long cables. Since the cables' input impedances shunt the feedback for the op-amp, the cable capacitance and the resonances in the cable can make the op-amp unstable. The instability can be eliminated by the addition of series resistors at the input and output of the amplifier, as shown in Figure 6.36. The resistances cannot be too large since the input resistor introduces noise into the circuit, and the output resistor increases the amplifier output impedance. The resistances are commonly chosen from 1 to 2 kΩ.

An oscilloscope should be connected to the breadboard while the circuit is tested, tuned, or troubleshooted. Without an oscilloscope, many possible problems can be misunderstood. For example, some power amplifiers with f_T of only 2 MHz can oscillate at 50 MHz if the wires connecting the IC are too long and their inductances form a resonance tank amplifying the signal in the parasitic feedback loop. Without an oscilloscope, when a signal analyzer is used, this oscillation might manifest itself by only a small change in the gain and some reduction of the output power. (But, at the same time, the IC gets hot, the power supply current increases, and the output signal decreases.)

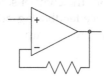

FIGURE 6.35
Using a feedback resistor in the follower.

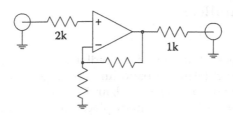

FIGURE 6.36
Reducing the effect of cables on an op-amp Feedback loop.

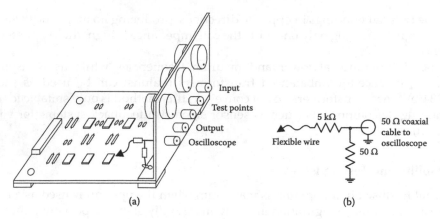

FIGURE 6.37
(a) Breadboard with connectors and (b) oscilloscope matching termination.

On the breadboard, the input, the output, and some spare connectors should be firmly attached, as shown in Figure 6.37(a). The spare connectors will be used to test the signals at various nodes of the circuit for troubleshooting.

Signal reflections from mismatched ends of a coaxial cable going to an oscilloscope can corrupt the measurements. To eliminate the distortions, the cable should be matched at least at one side. The test fixture shown in Figure 6.37(b) is convenient to have implemented on the board. It matches the 50 Ω cable and, at the price of attenuating the measured voltage 100 times, makes the input impedance of the probe 5 kΩ, which is sufficiently high for most troubleshooting tasks.

While designing an experimental breadboard, it is a good practice to leave one of the op-amps as an extra available inverter for the purpose of changing the phase of the signal from the sensor—since it is quite common that the sensor output signal polarity is not known or will be changed during experimentation.

While closing control loops over some physical plant, especially when the plant is mechanical and can be damaged by high-amplitude oscillation, one should introduce potentiometers either in the paths from the compensators to the actuators, or in the paths from the sensors, to be able to introduce feedback gradually and to reduce it rapidly if an oscillation starts. Such potentiometers greatly improve the troubleshooting options when multi-loop systems are tested.

6.6 PID Tunable Controller

6.6.1 *PID* Compensator

Figure 6.38 shows a block diagram of a feedback system with a *PID compensator* $I/s + P + Dqs/(s+q)$. The third term is a band-limited differentiation path approximating Ds for $\omega < q$. Here, the scalar parameters P, I, and D are to be determined (tuned) for a specific plant. A saturation link is commonly placed in front of the integral term (and, if the plant is a double integrator, also in front of the proportional term) to prevent wind-up (see Chapter 13).

Figure 6.39 shows Bode diagrams for each path of the compensator C_1 and the entire compensator. The component P dominates at midrange frequencies, the component I/s, at lower frequencies, and the component Ds, at higher frequencies. Typically, the distance between the corner frequencies f_{IP} and f_{PD} is two octaves or more, and the compensator transfer function has two real zeros corresponding to these frequencies. The pole q in the differentiator is typically chosen to be from $3 \times 2\pi f_{PD}$ to $10 \times 2\pi f_{PD}$.

The Bode diagram of a typical plant is convex, monotonic, and with increased slope at higher frequencies. If the plant response at higher frequencies is too shallow, a low-pass filter C_2 is commonly added to attenuate the sensor noise and to gain stabilize the system at higher frequencies. The concave response of the *PID* compensator needs to be adjusted for the loop Bode diagram to have the desired slope.

In a system with the plant having a pole at the origin and two or more additional real poles at higher frequencies, typically, the P term dominates near f_b, the Ds term at frequencies over $4f_b$, and the I/s term at lower frequencies up to $f_b/4$, as shown in Figure 6.40. Thus, P can be tuned for the feedback bandwidth, D, for the phase stability margin (D-term phase advance partially compensates the lags of the high-frequency poles of P and C_2), and I for disturbance rejection at lower frequencies. When a step test command is used for closed-loop tuning, P is tuned for the rise time/overshoot trade-off, D for the overshoot reduction, and the coefficient I is increased until it starts affecting the overshoot too much.

The *PID* controller for a double-integrator plant uses the term Ds to reduce the slope of the loop Bode diagram at the crossover. When the plant also has several high-frequency

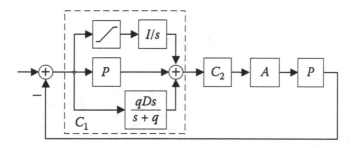

FIGURE 6.38
PID controller block diagram.

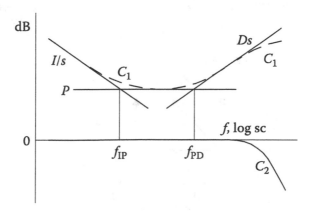

FIGURE 6.39
Bode diagrams for the *PID* compensator.

poles or a large n.p. lag, the *PID* controller may be augmented by a fourth parallel branch with a band-limited double differentiator.

The *PID* controller is very popular. In fact, it is the industry standard for tunable controllers. The controller is easily tuned to provide robust and fairly good performance for a great variety of plants. It typically provides an acceptable transient response without a prefilter or command feedforward. Several automatic tuning procedures based on tuning the transient response are successfully and widely used. However, the performance of the controller is not optimal. With the same average loop gain at higher frequencies, the Bode step responses provide larger feedback, and prefilters can reduce the overshoot.

6.6.2 *TID* Compensator

In the *PID* controller, the vector diagram at the corner frequencies f_{PD} is as shown in Figure 6.41(a). The output is larger by 3 dB than each of the components.

As shown in Figure 6.42, the zero-slope *P*-path can be replaced by the *tilted* response path $Ts^{-1/3}$ with constant gain slope −2 dB/octave, where T is a scalar tunable parameter. Implementation of the transfer function $s^{-1/3}$ was already described in Section 5.3 and exemplified in Figure 6.11. Bode diagrams for the T, I, and D path gains are shown in Figure 6.43. (For simplicity, the band-limiting pole in the differentiator is not shown in the diagram.)

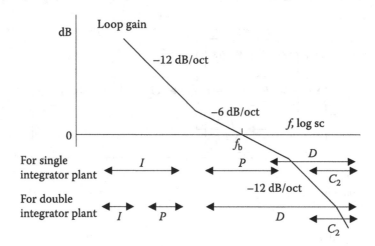

FIGURE 6.40
Ranges of dominant terms in *PID* compensators.

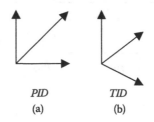

FIGURE 6.41
Vector diagrams for corner frequencies for (a) *PID* and (b) *TID* compensators.

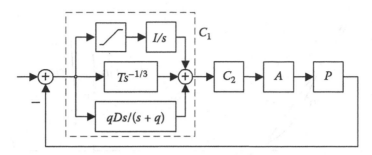

FIGURE 6.42
TID controller compensator.

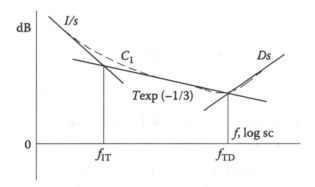

FIGURE 6.43
Bode diagrams for *TID*.

The vector diagram in Figure 6.41(b) shows the components forming the output signal of the *TID compensator* [38] at the corner frequency f_{TD}. The output signal amplitude is the same as the amplitudes of the components. Because of this, the f_{TD} corner of the *TID* gain frequency response is sharper than the similar corner in the *PID* compensator. Due to this and the slope of the *T*-term gain response, the controller provides a better loop response.

Example 6.11

Consider *PID* and *TID* controllers for a first-order plant using a sensor whose transfer function possesses a triple real pole at 80 Hz. Limited by the sensor noise, the crossover frequency is $f_b = 20$ Hz. The Bode diagrams for the loop gain achieved with *PID* and *TID* compensators are shown in Figure 6.44(a). The controllers have the same coefficient D (i.e., the same gain at higher frequencies, and therefore nearly the same level of high-frequency noise).

At the typically critical frequencies of about half f_b, the feedback in the *TID* controller is 4 dB larger than that in the *PID* controller (although the *TID* feedback is still smaller than the feedback achievable with a Bode step).

The *L*-plane open-loop diagrams for *PID* and *TID* control are shown in Figure 6.44(b). It is seen that the *PID* phase stability margin near f_b is too large, 70°, and the phase margin is not excessive in the *TID* compensator.

PID and *TID* controllers are both easy to tune when the plant parameters change, for example for a temperature control of an industrial furnace with variable payload. When the controller is not supposed to be tuned for each individual plant, preference should be given to a high-order compensator with a Bode step.

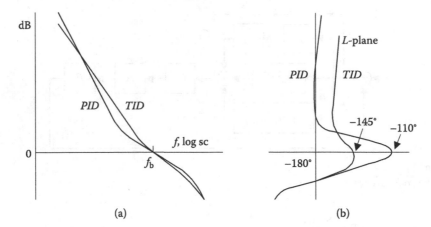

FIGURE 6.44
(a) Bode and (b) Nyquist diagrams for *PID* and *TID* controllers.

6.7 Tunable Compensator with One Variable Parameter

6.7.1 Linear Fractional Transfer Function

Plant characteristics may depend on the varying environment. Examples are the dependence of an aircraft's dynamics on altitude, and the dependence of a telecommunication cable's attenuation on temperature. For the loop response to be optimal for each of the intermediate values of the environmental parameter, the compensator must be variable, as shown in the block diagram for the resulting adaptive system in Figure 6.45 (and in Figure 9.2). We assume we know how the plant response varies with the changing environment (although not precisely, so that cancellation of the effects in an open-loop manner cannot be done).

It would be convenient to modify the transfer coefficient (or immittance; this term defines both impedance and admittance) W of a linear system by changing only one scalar parameter w. Here, w stands either for the variable transfer function of a unilateral link or for immittance of a variable two-pole.

The function $W(w)$ is a linear fractional function (referred to as "bilinear" in [6]) and as such can be expressed as:

$$W(w) = \frac{w_1 W(0) + w W(\infty)}{w_1 + w} \tag{6.9}$$

If w represents the variable impedance of a two-pole, w_1 is the driving point impedance between the terminals to which the two-pole w is connected. If w designates the transfer coefficient of an amplifier, then $-1/w_1$ is the feedback path transmission coefficient for this amplifier.

The flowchart corresponding to Equation 6.9 is shown in Figure 6.46.

The responses of the variable compensator in Figure 6.46 should be regulated smoothly. For this purpose, not all functions of the form in Equation 6.9 will do. For example, if it is desired to gradually change the slope of a Bode diagram from −6 dB to 6 dB, and we correspondingly choose $W(0)$ to be an integrator and $W(\infty)$ a differentiator, as shown in Figure 6.47, and only use a gain block w in series with the differentiator, then the response

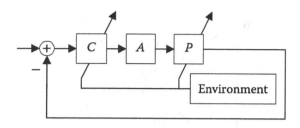

FIGURE 6.45
Adaptive system block diagram.

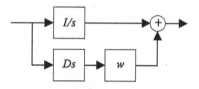

FIGURE 6.46
Flowchart for transfer function dependent on a parameter.

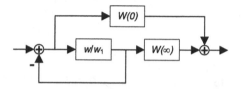

FIGURE 6.47
Block diagram of a regulator.

will be changed from that of an integrator to that of a differentiator with w changed from 0 to a large value, but, as is seen in Figure 6.48, the frequency response of W in the intermediate position possesses a zero; i.e., the regulation is not smooth.

6.7.2 Symmetrical Regulator

Smooth regulation can be obtained with Bode symmetrical regulators. Regulation is called symmetrical with respect to the nominal value w_o of the variable parameter w when the maximum relative deflections of w from w_o, up and down, cause symmetrical (in dB) variations in W, as shown in Figure 6.49, i.e., when the regulation has the following property:

$$Q^2 = \frac{W(\infty)}{W(w_0)} = \frac{W(w_0)}{W(0)} \qquad (6.10)$$

By substituting this expression into Equation 6.9 we have:

$$w_1 = w_0 Q \qquad (6.11)$$

and

$$W = W(w_0)\frac{1+(w/w_0)Q}{(w/w_0)+Q} \qquad (6.12)$$

The flowchart for the symmetrical regulator is shown in Figure 6.50.

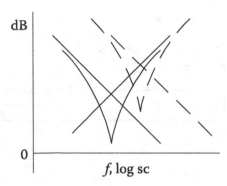

FIGURE 6.48
Frequency responses of a regulator.

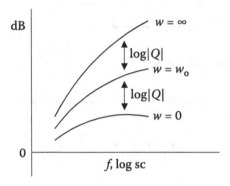

FIGURE 6.49
Regulation frequency responses of a symmetrical regulator.

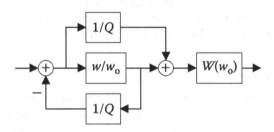

FIGURE 6.50
Flowchart for a symmetrical regulator.

The gain of the regulator changes gradually with variations in w. When $w = w_o$, the second component is 0. The second component retains the value but changes the sign when w_o is switched from 0 to ∞, as illustrated in Figure 6.49. The regulation is exactly symmetrical for these maximum variations of the gain response. It can be shown that in the Taylor expansion of the second component, the first-order term dominates to the extent that the gain $20 \log|W|$ depends on w/w_o nearly linearly over the regulation range more than 20 dB.

The regulator can be used in the compensator for changing the loop Bode diagram in accordance with changing system requirements, for solving the trade-offs between the available disturbance rejection and the output noise. It can be used to compensate the effects of known plant parameter variations in adaptive systems (adaptive systems are studied in Chapter 9). Also, a nonlinear element can be used in place of the variable element w. In this case, a nonlinear dynamic link with desirable properties can be built. Such nonlinear links can be employed in compensators for enhancing the system performance, as will be discussed in Chapters 10, 11, and 13.

6.7.3 Hardware Implementation

A hardware implementation of such a regulator is shown in Figure 6.51(a). The possible implementations of the block $1/Q$ as low-pass and high-pass RC filters are shown in Figure 6.51(b) and (c).

The regulator responses with the high-pass filter are shown in Figure 6.52. The regulator can be used to vary the compensator response when the plant response varies mostly at higher frequencies. Additional examples are given in Sections 11.7 and 13.5.

Regulation of frequency responses can also be performed with digital finite impulse response (FIR) filters, which can be made to approximate any desired frequency response. However, using the filter for gradual response changes requires changing several coefficients, which is less convenient.

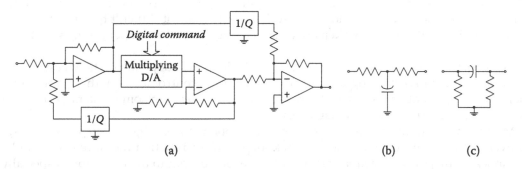

(a) (b) (c)

FIGURE 6.51
(a) Symmetrical regulator implementation with (b) low-pass and (c) high-pass RC filters as $1/Q$.

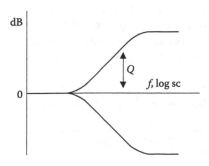

FIGURE 6.52
Frequency responses for the symmetrical regulator.

6.8 Loop Response Measurements

To design and verify the loop frequency response, one has to calculate or measure the plant response. The choice between the plant simulations and the actual plant measurements might depend on the plant size and accessibility. For example, if the plant is big or expensive (like a spacecraft or heavy machinery), or if it is very small, like a microwave feedback amplifier or a micromachined mechanical device that is difficult to access, then simulation of the plant should play a major role in the feedback loop design. In this case, the plant model needs to be accurate, and substantial efforts are justified to develop this model. At the beginning of the project, when this model is not yet completed, simpler models must be developed for preliminary control design, to be able to answer the questions crucial for the plant design. On the last stages of the project, when the plant is built, it can be measured and the feedback loop tested. In all cases, the compensator must be tested first, before testing the entire loop.

On the other hand, when the plant is inexpensive and easily measured, the compensator (analog or digital) can be well designed directly in the laboratory while working with the real plant, by measuring the loop response and adjusting the elements of analog compensators or coefficients in digital compensators.

The loop response can be measured when the feedback loop is open, as in Figure 6.53, or while the system is in the closed-loop configuration. While the loop response is being tested, the loop can be broken, theoretically, at any cross section. In practice, the cross section must be carefully chosen such that the signal at this cross section is large enough for convenient measurements but not excessively large to make it difficult to generate such a signal. Typically, the loop is broken after the first stage C_1 of the compensator, as shown in Figure 6.53.

The choice of the cross section for breaking the loop also depends on how easy is it to simulate the load for the loop output. For conventional control loops with separate compensators this is not an issue, but it can be a difficult problem in some systems, especially at microwave frequencies or in magnetics.

Measurements of the open-loop response in the open-loop configuration are not always convenient or even possible. The feedback loop transfer function measurements can be complicated by parasitic feedback loops from the loop output to the loop input (especially at higher frequencies). Also, some plants cannot be used without feedback—they could be unstable, self-destructive, or dangerous to deal with. The control loops about such plants need to be measured in the closed-loop configuration, but perhaps with reduced feedback.

Loop response measurements in the closed-loop configuration require that the closed-loop system be stable. If it is not, it still might be stable when the feedback is reduced by an extra attenuator; i.e., the system is closed with smaller feedback. Injection of the signal into the feedback loop for the purpose of the open-loop response measurements with the loop

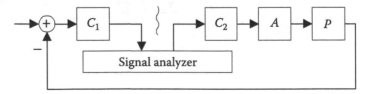

FIGURE 6.53
Breaking the feedback loop.

kept closed is shown in Figure 6.54. The loop transfer function is the ratio of the signal at input 2 to the signal at input 1 (the latter is applied to the "reference" input of the analyzer).

A practical circuit for the signal injection is shown in Figure 6.55. The variable resistor allows reducing the loop gain, and the *RC* lead is used to increase the test signal level at higher frequencies—as will be discussed below. The capacitance value $C = 2 \times 10^{-5}/f_b$ places the zero of this lead at approximately $0.08f_b$, and the pole at approximately $4f_b$.

If an analog compensator already includes an inverting op-amp, the signal injection can be implemented with a few passive elements, as shown in Figure 6.56, using the op-amp as a summing amplifier.

Signal analyzers typically use two types of test signals: swept-frequency sinusoidal and pseudorandom with subsequent fast Fourier transform (FFT) and averaging. Both methods are quite suitable for the loop response measurements. At higher frequencies, sinusoidal excitation is easy to implement and is appropriate. At low frequencies, FFT might provide better accuracy or shorter time of measurements. Still, low-frequency response

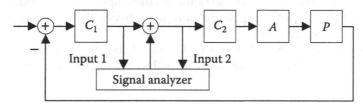

FIGURE 6.54
Stability margin boundary and the loop transfer function uncertainty at Nyquist frequency.

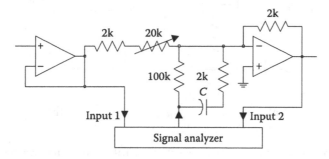

FIGURE 6.55
Summer for injecting test signals into the closed loop.

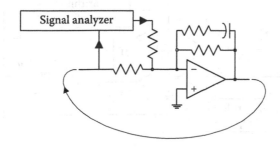

FIGURE 6.56
Signal injection using an already existing inverting amplifier.

measurements often take longer than desired, especially when the test is performed in the field, or some equipment normally in operation needs to be shut down to perform the test. Part of the problem is that when the test signal is employed with constant amplitude of sinusoidal signals at all frequencies, or with constant spectral density of pseudorandom signals, the dynamic range of the signals at the output of the loop becomes very large, since the loop gain is many tenths of dB in the functional feedback band and −20 dB at higher frequencies still of interest for the stability margin test. Then, the signal amplitude chosen so as not to overload the actuator at lower frequencies becomes too small at higher frequencies, and data recovery from the noise requires multiple runs and averaging.

An extra linear link with the gain increasing with the frequency, such as that shown in Figure 6.57, can be introduced at the signal analyzer generator output to reduce the dynamic range of the loop output signal, and correspondingly to reduce the required time for accurate measurements.

In the absence of a signal analyzer, the loop gain response can be measured point by point, at discrete frequencies, with a signal generator and a two-input oscilloscope. The input signal of the loop is applied to one of the oscilloscope Y-inputs, and the loop output is applied to the other Y-input. The phase difference between the signals can be seen on the scope screen with sufficient accuracy.

With a sweep generator and a scope, an addition of the circuitry shown in Figure 6.58 can mechanize the loop response measurements. Here, two amplifiers (optional) increase

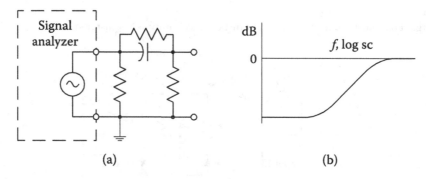

FIGURE 6.57
(a) Gain response corrector to be placed in cascade with the signal source to improve the measured signal-to-noise ratio, and (b) the corrector gain frequency response.

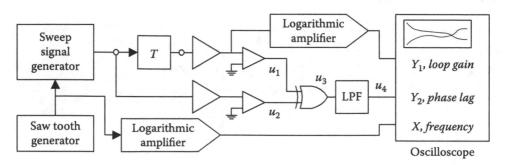

FIGURE 6.58
Set for loop response measurements.

the levels of the signals at the input and output of the feedback loop. The amplifier gain can be made to increase with frequency in order to reduce the dynamic range of the signals.

The two single-supply comparators convert the sinusoidal signals to rectangular single-polarity signals u_2 and u_1, as illustrated in Figure 6.59. The XOR gate (exclusive or) serves as phase detector. Its output signal u_3 is nonzero only when only one of the input signals is nonzero, as shown in Figure 6.59. The average value u_4 of the signal u_3 is proportional to the phase lag between u_2 and u_1. The low-pass filter producing u_4 can be a Butterworth active RC filter.

When the plant transfer function is relatively simple, the plant can be identified by its response to the step function applied to its input. This method is often employed for tuning *PID* controllers for chemical processes, as is discussed below.

While the process is going on open loop, the output is made steady by adjusting the command manually. Then, a small step is applied directly to the plant input as shown in Figure 6.60(a). This step produces an increment in the plant's output (controlled variable), as shown in Figure 6.60(b). From the response, the delay time t_d and the rate r are determined. The parameters in the *PID* controller written in the form $a[1 + 1/(bs)](1 + cs)/(1 + cs/d)$ are set to be $a = 1/(rt_d)$, $b = 5t_d$, $c = t_d/2$, and $d = 4$, and then further adjusted experimentally for good closed-loop performance. This open-loop Ziegler-Nichols tuning procedure is not universal and does not necessarily lead to good tuning of PID controllers for all possible plants.

A more general approach is to create a plant model with a transfer function having a couple of real poles and zeros, and adjust the values of the model poles and zeros with MATLAB to obtain the same transient response as that of the real plant. Then, using this plant model, a compensator can be designed that provides the desired loop response.

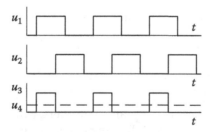

FIGURE 6.59
Time diagrams for the phase detector signals.

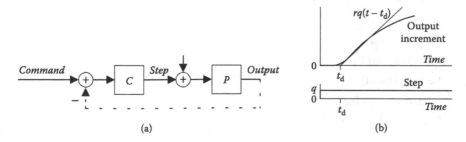

FIGURE 6.60
Plant transient response measurements: (a) block diagram and (b) transient response.

PROBLEMS

1. a. Design an amplifier with gain coefficient –25 using an op-amp in the invert-
      ing configuration. The amplifier will be employed in the compensator for a
      control system with 1 kHz feedback bandwidth. Choose f_T, and choose resis-
      tors such that the power consumption by the feedback resistor from the 12 V
      power supply does not exceed 3 mW.

   b. Do the same for the gain coefficient –100, $f_b = 100$ kHz, power not exceeding
      0.2 mW.

2. a. Discuss the accuracy of implementation of a transfer function as an op-amp
      with RC-impedances.

   b. Define the required tolerances of resistors and capacitors.

3. Draw a schematic diagram for an op-amp circuit implementing the following
   functions:

   a. $-ax - by - c$

   b. $+ax + by + c$

   c. $-ax + by - c$

   d. $-ax + b\int y\,dt - c$

   e. $-ax + b\,dy/dt - c$

   f. $\int(-ax - by - c)dt$

4. Draw a schematic diagram and specify the values of the circuit elements (use a
   100 kΩ feedback resistor) for a practical op-amp circuit implementing the following functions:

   a. $-0.5x - 0.2y - 3$

   b. $+2x + 8y + 0.2$

   c. $-3x + 6y + 1$

   d. $-4x + 4\int y\,dt + 0.3$

   e. $-2.72x + 0.5\,dy/dt + 0.1$

   f. $\int(-6x + 0.4y - 2)dt$

5. Using 20 kΩ feedback resistors in op-amp summers, draw a schematic diagram for
   an analog implementation of the following decoupling matrices:

   a. $x = 0.5038x' + 0.0443y' + 0.0772z'$

      $y = -0.0235x' + 0.4796y' - 0.0215z'$

      $z = -0.0094x' - 0.0243y' + 0.5291z'$

   b. $x = -0.2x' + 0.05y' - 0.1z'$

      $y = 0.1x' + 0.6y' + 0.01z'$

      $z = 0.001x' - 0.002y' - 0.7z'$

   c. $x' = 3x + 0.4y + 0.3z$

      $y' = 0.3x - 2.1y + 0.2z$

      $z' = 0.04x + 0.1y + 1.9z$

  d. $x' = 2x + 0.1y + 0.1z$

    $y' = 0.1x - 3.1y + 0.1z$

    $z' = 0.04x + 0.4y + 1.9z$

  e. $x' = x + y - z$

    $y' = -x + y + z$

    $z' = x - y + z$

  f. $x' = -x + y + 0.5 \int y dt + z$

    $y' = x + y + z$

    $z' = x - 0.1 \int x dt + y - z$

6. Choose the type of op-amp for applications requiring low noise in the following bandwidths:

  a. 10 Hz

  b. 100 Hz

  c. 1 MHz

7. Choose the type of op-amp to use when the output power needs to be up to 3 W and the frequency band of interest is the following:

  a. 0–100 kHz

  b. 0–1 MHz

  c. 100 Hz–1 kHz

  d. 1–10 kHz

8. In the compensators using the inverting op-amp configurations in Figure 6.10(a) and (c), the maximum gain (at lower frequencies for the lag and at higher frequencies for the lead) must be 40 dB. Choose $R_2 = 1$ MΩ and find the remaining element values for the following:

  a. Pole $f_p = 5$ Hz, zero $f_z = 10$ Hz

  b. Pole $f_p = 12$ Hz, zero $f_z = 20$ Hz

  c. Pole $f_p = 300$ Hz, zero $f_z = 1,000$ Hz

  d. Pole $f_p = 15$ Hz, zero $f_z = 40$ Hz

  e. Pole $f_p = 27.2$ Hz, zero $f_z = 120$ Hz

  f. Pole $f_p = 800$ Hz, zero $f_z = 3,000$ Hz

    Find the elements analytically (start with resistors), with MATLAB, or with the chart in Figure 6.23.

9. Use a noninverting op-amp with RC feedback according to Figure 6.9(b) to implement a lead link, assuming $R_2 = 500$ kΩ and dc gain 20 dB with the following:

  a. $f_z = 100$ Hz, $f_p = 300$ Hz

  b. $f_z = 50$ Hz, $f_p = 150$ Hz

  c. $f_z = 10$ Hz, $f_p = 100$ Hz

  d. $f_z = 150$ Hz, $f_p = 400$ Hz

e. $f_z = 27.2$ Hz, $f_p = 100$ Hz

f. $f_z = 60$ Hz, $f_p = 600$ Hz

10. Use a noninverting op-amp with RC feedback according to Figure 6.9(a) to implement a lag link, assuming $R_1 = 10$ kΩ and dc gain 30 dB with the following:

   a. $f_p = 300$ Hz, $f_z = 1{,}000$ Hz

   b. $f_p = 150$ Hz, $f_z = 500$ Hz

   c. $f_p = 3$ Hz, $f_z = 10$ Hz

11. Find a transfer function with three poles and two zeros approximating the gain response slope −4 dB/octave over the range 3 to 30 Hz. The gain at dc must be 26 dB, and the high-frequency asymptote must be −6 dB/octave. Design a compensator with this transfer function using an op-amp in the inverting configuration. Use an initially asymptotic Bode diagram, then (a) simulate the response in MATLAB or SPICE and (b) adjust the positions of poles and zeros (if using MATLAB), and then calculate the element values, or directly adjust the values of the circuit elements (if using SPICE).

12. Use the op-amp with RC feedback shown in Figure 6.12(a) to produce a broad notch, using a real pole and a real zero, and a pair of complex zeros. Make simulations with SPICE.

13. Use the FILTER1 program (available from Burr-Brown) to design the following Sallen–Key filters:

   a. A low-pass second-order link with cutoff frequency of 200 Hz and $\zeta = 0.3$

   b. Butterworth third-order link with cutoff at 1 kHz

   c. Chebyshev fifth-order link with 0.4 dB peak-to-peak ripples in the passband and cutoff at 10 kHz

   d. Butterworth fifth-order link with cutoff at 30 Hz

   Find filter elements, simulate the filter performance, and plot the gain response.

14. Use the FILTER2 program (available from Burr-Brown) to design the following multiple feedback filters:

   a. A single second-order link with 200 Hz cutoff frequency and $\zeta = 0.2$

   b. Butterworth third-order link

   c. Chebyshev fifth-order link with 0.4 dB peak-to-peak ripples in the passband

   d. Butterworth fifth-order link

Find filter elements, simulate the filter performance, and plot the gain response. Find sensitivities to the filter elements.

15. Some antialiasing filter ICs pass the dc input directly to the output, without contaminating the dc signal by the dc drift of the employed op-amp. For what applications are such filters especially suitable?

16. a. Implement a Bode step with the Sallen–Key low-pass filter shown in Figure 6.14 with the data in Example 1 in Section 6.9a, using the method illustrated in Figure 6.16(a).

b. Do it for the loop with $f_b = 500$ Hz, main Bode diagram slope -10 dB/octave, gain stability margin 10 dB, high-frequency asymptotic slope -18 dB/octave.

c. Design a Nyquist-stable loop Bode diagram, with the lower and upper Bode steps at -10 and 15 dB levels, respectively, using Sallen–Key filters.

Use the FILTER1 program from Burr-Brown, or use MATLAB simulation for second-order links with appropriate damping.

17. Use the FILTER1 program to study sensitivity of Sallen–Key filters. Show that plenty of margin exists when using 1% resistor tolerances, when the filter is used to implement a Bode step.

18. Prove that the equivalent resistance of a charge pump is $R = 1/(Cf_s)$.

19. Use SPICE to simulate the circuits in Figures 6.18 and 6.19. Discuss how to make these simulations with MATLAB and Simulink.

20. a. Using the plot in Figure 6.23, draw the frequency response of the magnitude of the impedance of the RC circuit in Cauer form in Figure 6.20(b), with capacitances (starting from input) 0.1 mF, 10 nF, and resistances 100 kΩ, 5 kΩ (you can draw the response directly on a copy of the chart).

b. Make the plot with SPICE.

c. Make the plot with MATLAB after deriving the equations describing the circuit.

21. Using the plot in Figure 6.23, find the following:

a. Where is the pole of impedance of the parallel connection of 1 mF and 1 kΩ?

b. Where is the zero of impedance of the series connection of 10 nF and 10 kΩ?

c. What is the impedance of the series connection of 20 kΩ and 4 µF at 300 Hz?

d. What the capacitor needs to be connected in series with a 50 kΩ resistor to make a zero at 800 Hz.

e. What the resistor needs to be connected in parallel with a 2 µF capacitor to move the pole from 0 Hz to 30 Hz.

f. R and C for the series RC two-pole connected in parallel to a 100 kΩ feedback resistor of an op-amp in inverting configuration; the transfer function pole is at 200 Hz and the zero at 800 Hz.

22. Choose ICs for the circuit in Figure 6.58, then build and test the device.

23. Simulate a control loop for pointing a spacecraft with flexible solar panels using thrusters (in pulse-width modulation mode, i.e., with a saturation type of characteristic) using the compensators in Figure 6.12.

24. Make SPICE simulation for the circuit in the following:

a. Figure 6.12(a)

b. Figure 6.12(b)

c. Figure 6.13

25. Use MATLAB or Simulink to solve Problem 24.

ANSWERS TO SELECTED PROBLEMS

1a. At 1 kHz, the closed-loop gain coefficient is −25, and the feedback must be 30 dB; i.e., the open-loop gain coefficient must be 750. Therefore, $f_T > 750 \times 1 \text{ kHz} = 750 \text{ kHz}$, $R_2 > 50 \text{ k}\Omega$, and $R_1 = 50/25 = 2 \text{ k}\Omega$.

2a. When impedance of a series feedback branch changes by +5%, and of the parallel branch by −5%, the gain coefficient increases by approximately 10%. The angle of an RC-impedance is within the $[0, -90°]$ interval. Because of this, the sensitivity of the modulus of the impedance to a resistance or a capacitance within the circuit is less than 1.

3d. The summer is shown in Figure 6.61.

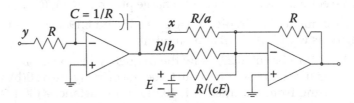

FIGURE 6.61
Implementation of function $-ax + b \times$ [integral of (y)] $- c$.

5a. The schematic diagram is shown in Figure 6.62. The resistances are 20 kΩ/(coefficient in the relevant equation).

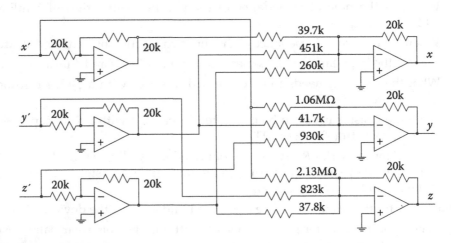

FIGURE 6.62
Schematic diagram for a decoupling matrix implementation.

8a. Since $f_z > f_p$, this is a lag link and the schematic in Figure 6.10(c) can be used, and the gain coefficient at zero frequency must be 100. Then, $R_1 = 10 \text{ k}\Omega$. At infinite frequency, the gain coefficient is $100 f_p / f_z = 50$. From here, $R_3 \# R_2 = 500 \text{ k}\Omega$, i.e., $R_3 = 1 \text{ M}\Omega$. The capacitance $C = 16 \text{ nF}$ is found from the condition $1/(2\pi f_z C) = R_3$.

9a. From the required gain coefficient 10 at zero frequency, $R_1 = R_2/11 = 45.5$ kΩ. At infinite frequency, the gain coefficient is three times larger (because the ratio of the pole to the zero frequencies is 3); i.e., it is 30. From this, the parallel connection of resistances R_1 and R_3 is $R_1 \# R_3 = R_2/31 = 16.1$ kΩ and $R_3 = 16.1R_1/(16.1 + R_1) = 25$ kΩ (we are designing a forward path compensator so there is no need for extreme accuracy of calculations). The capacitance can be found from the equality $1/(\omega C) = R_3$ at the frequency of the pole of the compensator transfer function, which is the frequency of the zero of the shunting impedance in the feedback path; $C = 21 \times 10^{-9}$ F, i.e., $C \approx 20$ nF.

10a. At zero frequency, the gain coefficient is $31.6 = (R_2 + R_1)/R_1$, wherefrom $R_2 = 306$ kΩ. At infinite frequency, the gain coefficient is $31.6 f_p/f_z = 9.48$. Thus, $9.48 = (R_3 \# R_2 + R_1)/R_1$. From here, $R_3 \# R_2 = 848$ kΩ and $R_3 = 477$ kΩ. The capacitance is found from the condition that at the pole frequency $R_3 + R_2 = 1/(2\pi f_p C)$, wherefrom $C = 400$ pF.

7

Linear Links and System Simulation

Two approaches are presented for modeling systems composed of electrical, mechanical, thermal, and hydraulic elements: describing the system elements and the topology of their connections, and deriving mathematical equations for the system.

It is emphasized that to reduce the plant uncertainty (and therefore, to increase the available feedback), the actuator output impedance needs to have a specific value. The use of local feedback loops to modify this impedance and, generally, the effect of feedback on impedance are considered, including parallel, series, and compound feedback.

Equivalent block diagrams are developed for the chain connection of two-ports, such as drivers, motors, and gears, to simplify system modeling and to make it structural.

Several issues are considered that are important for feedback control: flexible structures, collocated and non-collocated control, sensor noise, and the effect of feedback on the output noise.

The feedback-system equations are analogous to the equations describing the parallel connection of two links and the parallel connection of two two-poles. These analogies allow the theory developed for two-pole connections to be applied to feedback systems, as will be exploited in Chapters 10 and 12. The chapter ends with a brief demonstration of the specifics of linear time-variable systems.

For a shortened control course, Sections 7.2.2, 7.2.3, 7.4.2–7.4.5, 7.5, 7.8.2, 7.10, and 7.11 can be omitted.

7.1 Mathematical Analogies

7.1.1 Electromechanical Analogies

The first step in simulating a system behavior on a computer is to generate the equations describing the system. This may be done either by the user or, preferably, by a computer. In the latter case, parameters describing the system's elements and the topology of their connections constitute the simulation program input file.

Control engineers deal with systems that are part electrical and part mechanical, thermal, etc. Converting such a system to an equivalent system containing only one kind of variable and described by only one kind of physical law can facilitate both the preliminary back-of-the-envelope analysis and the write-up of the computer input file. Also, the conversion allows the use of programs that were initially supposed to be used for the analysis of electrical circuits, like SPICE, as universal control analysis and simulation tools.

There exist several mathematical analogies between mechanical and electrical systems. Among them, the most useful are those that preserve power; i.e., they convert the product (voltage × current) into the product (velocity × force). We will most often use the *voltage-to-velocity* analogy relating voltage U to velocity V, and current I to force F.

Example 7.1

Figures 7.1 and 7.2 show an example of the application of this analogy. It is seen that the topology of the diagram is conveniently preserved, the inductors are replaced by springs with spring coefficients k_1, k_3, and the capacitors, by rigid bodies specified by their masses M_2, M_4.

For nodes 2 and 4, Kirchhoff's equations correspondingly are:

$$I_1 + I_2 + I_3 = 0$$

$$U_4 = I_3 / (sC_4)$$

where

$$I_2 = U_2 s C_2$$

For bodies 2 and 4, Newton's equations are:

$$F_1 + F_2 + F_3 = 0$$

$$V_4 = F_3 / (sM_4)$$

where

$$F_2 = V_2 s M_2$$

The equality of the sum of currents to zero at a node corresponds to the equality of the sum of forces to zero at a rigid body (taking into account D'Alembert's force). Similarly, the equality of the sum of the voltages (relative potentials) about a contour to zero reflects the equality to zero of the sum of relative velocities about the contour. Zero voltage of the electrical "ground" most commonly translates to zero velocity of the inertial reference. Resistance converts into the inverse of the viscous damping coefficient, capacitance into mass, and inductance into the inverse of the stiffness coefficient. Generally, electrical impedance converts into mechanical mobility. Since mobility is the ratio of relative velocity to force, the mobility of a difficult to move or heavy load by a given force is small.*

Table 7.1 summarizes the analogy for translational and rotational motions.

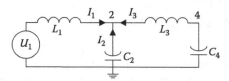

FIGURE 7.1
Electrical circuit schematic diagram.

FIGURE 7.2
Mechanical circuit schematic diagram. To accentuate the analogy, the springs are shown similar to inductors, and not similar to resistors, as is conventional in drawing mechanical diagrams.

---

* The term mechanical impedance is sometimes used to mean the mobility, and sometimes the inverse of the mobility.

Example 7.2

Consider the active suspension strut diagrammed in Figure 7.3(a). The strut consists of a spring, a dashpot (a device providing viscous friction), and a linear motor (voice coil) connected in parallel and driven by an electrical amplifier in accordance with the information obtained from the force sensor (load cell). The vibrations of the base should be prevented from shaking the payload. For the suspension strut, the mobility is defined as the ratio of the difference in velocities at its ends to the force. The mobility must be large at frequencies where the vibration should not pass through. At the same time, the mobility should be low enough at lower frequencies in order for the motion of the base to be conveyed to the body of interest. Thus, the notion of mobility is particularly applicable to the design of suspension and vibration isolation systems. Figure 7.3(b) exemplifies the modulus of the mobility with and without introducing feedback to make the isolation better.

For the analysis of mechanical systems with many degrees of freedom at the joints, such as ball joints, using the electromechanical analogy gets more difficult (although it is still possible). For complicated mechanical systems, specialized simulation programs should be used (like ADAMS® or SD FAST®). However, in the great majority of practical cases, control engineers only deal with sliding or pin joints, i.e., with the joints characterized by only two variables each, the position and the force, or the angle and the torque. For these systems, using the analogy of a mechanical joint to a joint of cascaded electrical two-ports is of great help.

The *voltage-to-velocity* (i.e., *current-to-force*) analogy described above is especially convenient when the system includes electromagnetic transducers like electrical motors

TABLE 7.1

Voltage-to-Velocity Analogy

Electrical	Translational	Rotational
Voltage U (or V)	Relative velocity V	Relative angular velocity Ω
Current I	Force F	Torque τ
Power $P = UI$	Power $P = VF$	Power $P = \Omega\tau$
Impedance $Z = U/I$	Mobility $Z = V/F$	Mobility $Z = \Omega/\tau$
At a node, $\Sigma I = 0$	At a rigid body, $\Sigma F = 0$	At a rigid body, $\Sigma\tau = 0$
About a contour, $\Sigma U = 0$	Along a contour through a sequence of connected bodies, $\Sigma\Delta V = 0$	Along a contour through a sequence of connected bodies, $\sum\Delta\Omega = 0$
Capacitance C	Mass M	Moment of inertia J
Inductance L	1/(stiffness coefficient k)	1/(angular stiffness)
Resistance R	1/(viscous damping coefficient B)	1/(angular viscous damping coefficient)

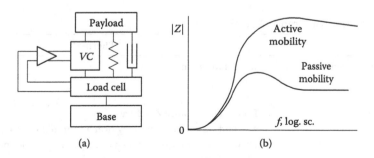

(a)   (b)

FIGURE 7.3
(a) Active suspension strut and (b) its mobility frequency response.

and solenoids, where *current* creates a magnetic field that in turn produces *force*. In piezo-electric and electrostatic transducers (the latter are used in micromachined devices), the applied voltage produces force, and the *voltage-to-force* analogy shown in Table 7.2 may appear more convenient.

7.1.2 Electrical Analogy to Heat Transfer

The electrical analogy to conductive heat transfer is shown in Table 7.3. The heat flow is measured in Watts or Btu/s, 1 kW corresponding to 1.055 Btu. The Fourier law for heat flow is analogous to Ohm's law.

Heat also can be transferred by radiation and convection. Since radiation increases as the fourth power of the temperature difference between the bodies (or between the body and the environment), it can be represented by a nonlinear current source. For heat transferred via convection, the equivalent thermal resistance depends on many parameters, including the geometry, the area, and the airflow.

The analogy is exemplified in Figure 7.4 by an equivalent electrical circuit employed to simulate the temperature control of the Cassini spacecraft narrow view photocamera. Here, H_2 and H_1 represent heaters attached to the secondary and primary mirrors, respectively. The temperatures T_2, T_c, and T_1 represent absolute temperatures of the secondary mirror, the case, and the primary mirror. The capacitances are the products of the bodies'

TABLE 7.2

Voltage-to-Force Analogy

Electrical	Translational	Rotational
Voltage U (or V)	Force F	Torque τ
Current I	Velocity V	Angular velocity Ω
Charge q	Displacement x	Angle θ
Power $P = UI$	Power $P = VF$	Power $P = \Omega\tau$
Impedance $Z = U/I$	Inverse of mobility $Y = F/I$	Inverse of mobility $Y = \tau/\Omega$
At a node, $\Sigma I = 0$	Along a contour through a sequence of connected bodies, $\Sigma\Delta V = 0$	Along a contour through a sequence of connected bodies, $\Sigma\Delta\Omega = 0$
About a contour, $\Sigma U = 0$	At a rigid body, $\Sigma F = 0$	At a rigid body, $\Sigma\tau = 0$
Capacitance C	$1/$(stiffness coefficient k)	$1/$(angular stiffness)
Inductance L	Mass M	Moment of inertia J
Resistance R	Viscous damping coefficient B	Angular viscous damping coefficient

TABLE 7.3

Voltage-to-Temperature-Difference Analogy

Electrical	Thermal
Voltage (potential difference) U (or V)	Temperature difference ΔT
Current I	Heat flow g
Power $P = UI$	ΔTg
Impedance $Z = U/I$	Thermal impedance $Z_T = \Delta T/g$
At a node, $\Sigma I = 0$	At a body, $\Sigma g = 0$ (not counting heat accumulation in C_T)
Capacitance C	Thermal capacitance C_T, which is the product of the specific heat c and the mass M
Resistance $R = dU/dI$	Thermal resistance $R_T = dT/dg$

specific heats and masses, and the resistors characterize the thermoconductivities between the bodies. The nonlinear sources represent the heat radiation into free space from the mirrors and the case:

$$G_2 = 0.507 \times 10^{-9} \times T_2^4, G_c = 0.5 \times 10^{-9} \times T_c^4, G_1 = 0.1 \times 10^{-9} \times T_1^4$$

This circuit was simulated using SPICE—the model also incorporated the driver and compensator. With this approach, the frequency and time responses can be plotted without derivation of the plant equations.

7.1.3 Hydraulic Systems

Example 7.3

A hydraulic system is exemplified in Figure 7.5. Here, the reservoir of volume v_1 is filled with a liquid kept at pressure h_1. The sum of the liquid flow at each node is 0. The pipes connecting the reservoir to the cylinder with current volume v_2 and pressure h_2 present resistances R_1, R_2 to the liquid flow (we consider the volume of the liquid in the pipe insignificant, or else the system must be considered a system with distributed parameters). The valve controls the liquid flow by introducing extra resistance in the pipe, or switching it on and off. The valve is operated by an electromagnet fed from the driver amplifier of the controller. The plunger stem moves a mechanical load with mobility Z_L.

The dashed lines separate the areas on the drawing for hydraulic, electrical, and mechanical variables. The power is the same to the left and to the right sides of the vertical dashed line. The hydraulic and mechanical diagrams can be converted to their electrical equivalents and connected by an ideal transformer that preserves the power and provides appropriate correspondence between the hydraulic and the mechanical variables. Then, this system can be simulated with a program for electrical circuit analysis.

An electrical analog to a hydraulic system is described in Table 7.4.

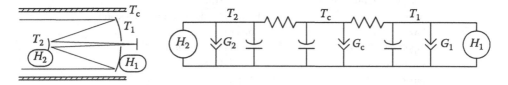

FIGURE 7.4
Heat transfer analysis for a spacecraft photocamera.

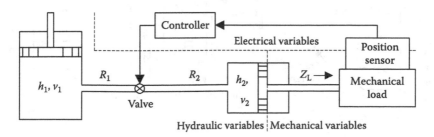

FIGURE 7.5
Hydraulic system example.

When the flow of liquid in a pipe is laminar, q is proportional to h, and the resistance

$$R = dh / dq$$

does not depend on h. However, in most applications, the flow of liquid is turbulent. In this case, the dependence of q on h is nonlinear, and the resistance is:

$$R = \frac{dh}{dq} = \frac{2(\Delta h)^2}{q} \tag{7.1}$$

The design problems and constraints are often similar for various systems and devices.

Example 7.4

Figure 7.6 demonstrates the analogy of an electrical amplifier (a) to a hydraulic amplifier (b). The electrical amplifier contains two parallel branches, each passing a single-polarity signal (which is the *class B* mode of operation). A small dead zone in input voltage prevents the waste of power supply current in the quiescent mode, i.e., when the input signal is 0. The summer combines the output currents of the branches. The input–output characteristic for the amplifier shown in Figure 7.6(c) is not quite linear, and in particular contains a dead zone. By application of large feedback (not shown in Figure 7.6), the input–output characteristic is made much closer to linear, and the dead zone is reduced many times (recall Section 1.7).

The hydraulic amplifier input is the position of the *spool valve*. The valve directs the liquid under pressure to the appropriate side of the plunger, thus providing output force at the plunger stem. The dead zone should be sufficiently large to reliably stop the

TABLE 7.4

Voltage-to-Difference-in-Pressure Analogy

Electrical	Hydraulic
Voltage (potential difference) U	Pressure difference Δh
Current I	Flow of liquid, volume per second, q
Power $P = UI$	Power $P = \Delta h q$
Impedance $Z = U/I$	Impedance $Z = \Delta p/\Delta h$
At a node, $\Sigma I = 0$	At a node, $\Sigma q = 0$
Capacitance C	Volume v
Resistance $R = dU/dI$	Resistance $R = dh/dq$

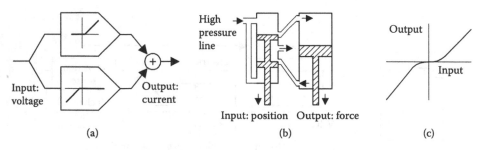

(a)   (b)   (c)

FIGURE 7.6

(a) Electrical class B amplifier, (b) spool valve hydraulic mechanical amplifier, and (c) the electrical amplifier's input–output characteristic.

power waste when the input signal is zero. Still, the zone must be small enough for the input–output characteristic to be close to linear.

This provision is especially difficult to implement in hydraulic amplifiers. The input–output characteristic of the spool valve amplifier is different for different speeds of the action. When the output motion is slow, even a small opening suffices to provide nearly full available pressure on the plunger, but if the output motion is fast, a small opening is insufficient to supply the liquid to the output cylinder. Thus, the input–output characteristic is steep for slow motion, but the gain decreases for faster motions. These variations in the actuator nonlinearity and dynamics limit the available accuracy of a control system using such an actuator. The dynamics and linearity can be improved by using local feedback about the actuator. This feedback can be implemented using a pressure sensor (or a position sensor, depending on the application) with electrical output, a compensator, a power electrical amplifier (motor driver), and an electrical motor moving the spool valve shaft. Such integrated actuators are commercially available, and their input is an analog or digital electrical signal.

7.2 Junctions of Unilateral Links

7.2.1 Structural Design

With structural design, the system is composed of subsystems that are relatively large functional parts with rather simple interconnections between them. Structural design simplifies the analysis and design, and facilitates having several people work together on the same project. Since the specifications for each of the subsystems are tailored such as to minimize the requirements, the subsystems' overdesign (for example, the accuracy requirements for the links in the forward path can be much relaxed from the required accuracy for the summer and feedback path links) are prevented. It simplifies verifying system performance, making design trade-offs, and troubleshooting. It facilitates redesigning the system, since, as long as the requirements for the subsystem are defined, each subsystem can be refined and redesigned independently of the others.

The subsystems are composed of smaller subsystems, etc. For example, the compensators themselves are built from simpler links cascaded or connected in parallel.

The interface between the subsystems should be as simple as possible. Between cascaded blocks in the block diagrams, the interface is the simplest: a single variable serves as the output of the preceding link and the input of the following links. However, in most physical links, we have to include into consideration the effects of *loading*, i.e., the effects of the following link on the output variable of the previous link. The effects of loading are conveniently described using the notion of impedance.

7.2.2 Junction Variables

Any electrical or mechanical source can be equivalently represented in Thevenin's and Norton's forms, shown in Figure 7.7. Here, E is the open-circuit voltage, or emf, I_S is the short circuit current, V_f is the free (unloaded) velocity, and F_B is the brake force. When the internal source impedance Z_S is small, the output voltage (or velocity) does not depend on the value of the load impedance, and therefore is called source of voltage U (or velocity V); when the source impedance is very large, the source is called source of current I or force F.

While the notion of driving point impedance is universally employed in the design of electrical circuits, in mechanical systems design, mechanical driving point impedance (or mobility) is used less frequently. This is probably so because at a mechanical junction, there could be more than two variables to deal with. For example, three angles and three torques need to be considered at a ball joint. Nevertheless, as mentioned before, the most common mechanical junctions are the simple ones, like a sliding joint or a pin joint with one position (or rotation) variable and one force (or torque) variable. These junctions are mathematically similar to connections of electrical two-node ports. The notion of mobility facilitates the analysis of such systems.

Figure 7.8 shows a preceding link (two-port) loaded at the input impedance of the following link. The voltage at the junction is:

$$U = \frac{EZ_L}{Z_L + Z_S} \tag{7.2}$$

and the current is:

$$I = \frac{I_S Z_S}{Z_L + Z_S} \tag{7.3}$$

For mechanical systems,

$$V = \frac{V_f Z_L}{Z_L + Z_S} \tag{7.4}$$

$$F = \frac{F_B Z_S}{Z_L + Z_S} \tag{7.5}$$

When the driver's output impedance (mobility) Z_S is much lower than the input impedance (mobility) Z_L of the following link, the effect of loading can be neglected and $U \approx E$. For instance, the output impedance of an operational amplifier is typically much lower than the load impedance due to the application of voltage feedback.

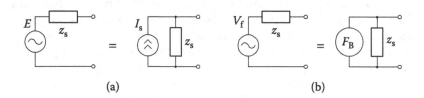

(a)   (b)

FIGURE 7.7
Thevenin and Norton representations of (a) electrical and (b) mechanical signal sources (for the voltage-to-velocity analogy).

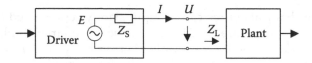

FIGURE 7.8
Variables at the links' joint.

7.2.3 Loading Diagram

The voltage across the terminals of the signal source with resistive load shown in Figure 7.9(a) is:

$$U = E - R_S I \qquad (7.6)$$

For a rotary mechanical actuator (motor), similarly, the angular velocity is:

$$\Omega = \Omega_f - R_S \tau \qquad (7.7)$$

where Ω_f is the *free-run* (no-load) relative angular velocity. When τ equals the *brake torque* (or *stall torque*) $\tau_b = \Omega_f / R_S$, the angular velocity becomes zero.

The *loading diagram* shown in Figure 7.9(b) reflects Equation 7.6. The resistance R_S determines the slope of the *loading line* $I = (E - U)/R_S$. The smaller the resistance, the steeper the line. The output power UI varies with the load resistance. It becomes zero when R_L is very large, and therefore the current is small, and also when R_L is very small, so that the output voltage is small.

Figure 7.9(c) shows the loading diagram (1) for a permanent magnet motor described by linear equations (we do not call the motor linear since the term *linear motor* is reserved for translational motion motors). Curve (2) exemplifies a nonlinear loading diagram for a motor with a flux winding. The slope of this curve defines the differential output resistance of the motor. The power, i.e., the product of the angular velocity and the torque, is defined similarly to the power in Figure 7.9(b).

The loading curve can be changed by application of feedback about the motor, although, certainly, the output power cannot be increased. Large torque feedback maintains the torque constant, as shown by curve (3). The torque does not depend on the angular velocity, and therefore does not depend on the load, within the range of the motor torque and velocity capabilities. The output mobility of the motor with torque feedback is large, as can be seen from the slope of the loading line. The loading line for a motor with velocity feedback is shown by curve (4); the output mobility of the motor is very low.

Output impedance (mobility) is an important parameter of drivers and motors, and its implications will be discussed in the next section.

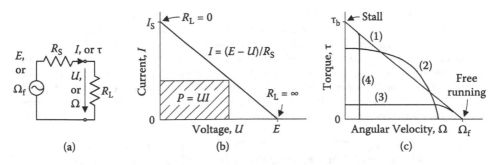

FIGURE 7.9
Schematic diagram (a) and loading diagrams for (b) an electrical linear signal source and (c) for motors: (1) permanent magnet motor, (2) inductive motor, (3) motor with torque feedback, and (4) motor with velocity feedback.

7.3 Effect of the Plant and Actuator Impedances on the Plant Transfer Function Uncertainty

Plant parameter uncertainty affects the accuracy of the closed-loop transfer function in two ways. The first is directly: the variations in the closed-loop transfer function are the feedback $|F|$ times smaller than the plant transfer function variations. The second is indirectly: the smaller the plant parameter variations, the larger is the feedback that can be implemented—as was discussed in Chapter 4 and will be further discussed in Chapter 9. In the hypothetical case that the plant is perfectly known (the so-called full-state feedback described in Chapter 8); infinite feedback is available. When, on the other hand, the plant is completely unpredictable (as in the extreme case described in *Alice in Wonderland*: "Yet you balanced an eel on the end of your nose"), no feedback control can be implemented. Reducing the plant transfer function uncertainty is an issue of high priority in feedback-system design.

As will be shown in the examples below, the plant transfer function uncertainty depends to a large extent on the value of the actuator output mobility. Therefore, an actuator with an appropriate output mobility must be used.

Example 7.5

Consider an actuator driving a plant whose mass M is uncertain. The control system output variable (controlled variable) is the plant velocity. Let's consider two extreme cases of the output mobility of the actuator shown in Figure 7.10(a) and (b), respectively. When this mobility is much larger than the load mobility, the actuator can be viewed as a force source, and when the actuator's mobility is much smaller than the load mobility, the actuator can be viewed as a velocity source.

When the actuator is a force source, the plant acceleration F/M and the plant velocity are uncertain due to the uncertainty of M. In contrast, when the actuator is a velocity source, the plant velocity equals the actuator velocity, and the plant transfer function is simply 1.

The actuator output mobility can be modified by the application of local feedback about the actuator.

Example 7.6

Consider the plant depicted in Figure 7.11. (The rotational version of this system is the problem of pointing a camera mounted on the end of a flexible boom attached to a spacecraft.) The actuator is placed between the second body and the spring, and the position of the second body is controlled. If the actuator is a force source, i.e., its output impedance is very large, the suspension resonance does not appear on the plant transfer function response: the second body only sees the force F, and this force does not depend on the spring parameters. On the other hand a velocity source actuator results in a profound resonance on the control loop response.

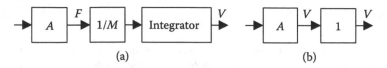

(a)   (b)

FIGURE 7.10
Actuators with (a) constant force and (b) constant velocity driving a rigid body plant.

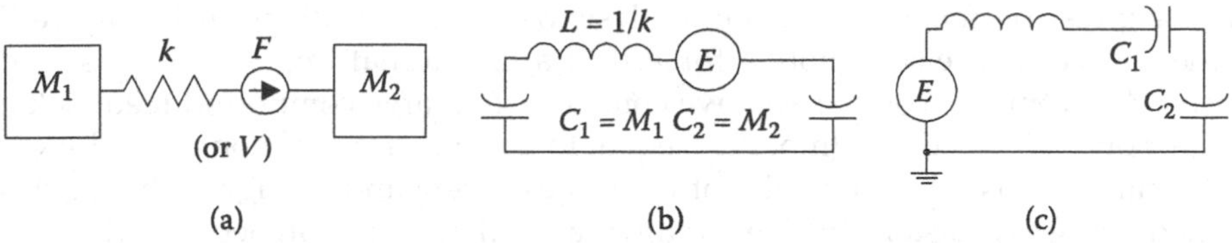

FIGURE 7.11
(a) Actuator and mechanical plant, (b) its equivalent electrical circuit diagram, and (c) its modification.

Notice that in the electrical circuit diagram, the order of the elements connected in series can be changed. Diagram (b) can be transformed into form (c), where one of the terminals of the source E is grounded and the capacitor C_1 is *floating* (not connected to the ground). Similar modifications of mechanical diagrams are not as evident. Conversion from a mechanical to the equivalent electrical circuit might therefore simplify the analysis.

While choosing the actuator output impedance, attention also must be paid to reduction of disturbance transmission from the base to the object of control, if this is a problem.

7.4 Effect of Feedback on the Impedance (Mobility)

7.4.1 Large Feedback with Velocity and Force Sensors

Examples already considered showed that feedback has a profound effect on impedance. The output impedance (mobility) is defined by the type of the sensor used. What is measured by the sensor is stabilized and made equal to the command or reference. Hence, if the sensor measures the output *voltage*, the voltage is stabilized, and the feedback-system output is a voltage source, which means its internal (output) impedance is 0 (certainly, not exactly 0, but very small). If a *current* sensor is employed, the feedback-system output represents a current source; i.e., the output impedance is high. If force feedback is employed, a force source is obtained. If velocity feedback is employed, a velocity source is obtained.

If both force and velocity sensors are used, and a linear combination of their readings is fed back, the feedback is called *compound*. Compound feedback is employed in most biological and many robotic motion control systems. With compound feedback, the output mobility of the control system at the joint of the actuator and the load can be made as desired, and damping can be introduced, which makes smoother the interaction between various feedback loops.

A feedback loop about an actuator is depicted in Figure 7.12. Two sensors are employed, one measuring the force and the other measuring the velocity. The outputs of the sensors are combined and fed back. A link Z is included in the force sensor path. The output of the link Z must have the dimension of velocity so that it can be summed with V. Therefore, the dimension of Z is mobility.

Since the velocity $V = FZ_L$, the return signal is $BF(Z_L + Z)$. If the feedback is large and therefore the error is small, the return signal equals the command U. From here,

$$F = U / [B(Z_L + Z)] \tag{7.8}$$

The same expression follows from the analysis of the circuit diagram in Figure 7.14(b), where the source with electromotive force U/B and internal impedance Z is loaded at impedance Z_L. From this comparison, we conclude that large compound feedback makes the actuator output impedance (mobility) equal to the transfer function of the link Z.

In electrical systems, a linear combination of the current and voltage at the junction can be obtained using Wheatstone bridge circuitry, as will be shown in Section 7.4.5.

7.4.2 Blackman's Formula

A two-wire connection where the current in one (direct) wire equals the negative of the current in the second (return) wire is called a *port*. The system shown in Figure 7.13 comprises a single-directional two-port network and a three-port linear system. Z_L is the impedance of an external load that can be connected to the port with the output impedance Z.

The feedback return ratio in the system certainly depends somehow on the impedance Z_L (examples will be provided later) so that $T \equiv T(Z_L)$. By this definition, when $Z_L = 0$, the return ratio is $T(0)$, and when $Z_L = \infty$, the return ratio is $T(\infty)$.

The input impedance of a feedback system can be calculated using *Blackman's formula*:

$$Z = Z_0 \frac{T(0)+1}{T(\infty)+1} \tag{7.9}$$

where Z_0 is the impedance without the feedback, i.e., with the feedback loop disconnected. (The proofs are given in Appendices 6 and 7.) The formula expresses the driving point impedance via three quantities that are much simpler to estimate and calculate: the impedance without the feedback, and the feedback in the cases when the system is simplified by setting the load impedance to 0 or ∞.

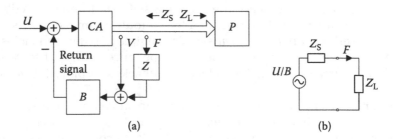

FIGURE 7.12
(a) Feeding back a linear combination of sensor readings and (b) an equivalent circuit diagram for the junction.

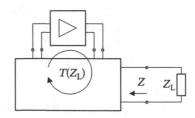

FIGURE 7.13
Driving point impedance of a feedback system.

7.4.3 Parallel Feedback

A *parallel feedback* block diagram is shown in Figure 7.14(a). Here, the output of the forward path, the input of the feedback path, and the load are all connected in parallel. The feedback signal is therefore proportional to the voltage across the load. This is why parallel feedback is also called voltage feedback.

In a parallel feedback circuit, reducing the load impedance (mobility) to 0 disconnects the feedback loop and makes $T(0) = 0$, and Blackman's formula reduces to:

$$Z = \frac{Z_0}{T(\infty)+1} \tag{7.10}$$

Hence, negative parallel feedback reduces the impedance. Large parallel feedback can be used to make nearly perfect voltage sources, as the one described in Chapter 1 and depicted in Figure 1.5.

The mechanical analog to parallel feedback is rate feedback. Such feedback is shown in Figure 7.14(b). It uses as a sensor a tachometer motor winding, an optical encoder, or some other velocity sensor. The feedback makes the actuator a velocity source, the velocity proportional to the input signal and independent of the loading conditions. Velocity feedback can be employed, for example, in the motor that propagates the tape in a tape recorder. The feedback maintains the tape velocity constant and independent of the tape tension.

Example 7.7

A driver amplifier with high-input and -output impedances has transconductance Y (i.e., when voltage V is applied to the input, the output current is YV). The output terminal is connected to the input terminal via feedback of two-poles with impedance Z_B. The signal source impedance is Z_S. The feedback path transfer function, from the output

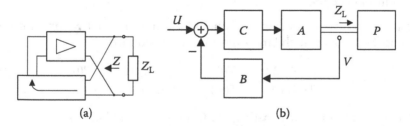

FIGURE 7.14
Parallel feedback: (a) electrical and (b) mechanical.

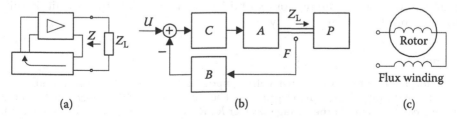

FIGURE 7.15
Series feedback: (a) electrical, (b) mechanical, and (c) series motor.

current to the input voltage, is YZ_S (when no load is connected to the output). Then, the output impedance of the amplifier is:

$$Z = Z_o / [T(\infty) + 1] = (Z_B + Z_S) / (YZ_S + 1)$$

7.4.4 Series Feedback

A *series feedback* diagram is shown in Figure 7.15(a). Here, three ports are connected in series: the load, the output of the forward path, and the input of the feedback path, so that the same current flows through these three ports. The feedback signal is proportional to this current, i.e., to the load current. This is why series feedback is also called current feedback. In this circuit, increasing the load impedance to ∞ disconnects the feedback loop and makes $T(\infty) = 0$. Then, Blackman's formula reduces to:

$$Z = Z_o[T(0) + 1] \qquad (7.11)$$

When series feedback is negative, it increases the impedance. Nearly perfect current sources can be made using large series feedback. In mechanical systems, a force or a torque source can be made using series feedback from a load cell or another kind of force or torque sensor. Torque sources are commonly used in pulleys, providing proper constant tension in the cable.

In series electrical motors, the flux winding is connected in series to the rotor as shown in Figure 7.15(c). The effect of the flux winding can be described by the following internal feedback mechanism. Increasing the load, i.e., increasing the torque, increases the rotor current, and therefore the flux winding current. This increases the voltage drop on the flux winding, and therefore reduces the voltage on the rotor winding, thus maintaining the torque nearly constant and independent of the load.

7.4.5 Compound Feedback

For *compound feedback* mentioned in Section 7.4.1, both $T(0)$ and $T(\infty)$ are nonzero. When both $|T(0)|$ and $|T(\infty)|$ are much larger than 1, the impedance

$$Z = Z_o \frac{T(0) + 1}{T(\infty) + 1} \approx Z_o \frac{T(0)}{T(\infty)} = Z_o \frac{CAPB(0)}{CAPB(\infty)} = Z_o \frac{B(0)}{B(\infty)} \qquad (7.12)$$

does not depend on the forward path gain. Compound feedback differs in this respect from series feedback and parallel feedback.

Compound feedback stabilizes neither the voltage nor current at the system's output, but the system's output impedance and the emf (and therefore the short circuit current). In mechanical rotational systems, compound feedback stabilizes the output mobility, the brake torque, and the free-run velocity.

Example 7.8

A circuit diagram for an amplifier with compound feedback at the output and series feedback at the input is shown in Figure 7.16. The output voltage is connected to the input of the feedback path via the voltage divider R_1, R_2. The current is sensed by the small series resistor R_3. The total signal at the input to the B-circuit is approximately $U_L R_2 / (R_1 + R_2) + I_L R_3$. Therefore, when the feedback is large, according to the rule developed when we were considering the system in Figure 7.12, or directly from Equation 7.11, the output impedance is $R_3(R_1+R_2)/R_2$.

The circuit in Figure 7.16 is often referred to as an implementation of *bridge feedback* since resistors R_1, R_2, and R_3 and the output impedance of the amplifier Z_A constitute a Wheatstone bridge. The load impedance is connected to one diagonal of the bridge, and the B-circuit input to another diagonal. In high-frequency amplifiers, bridge feedback with Wheatstone or transformer bridges is often employed to make the output impedance equal to 75 Ω or 50 Ω to match a cable, a filter, or an antenna.

Example 7.9

When compound feedback employs angular velocity and torque sensors, the output of a servo motor imitates a damper of desired value. In this way a flexible plant driven by this motor can be damped and the control accuracy improved. The drawback of this method is the need to use velocity and torque sensors that may be relatively expensive. Using only a rate sensor with resulting rate feedback and a driver amplifier with an appropriate output impedance is commonly sufficient for most practical applications.

Example 7.10

When power losses in a motor are small, the motor output mobility is proportional to the output electrical impedance of the driver amplifier, as will be shown in Section 7.6.2. Then, to obtain the required output mobility of the motor, it is sufficient to implement compound feedback in the driver amplifier, which only requires several resistors. The smaller the motor winding resistance, the better this method works.

Blackman's formula allows calculation of the driving point impedance at any specified port. Until now, we only considered the output port of the actuator, which is the most important for the control system designer. However, it can be any other port, for example, the input port of a feedback amplifier. While the impedance at a specified port is being calculated, Z_o is the impedance without feedback at this port, and $T(0)$ and $T(\infty)$ are the return ratios with, respectively, this port open or shorted.

7.5 Effect of Load Impedance on Feedback

We have already seen that the load impedance affects the feedback in the cases of series and parallel feedback. Generally, the return ratio can be expressed as a linear fractional function of Z_L [33] (see also Appendix 8):

$$T(Z_L) = \frac{Z_o T(0) + Z_L T(\infty)}{Z_o + Z_L} \tag{7.13}$$

It is easy to check that when $Z_L = 0$, then from Equation 7.13, $T(Z_L) = T(0)$, and when $Z_L = \infty$, then $T(Z_L) = T(\infty)$.

The feedback dependence on the load makes it more difficult to design systems where the load impedance is uncertain or could vary, especially when the plant is flexible and the resonance frequencies are not well known. It is preferable in this case to have the feedback independent of the load. The general condition for this is:

$$Z = Z_o \tag{7.14}$$

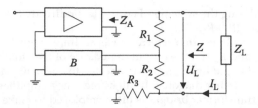

FIGURE 7.16
Resistive compound feedback at the output of an amplifier.

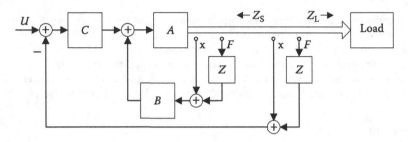

FIGURE 7.17
Balanced bridge feedback.

since in this case, according to Blackman's formula, $T(0) = T(\infty)$, and therefore from Equation 7.13, $T(Z_L) = T(0) = T(\infty)$. In order to implement this condition, a local loop about the actuator can be made. This loop provides the desired value of the output impedance of the actuator, this impedance (mobility) being Z_o for the main loop over the plant, as shown in Figure 7.17. Such a combination of local and main feedback is called *balanced bridge feedback*.

7.6 Flowchart for Chain Connection of Bidirectional Two-Ports

7.6.1 Chain Connection of Two-Ports

When two-ports are bidirectional, such as the electrical motors that can also be used as dynamos, or transformers and gears, the load for a link depends on the load for the following link. For example, the input impedance of an electrical motor is affected by the input mobility of the gear connected to the motor, and the gear input mobility is affected by its load.

General linear two-ports are described by a pair of linear equations relating their two input variables and two output variables, for example, the following equations:

$$I_2 = aI_1 + dU_2$$

$$U_1 = cI_1 + bU_2$$

(7.15)

Equation 7.15 can be represented by the block diagram in Figure 7.18. The inside of the diagram for each of the bidirectional two-ports is represented by four unidirectional links.

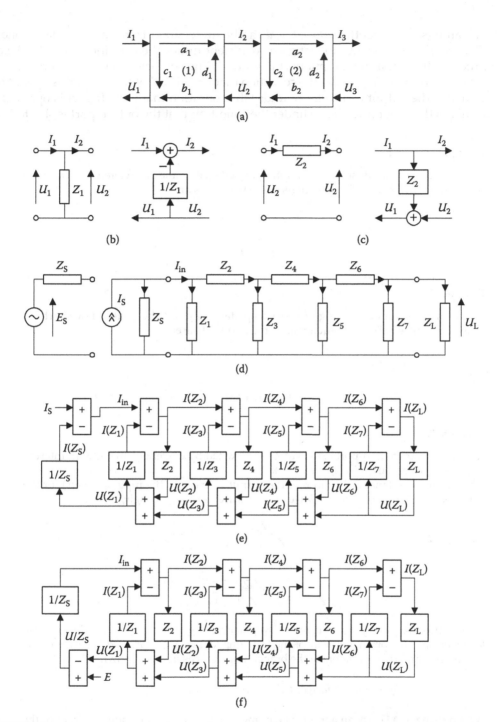

FIGURE 7.18
Flowcharts of (a) a cascade connection of two-ports, (b) a parallel impedance of two-ports, (c) a series impedance of two-ports, (d) a ladder network, and (e, f) block diagrams for the ladder network with current and voltage signal sources.

The meanings of the coefficients a, b, c, d can be understood from the boundary conditions. The coefficient a is the current gain coefficient under the condition that $U_2 = 0$, i.e., under the condition that the output port is shorted. The coefficient b is the reverse-direction voltage gain coefficient under the condition that $I_1 = 0$, i.e., that the left port is open. The coefficient d is the output conductance under the condition that the left port is open. The coefficient c is the input impedance under the condition that the output port is shorted.

Example 7.11

Electrical two-ports consisting of shunting impedance Z_1 and the related flowchart are shown in Figure 7.18(b). The equations of the two-ports are:

$$I_2 = I_1 - (1/Z_1)U_2$$

and

$$U_1 = U_2$$

Example 7.12

Electrical two-ports consisting of series impedance Z_2 and the related flowchart are shown in Figure 7.18(c). The equations of the two-ports are:

$$I_2 = I_1$$

and

$$U_1 = Z_2 I_1 + U_2$$

Example 7.13

A ladder electrical two-port loaded at Z_L and excited by current source I_S with internal impedance Z_S is shown in Figure 7.18(d).

The transimpedances (the ratios of the output voltages to the input currents) for the networks consisting of three and five branches are, correspondingly,

$$\frac{Z_1 Z_3}{Z_1 + Z_2 + Z_3}$$

and

$$\frac{Z_1 Z_3 Z_5}{(Z_1 + Z_2)(Z_3 + Z_4 + Z_5) + Z_3 Z_5}$$

For the most frequently encountered network consisting of three branches, the numerator *numtr* and denominator *dentr* of the transimpedance are expressed in terms of the numerators and denominators of the branch impedances n_i, d_i, as shown below:

$$numtr = n_1 n_3 d_1 d_2 d_3, \; dentr = n_2 n_3 + n_1 n_3 + n_1 n_2$$

(When using MATLAB, an appropriate number of zeros must be added in front of the vectors to be added.)

The flowchart of the network can be obtained by cascading flowcharts for parallel and series branches. The resulting block diagram is shown in Figure 7.18(e); here, $I(Z_n)$ is the current flowing through impedance Z_n and $U(Z_n)$, the voltage across impedance Z_n. As is seen, the summers in the upper row represent Kirchhoff's law for the currents in the nodes, and the summers in the lower row represent Kirchhoff's law for the voltages about the elementary contours. Bode diagrams of the flowchart transfer functions can be obtained with Simulink using the commands described in Section 7.7.

When the numerator order n of a link transfer function is higher than the order of the denominator, Simulink might not be able to find the solution. In this case, as^n can be added to the denominator with a sufficiently small coefficient a.

With current and voltage summers, any circuit (not only the ladder) can be modeled with a flowchart and modified, if required, by changing only a few links.

Example 7.14

In the previous example, the signal source is equivalently replaced by a source consisting of emf $E = I_SZ_S$ in series with impedance Z_S. The resulting block diagram is shown in Figure 7.18(f). A further example is given in Appendix 13.12.

7.6.2 DC Motors

An electrical dc motor is an electromechanical transducer. The current I and the voltage U characterize the electrical side of the motor model. The generated torque τ is applied to the mechanical load, which is rotated with angular velocity Ω.

In application to the motor, Equation 7.15 can be rewritten as follows:

$$\tau = kI - \Omega / Z_m$$

$$U = rI + k\Omega \tag{7.16}$$

and reflected in the flowchart in Figure 7.19. Here, k is the electromechanical conversion coefficient, i.e., the ratio of the break torque (i.e., the torque when the velocity $\Omega = 0$) to the applied current. As follows from the first equation, Z_m is the motor output mobility measured while the electrical winding is open, i.e., when the current is 0. This mobility reflects the viscous friction in the bearings and the moment of inertia of the rotor, $Z_m = 1/B + 1/(Js)$.

Z_L is the load mobility. The transfer function from the output of the link k in the forward path to the input of the second link k can be found using Equation 1.3 as:

$$\frac{Z_L}{1 + Z_L Z_m} = \frac{Z_L Z_m}{Z_L + Z_m}$$

This is the parallel connection of the load and motor mobilities.

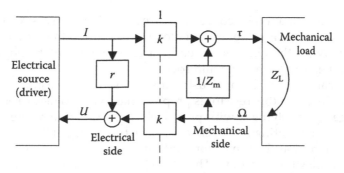

FIGURE 7.19
Electrical motor flowchart.

The input voltage U equals the sum of the voltage drop rI on the winding resistance r and the *back electromotive force* $k\Omega$. (The reactive component of the winding impedance is most frequently neglected since it is typically much smaller than the resistance, but in some cases this reactance must also be accounted for.)

If the mechanical losses, the rotor inertia, and the winding resistance are neglected, then $\tau = kI$ and $U = k\Omega$ so that $\tau\Omega = IU$; i.e., the power is converted from electrical to mechanical forms without losses.

In most low-power feedback control applications, brushless motors are used with a permanent magnet rotor. The stator windings (phases) are switched in accordance with the information about the angle position of the rotor obtained from position sensors based on the Hall effect or from optical encoders. The encoder consists of a disk on the motor shaft with specific patterns printed on it and several photosensors placed on the stator and separated by specific angles to read the printed information. The encoders can also serve as angle sensors for the feedback path. The sensors are described in more detail in Section 7.9.1.

A permanent magnet motor can be driven by sinusoidal ac current generated by a dc-to-ac inverter. The phase and the frequency of the ac current are controlled by a separate feedback loop, using either a rotor angle sensor or the information about the rotor angle extracted from the currents and voltages in the motor windings.

The periodic dependence of the coefficient k on the angle of rotation causes periodic variations in the motor torque. To model this effect in the system's block diagram, a parallel branch can be added to the k path. This branch includes a multiplier to whose second input the shaft angle is applied via the function $ak \sin n\phi$, where a is the relative amplitude of the torque variations, ϕ is the shaft angle, and n the number of torque ripples per rotation. A similar branch added in parallel with the path $1/Z_m$ in Figure 7.19 can be used to model the holding torque in step motors.

Higher-power dc motors have both rotor and stator windings. The stator winding is often referred to as a flux winding. The motor can be controlled by varying the current in either winding or in both windings.

7.6.3 Motor Output Mobility

Using the flowchart in Figure 7.19, the motor output mobility $Z_{\text{out mot}}$ can be calculated as the inverse of the transmission from Ω to τ. The link $1/Z_d$, which is the inverse of the output impedance of the driver amplifier, should be connected from U to I. The mobility is:

$$Z_{\text{out mot}} = (Z_d + r) / k^2 \tag{7.17}$$

(The derivation of this equation is requested in Problem 21.) When r is relatively small and a voltage driver is used, the actuator output mobility is low and the actuator approximates a velocity source. When the plant is driven by a velocity source, the angular velocity is constant and independent of the load and friction.

The effect of r can be compensated using a small sensing resistor in series with the motor. The voltage drop on this resistor is proportional to the voltage drop on r. By amplifying this voltage and applying it with proper phase to the input of the driver amplifier (i.e., by making a compensating feedback loop), an extra voltage at the output of the driver amplifier is created that has the same amplitude and opposite phase than the voltage drop on r, thus compensating the effect of r. The sensing resistor should be temperature dependent, or some additional circuitry should be added to compensate for the temperature dependence of the winding resistance.

The same circuit can be analyzed and designed with Blackman's formula by creating a driver amplifier with the output impedance equal to $-r$, thus making $Z_d + r = 0$ and $Z_{\text{out mot}} = 0$.

The feedback bandwidth in the compensating loop is limited by the winding inductance. The system stability can be analyzed with the multi-loop Bode–Nyquist criterion, or with the Bode–Nyquist criterion for connections of two-poles [6].

7.6.4 Piezoelements

Piezoelement actuators can be analyzed using the following equations:

$$F = aU + dV$$
$$I = cU + bV$$
(7.18)

reflected in the flowchart shown in Figure 7.20. The coefficient a is measured under the condition of zero output velocity, i.e., when the output is clamped. It is the ratio of the clamped force to the incident voltage. The coefficient b is the reverse-direction transmission coefficient from V to I while the input port is shorted. The coefficient d is the inverse of the output mobility under the condition that the input port is shorted. The coefficient c is the input port electrical conductance under the condition that the output port is clamped.

Both the electromagnetic actuators and the piezoactuators possess some hysteresis due to the ferromagnetic and piezoelectric material properties. Most often, the hysteresis is small and can be neglected. If not, it can be modeled by introducing hysteresis links (described in Chapter 11) into the elementary links in the diagrams in Figures 7.19 and 7.20.

7.6.5 Drivers, Transformers, and Gears

The input impedances of driver amplifiers are typically high; i.e., the amplifiers are *voltage controlled*. The amplifiers are characterized by their transconductance $Y_T = I_S/U_1$ (measured with zero impedance load) or their voltage gain coefficient $K = E/U_1$ (measured with no load); here, U_1 is the voltage at the driver's input, and U_2 is the voltage at the driver's output. Evidently, $K = Y_T Z_d$, where Z_d is the output impedance of the driver. The driver amplifier flowcharts are shown in Figure 7.21.

Example 7.15

In Figure 7.21(b), the emf at the output of the driver $E = U_1 K$. The motor angular velocity can be calculated from the equations obtained by cascading the flowcharts for the driver

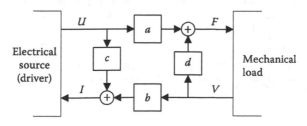

FIGURE 7.20
Piezoactuator flowchart.

and the motor in Figure 7.19. With the motor mobility neglected or included into the load, the angular velocity Ω and its sensitivity to variations in the load mobility Z_L are:

$$\Omega = \frac{\dfrac{E}{k}}{1 + \dfrac{Z_d + r}{Z_L k^2}}$$

and

$$S_\Omega = \frac{\dfrac{d\Omega}{\Omega}}{\dfrac{dZ_L}{Z_L}} = \frac{1}{1 + \dfrac{Z_L k^2}{Z_d + r}}$$

It is commonly desired that the velocity in velocity-controlled systems be less dependent on the load mobility Z_L (which includes the uncertain friction and load dynamics); i.e., $|S|$ needs to be small. The sensitivity depends on Z_d. Therefore, it is important to choose and implement Z_d properly. The sensitivity becomes small when $|Z_d + r|$ is small (recall Example 7.5, the description of the effect of r in Section 7.6.3, and the effect of the actuator impedance described in Section 7.5).

A flowchart for an electrical transformer with the turn ratio n is shown in Figure 7.22(a), and a flowchart for a mechanical gear with the velocity ratio n in Figure 7.22(b). With the resistance of the primary winding being r_1 and of the secondary, r_2, the total equivalent resistance of the primary is $r = r_1 + r_2/n^2$. For a mechanical gear box, the equivalent viscous friction coefficient for the motor output motion is $B = B_1 + B_2/n^2$.

The composite flowchart of a driver and a motor with an attached gear box is a cascade connection of the three corresponding flowcharts.

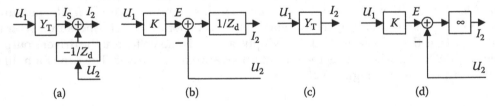

(a)   (b)   (c)   (d)

FIGURE 7.21
Flowcharts for (a, b) a driver amplifier with output impedance Z_d, (c) a current driver, and (d) a voltage driver.

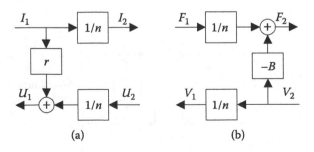

(a)   (b)

FIGURE 7.22
Flowcharts for (a) an electrical 1:n transformer and (b) a mechanical gear.

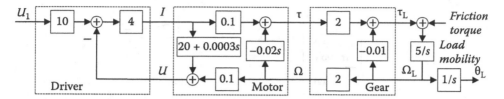

FIGURE 7.23
Block diagram of cascade connection of driver, motor, gear, and load.

Example 7.16

The driver has a voltage gain coefficient of 10 and an output impedance of 0.25 Ω. The motor winding impedance is $20 + 0.0003s$ Ω, and the motor constant is 0.1 Nm/A. The motor rotor's moment of inertia is 0.02 Nm². The gear amplifies the motor torque twice, which is the gear ratio, the load angle to the motor angle (or the load angular velocity to the motor angular velocity) is 1:2. The losses in the gear and bearings make $B = 0.01$ Nm/(rad/s). The load's moment of inertia is 0.2 Nm²; i.e., the load mobility is $5/s$. The model for the driver, motor, gear, and load assembly shown in Figure 7.23 is a cascade connection of the models in Figures 7.21, 7.19, and 7.22, and the load.

Since the output impedance of the driver and the impedance of the motor winding are connected in series, the model can be simplified by changing the value of the link $1/0.25 = 4$ to $1/(20.25 + 0.003s)$ and eliminating the link $20 + 0.003s$.

Example 7.17

The actuator of the previous examples is applied to the plant that is not rigid: the torque (τ_L—*friction torque*) is applied to the body with moment of inertia J_1 (which reflects the inertia of the gear and the motor), and an antenna with moment of inertia J_3 is connected to the gear via a shaft with torsional stiffness coefficient k_2.

Therefore, the load mobility $5/s$ in Figure 7.23 must be replaced by the mobility Z, which is the input impedance of the equivalent electrical ladder Π-type network:

$$Z = \cfrac{1}{J_1 s + \cfrac{1}{\cfrac{s}{k_2} + \cfrac{1}{J_3 s}}} = \frac{\frac{1}{J_1}\left(s^2 + \frac{k_2}{J_3}\right)}{\left[s^2 + k_2\left(\frac{1}{J_1} + \frac{1}{J_3}\right)\right]s}$$

This mobility has two zeros, $\pm j\sqrt{k_2/J_3}$, and three poles: $\pm j\sqrt{k_2(J_1 + J_3)/J_1 J_3}$ and 0. The pair of purely imaginary poles will bring about a pair of complex poles in the transfer function θ_L / U_1, where θ_L is the angular velocity of the output of the gear. These poles will be substantially damped by the output mobility of the actuator.

7.6.6 Coulomb Friction

Coulomb friction is modeled by the block shown in Figure 7.24(a). The dependence of the friction force F_{Coulomb} on the velocity V is shown in Figure 7.24(b).

The friction model is commonly incorporated in the plant model as a feedback path shown in Figure 7.25. In this composite plant model, F is the force applied to the plant, and the difference between this force and the Coulomb friction force is applied to the plant

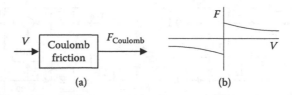

FIGURE 7.24
Coulomb friction model and characteristic.

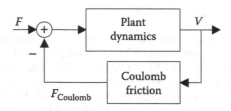

FIGURE 7.25
Model of a dynamic system with friction.

dynamics (the summer is also shown in Figure 7.23). When the actuator is not a pure force source, the inverse of the actuator output mobility can be included in the model as a feedback path in parallel with the Coulomb friction link.

In Figure 7.22, the model of a gear was shown with viscous friction B. This block can be replaced by a Coulomb friction link, or such a link can be added in parallel. The Coulomb friction link can be also placed parallel to the link $1/Z_m$ in Figure 7.19. It is seen that when Z_m is small, the link $1/Z_m$ will dominate and the effect of Coulomb friction will be negligible. It is therefore seen that the effect of Coulomb friction greatly depends on the actuator's output mobility.

Example 7.18

When a rigid body is dragged over another rigid body with a rough surface by a soft spring, Coulomb friction causes oscillation (with some rearrangement of coordinates, this is the case for the bowing of a violin string). In this case, the actuator's output impedance is large, and the actuator is a nearly pure force source. However, when the source is that of velocity, evidently no oscillation occurs.

7.7 Examples of System Modeling

A system model can be described either by the system elements and the topology of their connections (as in SPICE) or by mathematical equations (as in MATLAB and many other computer languages). The use of flowcharts simplifies the organization of the mathematical description of the system. The equations expressed in the flowcharts can be entered in the input file of a computer simulation program via a graphical interface (as in Simulink and some other control software packages).

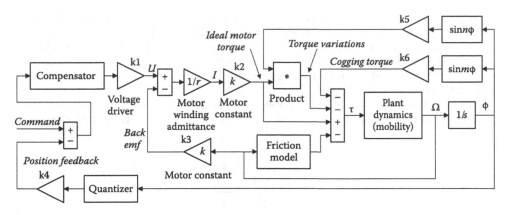

FIGURE 7.26
Block diagram of a control system.

Example 7.19

A Simulink-like model of a control system using a brushless dc permanent magnet motor is shown in Figure 7.26.

The system includes an input summer, a compensator with voltage output, a voltage driver as a voltage-controlled source of voltage U, a brushless permanent magnet dc motor with motor constant k, a plant with specified mobility, a friction model, a position (angle) feedback including a quantizer (since an optical encoder is used as the angle sensor), and the models of the motor torque variations and cogging as periodic dependencies on the output angle ϕ. The torque τ is applied to the plant, and the output angular velocity is Ω. The back electromotive force is subtracted from the driver's output voltage U and, divided by the motor winding resistance r, produces the winding current I. The torque τ is the ideal motor torque (current multiplied by the motor constant) from which the torque variations, cogging torque, and friction torque are subtracted. Simulink can provide the output time response to the input signal. The Simulink analysis tools (signal source, oscilloscopes, plotters, and multiplexers to provide data for the workspace) are not shown in Figure 7.26. Their use is described in the manuals.

A loop Bode diagram can be found as follows: disconnect the feedback path, attach inport 1 (from the Simulink "Connections" library) to the command input, outport 1 to the loop output, and type in the MATLAB command window:

```
[A,B,C,D]=linmod('file_name'); bode(A,B,C,D,1)
```

The closed-loop frequency response can be obtained by applying the same program to the system with the feedback path reconnected. The meaning of the matrices A, B, C, D is explained in Chapter 8. The Linear Analysis tool may also be used.

Example 7.20

The two methods of system description are illustrated below with the vibration isolation system shown in Figure 7.27. A voice coil actuator with a load cell sensor is placed between two flexible bodies with mobilities Z_1 and Z_2. The vibration source is on the second body, and the actions of the voice coil should reduce the vibrations of the first body. The load cell together with its amplifier has sensitivity 1 V/N. The voice coil is characterized by the coil resistance $r = 4$ Ω and the electromechanical coefficient $k = 3$ N/A. The coefficient's equivalent is a lossless down-transformer with the turn ratio of 3 (the ratio of the primary to the secondary windings).

Figure 7.28 shows the equivalent electrical schematic diagram. All currents and voltages to the left of the vertical dashed line represent physical electrical variables in amperes and volts; to the right of the line, they represent forces in newtons and velocities in m/s, correspondingly. The load cell is therefore represented by an amperometer, in this case, a current-controlled voltage source. The schematic diagram can now be coded into the input file of SPICE.

Figure 7.29 shows a flowchart description for the same system. It can be used, for example, with Simulink. Here, Z_{out} is the output impedance of the driver amplifier.

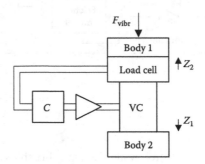

FIGURE 7.27
Vibration isolation system.

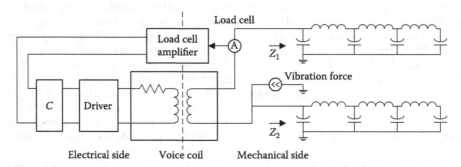

FIGURE 7.28
Equivalent schematic diagram for a vibration isolation system.

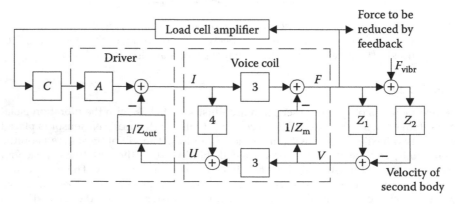

FIGURE 7.29
Flowchart description of a disturbance isolation feedback system.

The choice of the modeling method depends on the availability of simulation software, on the problem specifics, on the system links, and to a large extent, on the designer's personal preferences. If, for example, the driver amplifier is already designed and should not be changed, the compensator C is implemented in software, and the designer is a mechanical engineer, then probably flowchart simulation with Simulink will take less time than simulation with SPICE. If, however, the driver amplifier needs to be designed and optimized simultaneously with the compensator, the compensator is analog, and the designer is an electrical engineer, then probably the design and simulation is easier to perform using an equivalent electrical schematic diagram and SPICE.

7.8 Flexible Structures

7.8.1 Impedance (Mobility) of a Lossless System

Foster's theorem states that the zeros and poles of a driving point impedance (mobility) of a lossless system are purely imaginary and alternate on the $j\omega$-axis, as shown in Figure 7.30. At zero and at infinite frequencies, there could be either a pole or a zero.

Mobility frequency responses of nondissipative flexible plants are exemplified in Figures 4.38 and 7.31. The angle of the mobility is −90° where $|Z|$ is falling, and 90° where $|Z|$ is rising. The low-frequency asymptote and the high-frequency asymptote reflect $|Z|$ being either proportional or inversely proportional to frequency. It is seen that in Figure 7.31(a), the high-frequency asymptote is shifted from the low-frequency one. The low-frequency

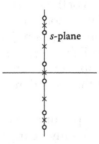

FIGURE 7.30
Poles and zeros of a lossless two-pole impedance.

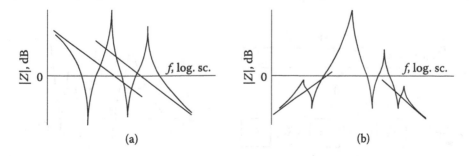

FIGURE 7.31
Driving point impedance (mobility) of a lossless system, logarithmic scale: (a) having a pole at zero frequency and a zero at infinite frequency, and (b) with zeros at zero frequency and infinite frequency, and a prominent suspension mode.

asymptote is determined by the rigid body mode, i.e., by the masses of all the bodies connected together by springs that are considered stiff at very low frequencies. The high-frequency asymptote is determined solely by the body to which the incident force is applied. All other bodies are disconnected from this body since the springs' mobilities become very large at high frequencies.

It can be calculated that each additional zero-pole pair lifts the high-frequency asymptote by the square of the ratio of the pole to the zero frequencies. This ratio commonly depends on the mass participation of the flexible mode. If, for example, an actuator drives a massive main body from which a small additional mass is suspended on a spring, then the pole-to-zero ratio of the flexible mode equals the square root of the ratio of the sum of the masses to the mass of the main body. The smaller the mass of the smaller body, the closer is the pole to the zero and the smaller is the shift in the asymptote.

Structural damping displaces the poles and zeros to the left of the $j\omega$-axis. As a result, the peaks and valleys on the frequency response of $|Z|$ become smoothed. Still, the losses in mechanical systems without special dampers can be so small that the peaks reach 40 and even 60 dB over the smoothed response of the mobility.

7.8.2 Lossless Distributed Structures

Distributed structures can be approximated by lumped element structures with a large number of elements, as shown in Figure 7.32(a) and (b).

Correspondingly, the impedance (mobility) of a distributed structure $Z(s)$ possesses an infinite number of poles and zeros, as illustrated in Figure 7.32(c).

These poles and zeros can be viewed as produced by interference of the incident signal and the signal reflected back from the far end of the structure. The frequencies of high-frequency resonances are very sensitive to small variations of the structure parameters, and therefore, in physical systems, are to a large extent uncertain. This uncertainty becomes a critical factor in limiting the bandwidth of the available feedback, as has been discussed in Chapter 4.

The *wave impedance (mobility)* is the input impedance (mobility) of the structure extended to infinity so that no signal reflects at the far end and returns back to the input. The wave impedance of a uniform lossless distributed structure is the resistance $\rho = 1 / \sqrt{kM}$, where k and M are the stiffness and the mass per unit length of the structure, respectively. The plot of $|Z|$ in logarithmic units is commonly nearly symmetrical about the wave impedance, as shown in Figure 7.32(c).

Matching at the far end means loading the structure at some resistor (or, if mechanical, at a dashpot) with the impedance (mobility) equal to the wave impedance (mobility) of the

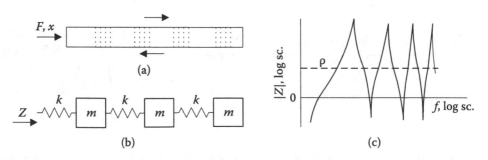

FIGURE 7.32
(a) Distributed structure. (b) Its equivalent representation with lumped parameters. (c) Its driving point mobility.

structure. In this case, we can think of the structure as extending to infinity, and the input impedance (mobility) of the structure equaling the wave impedance (mobility).

Matching between the signal source and the structure is provided when the signal source output mobility equals the wave mobility of the structure. Matching at either end of the structure fully damps it (since the resonances can be considered a result of interference of the waves reflected at the ends of the structure, and if at either end no reflection occurs due to matching, no resonances take place).

Example 7.21

A motor with motor constant k is employed to rotate a spacecraft solar panel having moment of inertia J to keep the panel perpendicular to the direction to the sun. The solar panel torsional quarter-wave resonance frequency is f_r. The spacecraft attitude control can be improved by damping this resonance. This can be achieved by making the motor output mobility approximately equal to the solar panel wave mobility ρ. The desired motor output mobility can be created by compound feedback in the driver.

We calculate the mobility using the voltage-to-velocity electromechanical analogy. In electrical transmission lines, the phase is $2\pi f x (lc)^{1/2}$, where x is the length and l and c are the inductance and capacitance per unit length. At the quarter-wave resonance, $2\pi f_r x(lc)^{1/2} = \pi/2$ wherefrom $(lc)^{1/2} = 4xf_r$. Then, the wave impedance is $\rho = (l/c)^{1/2} = 1/(4cxf_r)$. cx is the full capacitance of the line, which is analogous to the moment of inertia J. Hence, $\rho = 1/(4Jf_r)$ and the driver output mobility is $1/(4Jk^2f_r)$.

7.8.3 Collocated Control

Frequently, the actuator is a pure source of force (or torque or current), and the feedback sensor is connected to the same port of the plant but reads the variable that is related to the actuator variable by the plant driving point impedance (mobility). The plant transfer function is in this case the plant driving point impedance or admittance (mobility or the inverse of mobility). Such control is called *collocated*, as was already mentioned in Section 4.3.6.

The driving point impedance (mobility) of a passive plant is positive real (p.r.), and the phase angle of the function is constrained within the [–90°, 90°] interval. (Properties of p.r. functions are reviewed in Appendix 3.) This feature greatly simplifies the controller design and the provision of stability.

Example 7.22

In the mechanical arrangement shown in Figure 7.33(a), the actuator is a torque source and the sensor measures angular velocity $d\phi_1/dt$ (or a linear operator of the velocity, such as position or acceleration).

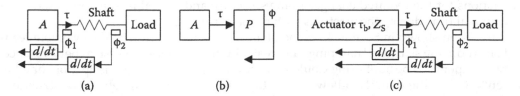

(a)   (b)   (c)

FIGURE 7.33
(a) Actuator is a torque source, and control is collocated with any of the sensors, ϕ_1 or ϕ_2, providing the fed-back signal. (b) Block diagram of collocated control. (c) The case is considered in Section 7.8.4; the output mobility of the actuator is finite; control is collocated when sensor ϕ_1 is used and is non-collocated when sensor ϕ_2 is used.

In the related block diagram shown in Figure 7.33(b), the plant transfer function $P = \phi/\tau$ is the mobility of the mechanical system.

Since the torque is the same before and after the shaft flexibility, the control remains collocated even when the sensor ϕ_2 is used. (The control will not be collocated if there will be an additional rigid body with substantial moment of inertia at the point of the torque application.)

7.8.4 Non-Collocated Control

Contrary to the driving point lossless mobility, transfer functions between different ports of a passive lossless structure may have several consecutive poles or zeros (sometimes referred to as *pole-zero flipping*), and the phase shift of these transfer functions is not constrained. As has been shown in Chapter 4, this phase shift reduces the available feedback. Right half-plane zeros may be present, usually real but less commonly in complex pairs. The attitude control of launch vehicles is typically non-collocated, with propellant sloshing and structural bending modes often resulting in n.p. transfer functions—refer to the Saturn V autopilot discussed in Example 4.20 and A13.9.

The sensor and the actuator cannot always be placed exactly at the same location. For example, for better control accuracy, the sensor might be placed closer to the load. If the plant were an ideal rigid body, or the actuator an ideal force (or torque) source, the control would still be collocated. Otherwise, as shown in Figure 7.33(c), the control is collocated when the first sensor is used and is not collocated; i.e., the actuator and plant transfer function is not generally p.r., when the second sensor is used. In this case, the control can be called collocated only as an approximation of reality, the approximation being valid only over a limited frequency bandwidth.

With respect to the flexibility of the shaft connecting the motor and the load, misunderstanding frequently arises between the designer of the mechanical structure and the control loop designer about the meaning of the statement: "Higher frequency structural modes are *much* higher than the frequency range of operation." The structural designer is mostly concerned with the structural soundness over the working frequency range and might underestimate the following factors: (1) the structural mode frequency is proportional to only the square root of the stiffness, and (2) the control feedback loop bandwidth must be *many times* wider than the functional band of the system (as has been shown in Chapters 4 and 5).

Trying to make the structure as inexpensive and as lightweight as possible, structural designers tend to design structures with structural modes falling within the feedback bandwidth, which prevents the control system from achieving high accuracy. This is why the control system designer/dynamicist needs to be involved in the design process before the mechanical structure design is completed, and s/he must be able to give their recommendations for the required changes in the structure and to evaluate the proposed solutions in real time during the meetings with the structural designer.

As has been shown in Section 4.3.6, the plant flexible modes restrict the available feedback. Introducing structural damping can greatly improve the control system performance.

The dampers can be built using Coulomb friction, hydraulic energy dissipation, or eddy currents. Using the dampers allows a substantial reduction of the effects of structural modes. The dampers are, however, relatively expensive, and to constrain the system cost, weight, and dimensions, the use of the dampers must be well justified by the available performance improvement.

7.9 Sensor Noise

7.9.1 Motion Sensors

7.9.1.1 Position and Angle Sensors

For position (or angle) variables' measurements, three categories of sensors are commonly employed: (1) position sensors, (2) rate (i.e., velocity) sensors, and (3) accelerometers.

Representative position and angle sensors are: potentiometers, linear variable differential transformers (LVDTs), resolvers, optical encoders, laser interferometers, and star trackers.

An electrical potentiometer with the tap moved mechanically by the plant can provide high resolution and a position reading accuracy and linearity of 0.1%. Less accurate potentiometers are universally used in small servos for radio-controlled toys, with resolution still better than 0.5°.

The LVDT shown in Figure 7.34(a) measures the position of the plant with respect to the base. It has three windings: the two symmetrical windings to which a signal from a generator is applied in opposite phases, and the winding providing the output voltage. The voltage is amplified and applied to a synchronous detector, the output of which is a product of this voltage and the generator voltage. The dc component of the detector's output is proportional to the third winding displacement from the central position.

The maximum stroke of common LVDTs is from 0.1 to 1 inch, with pretty good linearity, and the resolution of precision LVDTs can be 1 microinch. There also exist rotary versions of the device. Specialized ICs are available that incorporate all necessary electronics.

The *resolver* shown in Figure 7.34(b) is a rotating transformer. It has two stator windings and two rotor windings. To a pair of these windings an ac signal is applied in quadrature. From the signal induced on the two other windings, the angle of rotation can be determined with high accuracy.

The optical encoder was already briefly described in Section 7.6.2. The encoders can be *absolute*, with complicated patterns on the optical disk that at any time give full information about the shaft angle, or *incremental*, with simple patterns of alternating transparent and black lines. The incremental encoders must be accompanied by some electronics keeping track of the counts. *Quadrature* incremental encoders have two 90°-shifted readers. This improves the accuracy by a factor of 2 and also gives information about the direction of

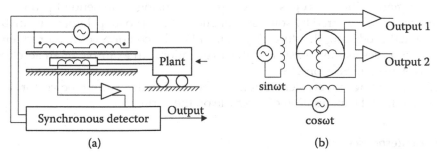

(a)　　　　　　　　　　　　　　(b)

FIGURE 7.34
Position sensors: (a) LVDT and (b) resolver.

rotation. Interpolation of the data from the readers additionally improves the angle reading accuracy by a factor of 2.

The *laser interferometer* compares the phases of the incident laser light with the light reflected from a mirror placed on the target, and counts interference fringes when the distance changes gradually. The interferometer can measure large distances with nanometer accuracy. Less accurate and less expensive interferometers use modulated light beams.

A *star tracker* is a small precision telescope equipped with an image recognition system.

7.9.1.2 Rate Sensors

The most often used rate (velocity) sensors are the *tachometer* and the *gyroscope*. The tachometer is a dynamo mounted on the same shaft as the motor (or there can be only a separate tachometer winding on the motor's rotor or stator; the motor windings in brushless motors that are disconnected from the driver during certain rotation angle intervals can be used as tachometer windings). The emf on the tachometer winding is proportional to the motor angular velocity. In contrast to the resolver, the tachometer is not able to detect the angle of rotation when the rotation rate is very low, since, in this case, the signal on the tachometer winding is below the noise level.

The *rate gyro* is a gyroscope with a position servo loop. An electrical winding generates a torque preserving the relative position of the gyro to the base. The current in the winding is the output signal of the gyro; this current is proportional to the base angular velocity (i.e., the rate of the base rotation). The rate signal is analog. A/D conversion simplifies the data interface with the rest of the system, but it loses some higher-frequency information. For these reasons, it is common to use both the analog and digital outputs from the gyro. *Integrating gyros* output angular increments.

7.9.1.3 Accelerometers

Accelerometers use electromagnetic, piezoelectric, piezoresistive, and tunnel effect devices.

In an electromagnetic accelerometer, the magnetic *proof mass* is suspended on a spring. Motion of the base relative to the proof mass produces electromotive force in a coil surrounding the proof mass, which is mechanically connected to the base. The proof mass position relative to the base is kept constant by a servo feedback loop that applies electromagnetic force to the proof mass. This force creates acceleration nearly equal to the base acceleration so the proof mass remains still relative to the base. The value of the compensating force in accordance to Newton's second law is the measure of the acceleration.

There exist many different types of accelerometers using a suspended proof mass, the position of which is measured by some sensor and, via a servo loop, kept constant by some force. The signal producing this force (e.g., current in the coil or a voltage-producing electrostatic force) serves as the measure of the acceleration. An example of an accelerometer control loop is given in Section 11.9.

A set of three orthogonal accelerometers can be used to determine the vector of gravity force, and therefore the inclination of a vehicle on the surface.

7.9.1.4 Noise Responses

Ideally, when the output data are postprocessed, motion sensors are interchangeable: suitable integrations and differentiations allow us to translate between positions, rates, and accelerations. The principal difference between them is their dynamic accuracy

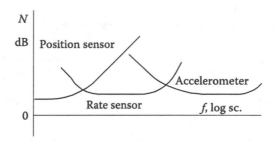

FIGURE 7.35
Frequency response of sensor noise normalized to the sensor input.

characteristics, or, with the frequency-domain characterization, their noise spectral density responses. The noise power and the mean square error can be calculated by integration of the spectral density with linear frequency scale over the bandwidth of interest.

Position sensors typically give an accurate steady-state value of the position, but their accuracy decreases when the position changes rapidly. For example, it takes considerable time to accumulate enough light to identify dim stars in a star tracker so that this sensor cannot react quickly to the rotation changes of the spacecraft on which the star tracker is mounted. On the other hand, gyros have *drift* (slow and gradual change in the reading caused by device imperfections), and therefore they do not determine the position accurately after some time passes from the initial setup. Gyros are worse in position determination than the star trackers at frequencies from 0 to 0.01 Hz, but they get better at higher frequencies. At even higher frequencies (say, from 15 Hz and up) the gyro noise increases and the accelerometers become superior.

The frequency responses of typical sensor noise spectral density converted to position data are shown in Figure 7.35.

7.9.2 Effect of Feedback on the Signal-to-Noise Ratio

The most common noise sources in control systems are those of the error amplifier (which is the first amplifier after the feedback summer), resistors, and sensors. The noise is commonly characterized by its spectral density.

A feedback loop that reduces both the signal source and the noise source effects does not change the ratio of the signal to the noise at the system output. Thus, it is possible and often convenient to examine the signal-to-noise ratio as if there were no feedback.

There is a caveat here: when the system performance is compared with and without feedback, the system should not be changed in any other aspect. Particularly, when the output effect of the sensor noise is calculated, the transmission coefficient from the sensor to the system output should remain unchanged. Therefore, the feedback loop should be disconnected between the system output and the sensor input, not between the sensor output and the feedback path. The difference is shown in Figure 7.36.

The loading for the disconnected ports must be preserved while disconnecting the feedback path for appropriate comparison of the system with and without feedback.

Example 7.23

Figure 7.37 shows an amplifier with the feedback path from the emitter of the outer stage to the emitter of the input stage. The feedback loop can be disconnected by setting $R_2 = \infty$. This reduces the signal-to-noise ratio—not because of change in the

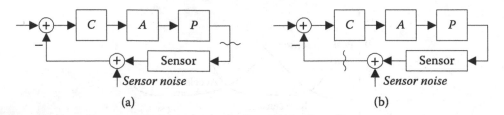

FIGURE 7.36
Disconnecting the feedback loop while (a) preserving the signal-to-noise ratio at the system output and (b) changing the ratio.

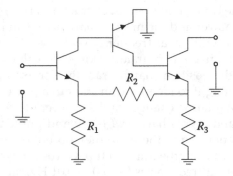

FIGURE 7.37
Amplifier with an emitter feedback path.

feedback, but because the resistance in the input contour increases, which reduces the signal and increases the noise. When the feedback is eliminated by setting $R_1 = 0$, the signal-to-noise ratio improves since in this case the resistance in the input contour decreases.

If, however, both disconnected ports of the feedback path are loaded onto loads equal to those that each port sees when the feedback path is closed, the signal-to-noise ratio remains the same with or without the feedback.

7.10 Mathematical Analogies to the Feedback System

7.10.1 Feedback-to-Parallel-Channel Analogy

The summer in Figure 7.38(a) implements the equation $E = U - TE$. From here it follows that $U = E + TE = (T+1)E = FE$, as re-diagrammed in Figure 7.38(b).

Equations 1.1–1.3 remain valid for this second system, which has no feedback and represents the parallel connection of two paths. When $|T+1| > 1$, introducing the channel T increases $|U|$ and reduces the ratio E/U. When $|T| \gg 1$, then $U \approx ET$ and $U_2/U \approx -1/B$. In this way, all the features of the feedback equations (Equations 1.1–1.3) are apparent.

This analogy can be employed to analyze or simulate responses of system (a) when this system is unstable but the system representation (b) is stable. We will use this analogy in Chapter 12.

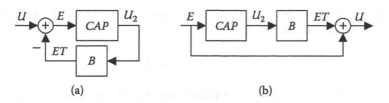

FIGURE 7.38
Analogy between (a) a feedback system and (b) a system with two parallel forward paths.

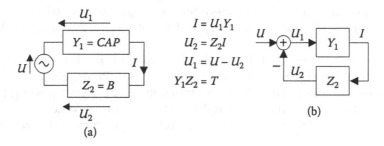

FIGURE 7.39
Analogy between (a) a two-pole connection and (b) a feedback system.

7.10.2 Feedback-to-Two-Pole-Connection Analogy

Equations 1.1–1.3 also describe the signals in the two-pole connections depicted in Figure 7.39(a). The transfer functions of the links in block diagram (b) follow Ohm's law: voltage U_1 applied to the first two-pole causes current I, and this current applied to the second two-pole produces voltage drop U_2. The contour equation $U_1 = U - U_2$ reflects the summer in the feedback loop.

The large feedback condition here, $|T| \gg 1$, is that $|Z_2| \gg |1/Y_1|$; i.e., the impedance of the second two-pole is much larger in magnitude than the impedance of the first two-pole. As a result, the second two-pole can be neglected when considering the current calculations: $I = U/(1/Y_1 + Z_2) \approx U/Z_2$. This is analogous to the closed-loop transmission of a system with large feedback where the output is the input divided by the feedback path transfer function.

The analogy can be employed to use the passivity condition of a network of electrical two-poles for stability analysis of feedback systems. We will use this analogy in Chapter 10.

7.11 Linear Time-Variable Systems

Linear time-variable (LTV) links are described by linear equations whose coefficients explicitly depend on time. When the signal applied to the input of an LTV link is sinusoidal, the output, contrary to the case of a linear time-invariant (LTI) link, is not necessarily sinusoidal, and might contain higher harmonics. When several sinusoidal components are applied to the link input, the output contains intermodulation products.

LTV links of digital compensators have already been analyzed in Section 5.10.7. In this section we will consider Mathieu's equation:

$$d^2y / dt^2 + (a + 2\varepsilon\cos(t))y = 0 \qquad (7.19)$$

which is representative of some LTV systems that might be encountered in practice. If $\varepsilon = 0$, this equation describes an LTI lossless resonator, the solution being a sinusoid with the angular frequency $\sqrt{a}$. The solution is on the boundary between self-oscillation and exponential decay. The time-variable coefficient $2\varepsilon\cos t$ changes the system behavior: some combinations of ε and a lead to solutions that are exponentially rising with time, and other combinations introduce damping into the system. The Ince-Strutt stability diagram shown in Figure 7.40 depicts the areas of stability and instability in the plane of the equation parameters.

A feedback system described by Mathieu's equation is shown in Figure 7.41(a). Intermodulation of the signal harmonics in the LTV link $2\varepsilon\cos t$ produces certain components at its output. Addition of these components to the signal passing through the LTI link a alters the phase of the signal at the summer's output. When the coefficient ε is large, the system is unstable with nearly all possible a, as seen from the stability diagram in Figure 7.40.

An electrical circuit equivalent to the feedback system is shown in the diagram in Figure 7.41(b), and an equivalent mechanical system in Figure 7.41(c). The time-variable term periodically changes the resonance frequency of the resonator. This pumps the energy, on average, over the length of the cycle, in or out of the resonator. For example, when the resonator capacitance is reduced while the charge is preserved, the voltage on the capacitor increases and the stored energy increases (the energy is proportional to the square of

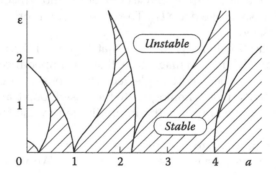

FIGURE 7.40
Ince-Strutt diagram.

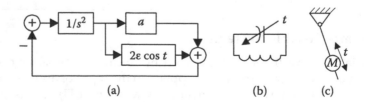

FIGURE 7.41
(a) Feedback system, and (b, c) resonators described by Mathieu's equation.

the voltage). The resonance frequency can be changed and the energy pumped into the pendulum resonator (c) by moving the center of mass of the swinging body up and down.

The phase shift for the passing signal in periodical LTV links depends on the phase of the incident signal. Therefore, while analyzing stability conditions, the worst possible case needs to be considered among all possible phases of the incident signal. This results in some uncertainty range in the link phase shift. However, the uncertainty of the loop transfer function reduces the potentially available feedback. Therefore, when controlling LTI plants, it is generally appropriate to use LTI compensators. When the compensators are LTV, as in the digital systems described in Chapter 5, or in the systems where the compensator parameters are varied in search for a maximum of a certain performance index, the available feedback is reduced.

To increase the feedback about an LTV plant, a controller can be chosen to be LTV in such a manner that the loop transfer function is less dependent on time (i.e., when the plant gain increases, the compensator gain decreases accordingly).

In adaptive systems (see Chapter 9), the compensators are LTV, but typically, the rate of the compensator variations in time is chosen much lower than the rate of changes in the critical part of the signals and the system dynamics (i.e., the dynamics that limit the value of the feedback and affect substantially the stability margins). In this case, the LTV link for the purpose of stability analysis can be considered LTI. Such links are called quasi-static.

For small-signal deviations from the current value, nonlinear links can be seen as LTV links. We will use this approximation in Chapter 12 for the analysis of process stability.

PROBLEMS

1. Why is it convenient to preserve power while choosing the type of electrome-chanical analogy?

2. How many independent variables are at the following:
   a. A ball joint
   b. A pin
   c. An x–y positioning system
   d. Sliding planes (e.g., one plate sliding arbitrarily on the surface of another plate)
   e. A two-wire junction of two electrical circuits
   f. A three-wire junction of two electrical circuits
   g. A four-wire junction of two electrical circuits

3. What are the equivalents of Ohm's law for translational and rotational mechanical systems? For thermal systems?

4. What are the equivalents of Kirchhoff's laws for mechanical systems (consider two analogies) and for heat transfer systems?

5. Draw the electrical analog circuit for the translational mechanical system in Figure 7.42(a), the rotational system in Figure 7.42(b) with torsional stiffnesses of the shafts k_1 and k_2, and for the thermal system in Figure 7.42(c).

6. Plot frequency responses of the following transfer functions:
   a. V_2/F for the system in Figure 7.42(a) for the case $M_1 = 100$, $k_1 = 2$, $M_2 = 5$
   b. V_2/F for the system in Figure 7.42(a) for the case $M_1 = 50$, $k_1 = 0.2$, $M_2 = 50$

c. Ω_3/t for the system in Figure 7.42(b) for the case $J_1 = 20$, $J_2 = 3$, $J_3 = 12$, $k_1 = 0.1$, $k_2 = 0.03$. Derive the function using Lagrange equations or equations corresponding to an equivalent electrical circuit, and use MATLAB or SPICE

d. Same as (c) for the case $J_1 = 10$, $J_2 = 24$, $J_3 = 2$, $k_1 = 0.01$, $k_2 = 0.02$

e. T_c/P for the system in Figure 7.42(c) for the case $R_T = 2.72$, $C = 100$

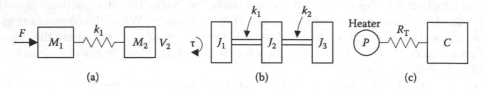

(a)   (b)   (c)

FIGURE 7.42
Examples of dynamic systems.

7. Draw an equivalent electrical circuit for cooling a power IC with a heat sink.

8. Using Equations 7.2–7.5, show that if $Z_L = 0$, then $I = I_S$, and when $Z_L = \infty$, $U = E$.

9. In Figure 7.9, the loading curve is expressed as $I = (E - U)/R_S$. Express I as a function of R_L.

10. The plant is a rigid body, $M = 50$ kg. The viscous friction coefficient is 0.01. The actuator output mobility is (a) 1(m/s)/N, (b) 5 (m/s)/N, or (c) 10 (m/s)/N. Use MATLAB to plot the frequency response of the actuator together with the plant. What is the plant transfer function uncertainty at 10 Hz?

11. The plant transfer function is the ratio of the output velocity to the force applied to a rigid body, $1/(sM)$, 20 kg < M < 50 kg. The actuator output mobility is (a) 1(m/s)/N, (b) 5(m/s)/N, or (c) 10(m/s)/N. Use MATLAB to plot the frequency response of the actuator with the plant. What is the response uncertainty?

12. The actuator is driving a rigid body, 20 kg < M < 50 kg, via a spring with stiffness coefficient $k = 1$. The actuator output mobility is (a) 1(m/s)/N, (b) 5 (m/s)/N, or (c) 10 (m/s)/N. Use MATLAB to plot the frequency response of the plant velocity to the input of the actuator, for cases of the maximum and minimum mass of the plant. Make a conclusion about the effect of the actuator impedance on the plant uncertainty.

13. Apply Blackman's formula to the calculation of the actuator output mobility in Figure 7.12.

14. Calculate the input and output impedances of the circuits diagrammed in Figure 7.43; the amplifier's voltage gain coefficient is 10,000, its input impedance is

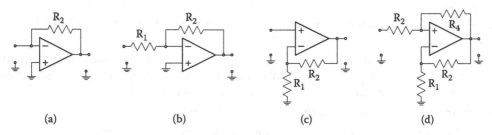

(a)   (b)   (c)   (d)

FIGURE 7.43
Examples of feedback amplifiers.

∞, and its output impedance is very low. For circuit (d), while calculating the input resistance, initially disregard R_3, calculate the impedance of the simplified circuit, and then simply add this resistance to the obtained result.

15. Determine the output mobility of the motor (without the main feedback loop about the plant) of the feedback systems shown in Figure 7.44. The driver's output impedance is small for voltage drivers and large for current drivers.

   Find the output mobility (at the load) for cases (c) and (d).

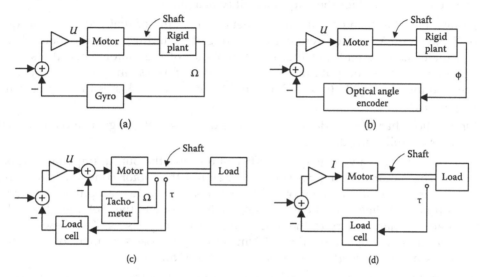

FIGURE 7.44
Examples of mechanical feedback systems.

16. The output impedance of the feedback amplifier shown in Figure 7.16 must be matched to the 50 Ω load. Find the resistors R_1, R_2, and R_3 such that signal losses in these resistors are not excessive, and at the same time, the attenuation of the circuitry in the direction of the feedback path is not too large (make an engineering judgment).

17. The velocity sensor gain coefficient is 1 V/(m/s), the force sensor gain coefficient is 3 V/N (the outputs of both sensors are in volts). The sensor outputs are combined to provide large compound feedback about the actuator. What is the output mobility of the actuator? What are the outputs of the sensors when the actuator output is:

   a. Clamped
   b. Unloaded

18. The output mobility of a translational actuator is 40 (m/s)/N. The return ratio is $100/(s+30)$ when only a force sensor is employed (i.e., when the output is clamped). The return ratio is (a) $300/(s+200)$, (b) $30/(s+10)$, and (c) $8/(s+0.3)$, when only a velocity sensor is employed (i.e., when no load is connected to the actuator). Find the loop transfer function when the load is a 20 kg mass. Plot the Bode diagram.

19. In the specifications of a brushless motor, it is written: "$k_t = 72$ oz×in./A, $k_e = 0.5$ V/s^{-1}." Make a good guess about what these parameters are, and how they correlate with the flowchart in Figure 7.19, where there is only one k.

20. a. A permanent magnet motor has $k = 0.2 \, N \times m/A$. The driver output impedance is $R_S = 15 \, \Omega$, the load mobility is $R_L = 0.8$ (rad/s)/(N×m), the free-run angular velocity is $\Omega_f = 100$ rad/s. What is the voltage of the source? The break torque? The torque?

    b. Same for $k = 0.1 \, N \times m/A$, $R_S = 25 \, \Omega$, $R_L = 1.8$ (rad/s)/(N×m), $\Omega_f = 200$ rad/s

    c. Same for $k = 0.24 \, N \times m/A$, $R_S = 6 \, \Omega$, $R_L = 0.6$ (rad/s)/(N×m), $\Omega_f = 400$ rad/s

    d. Same for $k = 0.5 \, N \times m/A$, $R_S = 2.5 \, \Omega$, $R_L = 0.4$ (rad/s)/(N×m), $\Omega_f = 850$ rad/s

21. Derive Equation 7.17 for the output mobility of a motor.

22. Draw a flowchart of a permanent magnet electrical motor that produces a torque of 10 Nm per 1 A of the current in the winding, the winding resistance being 10 Ω. Calculate the back emf and the output mobility of the motor when the source impedance (i.e., the output impedance of the driver) is 3 Ω and the current is 2 A, the load mobility is 0.12 (rad/s)/(Nm), and the mechanical losses in the motor are negligible.

23. Draw a flowchart of cascade connection of a driver, a motor, a gear, and an inertial load for the following data:

    a. The driver has a voltage gain coefficient of 5 and the output impedance is 0.5 Ω. The motor winding resistance is 2 Ω and the motor constant is 0.6 Nm. The motor rotor's moment of inertia is 0.5 Nm^2. The gear amplifies the motor torque four times, i.e., the gear ratio, the load angle to the motor angle (or the load angular velocity to the motor angular velocity), is 1:4. The losses in the gear and bearings make $B = 0.04$ Nm/(rad/s). The load's moment of inertia is 0.4 Nm^2; i.e., the load mobility is 2.5/s (rad/s)/(Nm).

    b. The driver has a voltage gain coefficient of 30 and the output impedance is 5 Ω. The motor winding resistance is 30 Ω and the motor constant is 0.15 Nm. The motor rotor's moment of inertia is 0.15 Nm^2. The gear amplifies the motor torque 10 times; i.e., the gear ratio, the load angle to the motor angle (or the load angular velocity to the motor angular velocity), is 1:10. The losses in the gear and bearings make $B = 0.08$ Nm/(rad/s). The load's moment of inertia is 0.05 Nm^2; i.e., the load mobility is 20/s (rad/s)/(Nm).

24. a. Design a control system for a motor similar to that shown in Figure 7.26, where $k = 0.3$, the motor winding resistance is 4, the plant transfer function is 55/s, the sensor is a 12-bit encoder, the torque variation model path is $0.1 \sin 4\theta$, the cogging model path is $0.05 \sin 4\theta$, and the crossover frequency is 12 Hz. (The values are in m, kg, s, rad, N, and Ohm.) Use a current driver. Use Simulink. Design the compensator and plot Bode and Nyquist diagrams with the function linmod. (Make the design using a PID controller and the prototype with the Bode step given in Chapter 4.) Plot the output time history in response to step and ramp commands in position using different scales, for the initial part of the responses and for estimation of the accuracy of the velocity during the ramp command. Study the effects of Coulomb and viscous friction, the cogging, and the motor torque variations on the accuracy of the output position and velocity.

    b. Do the same when $k = 0.2$, the sensor is a 16-bit encoder, the torque variation path is $0.06 \sin 8\theta$, the cogging path is $0.02 \sin 8\theta$, the crossover is 20 Hz, and the driver is a current source.

c. Do the same when $k=0.9$, the sensor is a 10-bit encoder, the torque variation path is $0.04 \sin 16\theta$, the cogging path is $0.05 \sin 16\theta$, the crossover is 6 Hz, and the driver is a voltage source.

25. What is in common between the driving point impedance and the transfer functions of a linear system: the poles, the zeros, or both?

26. What are the properties of the driving point impedance of a passive lossless system? Can the derivative of the modulus of the impedance on frequency be negative? What happens when the losses are small but not zero?

27. Indicate which of the following functions can be an impedance of a lossless two-pole, and plot the frequency response with MATLAB:

a. $(s^2+2)(s^2+4)/[(s^2+3)(s^2+5)s]$

b. $(s^2+2)(s^2+3)/[(s^2+4)(s^2+5)s]$

c. $(s^2+2)(s^2+40)/[(s^2+3)(s^2+5)s]$

d. $(s^2+4)(s^2+40)s/[(s^2+2)(s^2+20)]$

e. $(s^2+4)(s+40)s/[(s+2)(s^2+20)]$

28. Will the signal-to-sensor-noise ratio at the system output change when the feedback path is disconnected? How much?

29. a. A capacitor and an inductor are connected in parallel. What is the equivalent of the return ratio?

b. Same question, about the parallel connection of a capacitor and a resistor.

c. Same question, about the parallel connection of a resistor and an inductor.

30. In the system composed of two parallel paths in Figure 7.39(a), what is the analogy to the closed-loop transfer function for the system in Figure 7.39(b)?

31. What is the condition equivalent to large feedback, in the connection of two two-poles in parallel? How do we explain the effect without using feedback theory, but using Ohm's law?

32. Redo the equations and block diagrams in Figures 7.38 and 7.39 for the case of $B=1$.

ANSWERS TO SELECTED PROBLEMS

1. Since power losses in good motors are small, we can directly relate mechanical variables at one port to electrical variables at the opposite port, and the two-port output impedance is nearly proportional to the output impedance of the preceding link, and its input impedance is nearly proportional to the input impedance of the following link.

2a. Three angles (three degrees of freedom) and three torques.

7. The diagram is shown in Figure 7.45. The collector temperature must be below a certain specified temperature. When one is calculating the average collector temperature, thermal capacitances of the case and heat sink can be neglected, but they need to be taken into account when calculating the collector temperature during power bursts. The nonlinear resistance of radiation and convection cooling of the

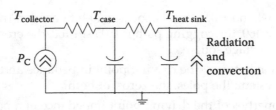

FIGURE 7.45
Electrical equivalent to heat sink.

heat sink is depicted by a nonlinear power source that can be specified in SPICE by the required mathematical expression.

14a. The input impedance is $Z_o = R_2$, $T(0) = 0$, $T(\infty) = 10{,}000$, $Z \approx R_2/10{,}000$.

27a. The zeros and poles alternate; therefore, this function can be a driving point impedance.

8

Introduction to Alternative Methods
of Controller Design

This chapter surveys several important design methods and compares them with the classical Bode approach presented in the previous chapters. The methods discussed in this chapter use linear time-invariant compensators and produce linear control laws, which is optimal according to some performance indices. These alternative methods may be encountered in industry, and software packages for many of them are readily available. The treatment here is cursory, with brief developments of the basic ideas.

8.1 QFT

The term *quantitative feedback theory* (QFT) was coined by Isaac Horowitz, the major contributor to the theory. (Some of his contributions to control theory have already been reflected in this book.) It is a frequency-domain design methodology that is based on Bode methods. QFT relies on simplified relationships between the frequency and time domains, uses prefilters and loop compensation to provide the desired closed-loop responses, considers sensor noise issues and actuator nonlinearities, and provides sufficient stability margins. Most of these issues have already been addressed in the previous chapters of this book. The QFT theorists aim to extend Bode methods to handle performance issues more precisely, and they augment them with additional formalizations, somewhat different problem statements, and extensions to cover MIMO cases, linear time-variable plants, and nonlinear problems.

For simplicity we consider the QFT design of a single-loop tracking system. The design begins by determining an acceptable set of input–output transfer functions that satisfy the tracking performance requirements. This set is defined by upper- and lower-bounding frequency responses. The idea is to design loop compensation and a prefilter so that the input–output transfer function remains between these bounding responses for all possible plant parameter variations. (Disturbance rejection requirements can be handled similarly.) Since the loop compensation and a prefilter can be implemented with negligible uncertainty, the design focuses on the *variations* of the closed-loop gain due to plant parameter variations. The QFT specification for the design of the loop compensation takes the following form: at each frequency ω_i of a certain set, the variation in the closed-loop gain should not exceed a_i dB for all possible plants defined by the uncertainty ranges of the plant parameters. The tolerances a_i are the gains spanned by the upper- and lower-bounding responses at the frequencies ω_i.

To satisfy the specification, it is first necessary to calculate the plant transfer function for all possible parameter values at each of the frequencies ω_i. With the allowable parameter variations, the plant transfer function maps to an area on the L-plane, which happens to be P-shaped, and the same at all frequencies in the example shown in Figure 8.1. The shape is characteristic of the effects of parameter variations on the plant transfer function and is referred to as the *plant template*. (The actuator is included in the plant.) The compensator transfer function at each frequency is defined by shifting the plant template to a proper location. With the template in a particular location on the Nichols chart, the gain curves indicate whether the variations in $|M|$ satisfy the QFT design requirement. If not, the template is shifted until the difference between the minimum and maximum gain is exactly a_i. In the example shown in Figure 8.1, the original gain variation is 6 dB. Suppose that the tolerance is $a_i = 1$ dB. From the lines on the Nichols chart, it is evident how the template must be shifted. There is a continuum of such shifted templates which satisfy the design requirement, and the edges or corners of the shifted template with minimum closed-loop gain form the *minimum performance boundary* $B(\omega_i)$, as shown in Figure 8.1.

For each of the frequencies ω_i at which the system requirements are specified, the boundaries $B(\omega_i)$ must be plotted on the L-plane, as shown by the dashed lines in Figure 8.2.

An additional high-frequency L-plane bound is included to guarantee system stability and robustness. With the boundaries in place, the next step is to search for a rational compensator transfer function such that the loop gain at each of the frequencies ω_i will be just over the minimum performance boundary. At frequencies near the zero-dB crossing and higher, the compensator gain is shaped to follow the stability boundary. The design is performed by trial and error or by using specialized software. Finally, a prefilter is synthesized that corrects the input–output response to achieve the desired response. (Remember, the prefilter's uncertainty contribution is negligible.)

It can be shown that a solution to the QFT problem always exists; although, the resulting feedback bandwidth may be unacceptably large. Generally, the best design is taken as that

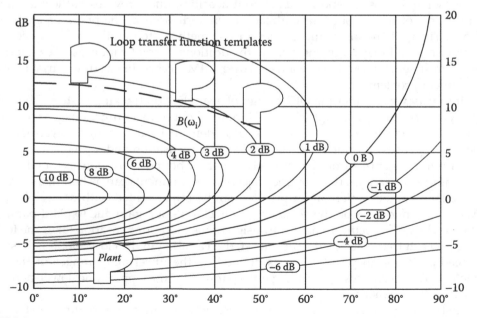

FIGURE 8.1
Plant templates on the Nichols chart forming the minimum performance boundary $B(\omega_i)$.

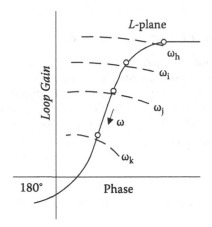

FIGURE 8.2
Boundaries on the *L*-plane.

which has the smallest feedback bandwidth while satisfying the minimum performance and stability boundaries.

The QFT design philosophy departs from the Bode approach in the following areas:

- The Bode approach is to maximize the performance (feedback) while satisfying the constraints on the high-frequency loop asymptote due to plant features and parameter variations, and high-frequency noise. QFT pursues the inverse problem of providing the minimum acceptable performance while minimizing the feedback bandwidth. The QFT-designed system is not the best possible, but rather is just good enough to satisfy the closed-loop response and disturbance rejection specifications. We prefer the Bode approach for the following reason: the cost differential between a substandard controller and the very best available is generally insignificant compared with the cost of the system. The controllers differ only by several resistors and capacitors or a few lines of code, and perhaps a few days of work by the control engineer (if he uses the Bode approach). Improving the control law might also relieve some of the requirements on other components of the device, making the entire system better and cheaper to manufacture. This may in turn affect decisions made about the development of the next generation of the system. It makes little sense to lower the system performance just to reduce the feedback bandwidth. The Bode approach identifies the constraints on the bandwidth upfront, rather than minimizing the bandwidth and determining later whether it is still too high. (Whatever the philosophical difference, there is little doubt that engineers well-trained in QFT can resolve these trade-offs and design high-performance controllers.)
- The QFT design is focused on satisfying the performance specifications for the worst-case plant parameters, and neglects optimizing the nominal performance. This may or may not be an advantage.
- With the multiple templates to be calculated and plotted, QFT design is far more complex than Bode design.

QFT methods have been developed to handle the design of MIMO systems with stable and unstable, time-invariant and time-variable, linear and nonlinear plants.

8.2 Root Locus and Pole Placement Methods

Another category of controller design methods focuses on the location in the complex plane of the roots of the closed-loop transfer function.

The *root locus method* uses plots of the root loci to choose the constant multiplier of the loop gain coefficient and to design additional loop compensation. An elaborate set of rules exists for constructing the root loci from the open-loop transfer function. Today, root locus *analysis* is usually performed by computer.

Example 8.1

Consider a control system with plant $P(s) = 100/s^2$, actuator $100/(s + 100)$, and lead compensator $C(s) = 10(s + 3) / (s + 30)$.

With the loop open, the poles of the system are just those of $T = CAP$, i.e., a double pole at the origin and real poles at -30 and -100 rad/s. For the purpose of this analysis, suppose that there is a variable gain coefficient k in the loop that is gradually increased from 0 to 1. As the gain coefficient is increased, the poles move from their open-loop positions to their closed-loop positions. Figure 8.3 shows the root loci for our example.

The loci can be continued by increasing the gain coefficient past the nominal closed-loop value of 1. In the example, when k reaches 3.56, some poles cross over into the right half-plane. The gain stability margin is therefore $20 \log(3.56) = 11$ dB. The robustness of the system is still difficult to determine from the root loci. The guard-point gain stability margin is apparent, but the phase stability margin and even the guard-point phase margin are not. It might be guessed that the distance of the roots from the $j\omega$-axis would be a good indicator of system robustness, but this is not always the case. A practical counterexample is an active RC notch filter, where the root locus passes very close to the $j\omega$-axis but the system is quite robust.

It is also not evident from the root locus whether the system is well designed. In fact, it is not. It would be instantly seen from the Bode diagram that the pole and the zero in the compensator are in the wrong places, and the phase and gain guard-point stability margins are not balanced.

The compensator design proceeds by trial and error, searching for compensation and a suitable loop gain that brings the closed-loop poles into desirable locations on the s-plane. What are desirable pole locations? Usually the system is examined for a pole or pole pair that is

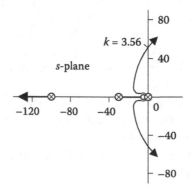

FIGURE 8.3
Root loci for a feedback system.

"dominant," meaning that the step response of the closed-loop system resembles the step response of a system with just this pole or these poles. The design goal is to move the dominant poles into areas on the *s*-plane with sufficient distance from the origin for the transient response to be fast, and with sufficient damping (i.e., not too close to the *jω*-axis) to prevent excessive overshooting. (The possibility of a prefilter is not factored into the design.) Meanwhile, other system poles must be monitored for stability.

When the designer makes an a priori decision about the precise location of the system poles, the method is sometimes referred to as *pole placement*; although, this label is often reserved for state-variable feedback control of MIMO systems, as will be discussed below. A common choice is to place the closed-loop poles in a Butterworth filter configuration.

A major inadequacy of the root locus design method is that it does not allow the designer to judge how close the system performance is to the best available. In addition, no convenient rules exist for designing good high-order compensators. Another problem is the complete lack of visibility into low-frequency disturbance rejection. Finally, system performance in the nonlinear mode of operation is difficult to determine from the root loci. (As we shall see in Chapters 10 and 11, the Bode and Nyquist diagrams enable the designer to deal effectively with common nonlinearities.) Because of these deficiencies, the root locus method is not recommended for control system design.

The root locus method can be valuable for the analysis of the effects of certain parameter variations on stability, and on the non-minimum phase lag in the link composed of several parallel minimum phase links (see Sections 3.13 and 4.5). Also, root locus plots make very impressive presentations for high-order systems that have already been well designed using other methods.

8.3 State-Space Methods and Full-State Feedback

From the classical control perspective, the linear control system is a block diagram of transfer functions, i.e., Laplace transforms. Of course, the system can also be represented by one or several linear differential equations.

The system of equations can be transformed into a set of first-order differential equations by introducing intermediate variables where necessary. The following *state-space* system description is standard:

$$\dot{x} = Ax + B(u + r) \tag{8.1}$$

$$y = Cx \tag{8.2}$$

where *x* is a vector (column) of *state variables* (or *states*), *u* is the input or *control vector*, *r* is the *reference*, and *y* is the *output vector*. The square matrix *A* is referred to as the *system matrix*. It describes the dynamics of the system without feedback (i.e., dynamics of the actuator and plant). *B* is the *control input matrix*, and *C* is the *output matrix*.

It may be helpful to think about how a SISO system would fit into this format. The system matrix *A* would be $n \times n$, where *n* is the order of the combined actuator-plant transfer function. The control input matrix *B* would be a column matrix of length *n* that distributes the scalar control input among the state derivatives. The output matrix *C* would be a row matrix of length *n* that reassembles the scalar output (which is a function of time) from the states. Note that the representation is not unique but depends on the choice of states. It is customary to try to choose states that correspond to some physical variable of the system.

An advantage of the state-space notation is that it is easily generalized to multi-input multi-output systems by changing the matrix dimensions. For example, a two-input three-output system would have a control input matrix B that is $n \times 2$ and an output matrix C that is $3 \times n$.

The feedback loops are closed when the second component in Equation 8.1 is added to the state vector. With linear state feedback the control u is a linear combination of the states:

$$u = -Kx \tag{8.3}$$

where K is the *gain matrix*. The closed-loop system is then described by the equations:

$$\dot{x} = (A - BK)x + Br \tag{8.4}$$

$$y = Cx \tag{8.5}$$

To be more general, and to conform to the convention adopted by MATLAB®, we can allow the control to affect the output directly by introducing the matrix D and rewriting Equation 8.5 as:

$$y = Cx + Du \tag{8.6}$$

The state-space block diagram of the feedback system is shown in Figure 8.4.

The open-loop system corresponds to $u = 0$ in Equation 8.1. The open-loop system becomes:

$$\dot{x} = Ax + Br \tag{8.7}$$

$$y = Cx \tag{8.8}$$

Example 8.2

Suppose that the open-loop plant is the pure double integrator:

$$P(s) = \frac{1}{s^2} \tag{8.9}$$

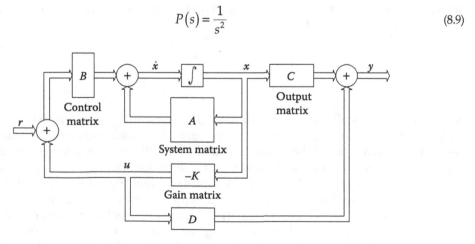

FIGURE 8.4
State-space block diagram of a feedback system.

(and that the actuator transfer function is unity). The open-loop system could be represented as follows:

$$y = x_1 \tag{8.10}$$

$$\dot{x}_1 = x_2 \tag{8.11}$$

$$\dot{x}_2 = u \tag{8.12}$$

Here u is the input and y is the output, both scalars (functions of time). The state vector consists of a position-like and a velocity-like state $x = [x_1 \ x_2]^T$. Per our notation, $n = 2$, and A, B, C, and D are as follows:

$$A = \begin{bmatrix} 0 & 1 \\ 0 & 0 \end{bmatrix} \quad B = \begin{bmatrix} 0 \\ 1 \end{bmatrix} \quad C = \begin{bmatrix} 1 & 0 \end{bmatrix} \quad D = 0 \tag{8.13}$$

The nomenclature is important since systems are represented this way in MATLAB. Suppose that we had manually entered the A, B, C, and D matrices for our example into MATLAB. The following command would then produce the open-loop frequency response plot:

```
bode(a,b,c,d,1)
```

The last argument is letting the MATLAB function know that we're interested in the response of the output (all outputs in the general case) to the *first* input (the system in case has only one input and one output). This may seem like a lot of overhead to calculate the frequency response of a double integrator. Fortunately, the matrices are usually created by other programs. For instance, the block-diagram-oriented Simulink has a function `linmod`, which creates the appropriate A, B, C, and D matrices for further analysis:

```
[a b c d] = linmod('model_name')
```

After the gains k_1 and k_2 are chosen, the closed-loop frequency response can be obtained using MATLAB commands to manipulate the system matrices, or by making the appropriate connections in the Simulink block diagram and rerunning `linmod`.

The state-space closed-loop design problem is to choose the control matrix K to obtain the desired closed-loop transient response. (We might already disagree with the practicality of such an approach since obtaining the desired closed-loop response is not the only or the main purpose of closed-loop control in practical systems.)

Before we discuss the possible strategies for choosing K, some implications of the state-space notation should be noted. An implicit assumption is that the states x are somehow available to be plugged into Equation 8.3 and fed back to the input of the system. For this reason, Equation 8.3 is often referred to as *full-state feedback*. In a typical control system the order of the actuator-plant combination exceeds the number of sensed outputs, making full-state feedback unrealistic. The missing states must be estimated using the available ones; this is discussed in the next section. Another feature of the state feedback framework is that it does not allow compensators whose order exceeds the order of the actuator/plant. In our example above, Equation 8.3 restricts the compensator function to consist of a single unrealizable zero: $C(s) = k_1 \, k_2 s$. A work-around is to expand the state vector to include some of the compensator dynamics, as is typically done to add integral control in the state-space version of the *PID* controller.

A more insidious problem with the state-space approach is inherent to the representation of the system by a set of linear matrix differential equations rather than a block

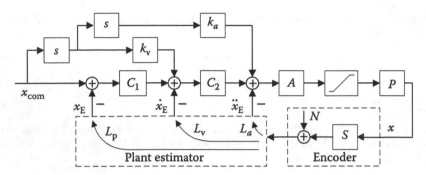

FIGURE 8.5
Block diagram of a position control system with position, velocity, and acceleration loops.

diagram of transfer functions. This draws the designer's attention away from the physical elements of the control system, along with their limitations and imperfections, and instead focuses on matrix algebra.

The state-variable approach is sometimes mixed with conventional block diagram design methods, as in the following example.

Example 8.3

The block diagram in Figure 8.5 has been employed for the control of position x (or the attitude angle) in many space systems, especially those having rigid-body plants with small parameter uncertainty. The feedback bandwidth in these systems is typically limited by the sensor quantization noise. The plant is considered rigid, the acceleration is proportional to the actuator force (or torque, for attitude control), the actuator transfer function A is a constant, and the plant P is seen as a double integrator.

The position command x_{com} and the velocity and acceleration commands obtained by differentiation are forwarded with appropriate gain coefficients to, respectively, position, velocity, and acceleration summing points.

A plant estimator (filter) generates the plant variable estimates x_E, etc., from the noisy readings of the sensor. The transfer functions L_p, L_v, and L_a are m.p., and the phase lag responses are related to the gain responses by the Bode integral (Equation 3.5). The bandwidth of L_v is wider than the bandwidth of L_p and smaller than that of La. The filter cutoff frequencies must be sufficiently low to extensively attenuate high-frequency sensor noise components, but not too low since, first, the filter distorts the output signal and, second, the filter phase lag reduces the available feedback and the disturbance rejection.

The errors in position, velocity, and acceleration are formed by subtraction of the plant variable estimates from the signals arriving to the summing points. The errors are reduced by the three feedback loops. It is seen that when the position error is 0, the output of the compensator C_1 is 0. When the velocity error is also 0, the output of C_2 is 0. When the acceleration error is also 0, the signal at the actuator input is 0. This control scheme can be perceived as multivariable.

8.3.1 Comments on Example 8.3

The three feedback loops are coupled. Still, the design can be made by iterative adjustments, one loop at a time, since (1) the compensators are typically low order (*PD*), and (2) the three loops substantially differ in bandwidth: the bandwidth of the velocity loop is wider than that of the position loop, and the bandwidth of the acceleration loop is still wider.

When the plant is flexible, the compensators' order must be much higher than that of the *PD* compensators shown in the block diagram, but the higher-order compensators are not easy to fit within this design.

The differentiators in the feedforward paths are implemented in practice as lead links whose frequency responses approximate the response of the ideal differentiator over the required frequency band. The effects of saturation in the actuator limit the useful bandwidth of the feedforward.

The torque source actuator (using a driver with high-output impedance) simplifies the analysis. On the other hand, a velocity source (a motor driven by a driver with low-output impedance) may improve the system accuracy, especially when the plant is flexible with Coulomb friction.

The controller can be augmented with inclusion of nonlinear links to reduce the wind-up and to improve the transient response for large-level commands.

This system is multivariable only formally since it has only one sensor, the plant is rigid, and the actuator is a force source. Since the position, rate, and acceleration have unique and simple interrelations, such a system can be equivalently and better described as a single-loop SISO system.

The complex diagram in Figure 8.5 can be equivalently transformed into the diagram in Figure 8.6(a), and further into the diagram in Figure 8.6(b), which follows the diagram in Figure 2.1 (the loop transfer function about the plant is the same in these diagrams, and the input–output transfer function without the feedback, i.e., with the sensor transfer function $S = 0$, is the same). The diagram in Figure 8.6(b) includes only two independent linear links

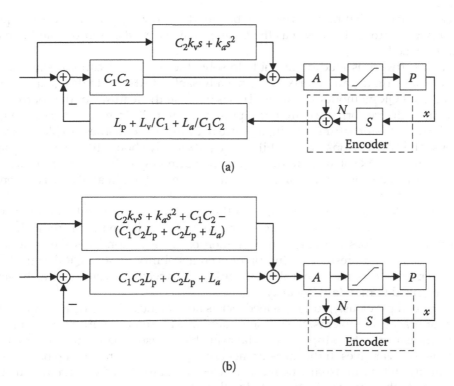

(a)

(b)

FIGURE 8.6
Single-loop equivalents of the block diagram shown in Figure 8.5.

whose transfer functions are defined by the designer: the feedback compensator and the feedforward path.

Therefore, the performance of the system shown in Figure 8.5 cannot be superior to a conventional well-designed system with a prefilter or a feedforward path.

8.4 LQR and LQG

The general plan of the so-called *modern control theory* is to take the state-space description of the control system literally, set up some scalar performance index that quantifies the desirable features of the closed-loop system, and then find the gain matrix K that is optimal for this index. One such approach is to minimize a quadratic functional J of the state and control history for the system's step response:

$$J = \int_{t=0}^{\infty} \left(x^{\mathrm{T}}Qx + u^{\mathrm{T}}Ru \right) dt = \min \tag{8.14}$$

where Q and R are *weighting matrices*. It is assumed that the desired state is $x = 0$, but the initial condition is nonzero, so the matrix Q penalizes the state error in a mean square sense. Similarly, the matrix R penalizes the control effort, i.e., limits the control signals' magnitude.

The gain matrix K that minimizes J can be found by solving a matrix Riccati equation. The resulting controller is known as the *linear quadratic regulator*, and the methodology is referred to as *LQR*.

Although software is readily available to solve the matrix Riccati equation and thus determine the optimal gain matrix K, it is not advisable to attempt to design a control system using the LQR methodology alone. This is because the features of the control system that constrain the performance are not captured in the LQR framework. There has been no mention of actuator saturation, disturbance rejection, or robustness to plant parameter variations. The only reasonable possibility is to judiciously choose the weighting matrices Q and R, run the LQR software to determine the optimal gain matrix K, and then examine the resulting control system using classical frequency-domain analysis. This generally entails several iterations.

The inability to address nonlinearities in the LQR framework is difficult to overcome. A common design strategy is to increase the control penalty matrix R until the largest expected transient does not result in saturation of the actuator. This seems wise since actuator saturation can result in wind-up or even instability for an LQR design. The system design may be very inefficient since, to achieve the specified performance, the actuator will be oversized to maintain linearity.

As mentioned previously, full-state feedback is not practical. If the LQR framework is to be used for practical problems, the missing states must be continually estimated from the available measurements. Suppose that the available measurements are linear combinations of the state variables. If the measurements are perfect, and the plant model is perfect, the remaining states can usually be reconstructed by repeated differentiation. In fact, the entire future of the state can be predicted with certainty. To make the estimation process nontrivial, the state-space formulation has to be augmented by introducing errors in sensing and modeling. An analytically tractable approach is to assume that the measurements are corrupted by white noise, and that the actual plant differs from the plant model by an additional white noise input. The noise in the measurements is referred to as the *sensor noise* and denoted w. The noise added to the plant model is referred to as the *process noise* and denoted v. Note that w and v are generally vectors. Let the measurements be z, so that the system description becomes:

$$\dot{x} = Ax + Bu + Gw \tag{8.15}$$

$$z = Hx + v \tag{8.16}$$

where G is the *plant noise distribution matrix* and H is the *measurement matrix*.

The *state estimate* x_E is to be propagated as:

$$\dot{x}_E = Ax_E + Bu + K_E(z - Hx_E) \tag{8.17}$$

where K_E is the *estimator gain matrix*. Given the second-order moments of the white processes w and v, the optimal estimator gain can be found that minimizes the mean square error in x_E. This estimator is referred to as *linear quadratic Gaussian (LQG)*. When these estimates are used in conjunction with an LQR controller, the combined approach is referred to as an LQR/LQG regulator.

The LQR/LQG regulator theory was intended to resolve the trade-off between the sensor noise and the disturbance rejection. Since this method by itself does not address the robustness issue, it does not provide the best solutions to most practical problems. However, when the plant is known pretty well (say, with 1% accuracy) and the feedback bandwidth is limited by the sensor noise, LQG provides a loop response that is well shaped in the crossover area of the frequency band. This response can be later modified with classical methods for better disturbance rejection at lower frequencies.

The addition of the *loop transfer recovery (LTR)* method to LQR/LQG allows the system robustness to be addressed. The LTR method recalculates the frequency-domain loop responses of the system designed in state-space time domain with the LQR/LQG method, and allows adjusting the responses to provide the desired stability margins [16]. However, the design process is not simple, reflecting the fact that the quadratic norm of the closed-loop response is not appropriate for stability analysis.

8.5 H_∞, μ-Synthesis, and Linear Matrix Inequalities

The state-space approach to control system design and the state-space performance indices are difficult to use during the conceptual design. G. Zames, who initiated the H_∞ method, often said that the processes of approximation in model building and obtaining state-space models do not commute [78]; i.e., input–output (black box) formulations are the preferred framework for uncertain (practical) system modeling, and state-space models should come into picture only as internal models at the level of computation and at the level of implementation of control systems.

The computational aspects of control system design have been already advanced to the degree when they cease to be critical for the design of most practical systems. However, building the system model and designing optimal compensators still present a challenge, and are easier to accomplish with the input–output formulations.

In other words, control system engineers should structure the systems as sets of physical blocks interconnected via ports, instead of structuring the systems mathematically in sets of linear matrices. Mathematically, the input–output formulations mean separating the system variables into the sets of local (internal for the blocks) and global (at the blocks' ports) variables, and typically, the number of the global variables is much less than the number of the local variables.

For the input–output formulations, frequency-domain characterizations are in many aspects more convenient than the time domain ones. As was exemplified in Chapter 7 with the two-ports, linear black boxes can be described by the matrices of their transfer functions and impedances (mobilities), and the entire system, as a conglomerate of linear and nonlinear multi-ports interconnected via their ports.

H_∞ is an extension of the classical frequency-domain design method. It solves in one operation the two problems that are sequentially solved with the Bode approach: maximization of the available feedback bandwidth with related shaping of the loop response over the frequency region of crossover frequency and higher, and distribution of the available feedback over the functional bandwidth, as was described in Chapter 4. The method is formulated such that it is directly applicable to multivariable control systems.

The H_∞ *norm* is the limit on the magnitude of a vector in the Hilbert space. This norm is an extension of the Chebyshev norm widely used in frequency-domain network synthesis. The H_∞ feedback control design method applies this norm to the closed-loop frequency responses from the disturbance sources to the system output.

With the H_∞ method, frequency responses of the disturbance rejection are first specified with weight functions. The weight functions define at which frequencies disturbance rejection should be higher than that at other frequencies, and by how much. The weight functions should be calculated from the known disturbance spectral densities. For the functional feedback bandwidth, the norm on F is nearly the same as the norm on T.

Since it is not easy to properly shape the crossover area of the loop Bode diagram with the H_∞ method, the method may lead to an overly conservative system. A less conservative solution can be achieved with μ-*synthesis*, which combines the H_∞ design and μ-*analysis* in an iterative procedure. The μ-analysis method introduces into the loop special links that imitate the plant uncertainty. It is required that with these links added, the nominal system should still be stable and perform well.

The H_∞ design method is for linear control system design. It optimizes the system performance without paying special attention to the system global stability. Because of this, the H_∞ design often results in Nyquist-stable systems that are not absolutely stable and can burst into oscillation after the actuator becomes overloaded. The solutions to this stability problem are either making several iterations by relaxing disturbance rejection requirements and modifying the weight functions such that the resulting loop response is of the absolutely stable type, which is easy to do when the designed system is single loop, or better, using the nonlinear controllers that will be studied in Chapters 10 and 11. The nonlinear controller design methods should be also employed to further improve or optimize the system performance in the nonlinear state of operation when certain commands or disturbances overload the actuators.

H_∞ control and many other linear control and stability analysis problems can be formulated in terms of *linear matrix inequalities (LMIs)*. The LMI is the algebraic problem of finding a linear combination of a given set of symmetric matrices that are positive definite. LMIs find applications outside of control, in such diverse areas as combinatorial optimization, estimation, and statistics. Although it has long been recognized that LMIs are important in control, it was only with the advent of the efficient algorithms (based on the interior point methods) that their popularity has increased among researchers in recent years.

9

Adaptive Systems

Large plant uncertainty reduces the available feedback. One remedy is to use an adaptive control law that changes the transfer functions of the compensator, the prefilter, and the feedback path on the basis of accessible information about the plant.

On the basis of the information used for the adaptation, the adaptive controllers can be divided into three types. The first type uses sensor readings of environmental parameters (temperature, pressure, time, etc.), and plant parameter dependencies on these environmental factors to correct the control law. (Obviously, the dependencies must be known a priori.) The second uses the plant response to the command or disturbances to correct the control law. The third type uses the control loop response to specially generated pilot signals to correct the control law.

The first and third types of adaptation schemes substantially improve the control when the plant changes at a much slower rate than the control processes. The second method can be useful when the command profile is well known in advance. If this is not the case, the second type may result in a system with rapidly varying parameters whose stability analysis represents a formidable problem.

It is shown that the plant is easier to identify in the frequency bands where the feedback in the main loop is not large. This identification provides most of the available benefits in the system performance.

A brief description is provided for adaptive systems for flexible plants, for disturbance rejection and noise reduction, and for dithering systems. Examples of adaptive filters are described.

9.1 Benefits of Adaptation to the Plant Parameter Variations

Plant parameter uncertainty impairs the available feedback. The uncertainty can be reduced by a *plant identification* procedure that gives the plant transfer function estimation as P'. The improved knowledge of the plant should then be used for controller *adaptation*, i.e., adjustments of the compensator, feedback path, and prefilter to reduce the output error. The rate and dynamics of the adaptation are defined by some adaptation law.

Disturbance sources and plant parameter uncertainty both contribute to the feedback-system output error. Increasing the feedback reduces both error components, and as explained in Chapter 2, the second component can be additionally reduced by using an appropriate prefilter.

Correspondingly, the benefit of good plant identification, $(P' \approx P)$ is twofold: the feedback can be increased by appropriate modifications to the compensator C, and the prefilter can be made closer to the ideal. When plant parameter variations are large, adaptation can significantly improve the system performance.

Adaptation schemes where the plant is identified first are called *indirect*.

Example 9.1

Assume the plant gain varies by 20 dB in a loop designed as shown in Figure 9.1. The feedback bandwidth is limited by sensor noise or plant resonances, $f_b < f_{bmax}$. The loop response with maximum available feedback bandwidth is the one when the plant gain is largest. When the plant gain drops, the feedback and the disturbance rejection reduce and the sensitivity of the output to plant parameter variations increases. To alleviate the problem, plant gain can be monitored and the compensator gain coefficient continuously adjusted for the loop gain to remain equal to the loop gain of the case of the maximum plant gain. This is an example of indirect adaptation.

Example 9.2

Assume the loop in the previous example is designed as Nyquist stable to increase the feedback in the functional band. When the plant gain decreases, the system can become unstable. Therefore, the plant gain needs to be monitored and the compensator gain modified not only to maintain the desired level of disturbance rejection, but also to just preserve stability.

Example 9.3

In the plant, a real pole position can go down by a factor of four from the initial position at $2f_b$. The phase lag increases correspondingly, and the system can burst into oscillation. To counteract this pole motion, a zero can be introduced in the compensator transfer function and, after the plant identification, kept close to the pole position to cancel its effect. The accuracy of the pole position identification need not be very high: if the zero-to-pole distance is within 5% of the pole magnitude, the resulting loop gain uncertainty will be less than 0.5 dB.

Example 9.4

In the previous example, not a real pole but a complex pole related to a plant flexible mode with high Q is uncertain. In this case, if the same remedy is contemplated—compensation of the plant pole with a compensator zero—the plant identification needs to be Q times more accurate than in the previous example, which might be difficult to accomplish.

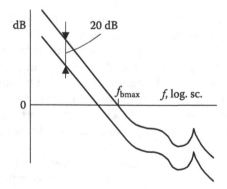

FIGURE 9.1
Loop gain responses with maximum and minimum plant gain.

Direct adaptive algorithms do not use plant identification explicitly, but merely tune the compensator to improve the closed-loop performance. For tuning the most popular *PID* controllers, there exist many adaptive algorithms. Time domain performance adjustment is most often used: the time response parameters are measured, and *P*-, *I*-, and *D*-coefficients of the controllers are adjusted correspondingly. The algorithms typically work well as long as the plant's responses are smooth. The typical goal for such adaptive control is improving the responses to the commands in the system where the plant parameters are varying slowly, and not improving the higher-frequency disturbance rejection or the system tolerance to fast plant parameter variations.

9.2 Static and Dynamic Adaptation

Plant identification presents no problems and can be performed rapidly when the dependence of the plant transfer function on the environmental variables is well known and the environmental variables are measured accurately. Such an adaptive control system is shown in Figure 9.2. The plant is identified, and then the compensator and prefilter transfer functions are adjusted. Such a process is sometimes called *gain scheduling*. An example of such a system has been given in Section 6.7. Commonly, the environment varies much more slowly than the main control loop dynamics, i.e., the adaptation is *quasi-static*. Therefore, during the system analysis, the main feedback loop dynamics can be assumed to be independent of the process of adaptation, and the prefilter and the main loop links can be considered time invariant.

The rate of the plant identification is limited by effects of noise and disturbances, and by the rate of the information processing. The rate of adaptation—which can be critical when the plant parameters vary rapidly—is limited by the plant identification rate and also by the system stability conditions.

Increasing the speed of the adaptation makes the adaptation *dynamic*. Stability analysis of such systems presents a formidable problem, and ensuring the system stability requires increasing stability margins (and reducing the feedback) in the main control loop. This is why the adaptation laws are commonly designed as quasi-static, i.e., much slower than the processes in the main control loop. In this case, the main loop can be designed as a linear time-invariable (LTI) system.

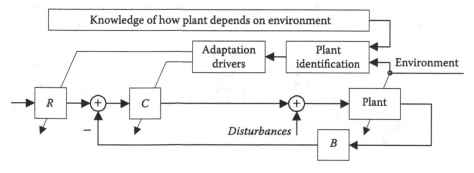

FIGURE 9.2
Prefilter and compensator regulation.

9.3 Plant Transfer Function Identification

Figure 9.3 illustrates three methods for plant transfer function identification that are used with indirect adaptation:

1. Using sensors of environmental variables and the known dependencies of the plant transfer function on these variables.

2. Using the input and output signals of the plant that are generated during normal operation of the control system as the result of commands and disturbances.

3. Using *pilot signals* generated and sent to the plant for identification purposes. The pilot signal amplitude should be sufficiently small so as not to introduce substantial error in the system's output.

The plant is identified from its measured input and output signals. The measurements are corrupted with the disturbance signals and sensor noise. This limits the accuracy and speed of the adaptation.

Plant identification is ill-conditioned over the bandwidth of large feedback where the error frequency components are small, and therefore, the pilot signals must be very small. This increases the requirements on the sensors' sensitivity and accuracy, and makes the identification complicated. Also, the frequencies where the feedback is large are relatively low, which makes the plant identification process slow. Further reduction of the control error components at these frequencies can be achieved more economically by using nonlinear dynamic compensation, as explained in Chapters 10 and 11.

In contrast, the problem of plant identification is well conditioned over the frequency ranges of positive feedback and small negative feedback, i.e., from $f_b/4$ up to $4f_b$. This bandwidth is important for feedback bandwidth maximization. The plant identification over this frequency range provides tangible benefits and is relatively easy to implement since at these frequencies the pilot signal amplitudes can be larger and the measurements much faster.

9.4 Flexible and N.P. Plants

Plant identification might especially improve the control when the plant is flexible and uncertain. For example, when the torque actuator and the angular velocity sensor are collocated in a structural system with flexibility, the plant transfer function is that of a passive

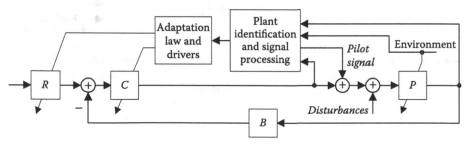

FIGURE 9.3
Adaptive feedback system with plant identification.

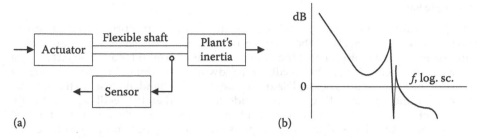

FIGURE 9.4
(a) Plant with flexibility and (b) the uncompensated loop response.

two-pole impedance, and the phase variations do not exceed 180°. At higher frequencies, however, due to the flexibility of the actuator-to-plant shaft, as shown in Figure 9.4, the control ceases to be collocated and two poles in a row follow in the plant transfer function. Plant identification makes possible the design of a compensator whose transfer function has zeros to compensate for the unwanted plant poles.

For this application, using frequency-domain identification is also economical since the plant response needs to be identified only at higher frequencies, in the vicinity of the structural modes.

N.p. lag caused by parallel paths of the signal propagation, such as that of rotation and translation (due to the thruster's position and the propellant slosh in the tank of a spacecraft, for example), can be removed from the main feedback loop by the addition of appropriate sensors for the plant identification. The n.p. lag increases with frequency, and the best way to measure it is at higher frequencies, i.e., within the bandwidth of positive feedback.

We conclude that for the flexible plants and n.p. plants, the plant identification is less complicated and most beneficial over the frequency band where the feedback is either positive or small negative.

9.5 Disturbance and Noise Rejection

Disturbance rejection can be improved by using adaptive loops within the compensator, as diagrammed in Figure 9.5. The disturbance identification allows modifying the loop gain response to match the disturbance spectral density, under the limitation of Bode integrals (Equations 3.11 and 3.12). The disturbances can also be compensated in a feedforward manner.

Example 9.5

Using an adaptive *comb filter*, i.e., a filter with multiple narrow passbands, allows the provision of substantial feedback at the harmonics of a periodic disturbance, as shown in Figure 9.6, instead of broad-band feedback with the response shown by the dashed line, the feedback in both of the responses being limited by the Bode integrals. Such a filter tunes by tracking the disturbance frequency f_d using a phase-locked loop (PLL), a frequency-locked loop (FLL), or some other adaptive algorithm.

Example 9.6

Diagram 1 in Figure 9.7 was found optimal for a spacecraft attitude control system that uses reaction wheels in the tracking regime, when the feedback bandwidth is limited by the gyro noise. (This kind of diagram was also used for a spacecraft camera temperature control where the feedback bandwidth was limited by the quantizing noise of the temperature sensor.) Diagram 2, on the other hand, is closer to the optimal for the regimes of retargeting of the attitude control, when the feedback bandwidth needs to be wider and the gyro noise is of small importance. Therefore, for the best performance, the shape of the loop response can be changed from (1) to (2) by an adaptation process.

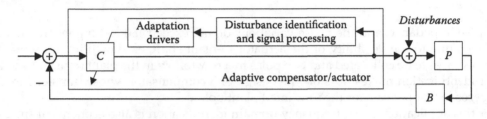

FIGURE 9.5
Loop response adaptation for disturbance rejection.

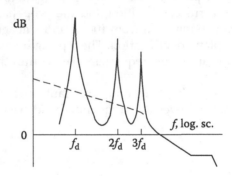

FIGURE 9.6
Loop gain Bode diagram with adaptive equalizer.

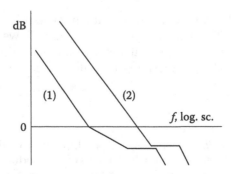

FIGURE 9.7
Loop gain Bode diagrams for systems with different bandwidths.

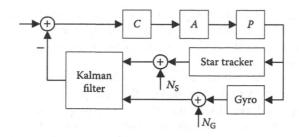

FIGURE 9.8
Adaptive control system with a Kalman filter in the feedback path.

Example 9.7

A spacecraft attitude control system using star tracker and gyro attitude sensors is shown in Figure 9.8. The measurements are corrupted by the noise represented by the sources N_S and N_G. At any moment, the signal from the star tracker depends on the particular stars that happen to be in its field of view. When several bright stars are in view, they are easily recognized compared with the star map in the computer memory, and the signal-to-noise ratio of the tracker becomes excellent, but when the camera turns in a direction where no bright stars exist, the tracker output is noisy and the attitude control should depend more on the gyro. In the latter case, the bandwidth of the filter in the feedback path from the star tracker must be reduced. A *Kalman filter* (a widely used algorithm minimizing the mean square error) performs this adaptation task continuously.

9.6 Pilot Signals and Dithering Systems

Without pilot signals, *passive identification* relies on the feedback system's responses to noise, distortions, and commands, which are all far from optimal test signals for the plant identification. Consequently, passive identification of the plant is slower, less accurate, and less reliable than *active* plant identification with specially generated pilot signals. When pilot signals are used, there always is a trade-off between the accuracy of the plant identification, which increases with the level of the pilot signals, and the error the pilot signals themselves introduce at the system's output.

When the plant changes rapidly, the adaptation must be faster. This requires an increased rate of information transmission by the pilot signals, and the pilot signals must have larger amplitudes or a broader spectrum.

When only the constant multiplier of the gain coefficient of the plant is changing, and this change is slow, a sinusoidal pilot signal with the frequency in the crossover region and very small amplitude suffices to identify the plant accurately. The small-amplitude pilot signal can be superimposed on the command. Or, the pilot signal can be generated in the system itself by self-oscillation. High-frequency oscillation of small amplitude is called *dither*. A *dithering system* block diagram is shown in Figure 9.9.

The Nyquist and Bode diagrams for the system are shown in Figure 9.10.

The dithering system oscillates at the frequency where the phase stability margin is zero. In a conventional system, at this frequency the loop gain must be −8 to −10 dB to provide a sufficient gain stability margin. In the dithering system, the loop gain at this frequency is 0 dB, which allows an increase in the loop gain by 8–10 dB, i.e., a provision of approximately three times the additional rejection of the disturbances within the functional frequency range.

The gain in the feedback loop needs to be adaptively adjusted in order for the amplitude of the oscillation to remain small. For this purpose, the dither is selected by a band-pass filter, amplified, rectified, smoothed by a low-pass filter, and compared with a reference. The difference between the measured dither and the reference, the error, is amplified and slowly and continuously regulates a variable attenuator in the forward path of the main feedback loop. The attenuator varies the main loop gain so that the dither level becomes as small as the reference. The dither loop compensator C_d implements the desired adaptation law, resulting in the desired time response of the adaptation.

This system has certain similarities with the automatic level control system for AM receivers described in Section 1.10.

(In some systems, the dither signal reduces the loop gain by partially saturating the actuator nonlinear link. These systems are simpler since they have neither a dither detector nor a variable attenuator, but their performance is inferior, with smaller range of adaptation, larger error introduced by the dither, and reduced power available from the actuator to drive the plant.)

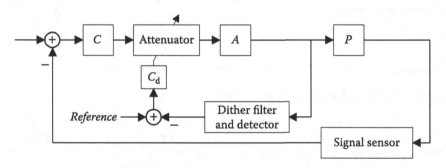

FIGURE 9.9
Dithering control system.

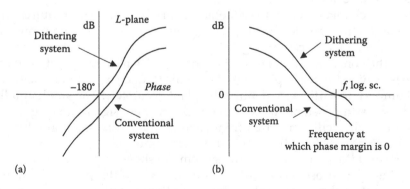

FIGURE 9.10
Dithering and conventional control systems: (a) Nyquist diagrams and (b) Bode diagrams.

9.7 Adaptive Filters

Adaptive controllers use linear links, also called *linear filters*, which are varied by some adaptation algorithm. For these filters, the symmetrical regulators described in Chapter 6 can be used. Two additional implementation examples are given below.

Example 9.8

A *transversal filter* compensator is shown in Figure 9.11. It consists of delay links τ and variable gain links, the *weights* w_i. With appropriate weights, any desired frequency response of the adaptive filter can be obtained.

The weights are adjusted by an adaptation algorithm (defining the dynamics of the adaptation loops) on the basis of the performance error produced by the performance estimator. Such a system can be classified as self-learning.

This system works well when the required frequency responses are relatively smooth. Responses with sharper bends and responses with resonance modes would require using too many sections in the transversal filters.

Example 9.9

Adaptive filters can be made with amplitude modulators, as for example in Figure 9.12. The first balanced pair of modulators (multipliers) with carriers in quadrature transfers

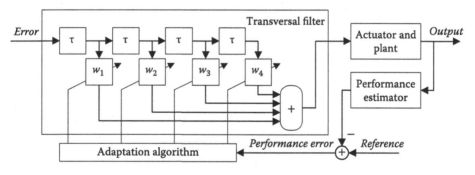

FIGURE 9.11
Transversal adaptive filter as a compensator.

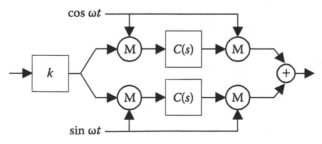

FIGURE 9.12
Adaptive filter with modulators.

the input signal from baseband to higher frequencies where the signal is filtered by filters $C(s)$, and the second pair of balanced quadrature modulators returns the signal to the baseband. The balanced quadrature modulators are used here to cancel most of the parasitic intermodulation products. The input–output frequency responses can be varied by changing $C(s)$ and using multi-frequency carriers.

   These adaptive filters are particularly useful for rejection of periodic non-sinusoidal disturbances.

10

Provision of Global Stability

Nonlinearities in the actuator, feedback path, and plant are reviewed. The following concepts are developed: limit cycles, stability of linearized systems, conditional stability, global stability, and absolute stability. The Popov criterion is discussed and applied to control system analysis and design. Nonlinear dynamic compensators (NDCs), which ensure absolute stability without penalizing the available feedback, are introduced.

10.1 Nonlinearities of the Actuator, Feedback Path, and Plant

In a nondynamic link, the current value of the output variable depends only on the current value of the input variable and not on its previous values. The link has no memory. It is fully characterized by the input–output function. Several examples of the input–output characteristics of nonlinear nondynamic links—hard and smooth saturation, hard and smooth dead zone, dead zone and saturation, and three-position relay—are shown in Figure 10.1(a)–(d), respectively. These nonlinear links represent properties of typical actuators.

A nonlinear link $y(x)$ can be placed in the feedback path to implement an inverse nonlinear operator $x(y)$, as shown in Figure 10.2. When the feedback is large, the error e is small and input $y + e$ approximately equals the signal fed back, $y(x)$.

Example 10.1

Inserting an exponential link (using a semiconductor pn-junction) in the feedback path of an op-amp, as shown in Figure 10.3, produces a logarithmic link. If the error is small, the input current $i_{in} = -i$. The range of i_{in} is about 10^4, and the range of the output voltage u is 4. The feedback path gain coefficient is 2500, i.e. 68 dB, when the input signal is at the maximum.

For the error $i_{in} - i$ to be small relative to i_{in}, the feedback must not be smaller than, say, 40 dB, even when y is small and the feedback path coefficient is relatively small. Therefore, the feedback reaches 108 dB when i_{in} is the largest.

The stability conditions limit the feedback bandwidth when the loop gain is largest (i.e., for the largest i_{in}). Therefore, assuming the loop Bode diagram slope is –10 dB/octave, the feedback bandwidth becomes 6.8 octaves smaller for the smallest i_{in}

To counteract this detrimental effect and to keep the functional feedback bandwidth wide enough when i_{in} varies with time, a nonlinear link with the characteristic approximately inverse to that in the feedback path can be installed in the forward path, which will increase the loop gain and the feedback bandwidth for low-level signals.

Two examples of plant nondynamic nonlinearities are heat radiation and turbulent liquid flow, which were mentioned in Section 7.1. These are *static nonlinearities* that can be characterized by the link's input–output function. In a link with *kinematic nonlinearities*, the relation between the input and output variables (forces, torques) depends on the output position. For example, the force-to-torque ratio in a robotic arm driven by a rotational motor depends on the elbow angle, as shown in Figure 10.4(a) and (b).

   Dynamic nonlinearities are those where the output nonlinearly depends on the current and the previous values of the input variable. They will be studied in Section 10.7 and Chapters 11–13.

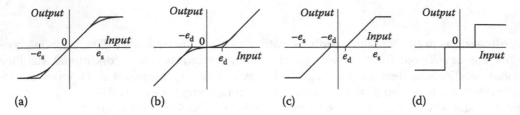

FIGURE 10.1
Characteristics of nonlinear nondynamic links: (a) soft and hard saturation, (b) soft and hard dead zone, (c) dead zone and saturation, and (d) three-position relay.

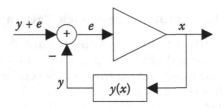

FIGURE 10.2
Inverse operator.

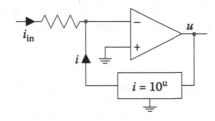

FIGURE 10.3
Logarithmic amplifier.

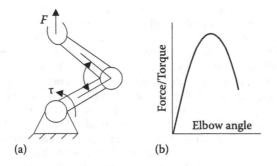

FIGURE 10.4
(a) Robotic arm and (b) its force-to-torque ratio as the function of the elbow angle.

10.2 Types of Self-Oscillation

The most frequently encountered oscillation $x(t)$ is periodic. The *phase plane* (x, x'), where $x'(t)$ is the signal's time derivative, is shown in Figure 10.5. The trajectories on the plane reflect the time history of the process. When the oscillation is sinusoidal, as in a second-order conservative system, the trajectory is a circle, as depicted in Figure 10.5(a). The unmarked phase-plane trajectories relate information about the shape of the oscillation but not about the frequency.

Figure 10.5(b) depicts a stable second-order system with some damping. Starting with any initial condition, the signal amplitude decreases with time and the trajectories approach the origin; i.e., x and x' approach 0 when time increases indefinitely. In the language of dynamic systems, the origin is a *stable fixed point* for the trajectories.

Self-oscillation in physical non-conservative nonlinear systems is initially aperiodic but often asymptotically approaches periodicity. This type of periodic oscillation is called *limit cycle*, which is a stable kind of *attractor*. In Figure 10.5(c), a nonlinear system with a limit cycle is shown. This circular limit cycle describes sinusoidal oscillation.

In many systems, $x(t)$ is rich in high harmonics, as, for example, in the oscillations that are shown in Figure 10.6. When, for example, the signal is triangular as in Figure 10.6(a), the limit cycle takes a rectangular shape as in Figure 10.5(d), with the rate x' changing by instant jumps.

A nonlinear system might possess several limit cycles, each surrounded with its own *basin of attraction* in the phase space. Different initial conditions lead to different limit cycles, and some initial conditions might lead to stability.

The limit cycle and the basin of attraction of a feedback loop that is stand-alone unstable can be modified with linear and nonlinear compensators so as to facilitate stabilizing the system with other feedback loops (see an example in A13.14).

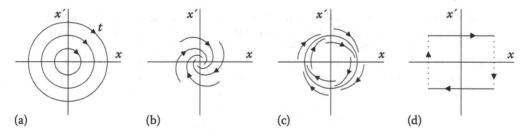

(a)   (b)   (c)   (d)

FIGURE 10.5
Trajectories on the phase plane, for (a) non-dissipative second-order system, (b) dissipative second-order system having a static attractor at the origin, (c) limit cycle for an oscillator of a sinusoidal signal, and (d) limit cycle for an oscillator of a triangular signal.

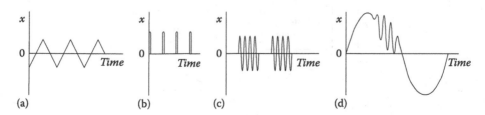

(a)   (b)   (c)   (d)

FIGURE 10.6
Periodic oscillations with high harmonic content.

Aperiodic oscillation usually does not arise in control systems of moderate complexity near the border of global stability. Because of this, we will not consider it in this book; although, aperiodic oscillations do happen in some engineering feedback systems such as in improperly designed microwave voltage-controlled oscillators (VCOs).

Phase-plane analysis can be utilized for the design of low-order relay control systems. For example, when the variable x needs to be constrained by $-a < x < a$, the *switching lines* $x = a$ and $x = -a$ drawn on the phase plane define the conditions for the relay to switch, and the x time history is derived from the trajectories on the phase plane. For double integrators such as the attitude control of a spacecraft using thrusters, the switch function is usually a combination of position and rate, and the switch lines may be curves, or the phase-space may be carved up into different regions to improve transients and also to control the characteristics of the inevitable limit cycle. For analyzing potential interaction with structural flexibilities, it is necessary to instead use a frequency-response approach using describing functions.

10.3 Stability Analysis of Nonlinear Systems

10.3.1 Local Linearization

The *first Lyapunov method* for stability analysis is applicable to nonlinear systems with differentiable characteristics. Lyapunov proved that stability of the system equilibrium can be determined on the basis of the system parameters linearized for small increments, i.e., *locally*.

Example 10.2

Consider a nonlinear two-pole (an electrical arc, neon lamp, or the output of an op-amp with in-phase current feedback) with an external dc bias voltage source u as shown in Figure 10.7(a). The differential dc resistance du/di depends on u, as shown in Figure 10.7(b). The resistance is negative on the *falling branch* between the two *bifurcation points*.

According to the first Lyapunov method, the locally linearized impedance of the two-pole

$$Z(u,s) = \frac{a_m s^m + a_{m-1} s^{m-1} + \ldots + a_0}{b_n s^n + b_{n-1} s^{n-1} + \ldots + b_0}$$

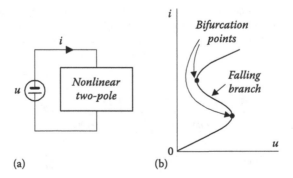

(a)   (b)

FIGURE 10.7

(a) Nonlinear two-pole device with (b) S-type current-on-voltage dependence.

with the polynomial coefficient functions of u, can be used to determine the system local stability. The system is locally unstable when the contour impedance, $Z(u, s)$, has a zero in the right half-plane of s.

Assume it is known from experiments that outside the falling branch such a device is passive, and therefore the coefficients a_o, b_o, a_m, and b_n are positive. Assume that it is also known that on the falling branch the device is stable if connected to a current source, i.e., $Z(s)$ has no poles in the right half-plane.

The differential dc resistance of the two-pole is:

$$Z(u,0) = \frac{a_o(u)}{b_o(u)}$$

As the bias point moves along the characteristic in Figure 10.7(b) into the falling branch, the resistance and therefore coefficient a_o pass through zero at the bifurcation point and both become negative on the falling branch.

On the other hand, a_m and b_n are determined by the high-frequency behavior of the circuit (by stray inductances and capacitances), and remain positive on the falling branch. With a_m positive and a_o negative, the numerator of Z must have a positive real root. Therefore, when the two-pole is shorted (connected to a voltage source), the contour impedance has a zero in the right half-plane of s, and the system is unstable. We will use this result in Section 12.3.

The first Lyapunov method justifies using the Nyquist and Bode methods of stability analysis to determine whether a nonlinear system is stable locally, i.e., while the signal deviations from the solution are small. For a system where the deviations can be big, the locally applied Nyquist and Bode stability conditions are necessary but not sufficient. In particular, a Nyquist-stable system is commonly stable when first switched on, but after being overloaded, can become unstable.

10.3.2 Global Stability

Signals that are initially finite and then, when the time increases, remain within an envelope whose upper and lower boundaries asymptotically approach zero are called *vanishing*. The system is called *asymptotically globally stable (AGS)* if its responses are vanishing after any vanishing excitation. Time exponents $\pm ae^{-kt}$, $k > 1$, are commonly used for the envelope boundaries. Asymptotic stability with such boundaries is called *exponential stability*.

An AGS system remains stable following all possible initial conditions, as opposed to a *conditionally stable* system, which is stable following some initial conditions but in which some other initial conditions trigger instability. Well-designed control systems must be AGS. However, to directly test whether a system is AGS one would be required to try an infinite number of different vanishing signals, which is not feasible. This necessitates devising convenient practical global stability criteria.

One such criterion is the *second Lyapunov method*. It uses the so-called *Lyapunov function*, which is a scalar function of the system coordinates, is equal to zero at the origin, is positive definite (i.e., positive and nonzero) outside of the origin, as illustrated in Figure 10.8 for the case of two coordinate variables, and has a negative time derivative. The function V continuously decreases with time and approaches the origin. As a result, each variable approaches zero and the system is AGS.

The Lyapunov function is often constructed as a sum of a quadratic form of the system variables and an integral of a nonlinear static function reflecting the system nonlinearity. Finding an appropriate Lyapunov function is simple when the system is low order and the stability margins are wide. Unfortunately, finding the function for a practical nonlinear system with a high-performance controller is typically difficult.

10.4 Absolute Stability

Many practical feedback systems consist of a linear link $T(s)$ and a nonlinear link (actuator) that can be well approximated by a memoryless (i.e., nondynamic) nonlinear link $v(e)$, as shown in Figure 10.9.

This system is said to be *absolutely stable (AS)* if it is AGS with any characteristic $v(e)$ constrained by

$$0 < v(e)/e < 1 \tag{10.1}$$

as illustrated in Figure 10.10. Hard and soft saturation, dead zone, and the three-level relay belong to the class of nonlinear characteristics defined by Equation 10.1.

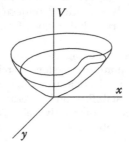

FIGURE 10.8
Lyapunov function example.

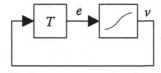

FIGURE 10.9
Feedback system with a nonlinear link.

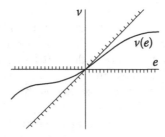

FIGURE 10.10
Characteristic of the nonlinear link in Figure 10.9.

10.5 Popov Criterion

10.5.1 Analogy to Passive Two-Poles' Connection

The stability criterion due to V. M. Popov [66] was obtained by applying Parseval's theorem to the absolute stability criterion previously obtained by A. I. Lur'e [32] using the second Lyapunov method. The Popov criterion can be readily applied to systems defined by plots of their frequency responses, and is instrumental in developing controllers with improved performance.

The Popov criterion can be understood through the mathematical analogy between the feedback system and the connection of electrical two-poles (refer to Section 7.10).

Consider a nonlinear inductor that in response to current i creates a magnetic flux $q\phi(i)$, where q is some positive coefficient and $\phi(i)/i > 1$, as shown in Figure 10.11(a). Since the flux has the same sign as the current, the energy $qi\phi(t)/2$ stored in the inductor is positive. Consider also a nonlinear resistor with the dependence of voltage on current $u = \phi(i) - i$. The power dissipated in the resistor $ui = i(\phi(i) - i)$ is positive for all i.

Let us connect this inductor, this resistor, and a passive linear two-pole with impedance $Z(s)$ (that is positive real) as shown in Figure 10.11(b). The resistor and the two-pole $Z(s)$ draw and dissipate the energy stored in the inductor. Since this energy cannot become negative, i.e., the energy cannot be overdrawn from the magnetic field, the current i decays with time and approaches 0. The system is AGS.

The voltage u in the circuit in Figure 10.11(b) is, according to the equations related to the upper branch,

$$u = \phi(i) - i + q\frac{d\phi(i)}{dt} \tag{10.2}$$

Using Laplace transforms $U = Lu$ and $I = Li$, the voltage

$$U = \ L\phi(i) - I + qsL\phi(i) = (1 + qs)L\phi(i) - I$$

The current, according to the Ohm's equation for the lower branch, is

$$I = -\frac{1}{Z(s)}U$$

These equations describe block diagram (a) in Figure 10.12.

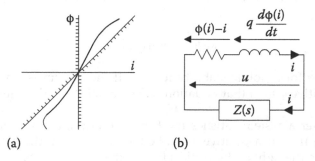

(a)   (b)

FIGURE 10.11
(a) Function $\phi(i)$ and (b) an AGS circuit.

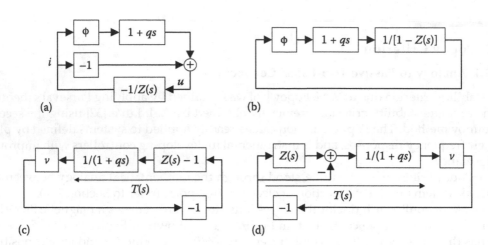

FIGURE 10.12
AGS equivalent feedback systems.

Since the link –1 can be viewed as the feedback path for the link –1/$Z(s)$, the diagram can be converted to that in (b). Next, reversing the direction of the signal propagation generates diagram (c), where the function v is the inverse of ϕ. Notice that such v satisfies Equation 10.1. Diagram (c) is equivalently redrawn with conventional clockwise signal transmission in (d).

The feedback system shown in Figure 10.12(d) is described by the same equations as the passive circuit in Figure 10.11, and is therefore AGS. The return ratio of the linear links of the loop is:

$$T(s) = [Z(s) - 1] / (1 + qs) \tag{10.3}$$

Therefore, a feedback system consisting of a nonlinear link v and the linear link $T(s)$ is AGS if $T(s)$ is representable in the form of Equation 10.3 with any p.r. $Z(s)$. In other words, the system is AS if there exists a positive q such that the expression

$$Z(s) = (1 + qs)T(s) + 1 \tag{10.4}$$

is positive real (p.r.). This means that $T(s)$ should have no poles in the right half-plane, and at all frequencies

$$\text{Re}\,[(1 + jq\omega)T(j\omega)] > -1 \tag{10.5}$$

The *Popov criterion* for absolute stability follows: if the system is stable open-loop and there exists a positive q such that Equation 10.5 is satisfied at all frequencies, the system is AS.

To check whether a system satisfies the Popov criterion, one might plot the Nyquist diagram for $(1 + qs)T(s)$. If a positive q can be found such that the Nyquist diagram for $(1 + qs)T(s)$ stays to the right of the vertical line –1, the system is AS.

The Popov criterion (which is sufficient but not necessary for AS) is more restrictive than the Nyquist criterion (which is necessary but not sufficient for AS).

Instead of the Nyquist diagram, equivalent Bode diagrams can be used.

Example 10.3

Consider a system with a loop gain response that is flat at lower frequencies and has a high-frequency cutoff with $-10\,dB$/octave slope. The phase shift in this system varies from 0 to $-150°$, as shown in Figure 10.13(a). By multiplying the loop transfer function by the Popov factor $1 + qs$ with large q, so that the Popov factor's angle is nearly $+90°$, we obtain the function $(1 + qs)T(s)$ with the angle from $+90°$ to $-60°$. The real part of this function is positive at all frequencies, and the absolute stability condition in Equation 10.5 is satisfied.

Example 10.4

Consider a band-pass system with a low-frequency roll-off slope of $10\,dB$/octave and with the associated phase shift of $150°$, as shown in Figure 10.13(b). The phase shift of the Popov factor $(1 + jq\omega)$ is within the $0–90°$ limits. Therefore, for the frequencies on the roll-off, the phase shift of the expression in the brackets in Equation 10.5 is within the $150–240°$ limits, and Equation 10.5 is not satisfied. This system therefore falls into the gap between the Nyquist and the Popov criteria, and no judgment can be passed on whether the system is AS.

Example 10.5

In a Nyquist-stable system, T must be real and less than -1 at some frequency, as shown in Figure 10.14. At this frequency, since T is real, q does not affect the left side of Equation 10.5, and since T is less than -1, the inequality is not satisfied. Hence, Nyquist-stable systems do not satisfy the Popov criterion.

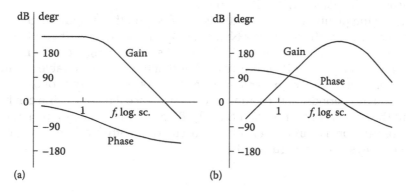

(a)   (b)

FIGURE 10.13
Loop responses for (a) an AS system and (b) a system for which AS cannot be proved with the Popov criterion.

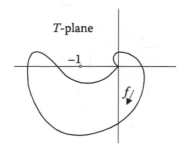

FIGURE 10.14
Nyquist-stable system.

10.5.2 Different Forms of the Popov Criterion

Equation 10.5 uses the real part of a function. The condition can be changed into an equivalent form that makes use of Bode diagrams, which is more convenient. Adding qs to $Z(s)$ does not change whether $Z(s)$ is positive real since $qj\omega$ is purely imaginary. From this follows the *second form* of the criterion: the system is AGS if a real positive q exists such that $(1 + qs)T(s) + 1 + qs$ is p.r., i.e.,

$$(1 + qs)F(s) \text{ is p.r.} \tag{10.6}$$

at all frequencies:

$$|\arg[(1 + jq\omega)F(j\omega)]| < \pi/2 \tag{10.7}$$

Therefore, AGS can be established by plotting the Bode diagram for $(1 + jq\omega)F(j\omega)$ and, using the phase-gain relations, deciding whether Equation 10.7 is satisfied.

The *third form* of the Popov criterion uses the *modified Nyquist plane* $(ReT, \omega ImT)$. (When drawn on this plane, the Nyquist diagram is vertically compressed at lower frequencies and expanded at higher frequencies.)

Equation 10.5 can be rewritten as $ReT(j\omega) - q\omega ImT(j\omega) > -1$ or

$$\omega ImT(j\omega) < q^{-1}[ReT(j\omega) + 1] \tag{10.8}$$

With the sign < replaced by =, Equation 10.8 represents the *Popov line* shown in Figure 10.15. It passes through the point (–1, 0) with slope $1/q$.

Therefore, the inequality in Equation 10.8 is satisfied and the system is AS if a Popov line can be drawn entirely to the left of the Nyquist diagram on the modified plane.

For example, the system characterized by the plot shown in Figure 10.15 is AS.

The plot shown in Figure 10.16 indicates neither that the system is stable nor that the system is unstable. This plot only guarantees the absence of certain types of oscillation. It has been proven that no periodic self-oscillation can take place if a Popov line can be drawn to the left of all the points of the modified Nyquist diagram that relate to the Fourier components of the oscillation. For instance, periodic oscillation with fundamental f_1 cannot exist in the system represented in Figure 10.16.

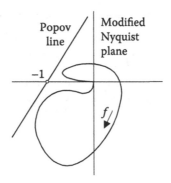

FIGURE 10.15
Popov line on a modified Nyquist plane.

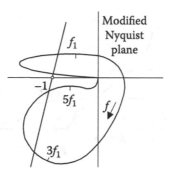

FIGURE 10.16
On the basis of this diagram a judgment can be passed that a limit cycle with fundamental f_1 is not possible, but no judgment can be made about whether the system is AS.

The Popov criterion can be extended using more complex circuits than that in Figure 10.12, for example circuits including several various nonlinear two-poles and several passive two-poles. It was proven, in particular, that the criterion is valid with the function $(1 + j\omega q_1)/(1 + j\omega q_2)$ replacing the Popov factor $(1 + qs)$.

10.6 Applications of Popov Criterion

10.6.1 Low-Pass System with Maximum Feedback

The Popov criterion is not a tool in everyday use like the Nyquist criterion or Bode diagrams. Its major application is to prove that the Nyquist and Bode criteria established for linear systems also guarantee absolute stability of most practical nonlinear systems, including those with maximized feedback.

The low-pass system with the Bode optimal cutoff is AS. This can be established with the second form of the Popov criterion employing $(1 + q_1 s)/(1 + q_2 s)$ instead of the usual Popov factor. With $1/q_1 \ll 1$ and $1/q_2 = 4\omega_c$, the inequality in Equation 10.7, expressed as

$$|\arg F + \arg(1 + j\omega q_1)/(1 + j\omega q_2)| < \pi/2$$

is satisfied, as seen in Figure 10.17.

10.6.2 Band-Pass System with Maximum Feedback

Consider a band-pass system with the feedback maximized over a large relative frequency band, with phase stability margin $y\pi$ in the high-frequency cutoff region and $y_1\pi$ in the low-frequency roll-up region, as shown in Figure 10.18(a) and (b). It is seen that the combination of $y = 0$ and $y_1 = 1/2$ satisfies Equation 10.7, i.e., $\arg(1 + jq\omega)T$ varies only from $-\pi/2$ to $\pi/2$, if $1/q$ is chosen close to the mean square frequency of the passband.

Therefore, stability can be proved with the Popov criterion only when $y_1 \geq 1/2$. However, the value of $y_1 = 1/6$ does well in practical design. Making y_1 excessively large to satisfy the criterion imposes little impairment to the available feedback but does requires an extension of the loop gain to lower frequencies.

In a narrow-band band-pass system, the Popov criterion requires that the roll-up and cutoff phase stability margins, $y_1 180°$ and $y180°$ exceed 90°. Satisfying this requirement dramatically reduces the maximum available feedback compared with using the typical

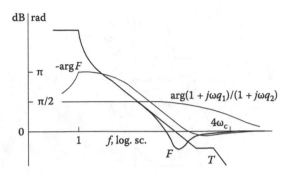

FIGURE 10.17
Application of the Popov criterion to the Bode optimal low-pass cutoff.

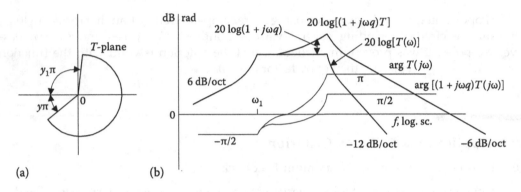

(a)   (b)

FIGURE 10.18
Band-pass system analysis with Popov criterion.

stability margin of $y/6$ in the band-pass optimal cutoff. The smaller the relative band-width, the more the feedback is limited by the Popov criterion because of the deficient filtering properties of the Popov factor $(1 + jq\omega)$. The Popov criterion is excessively restrictive in this case—phase stability margins of $\pi/6$ at both higher and lower frequencies suffice to provide AGS of a system with a saturation link, regardless of the relative bandwidth—although this has not yet been proven theoretically.

10.7 Absolutely Stable Systems with Nonlinear Dynamic Compensation

10.7.1 Nonlinear Dynamic Compensator

Larger feedback is available in Nyquist-stable systems; although, such systems are not AS when the compensators are linear. Still, they can be made AS by using *nonlinear dynamic compensators* (NDCs). The NDC can be built of linear and nondynamic nonlinear links.

In the system shown in Figure 10.19, the nonlinear link $1 - v(e)$ in the local feedback of the NDC uses the same nonlinear function $v(e)$ as the nonlinear link of the actuator. Typically, $v(e)$ is a saturation link so that $1 - v(e)$ represents a dead zone. The rest of the links in the block diagram are linear. For the AS analysis, it can be assumed the command is 0.

We denote by T_P the return ratio for the plant measured when the link v is replaced by 1 (and the link $1 - v$, by 0). Then, when $v(e)$ is saturation, the compensator transfer function for small-level signals is expressed as T_P/P. When the signal level is very large, the return ratio in the NDC's local loop becomes G.

10.7.2 Reduction to Equivalent System

The diagram shown in Figure 10.19 depicts a system that has two identical nonlinear links $v(e)$, with the same input signal e, and therefore the same output signal v. For the sake of stability analysis, the system can be modified equivalently into the one shown in Figure 10.20, which contains only one nonlinear link v. The linear links within the dashed envelope form a composite linear link. We denote the negative of its transfer function as T_E (equivalent return ratio). If T_E satisfies the Popov criterion, the system must be globally stable.

To find the expression for T_E, the diagram in Figure 10.20 is further redrawn as shown in Figure 10.21.

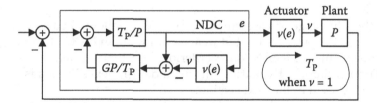

FIGURE 10.19
Feedback system with nonlinear link v in the actuator and link $1 - v$ in the feedback path of the NDC.

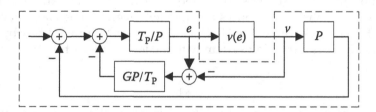

FIGURE 10.20
Equivalently transformed system containing a single nonlinear link v.

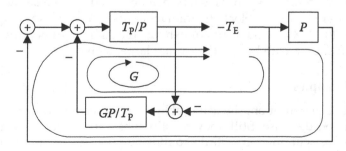

FIGURE 10.21
Calculation of T_E using two parallel paths and a loop tangent to both paths.

From this diagram, using Mason's rules, the negative of the transfer function from the output of the nonlinear link to its input is:

$$T_E = \frac{T_P - G}{1 + G} \tag{10.9}$$

Given T_E and T_P, the NDC linear link transfer function is:

$$G = \frac{T_P - T_E}{1 + T_E} \tag{10.10}$$

From Equation 10.10, the plant feedback is:

$$1 + T_P = (1 + G)(1 + T_E) \tag{10.11}$$

Equation 10.11 states that the plant feedback is the product of the local feedback in the NDC for large-level signals and the feedback in the equivalent system. Therefore, the areas of substantial positive feedback in the G and T_E loops should not overlap or else the positive feedback in the plant will be excessive and the phase stability margin in the plant loop will be, correspondingly, small. If positive feedback in each loop is substantial, the crossover frequency of $T_E(j\omega)$ must be either much smaller or much larger than the crossover frequency of $G(j\omega)$.

We can see the advantage of a system with an NDC as follows: in a conventional system, T_P must satisfy the Popov criterion, but in a system with an NDC, the only requirements are that T_P satisfies the weaker Nyquist criterion, and that T_E, defined by Equation 10.9, satisfies the Popov criterion. Exploiting this extra design flexibility leads to better performance, as will be shown in the next section.

Commonly, there is no need to maximize to the full extent the disturbance rejection while keeping specified stability margins, or maximizing the stability margins while providing specified disturbance rejection response. In these cases, there still remains some freedom in choosing the responses for G and T_E. This freedom can be utilized for the provision of desired transient responses for large-level signals. For a large-level signal, the dead zone in the feedback path of the NDC can be replaced approximately by a unity link, and the system becomes a tracking system with the loop transfer function:

$$\frac{T_P}{G + 1} \tag{10.12}$$

and saturation in the actuator. Such a system has response without an overshoot if the guard-point phase stability margin is 90° or so. To achieve this, the Bode diagram for Equation 10.12 must be rather shallow over a rather wide frequency interval, including the crossover frequency. Examples with such responses will be demonstrated in the next section.

The equations derived for the block diagram in Figure 10.19 are also valid for the system shown in Figure 10.22, where the actuator exhibits the same nonlinearity as the function $v(e)$ in the NDC feedback path but for the k-times larger signal.

10.7.3 Design Examples

With lower-order compensation, the achievable disturbance rejection and the benefits of an NDC application decrease. Still, they remain substantial, and in Examples 10.6–10.8 we will use systems with low-order links to illustrate the benefits of nonlinear dynamic compensation, rendering systems with rather steep Bode diagrams AS, and in some cases, improving the transient responses to large commands.

Example 10.6

In the homing system diagrammed in Figure 10.23, $v(e)$ is a saturation, and the plant P is a single integrator. The NDC forward path transfer function T_P/P is a lead-with-integration compensator and the plant return ratio for small-level signals is T_P:

$$\frac{T_P}{P} = \frac{2(s+0.5)}{s(s+2)}$$

$$T_P = \frac{2(s+0.5)}{s^2(s+2)}$$

The asymptotic Bode diagram for T_P shown in Figure 10.24(a) is typical for simple lead compensation. The logarithmic Nyquist diagram for T_P is plotted in Figure 10.24(b) with the nyqlog function from the Bode step toolbox described in Appendix A14:

```
np = [2 1]; dp = [1 2 0 0]; nyqlog(1,np,dp)
```

In the linear mode of operation (for small signals), the guard-point phase stability margin is 37° and the closed-loop transient response has a 43% overshoot.

The general rules for selecting T_E and G have been introduced in Section 10.7.2. We will consider two low-order choices (A and B) of T_E and G.

A. We will start the design by a guessed response for T_E:

$$T_E = \frac{2}{s(s+2)}$$

As shown in Figure 10.24, this response merges with T_P at higher frequencies but has a shallower slope at lower frequencies and, correspondingly, lesser phase lag. The system with such a T_E is AS (and even the *processes* in this system are stable as discussed in Chapter 12). From Equation 10.10,

$$G = \frac{1}{s(s^2+2s+2)}$$

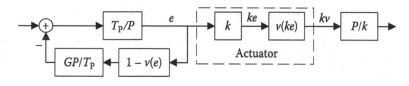

FIGURE 10.22
Scaling down the nonlinear link in the NDC by a factor of k.

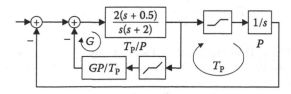

FIGURE 10.23
Block diagram of a system with an NDC.

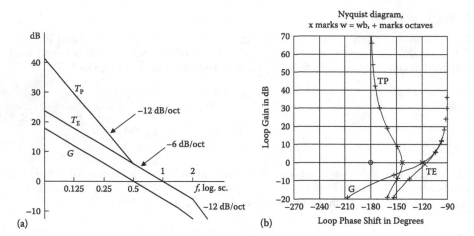

FIGURE 10.24
(a) Asymptotic Bode diagrams and (b) *L*-plane plots for Example 10.6, case A.

The asymptotic Bode diagrams for these T_E and G are shown in Figure 10.24(a). The *L*-plane Nyquist diagrams are plotted in Figure 10.24(b) with the following script (which includes the previously shown line for plotting T_P):

```
np = [2 1]; dp = [1 2 0 0]; nyqlog(1,np,dp); hold on
ne = 2; de = [1 2 0]; nyqlog(1,ne,de); hold on
ng = 1; dg = [1 2 2 0]; nyqlog(0.5,ng,dg); zoom off
gtext('TP');     gtext('TE');     gtext('G') %place labels with mouse
```

The transfer function of the link in the feedback path of the NDC is:

$$\frac{GP}{T_P} = \frac{s+2}{(s+0.5)(s^2+2s+2)}$$

The large-amplitude loop transfer function (Equation 10.12) is

$$\frac{T_P}{G+1} = \frac{2(s+0.5)}{s^2(s+2)}\frac{s^3+2s^2+2s}{s^3+2s^2+2s+1} = \frac{2s^3+5s^2+6s+2}{s^5+4s^4+6s^3+5s^2+2s}$$

The large-amplitude closed-loop response (not shown here) is dissimilar to the response of a Bessel filter; therefore, closed-loop transient responses to large commands are not expected to be acceptable.

B. If, while retaining the NDC forward path link, we interchange the functions for G and T_E, making them

$$T_E = \frac{1}{s(s^2+2s+2)} \quad \text{and} \quad G = \frac{2}{s(s+2)}$$

then Equations 10.9–10.11 are still satisfied since G and T_E enter these equations symmetrically. The transfer function of the link in the feedback path of the NDC is:

$$\frac{GP}{T_P} = \frac{2}{2s+1}$$

The stability margins for this T_E (i.e., for G in Figure 10.24) are smaller than in version A, but sufficient for the system to remain AS (although not process stable).

The large-signal loop transfer function (Equation 10.12) is:

$$\frac{T_P}{G+1} = \frac{2(s+0.5)}{s(s^2+2s+2)}$$

This frequency response and the closed-loop response are plotted in Figure 10.25(a). The step response is shown in Figure 10.25(b); it has no overshoot, indicating that the overshoot in the nonlinear regimes in response to large commands will be small or nonexistent.

Indeed, the system transient responses to step commands obtained with Simulink and shown in Figure 10.26 demonstrate that NDC improves the responses.

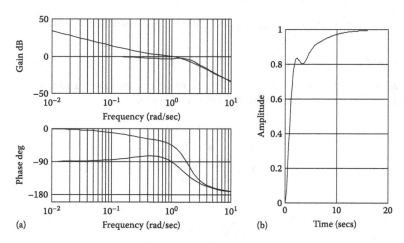

FIGURE 10.25

(a) Open-loop and closed-loop responses for large signals and (b) step response in Example 10.6.

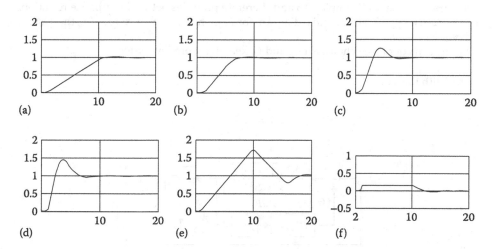

FIGURE 10.26

Transient response to unity step command in Example 10.6, case A: (a–c) systems with threshold of saturation and dead zone set to, respectively, 0.1, 0.2, and 0.5; (d) system without NDC; (e) wind-up in the system where threshold is set to 0.2, and the NDC feedback path is disconnected; and (f) actuator output for (a).

In the linear mode of operation (d) the overshoot is large, approximately 45%. In (a)–(c) the responses are shown for the systems with the unity step command and different saturation thresholds. The larger ratio of the command to the saturation threshold, the smaller the percentage of overshoot.

Figure 10.26(e) shows the wind-up response, which occurs when the feedback path in the NDC is disconnected and the actuator overloaded by a relatively large disturbance or command. The effects of NDCs on the transient response for large-level signals will also be considered in Chapter 13.

Figure 10.26(f) shows that the actuator output is constant over some time range and then rapidly drops to nearly zero. This indicates efficient utilization of the actuator output power capability.

The system exhibits performance that is desirable for many homing systems: the absolute value of the overshoot is limited to a small value. Large (percent-wise) overshoots occur in the responses to small disturbances but not in the responses to large disturbances.

The performance can be further improved with higher-order compensation.

Example 10.7

The system block diagram is shown in Figure 10.27. The transfer function of the NDC forward path and the plant small-signal return ratio are:

$$\frac{T_P}{P} = \frac{2s^2 + 1.2s + 0.1}{s^3 + 2s^2}$$

$$T_P = \frac{2s^2 + 1.2s + 0.1}{s^4 + 2s^3}$$

i.e.,

$$T_P = \frac{2(s + 0.5)(s + 0.1)}{s^3(s + 2)}$$

As seen on the asymptotic Bode diagram in Figure 10.28(a), $|T_P|$ is larger at lower frequencies than that in Example 10.6 and therefore provides better disturbance rejection. The system is type 3. The Nyquist diagram for T_P is shown in Figure 10.28(b); the system is Nyquist stable.

Let us consider three choices of T_E and G, keeping them low order.

A. With the same

$$T_E = \frac{2}{s^2 + 2s}$$

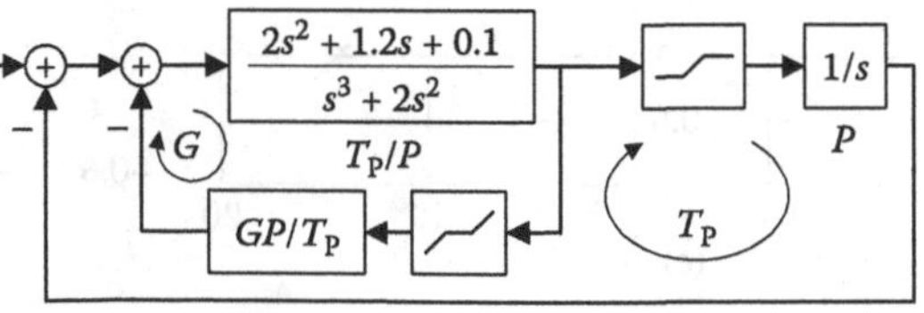

FIGURE 10.27
Block diagram of a system with an NDC.

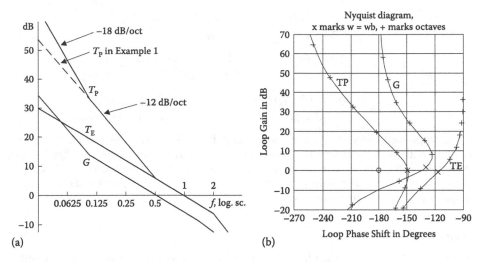

FIGURE 10.28
(a) Asymptotic Bode diagram and (b) *L*-plane plots for Example 10.7.

as in case A of Example 10.6, *G* is found from Equation 10.10 as:

$$G = \frac{1.2s + 0.1}{s^2\left(s^2 + 2s + 2\right)}$$

The asymptotic Bode diagrams for T_E and G are shown in Figure 10.28(a). The *L*-plane plots are shown in Figure 10.28(b). For large-level signals, the system is type $3 - 2 = 1$ since the NDC local loop is type 2. The NDC feedback path transfer function is:

$$\frac{GP}{T_P} = \frac{\left(0.6s + 0.05\right)\left(s + 2\right)}{\left(s^2 + 2s + 2\right)\left(s + 0.5\right)\left(s + 0.1\right)}$$

The system is globally stable and the stability margins in the equivalent system are large, but it does not qualify as a good homing system because the transient responses to large-amplitude signals (not shown here) are far from the best possible.

B. With the functions $T_E(s)$ and $G(s)$ interchanged, i.e.,

$$T_E = \frac{1.2s + 0.1}{s^2\left(s^2 + 2s + 2\right)} \quad \text{and} \quad G = \frac{2}{s\left(s + 2\right)}$$

the NDC feedback path becomes:

$$\frac{GP}{T_P} = \frac{2}{s\left(s + 2\right)} \frac{s^2\left(s + 2\right)}{2\left(s + 0.5\right)\left(s + 0.1\right)} = \frac{s}{2\left(s + 0.5\right)\left(s + 0.1\right)} = \frac{s}{2s^2 + 1.2s + 0.1}$$

The type of system for large-level signals is $3 - 1 = 2$. The transient responses for large-magnitude step commands (not shown here) still deserve improvement.

C. The numerator of the NDC feedback path transfer function is further adjusted by trial and error, in order to improve the transient response to a large-amplitude step command. With NDC feedback the path transfer functions are:

$$\frac{GP}{T_P} = \frac{1.6s + 0.16}{2s^2 + 1.2s + 0.1}, \qquad G = \frac{GP}{T_P}\frac{T_P}{P} = \frac{1.6s + 0.16}{s^3 + 2s^2}$$

$$G + 1 = \frac{s^3 + 2s^2 + 1.6s + 0.16}{s^3 + 2s^2}, \qquad \frac{T_P}{G + 1} = \frac{2s^2 + 1.2s + 0.1}{s(s^3 + 2s^2 + 1.6s + 0.16)}$$

$$T_P + 1 = \frac{s^4 + 2s^3 + 2s^2 + 1.2s + 0.1}{s^4 + 2s^3}, \qquad T_E = \frac{0.4s^2 + 1.04s + 0.1}{s^4 + 2s^3 + 1.6s^2 + 0.16s}$$

The large-signal open-loop and closed-loop Bode diagrams are shown in Figure 10.29(a). With such diagrams, the transient response shown in Figure 10.29(b) has no overshoot, and the nonlinear system transient response shown in Figure 10.30 becomes satisfactory. The Nyquist diagram for T_E is shown in Figure 10.31, where the crossover frequency $\omega_b = 0.65$. The system is AS.

Example 10.8

In Examples 10.6 and 10.7, the plant is $1/s$. Let us next consider the system with a double-integrator plant.

To retain the plant loop return ratio the same as before, let us multiply the NDC forward path transfer functions in Examples 10.6 and 10.7 by s, and to retain the same internal feedback in the NDC, let us multiply the feedback path by $1/s$. The system with the $1/s^2$ plant and the modified NDC has the same G and T_E, and is therefore AS.

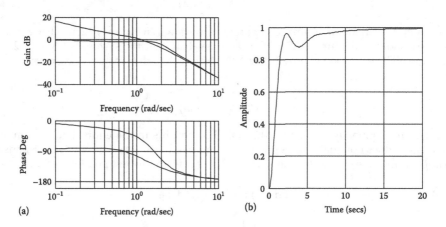

FIGURE 10.29
(a) Bode diagrams and (b) transient response for $T_P/(G + 1)[1 + T_P/(G + 1)]$.

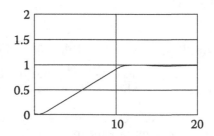

FIGURE 10.30
Transient response of the nonlinear system.

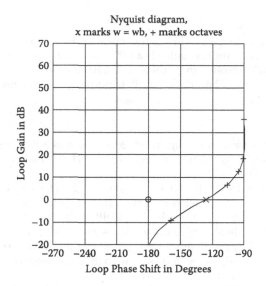

FIGURE 10.31
Nyquist diagram for T_E to large step commands in Example 10.7, case C, obtained with Simulink.

However, since the NDC reduces the system's type for large signal amplitudes by the number of poles of G at dc, the output transient responses to large-magnitude step commands undershoot. When good large-signal transient responses must be ensured in the nonlinear system, the Bode diagram for T_E should be made steeper. Acceptable performance can be obtained with:

$$P = \frac{1}{s^2}, \frac{T_P}{P} = \frac{2s^3 + 1.2s^2 + 0.1s}{s^3 + 2s^2}, \frac{GP}{T_P} = \frac{s}{1.5s^3 + 1.5s^2 + 0.15s + 0.002}$$

Better transient responses can be achieved by making the dead zone in the NDC somewhat smaller than the actuator's threshold of saturation (although in this case the system analysis with the Popov criterion is no longer valid, and the disturbance rejection for large signals will decrease, which may or may not be of importance).

In examples 10.6–10.8 we use explicit algebraic low-order transfer functions for the NDC's. The following three examples use graphical representations for the gain responses. The related phase responses can be calculated using either graphical methods described in Chapter 4, or the programs from Appendices 5 or 14. Then the responses for G and for the compensator T_P/P can be approximated by rational functions.

Example 10.9

Figure 10.32 shows the diagram for $G + 1$ as Bode optimal cutoff with $f_b = 8\,Hz$, 30° phase stability margin, and 10 dB amplitude stability margin, resulting in 40 dB of feedback over the bandwidth [0,1]. T_E is chosen with only 10 dB over the functional bandwidth, and having a phase stability margin of more than 90°. The attained feedback $|T_P + 1|$ of 50 dB exceeds the 40 dB available in the single-loop AS system. The system is Nyquist stable and AS.

At lower frequencies the responses can be reshaped, for example, as shown by the dashed lines. A rational function approximation for these responses can be found as described in Chapter 6.

Example 10.10

As indicated in Section 10.6, a band-pass transform of the low-pass Bode optimal cutoff with a phase stability margin smaller than 90° does not satisfy the Popov criterion. Application of an NDC resolves this problem. The band-pass transform of the Bode diagrams shown in Figure 10.32 produces an AS band-pass system with 50 dB of feedback.

Example 10.11

The Bode plot for T_E displayed in Figure 10.33 is chosen as a Bode optimal cutoff with 40 dB of feedback and a 30° phase stability margin. The available feedback in the AS system without the NDC is 40 dB.

A Nyquist-stable local loop in the NDC is chosen to reduce the loop crossover frequency so that the area of positive feedback is kept away from the area of positive feedback in the T_E loop. The Nyquist-stable plant feedback is 80 dB in the operational frequency range, much larger than the 40 dB allowable in the AS system without an NDC.

In Examples 10.9–10.11, no attention was paid to the closed-loop transient responses. These responses, if desired, can be improved with the command feedforward technique or with additional nonlinear feedback loops.

We can conclude that NDCs improve the performance of AS systems.

Another approach to designing feedback systems with nonlinear dynamic compensation will be discussed in Chapter 11.

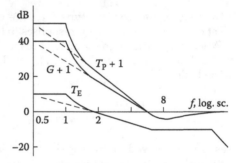

FIGURE 10.32
Bode diagrams for the AS system of Example 10.9.

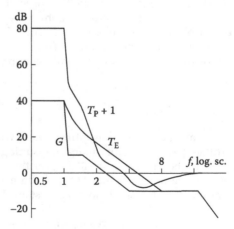

FIGURE 10.33
Bode diagrams for the AS system of Example 10.11.

PROBLEMS

1.
   a. Make an amplifier equalizing the attenuation of a span of a coaxial cable (the cable attenuation in dB is proportional to the square root of the frequency) using a passive two-port imitating the cable.
   b. Make an A/D converter from a D/A converter.
   c. Make a voltage-to-frequency converter from a frequency-to-voltage converter.

2. What is the input–output characteristic of a feedback system with a high-gain forward path and a feedback path where (a) saturation or (b) dead zone is placed?
3. Sketch the spectral density plots for the signals in Figure 10.6.
4. Find the point-to-point correspondence between the limit cycles in Figure 10.5 and the time responses for x and x'.
5. Draw phase-plane limit cycle curves for (a) a triangular symmetrical signal and (b) a triangular signal with different front and back slopes.
6. What is the current-to-voltage characteristic of the output of an amplifier with in-phase voltage feedback, if the amplifier gain coefficient is 20, the feedback voltage divider consists of two $10\,\text{k}\Omega$ resistors, and the power supply voltage is $\pm10\,\text{V}$? What external load resistance makes the circuit stable?
7. Is the system with saturation AS if the following? (Draw asymptotic Bode diagrams or use computer-generated Bode diagrams, and apply the Popov criterion.)

   a. $T(s) = 100(s + 3)(s + 6) / [(s + 10)(s + 200)s]$
   b. $T(s) = 100(s + 2)(s + 16) / [(s + 20)(s + 200)s]$
   c. $T(s) = 300(s + 1)(s + 5) / [(s + 100)(s + 200)s]$
   d. $T(s) = 300s(s + 1)(s + 5)(s + 5{,}000) / [(s + 100)(s + 120)(s + 130)]$

8. Are the systems from the previous problem AS when the saturation link is replaced by a dead-zone link? By a three-position relay?
9. A system is AS. Is it globally stable (yes, no, or no judgment can be passed) if the nonlinear link $v(e)$ is replaced by a nondynamic link described by the following equations:

   a. $v = e^2 / 2$

   b. $e = v^2 / 2$

   c. $v = e^3 / 2$

   d. $e = v^3 / 2$

   e. $v = e - e^2 / 2$

   f. $e = v - e^2 / 2$

   g. $e = 5v$

   h. $v = 5e$

   i. $e = v + 0.25v^3$

   j. $e = v + v^{25}$

   k. $e = v / (v - 1)$

   over the range $|v| < 1$, no solution outside this range

10. Draw the characteristic of the nonlinear link obtained by connecting in parallel a unity link and the negative of a three-position relay link.

11. Design an AS system with an NDC, with $f_b = 1\,\text{kHz}$, using the design prototype in Example 10.6, Figure 10.24(a). Make MATLAB or SPICE simulations. Plot the output time responses to step function input of small and large amplitudes.

12. Design an AS system with an NDC, with $f_b = 140\,\text{Hz}$, using the design prototype in Example 10.7, Figure 10.24(b). Make MATLAB or SPICE simulations. Plot the output time responses to step function input of small and large amplitudes.

13. Using the block diagram shown in Figure 10.22, scale down the saturation link for the NDC, when the plant saturation threshold is 20 V, and the maximum output signal of the driver is 2 V.

14. What is the meaning of the word *asymptotic* in the terms *asymptotic Bode diagram*, *AS*, and *AGS*? Can an asymptotic Bode diagram be used in determining AS?

15. Research projects:

    Design a Nyquist-stable system for a plant specified by the instructor, with the feedback bandwidth specified by the instructor, with saturation in the actuator. Verify the design with Simulink. Apply a large-amplitude vanishing signal to the system's input, and observe the following self-oscillation.

    Introduce a link with gain coefficient 100 in the loop after the compensator.

    Introduce local feedback about this link that makes the main loop response AS type. Introduce a dead zone into the local feedback path.

    Transform equivalently this system to a single-loop system. Modify the linear links, if necessary, for the system to satisfy the Popov criterion.

    Study the responses to commands of different shapes and amplitudes. Modify the linear links, if necessary, to improve the responses.

    Compare the disturbance rejection of the designed system with the disturbance rejection of an AS system without an NDC and with (a) Bode step response or (b) PID response.

16. Research project: Do a design similar to that of the previous problem for a system with a three-position relay in the actuator and an appropriate nonlinear link in the compensator.

17. Research project: Do the same as in Problem 13 for the system with the characteristic of the nonlinear link in the actuator described by the equation $e = -v - 0.25v^3$ and an appropriate nonlinear link in the compensator.

18. Research area: Design of nonlinear prefilters and feedforward paths complementing nonlinear feedback loops.

19. Research area: Design of NDCs with multiple nonlinear links.

20. All links in the system diagrammed in Figure 10.34 are linear except the two identical links v. Which of the following conditions allows stability analysis with the Popov criterion?

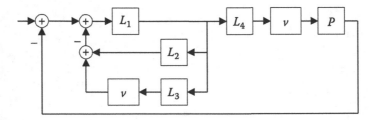

FIGURE 10.34
Feedback system.

a. $L_1 = L_2$

b. $L_2 = L_3$

c. $L_3 = L_4$

d. $L_2 = L_4$

e. $L_1 = 2.7L_4$

21. The noninverting amplifier is shown in Figure 10.35. Describe the (a) voltage-current ratio and (b) ratio behavior with the falling branches (like in Figure 10.7), and show (c) the direction of jumps while given the source of either a voltage or a current source and (d) the rotational sources of voltage and current, to measure these responses.

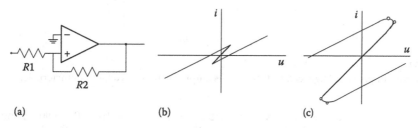

(a) (b) (c)

FIGURE 10.35

(a) Noninverting amplifier, (b) volt-amperes characteristic of its input resistance, (c) increased scale (slightly distorted) of the same volt-amperes characteristic showing the bifurcation points.

ANSWERS TO SELECTED PROBLEMS

3. The sketches of the spectral density for the oscillations are shown in Figure 10.36.

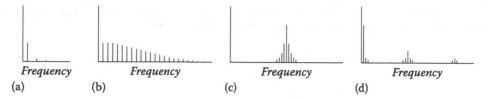

(a) (b) (c) (d)

FIGURE 10.36

Spectral density plots for oscillation shapes in Figure 10.6.

6. The circuit is depicted in Figure 10.37(a), and the characteristic is shown in Figure 10.37(b). The output impedance in the linear mode of operation can be calculated with Blackman's formula as $R = R_oF(\infty) = (1 - 10)20 \text{ k}\Omega = -8 \text{ k}\Omega$. When the voltage exceeds VCC, which is 10 V, the amplifier is saturated, and the output resistance is $R_o = 20 \text{ k}\Omega$ (we assume the amplifier output impedance remains infinite in the linear mode of operation and while it is saturated). The system is certainly stable when the load impedance is 0 since this disconnects the feedback. The negative impedance on the falling branch contains a positive real zero.

7a. The system is AS, as can be proved with $q = 0.1$, using MATLAB to verify Equation 10.7 or any of the other equivalent conditions.

20. The answer is (c) since in this case the input to both identical nonlinear links is the same and the outputs are therefore the same. Thus, for the purpose of stability analysis, one of the nonlinear links can be removed.

22. The input impedance of the amplifier is shown with input impedance as $R_1 - R_2/(K-1)$, where K is the amplifier gain on Figure 10.37 and R_2 the feedback resistor. The branch going through the central point has the input resistance R_1.

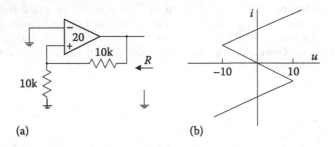

(a)   (b)

FIGURE 10.37
(a) Amplifier with in-phase voltage feedback and (b) its output current-to-voltage characteristic.

The direction of the rotational axis changes when we deviate from 0° to 90° using the converted voltage source. In this case none of the points correspond to the three-valued characteristic curve, Figure 10.38.

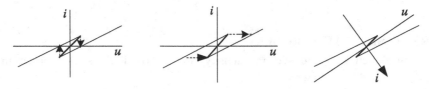

FIGURE 10.38
Jumps for a voltage source (left), a current source (center), and a current source with an appropriate internal conductance that create single-valued voltage-to-current characteristics (right).

11

Describing Functions

Stability analysis based on harmonic balance provides sufficient accuracy when applied to well-designed control feedback loops having low-pass filter properties.

Describing function (DF) stability analysis is simple and convenient. In this chapter, DFs are derived for most common nonlinear links: saturation, dead zone, three-position relay, and hysteresis. Simple approximate formulas are derived and used in the design of nonlinear controllers.

A bang–bang temperature controller with a hysteresis link is described, and spacecraft attitude control using thrusters is discussed.

Iso-f and iso-E responses are introduced. The responses characterize the dependence of DF on the signal amplitude and frequency.

Nonlinear dynamic compensators are links that provide phase advance for large-level signals. A loop is analyzed that includes a nonlinear dynamic compensation (NDC) and an additional nonlinear link (actuator). Stability margins for such systems are defined.

An NDC can be built of one nonlinear and several linear links. Conditions are discussed for maximizing the phase advance for large-level signals. NDCs are described with parallel nonlinear channels and with nonlinear local feedback paths.

It is known that SISO and MIMO Nyquist-stable systems provide larger feedback, thus enhancing the disturbance rejection. However, their stability is conditional when the only nonlinear links in the systems are the actuators' saturation. NDCs eliminate limit cycles in such systems. NDCs also improve transient responses for large command amplitudes and reduce the effects of process instability. For all these purposes, NDCs can be built in the same configurations as the NDCs designed for provision of global stability. NDCs are neither expensive nor complex—an analog NDC can be made with an op-amp and five to ten passive elements, resistors, and capacitors. A digital NDC is trivial to program. So, there is little excuse for lowering the system performance by using only linear controllers.

Nonlinear interaction between local and common loops is reviewed, and the reader is warned about practical cases where such interaction can lead to a limit cycle.

The effects of harmonics and intermodulation on system stability and accuracy are discussed. The chapter ends with a description of the procedure for testing whether a system is AGS (absolutely globally stable).

11.1 Harmonic Balance

11.1.1 Harmonic Balance Analysis

The condition for periodic self-oscillation (limit cycle) is unity transmission about the feedback loop; i.e., after passing about the loop, the same signal must return to any initially chosen cross section. Since periodic signals can be presented in Fourier form, each harmonic of the return signal is the same in amplitude and phase as the harmonic of the signal at the beginning of the feedback loop. This condition is called *harmonic balance*.

Since the harmonics interact in the nonlinear links, the superposition principle cannot be used here. We cannot consider a single harmonic in isolation from the others. For each of the harmonics, the transfer function's equality to 1 should be verified in the presence of all other harmonics. Although generally this procedure could be cumbersome, harmonic balance analysis is simplified when the Fourier series can be justifiably truncated to a few terms. To establish whether this is possible, let us consider a typical feedback system example.

Example 11.1

The feedback system diagrammed in Figure 11.1 contains a common type of plant with an integrator and two real poles. The compensator has a real gain coefficient k (the compensator is not optimally designed). The loop phase lag reaches π at frequency 0.5 Hz. A self-oscillation takes place when the loop gain coefficient exceeds 1 at this frequency, which happens when $k > 11$. The self-oscillation initially grows exponentially until, due to the saturation link in the loop, the signal stabilizes with some specific amplitude and shape.

The shapes of the signal $v(t)$ at the output of the saturation link are illustrated in Figure 11.1(b) for three particular values of k. When the gain is barely sufficient for the self-oscillation to occur, $v(t)$ is sinusoidal. With larger k, the signal $e(t)$ is clipped in the saturation link and $v(t)$ becomes nearly trapezoidal. When k is large enough to make the loop gain coefficient 20 or more, $v(t)$ becomes nearly Π-shaped.

Because of the saturation symmetry, $v(t)$ is symmetrical, and therefore contains only odd harmonics. The amplitudes of the harmonics increase when the shape of $v(t)$ approaches rectangular. In this case, the Fourier series for $v(\omega t)$ is:

$$v(\omega t) = (4/\pi)\left[\sin \omega t + (1/3)\sin 3\omega t + (1/5)\sin 5\omega t + (1/7)\sin 7\omega t + \ldots\right]$$

At the same time, as seen in Figure 11.1(c), the signal $e(t)$ at the input to the nonlinear link looks smooth. This happens because $v(t)$ is nearly doubly integrated by the loop linear links. As a result, $e(t)$ and its first and second derivatives are continuous. Another explanation is that higher harmonics of $v(t)$ are effectively filtered out by the low-pass properties of the loop's linear links. In fact, $e(t)$ does not differ much from the sinusoid. Because of this, the cross section at the input to the nonlinear link is, generally, the simplest choice for stability analysis.

11.1.2 Harmonic Balance Accuracy

During self-oscillation, the signal $e(t)$ at the input to the nonlinear link is not exactly sinusoidal, and the interference of its harmonics in the nonlinear element contributes to the fundamental V of $v(t)$. However, as mentioned above, we generally expect the effects of

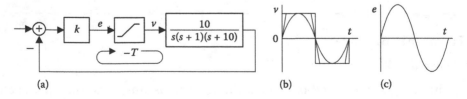

(a)   (b)   (c)

FIGURE 11.1
(a) Block diagram of a feedback loop and the shapes of self-oscillation at (b) the output and (c) the input of the saturation link.

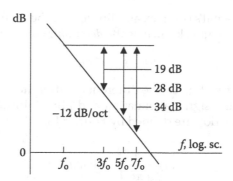

FIGURE 11.2
Filtering of higher harmonics by the loop linear links.

the higher harmonics of $e(t)$ on V to be small because of the following features of feedback control systems' typical nonlinear and linear links:

1. The nonlinear link characteristic $v(e)$ is such that small-amplitude harmonics of $e(t)$ have a negligible effect on V. This assertion is fair for common types of control system nonlinear links.

2. $T(j\omega)$ possesses properties of a filter attenuating the return signal harmonics. This statement is called the *conjecture of a filter*.

The filter conjecture is typically satisfied for well-designed control systems with relatively smooth response and a small amount of non-minimum phase. If a periodic self-oscillation appears in such a system, then, due to Bode phase-gain relations, the phase condition of the oscillation arg $T(j\omega) = -\pi$ occurs at some frequency where the average steepness of the Bode diagram is -12 dB/octave. Therefore, the third harmonic is attenuated by $12\log_2 3 \approx 20$ dB relative to the fundamental by the linear links of the loop. Since in the Π-shaped $v(t)$ itself the third harmonic is 10 dB below the fundamental, we conclude that in $e(t)$, the amplitude of the third harmonic is 30 dB (i.e., 30 times) lower than the fundamental. The higher harmonics in $v(t)$ are even smaller, and they are attenuated even more by the loop linear links, as indicated in Figure 11.2.

With conditions 1 and 2 satisfied, looking at the fundamental only and neglecting the harmonics gives a good estimate for whether the system is stable, and for the value of the stability margins. When, however, the slope of the gain response of the loop links is not steep and the phase condition of oscillation is met due to non-minimum phase lag or due to phase lag in nonlinear links, then harmonic analysis must involve not only the fundamental but also several harmonics of the signal $e(t)$.

11.2 Describing Function

Using harmonic stability analysis for nonlinear control systems was suggested almost simultaneously by several scientists—in 1947 by L. Goldfarb in Russia [22] and by A. Tustin in the UK [76], and in 1949 by R. Kochenburger in the United States [29] (who introduced the term *describing function*), along with several other scientists (in Germany and France).

The ratio of the fundamental's complex amplitude V to the amplitude E of the sinusoidal signal applied to the link input is known as the *describing function* (DF):

$$H(E, j\omega) = V / E \tag{11.1}$$

In general, the describing function can be different at different frequencies, and it is a complex number, describing the signal attenuation and phase shift. The real and imaginary parts of the describing function are defined by Fourier formulas as:

$$\mathrm{Re}H = \frac{1}{\pi E} \int_0^{2\pi} v \, \sin\omega t d\omega t \tag{11.2}$$

and

$$\mathrm{Im}H = \frac{1}{\pi E} \int_0^{2\pi} v \, \cos\omega t d\omega t \tag{11.3}$$

When the characteristic of the nonlinear link is symmetrical, the dc component of its output is zero. When, further, the harmonics at the input to the nonlinear links are negligible compared to the fundamental, the oscillation condition in a feedback system is, approximately,

$$T(j\omega)H(E, j\omega) = 1 \tag{11.4}$$

or

$$T(j\omega) = 1 / H(E, j\omega) \tag{11.5}$$

(When the characteristic is asymmetric, the dc signal component needs to be taken into account.)

Figure 11.3 illustrates two ways to do stability analysis. The trajectories formed by the DF $H(E)$ and *inverse DF* $1/H(E)$ at a specific frequency and varying E are shown by dashed lines. The oscillation conditions (Equations 11.4 and 11.5) become satisfied when the DF trajectory passes through the critical point in Figure 11.3(a), or equivalently, when the inverse DF line in Figure 11.3(b) intersects the Nyquist plot.

When H does not depend on frequency, analysis with the inverse DF appears to be easier. However, when the DF is different at different frequencies, which is typical for

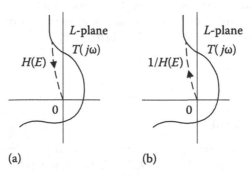

(a)   (b)

FIGURE 11.3
Using (a) DF and (b) inverse DF.

multi-loop nonlinear systems and systems with nonlinear dynamic compensation, the direct DF method is more convenient.

DF analysis replaces the nonlinear link by an equivalent linear link with transmission function $H(j\omega)$. The difference is that H depends on E, and generally, the superposition principle and Bode integral relationships do not apply. The integrals can be applied only in a modified form, as shown in Appendix 10.

11.3 Describing Functions for Symmetrical Piece-Linear Characteristics

11.3.1 Exact Expressions

Characteristics of hard saturation, dead zone, saturation with a dead zone, and a three-position relay are shown in Figure 11.4. The saturation threshold is e_s, and the dead zone is e_d.

The characteristics in Figure 11.4 are symmetrical. Therefore, when the inputs to the links are sinusoidal, the outputs are symmetrical, i.e., do not contain dc components. The output is not shifted in time relative to the input. Hence, the links' DFs are real (the DFs have no phase shift). These DFs do not depend on frequency.

Let us derive the DF for a dead-zone link. The input and output signals for the dead-zone link are shown in Figure 11.5. When e is positive, the output is 0 as long as $e < e_d$, i.e., when the angle $\omega t < \arcsin(e_d/E)$. Therefore, the output is $\sin\omega t - e_d/E$ up to the angle $\pi - \arcsin(e_d/E)$.

Since the output is not shifted in time relative to the input, the phase of the DF and the imaginary part of the DF are 0. (In other words, since the integrand in Equation 11.3 is an odd function, Equation 11.3 becomes 0.) Therefore, the DF can be found with only Equation 11.2. Because the integrand is an even function and the function is symmetrical, we can take the integral from $\arcsin(e_d/E)$ to $\pi/2$ and multiply the result by 4:

$$H = \frac{4}{\pi} \int_{\arcsin(e_d/E)}^{\pi/2} \left(\sin\omega t - \frac{e_d}{E} \right) \sin\omega t\, d\omega t =$$

$$\frac{4}{\pi} \left[\frac{\omega t - \sin\omega t\cos\omega t}{2} + \frac{e_d}{E}\cos\omega t \right]_{\arcsin(e_d/E)}^{\pi/2}$$

or

$$H = \frac{4}{\pi} \left[\frac{\pi}{2} - \frac{1}{2}\arcsin\frac{e_d}{E} + \frac{e_d}{2E}\cos\left(\arcsin\frac{e_d}{E}\right) - \frac{e_d}{E}\cos\left(\arcsin\frac{e_d}{E}\right) \right]$$

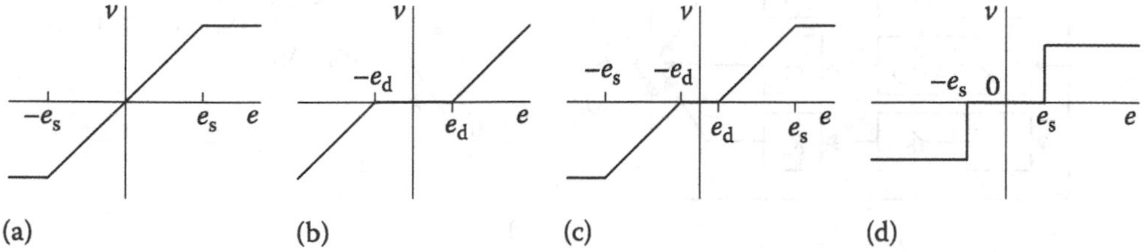

(a)   (b)   (c)   (d)

FIGURE 11.4
Characteristics of (a) saturation, (b) dead zone, (c) saturation with dead zone, and (d) three-position relay.

i.e.,

$$H = \frac{2}{\pi}\left[\pi - \arcsin\frac{e_d}{E} - \frac{e_d}{E}\sqrt{1 - \left(\frac{e_d}{E}\right)^2}\right]$$

This expression for the dead-zone DF is valid for the signal with the amplitude $E > e_d$. When $E < e_d$, the DF is 0.

Any piecewise-linear characteristic can be obtained by connecting a link k_o in parallel with several links with dead zones e_{d1}, e_{d2}, e_{d3}, etc., followed by linear links k_1, k_2, k_3, etc., as shown in Figure 11.6; here, a, b, c, etc. are certain constants. The DF of the total link can be obtained as the sum of the DFs of the parallel paths, since the integral in Equation 11.2 is a linear function.

For example, the *saturation* link DF is the DF for the link with the dead zone e_s subtracted from 1. Therefore, the saturation DF equals 1 for $E < e_s$, and for $E > e_s$ is given by:

$$H = \frac{2}{\pi}\arcsin\frac{e_s}{E} + \frac{e_d}{\pi E}\sqrt{1 - \left(\frac{e_s}{E}\right)^2} \qquad (11.6)$$

For a *three-position relay* actuator (forward, stop, reverse) with the characteristic shown in Figure 11.4(d), the output $v(t)$ is two-polarity pulses, shown in Figure 11.7. The amplitude of the pulses is 1. Therefore, from Equation 11.2 the describing function is:

$$H = \frac{4}{\pi}\int_{\arcsin e_d/E}^{\pi/2} \sin\omega t\, d\omega t = \frac{4}{\pi}\cos\left(\arcsin\frac{e_d}{E}\right)$$

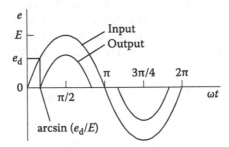

FIGURE 11.5
Input and output signals of the dead-zone link.

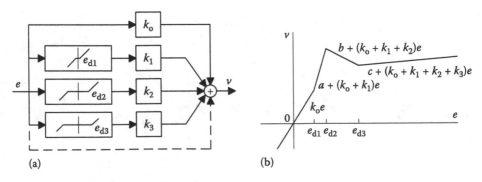

(a)   (b)

FIGURE 11.6
(a) Parallel connection of links with dead zones to implement the (b) piece-linear characteristic.

or

$$H = \frac{4e_d}{\pi E} \sqrt{1 - \left(\frac{e_d}{E}\right)^2} \tag{11.7}$$

When E/e_d is large, the second component under the square root can be neglected and $H \approx (4/\pi)e_d/E$.

Describing functions for saturation, dead zone, and three-position relay are plotted in Figure 11.8.

When E is only slightly larger than 1, then the pulses are short, the fundamental of $v(t)$ is small, the loop gain for the fundamental is small, and oscillation cannot take place. The oscillation is more likely to take place in practical systems when DF is close to the maximum. As seen in Figure 11.8, the relay DF is maximum when $E \approx 1.5$. In this case, the pulses are rather wide, as shown in Figure 11.7(b), the harmonics in $v(t)$ are therefore relatively small, and the DF analysis is sufficiently accurate.

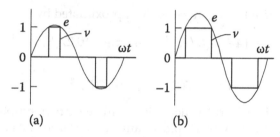

FIGURE 11.7
Output signal of a three-position relay for (a) $E = 1.05$ and (b) $E = 1.5$.

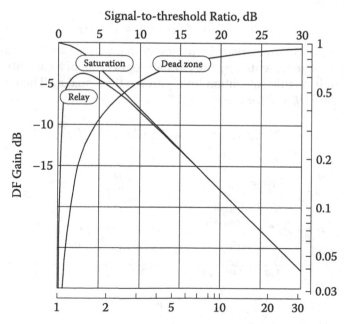

FIGURE 11.8
Describing function characteristic for saturation, dead zone, and three-position relay.

Example 11.2

The stability analysis of a system with a three-position relay and the loop transfer function,

$$T = \frac{50,000(s+500)}{s(s+20)(s+100)^2}$$

is shown in Figure 11.9. The Nyquist diagram and the inverse DF plot are shown in Figure 11.9(a). The direct DF analysis is presented in Figure 11.9(b). The frequency at which the return ratio phase lag is 180° is 5.1 Hz. The loop gain of the linear links at this frequency is 5.4 dB. From the diagram in Figure 11.8, for the DF of −5.4 dB, the signal-to-threshold ratio is 2, i.e., 6 dB. Either analysis concludes that the system can oscillate with the signal amplitude about twice the threshold, and the frequency of oscillation is approximately 5.1 Hz.

11.3.2 Approximate Formulas

Equation 11.6 for *saturation* DF is conveniently approximated by:

$$H \approx (4/\pi)(E/e_s)^{-1} - (4/\pi - 1)0.27(E/e_s)^{-4}$$

or

$$H \approx 1.27(E/e_s)^{-1} - 0.27(E/e_s)^{-4} \tag{11.8}$$

with the error smaller than 0.1 dB. Calculations can be further simplified by omitting the second term in Equation 11.8, which rapidly vanishes for large E. It contributes less than 2, 0.6, and 0.35 dB correspondingly for E larger than e_s, $1.5e_s$, and $2e_s$, respectively.

The link with the *dead-zone* characteristic of Figure 11.4(b) can be replaced by a parallel connection of a unity link and an inverting saturation link. Hence, the DF for the dead-zone link for $E > e_d$ is:

$$H \approx 1 - 1.27(E/e_d)^{-1} + 0.27(E/e_d)^{-4} \tag{11.9}$$

A nonlinear link with a characteristic including both *dead zone and saturation* is shown in Figure 11.4(c). It can be represented by the parallel connection of a saturation link with the threshold e_s and an inverting saturation link with the threshold e_d. Then, for $E \in [e_d, e_s]$, H is given by Equation 11.9, and for $E > e_s$,

$$H \approx 1.27[E/(e_s - e_d)]^{-1} - 0.27\left[(E/e_s)^{-4} - (E/e_d)^{-4}\right]$$

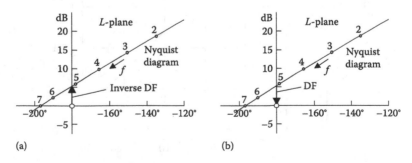

FIGURE 11.9
Nyquist diagram and the (a) inverse and (b) direct DF plots for a system with a three-position relay.

or

$$H \approx \{1.27 / [1-(e_d / e_s)]\}(E / e_s)^{-1} - \{0.27 / [1-(e_d / e_s)^4]\}(E / e_s)^{-4}$$

A *saturation link with frequency-dependent threshold* can be made by cascading a linear link *L*, a saturation link, and a link 1/*L* as shown in Figure 11.10. The threshold is $e_s |L(j\omega)|$. For example, if *L(s)* approximates *s* over the frequency band of interest, this is a *rate limiter*. A link in which the width of the dead zone is frequency-dependent can be made in a similar manner.

11.4 Hysteresis

Figure 11.11(a) shows the input/output characteristic of smooth saturation with hysteresis. The output *v(t)* in Figure 11.11(b) is found by using branch v_1 while the input *e(t)* is rising, and branch v_2 while *e(t)* is decreasing. The time delay of the output relative to the input indicates that the DF must have a negative imaginary component.

The DF phase shift is:

$$\arg H = \arcsin \frac{\text{Im} H}{H} \qquad (11.10)$$

From Equation 11.3,

$$\text{Im} H = \frac{1}{\pi E} \int_0^{2\pi} v d\sin\omega t$$

FIGURE 11.10
Nonlinear link with a frequency-dependent threshold.

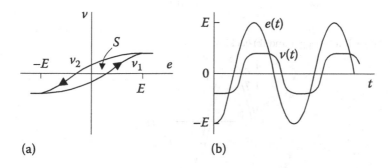

(a) (b)

FIGURE 11.11
(a) Output/input characteristic of smooth saturation with hysteresis and (b) time history of the input and the output.

After replacing $d\sin\omega t$ by $E^{-1}de(t)$ and taking the integral of the output from $-E$ to E using $v_1(e)$, and back to $-E$ using $v_2(e)$, we have:

$$\mathrm{Im}H = \frac{1}{\pi E^2}\int_{-E}^{E}v_1 de + \frac{1}{\pi E^2}\int_{E}^{-E}v_2 de = \frac{1}{\pi E^2}\int_{-E}^{E}(v_1 - v_2)de = -\frac{S}{\pi E^2} \tag{11.11}$$

where S stands for the area within the hysteresis loop. By substituting Equation 11.11 and $H = V/E$ into the right side of Equation 11.10, we have:

$$\arg H = -\arcsin\frac{S}{\pi EV} \tag{11.12}$$

Example 11.3

In Figure 11.11, $V \approx 0.7E$ and $S \approx 0.3$. Therefore,

$$\arg H \approx -\arcsin\frac{0.3}{0.7\pi} \approx 0.14\mathrm{rad}$$

Example 11.4

The Schmitt trigger has a rectangular hysteresis characteristic, shown in Figure 11.12(a). The characteristic for *backlash* shown in Figure 11.12(b) is typically caused by air gaps in gears. For backlash, the area of the hysteresis depends on the signal amplitude. The time history for the backlash link output when the input signal is sinusoidal is shown in Figure 11.13.

Example 11.5

If the width of the backlash is 2, and $e = 20\sin\omega t$, then the output signal amplitude is nearly the same as that of the input, and the phase lag is $\arcsin[2 \times 20/(400\pi)] = 0.064$ rad, i.e., 3.6°.

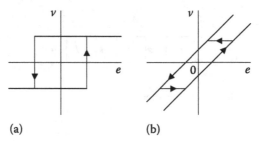

(a)   (b)

FIGURE 11.12
Characteristics of (a) Schmitt trigger and (b) backlash.

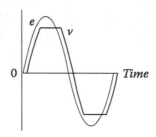

FIGURE 11.13
Time response of the backlash link output.

Because the phase lag reduces the available feedback, efforts are always made to eliminate or decrease backlash in gears and machinery. Reducing the backlash is important even for manually operated tools and equipment (like lathes) because of the man-machine feedback loop via the operator's tactile and visual sensors and the operator's brain (compensator). The backlash necessitates slowing the actions of the operator, or else the overshoots in the feedback system become large or the system becomes unstable.

Example 11.6

Hysteresis links are frequently used in oscillating feedback loops, of which the sawtooth signal generator shown in Figure 11.14 is representative. The feedback loop is composed of an inverting integrator with transfer function $-1/(R_3Cs)$ and a noninverting Schmitt trigger. The input thresholds of the Schmitt trigger shown by dashed lines in Figure 11.14(c) are $u_{th} = \pm [R_1/(R_1 + R_2)]VCC$, where VCC is the power supply voltage. The integrator output is constrained within the *dead-band* $[-u_{th}, u_{th}]$.

When the output voltage of the integrator arrives at u_{th}, the trigger output switches from $-VCC$ to VCC. After the switching, the integrator output begins decreasing with constant rate $VCC/(R_3C)$ (V/s) until the next switching occurs. The period is therefore:

$$T = 4(u_{th} / VCC)R_3C = 4R_1R_3C / (R_1 + R_2)$$

In Figure 11.14(b) the output signals' rates of rising and falling are the same. The rates can be made different by adding in parallel to R_3 an additional resistor R_4 in series with a diode, so that in one direction the resistance will be smaller and the integrator gain coefficient larger, thus increasing the rate.

It is seen in Figure 11.14 that the linear link and the nonlinear link both lag the signal by 90°, so that the signal comes back in phase after passing about the loop. It is also seen that during the analysis of such a system the harmonics cannot be neglected; i.e., DF analysis cannot be used: if the sawtooth signal at the input to the Schmitt trigger is replaced by its fundamental, the phase lag in the Schmitt trigger will be less than 90° and the condition of oscillation will be not satisfied.

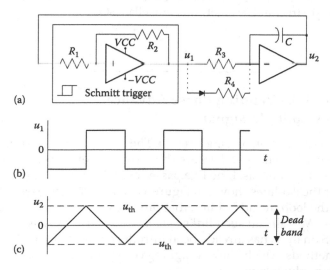

FIGURE 11.14
(a) Sawtooth signal generator and signal histories at (b) the Schmitt trigger output and (c) the integrator output.

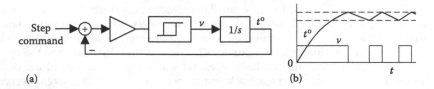

FIGURE 11.15
(a) Bang–bang temperature controller and (b) its output time response.

Example 11.7

An *on–off* (or *bang–bang*) oscillating temperature controller is shown in Figure 11.15(a). The actuator is a *two-position relay with hysteresis* that can be implemented either as an electromechanical device or electronically, for example, employing a Schmitt trigger. The temperature of the single-integrator plant oscillates within the dead band. The positive temperature rate depends on the power supplied by the heater, and the negative rate (the cooling rate) depends on the heat transfer and radiation conditions. The narrower the dead band is, the higher is the control accuracy, but also the higher is the oscillation frequency. The oscillation frequency cannot be chosen to be excessively high since each switching in the physical actuators consumes some energy or wears out the contact mechanisms.

When the plant in this feedback system contains extra high-frequency poles, the control law is often augmented by additional rules. For example, switching can be done not in the instant the output approaches the end of the dead band, but somewhat earlier in time or position, to counteract additional plant inertia. The accuracy of such an on–off controller, although sufficient for many applications, is typically inferior to the well-designed controllers using pulse-width-modulated (PWM) drivers. When the actuator is electrical, switching on and off with sufficiently high frequency does not present a problem, and using PWM is common. However, when the switching cannot be fast and causes noticeable power losses, bang–bang control may be preferred. Bang–bang controllers are also used for the attitude control of spacecraft using thrusters. PWM is less efficient due to thrust transients and excessive valve actuations. Although these controllers are often designed in the phase-plane, DF analysis is required to handle potential interactions with structural flexibility and propellant slosh.

11.5 Nonlinear Links Yielding Phase Advance for Large-Amplitude Signals

In general, the DF is a function of E and ω (or f). The DF *iso-f lines* are shown in Figure 11.16 by the dashed lines, while the *iso-E lines* are shown by the solid lines. Either set of lines can be used for the stability analysis. If no line passes over the critical point, the system is considered stable. For the iso-lines shown in Figure 11.16, arg DF increases with E. The system is stable because the loop includes a link that provides a certain phase advance for large-amplitude signals. We will call such links *nonlinear dynamic compensators* (NDCs). Design of NDCs using AS analysis has already been introduced in Chapter 10. In this chapter we will employ DF methods, which allow designing certain NDCs that cannot be analyzed or designed with AS methodology.

From the iso-*E* lines, *iso-E Bode diagrams* can be drawn as shown in Figure 11.16(b) (MATLAB functions described in Appendix 14 can be used to plot the iso-*E* diagrams for some typical NDCs). We should keep in mind, however, that these Bode diagrams do not uniquely define the phase shift. The Bode relations can only be used when constant *E* causes the DF of nonlinear links to be constant at all frequencies so that the DF can be equivalently replaced by a constant gain linear link.

Example 11.8

Figure 11.17(a) shows a simple proportional-integral (*PI*) NDC using a parallel connection of a linear link with a nonlinear link. Because of the saturation, the upper channel contributes relatively less to the output when *E* increases. The compensator iso-*E* Bode diagrams are shown in Figure 11.17(b). Consequently, the NDC lag decreases as *E* increases. An iso-*E* Bode diagram of the loop gain of a feedback system with such an NDC is shown in Figure 11.17(c).

This example shows that an effective NDC can be implemented rather simply, and that using iso-*E* Bode diagrams is convenient for certain practical classes of NDCs.

Before concentrating on the NDC design, we first consider the NDC performance in the loop, which also includes another nonlinear link, the actuator.

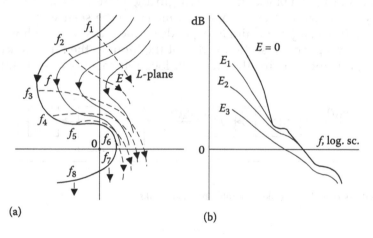

(a)   (b)

FIGURE 11.16
(a) Iso-*f* (solid) lines and iso-*E* (dashed) lines, as well as (b) iso-*E* Bode diagrams.

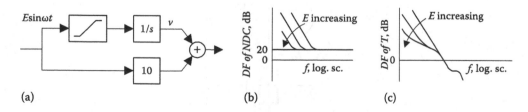

(a)   (b)   (c)

FIGURE 11.17
(a) NDC with parallel channels, (b) its iso-*E* Bode diagrams, (c) and the loop iso-*E* Bode diagrams. Phase shift can be calculated from these diagrams.

11.6 Two Nonlinear Links in the Feedback Loop

A feedback system including an NDC and a nonlinear actuator is shown in Figure 11.18. The NDC and the actuator are separated by some linear link L. For the purpose of stability analysis the loop can be cross-sectioned at the input to either of the nonlinear links, as shown in the figure. The loop return ratio DF can be measured by applying some test signal either to the input to the NDC, with some amplitude E_1, or to the input of the actuator with some amplitude E_2.

Although different, the iso-f lines for E_1 and E_2 have a common point when $|T| = 1$, as shown in Figure 11.19. This can be seen by considering that the condition $|T| = 1$ is satisfied with some specific values of $E_1 = E_{1C}$ and $E_2 = E_{2C}$. Then, if the loop is broken at the input to the NDC and E_{1C} is applied to the input of the loop, the value E_{1C} returns to the cross section, and the value E_{2C} appears at the input to the actuator. Also, when the loop is broken at the input to the actuator and E_{2C} is applied to the input of the loop, the value E_{2C} returns to this cross section, and the value E_{1C} appears at the input of the NDC. In both cases, the same signals appear at the inputs of both nonlinear links, and therefore, the loop DF is the same and the loop phase shift is the same. Hence, these two iso-E lines intersect when $|T| = 1$.

The nominal guard-point phase stability margin, therefore, does not depend on the position of the cross section. However, some confusion can be seen with the gain stability margin: this stability margin seemingly depends on the cross section chosen for examining the loop. The proper cross section is that at the input to the NDC, since in this case, uncertainty in the plant gain directly affects the loop return ratio: when plant gain reduces

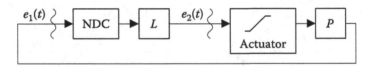

FIGURE 11.18
Feedback loop with two nonlinear links separated by linear links.

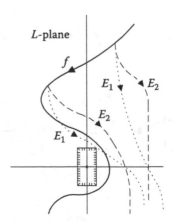

FIGURE 11.19
Iso-f lines in a system with two nonlinear links.

by a dB, then the Nyquist diagram and the iso-f lines simply sink down by a dB. Therefore, in Figure 11.19, the stability margin is just exactly satisfied.

The accuracy of DF analysis suffers when the loop incorporates two nonlinear links since in this case the linear links separating them, although commonly still low-pass filters, are not as frequency selective as the linear link in the loop with a single nonlinear link. Nonetheless, when the feedback in the loop is large and the Bode diagrams of the linear links are reasonably steep (they are quite steep when the system is Nyquist stable), the phase uncertainty caused by the harmonics typically does not exceed ±20°. This range seems large: generally, such uncertainty can make the entire difference between an unstable and a high-performance system. However, using DF enables an easy design of a simple NDC providing phase advance of more than 120°. Therefore, even with the ±20° phase error, the NDC gives the phase advance no less than 100°, which is certainly better than no phase advance at all, and is adequate for most practical applications.

11.7 NDC with a Single Nonlinear Nondynamic Link

For an NDC composed of several linear links and a single nonlinear nondynamic link with DF w, the normalized NDC transfer DF depends on w as a linear fractional function:

$$H(w) = \frac{w + M}{w + N} \tag{11.13}$$

where M and N are some functions of s.

When the signal amplitude changes, w changes and $H(w)$ changes. When w changes from 0 to ∞, H changes from M/N to 1, and therefore $\arg H$ changes by $\arg (M/N)$. To maximize the phase advance produced by the NDC, $\arg N$ and $-\arg M$ must be made as large as possible. This angle is, however, bounded by continuity considerations.

During the gradual change in the signal amplitude and in w, the transitions in H are desired to be monotonic and smooth. The most critical values of w, from this point of view, are those equal in modulus to either $|M|$ or $|N|$, as shown in Figure 11.20. For these values of w, if the angle of M (or, correspondingly, N) is π, the vector $w + M$ (or, correspondingly, $w + N$) vanishes, and in the neighborhood of this point, H becomes excessively sensitive to w. To avoid this situation with a $\pi/6$ *safety margin*, neither $|\arg M|$ nor $|\arg N|$ should be allowed to exceed $2\pi/3$. This requirement limits the phase shift of the NDC to $4\pi/3$, i.e., 240°, which is more than enough for all practical purposes.

A flowchart implementation of Equation 11.13 is shown in Figure 11.21(a). (Notice that when w is very large, the error signal at the beginning of the upper branch decreases and

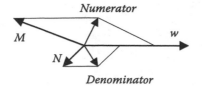

FIGURE 11.20
Vectors $(w + M)$ and $(w + N)$ with minimum moduli.

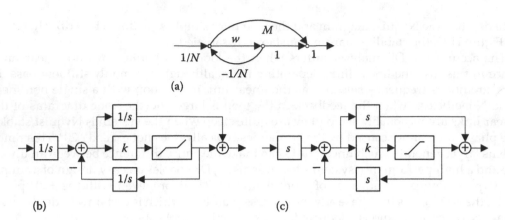

FIGURE 11.21
(a) NDC flowchart and (b,c) block diagrams.

the branch output signal becomes negligible.) Figure 11.21(b) exemplifies the case where w is a high-gain amplifier with the gain coefficient k and a dead zone, and $M = 1/s$, $N = s$. This NDC reduces the phase delay by 180° for large-amplitude signals. Figure 11.21(c) exemplifies the case when a saturation link is used, with $M = s$, $N = 1/s$ over the frequency range of interest, resulting again in a 180° phase advance for large-amplitude signals.

To shift the phase over the full available range of 240°, the NDC includes a forward path with w, a path in parallel with the forward path, and a feedback path. When the NDC includes only a parallel path or only a feedback path, i.e., when either M or N is zero, the available phase shift change must not exceed 120°, which still suffices for most applications.

In addition to iso-f and iso-E responses, *iso-w Bode diagrams* can be used. These are the frequency responses measured while holding constant the value of the signal amplitude at the input to the nonlinear element. Such a response can be calculated after replacing the nonlinear element with a constant linear element. The iso-f and iso-E responses can be further calculated using the set of iso-w responses, the DF for the nonlinear element, and the signal amplitude at the nonlinear element.

Example 11.9

In the NDC shown in Figure 11.21(b), $k = 10$. Calculate and plot the iso-w Bode diagrams for three values of $k \times$ (DF of the dead-zone link): 0.1, 1, and 10. The SPICE simulation input file is shown below and uses the schematic diagram in Figure 11.22. The iso-w Bode diagrams are shown in Figure 11.23.

```
**** ch9ex1.cir for iso-w simulation of NDC Figs. 11.21(b), ****
11.22
*** input integrator
G2 2 0 0 1 1
C2 2 0 1
R2 2 0 1MEG
*** feedback summer
G3 3 0 7 2 1
R3 3 0 1
*** kDF path: G5 =.1, 1, or 10
G5 5 0 0 3 10
```

```
R5 5 0 1
*** forward path, inverting
G4 4 0 3 0 1
C4 4 0 1
R4 4 0 1MEG
*** forward summer, the output is VDB(6)
G6 6 0 4 5 1
R6 6 0 1
*** feedback path
G7 7 0 0 5 1
C7 7 0 1
R7 7 0 1MEG
***
VIN 1 0 AC 1
RIN 1 0 1MEG
.AC DEC 20.001 10
.PROBE
.END
```

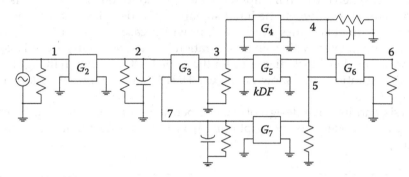

FIGURE 11.22
SPICE model for iso-*w* response.

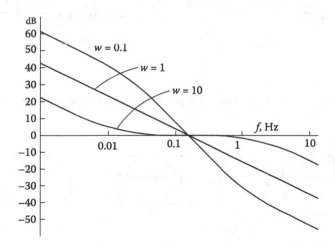

FIGURE 11.23
Iso-*w* Bode diagrams.

With increasing w, i.e., increasing the signal at the input to the NDC, the plot gradually changes from that of a double integrator to a constant gain response. Correspondingly, the phase lag decreases from π to 0.

When the system is linear and the loop gain is large, the signal amplitude after the summer is nearly constant at all frequencies where the loop gain is large, since the input block is an integrator, and also, the feedback decreases with frequency.

11.8 NDC with Parallel Channels

Simple examples of NDCs with parallel paths are the PID controller with saturation in the *I*-channel (or in the low-frequency channel, as described in Chapter 6) and the *PI* controller shown in Figure 11.17.

The feedback system in Figure 11.24(a) includes an actuator with saturation and an NDC with two parallel channels. Nonlinear nondynamic links are placed here in both channels. The first channel starts with a saturation link with unity threshold, and the second, with unity dead zone. When the signal amplitude is low, the loop transfer function is $T_{01} = C_1AP$. The second channel is off when the input signal amplitude is less than 1, and takes over when the first channel becomes overloaded, in which case $T_{02} = C_2AP$. The NDC can be equivalently implemented with a single saturation link as shown in Figure 11.24(b).

Three more equivalent block diagrams for the same NDC are shown in Figure 11.25. The versions with saturation links shown in Figure 11.24(b) and in Figure 11.25(a) are typically easier to implement.

The analysis and iterative design of the compensator can be performed by calculating and plotting iso-*E* Bode plots, or by plotting iso-*f* lines on the *T*-plane as exemplified in Figure 11.26.

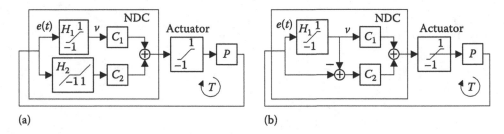

(a)   (b)

FIGURE 11.24
(a) Feedback system with two nonlinear elements in the NDC and (b) an equivalent diagram with a single nonlinear element in the NDC.

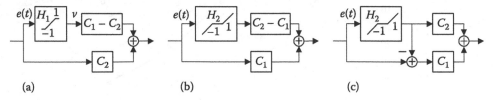

(a)   (b)   (c)

FIGURE 11.25
Three more NDC configurations.

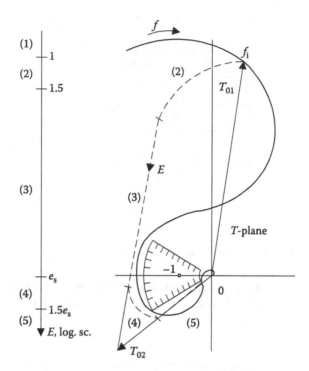

FIGURE 11.26
Intervals on *E*-axis and on the iso-*f* line for a system with two parallel channels.

An iso-*f* line for

$$T = T_{01}H_1(E) + T_{02}H_2(E) \tag{11.14}$$

is displayed in Figure 11.26. We can examine it piece by piece using simplified Equations 11.8 and 11.9 for the intervals of *E* shown in the left of the figure.

On the *first interval*, $E < 1$ and $T = T_{01}$.

The *second interval*, where $E \in [1, 1.5]$, is comparatively short. Over this interval, H_1 reduces from 1 to 0.8, and H_2 increases from 0 to 0.25, as can be verified with the plots in Figure 11.7. This segment of the iso-*f* line is curvilinear, and its exact shape is not important for the stability analysis.

On the *third interval*, where $E \in [1.5, e_s]$, Equation 11.14 reduces to:

$$T = T_{01}1.27 / E + T_{02} - 1.27T_{02} / E = T_{01} + T_{02}(1 - 1.27 / E) \tag{11.15}$$

This piece of the iso-*f* line is a segment of a straight line aimed at the end of the vector T_{02}. In the vicinity of the stability margin boundary the iso-*f* line can be approximated by the side of the parallelogram shown in Figure 11.26.

On the *fourth interval*, $E \in [e_s, 1.5e_s]$. This section is curvilinear, short, and does not deserve detailed analysis.

On the *fifth interval*, $E > 1.5e_s$ and

$$T = 1.27T_{01} / E + 1.27T_{02} / E = 1.27(T_{01} + T_{02}) / E \tag{11.16}$$

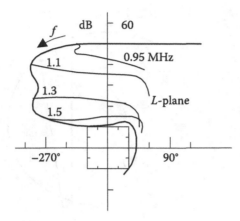

FIGURE 11.27
Nyquist diagram and iso-*f* lines for an experimental system.

so that the iso-*f* line is straight and directed toward the origin. From the plot we can conclude that, at this frequency, limit cycling is avoided with sufficient stability margins.

Example 11.10

Figure 11.27 shows the *L*-plane Nyquist diagram and iso-*f* lines measured in an experimental system designed with the method presented in this section. The system has no limit cycles.

11.9 NDC Made with Local Feedback

In many practical links the DF's modulus is the same as that for saturation, but the phase is nonzero and changes with the signal level. We will call such links *dynamic saturation* links. A typical case is presented in Figure 11.28(a). The phase shift in the nonlinear dynamic link varies with the signal level, while the output signal amplitude and therefore $|H|$, the magnitude of the DF of the composite link, does not vary after the saturation threshold is exceeded.

Figure 11.28(b) gives another implementation of dynamic saturation. Here, $|K(j\omega)|$ is large, although decreasing with frequency, and B is real. When the signal is small, the closed-loop transfer function is $1/B$. When the signal is large, the describing function of the forward path decreases, the feedback becomes negligible, and the phase lag becomes that of $K(j\omega)$. That is, the phase lag increases with the signal level.

Still another version is shown in Figure 11.28(c) using similar links $K(j\omega)$ and B. For small-signal amplitudes, the link transfer function is $K(j\omega)$. As the signal amplitude gets much larger than the dead zone, the dead-zone DF approaches 1, large feedback is introduced, and the output signal is limited as if by saturation. The phase shift of the link will then be determined by the feedback path, that is, the phase lag reduces with the signal level. This link is especially suitable for nonlinear dynamic compensation.

When the linear link loop gain is high, and the dead-zone link DF w changes from 0 to 1, the DF of the link changes from $1/B$ to K; i.e., the phase shift changes by arg (KB), i.e., by the angle of the local loop phase shift. As indicated in Section 11.7, this value must be limited to 120°, which is sufficient for most applications.

Example 11.11

Consider an NDC with dead zone in the local feedback path, as shown in Figure 11.29(a), with $B = s^{-4/3}$ (i.e., the Bode diagram for B has constant slope −8 dB/octave, and the phase shift is −120°), and $K = 200$. Since V and the dead-zone link DF w are uniquely related and this relation does not depend on frequency, the full set of iso-V Bode diagrams is the same as the full set of iso-w diagrams, and either one can be employed for the stability analysis. Figures 11.29(b) and 11.29(c) show a parallel-channel NDC implementation that could be advantageous in some circumstances.

The iso-V Bode diagrams for the NDC are shown in Figure 11.30(a), and the iso-V lines of the main loop on the L-plane might look as shown in Figure 11.30(b). The system has no limit cycle. An example of an NDC with local high frequency feedback is shown in Figure 11.29(b), (c).

Example 11.12

A tunnel effect accelerometer is shown in Figure 11.31(a). The proof mass and the soft springs that suspend the mass are etched of silicon. The position of the proof mass is regulated by electrostatic forces between the proof mass and the upper and lower plates. The accelerometer uses a tunnel effect sensor to measure the proof mass position. The tunnel current flows between the proof mass and the sharp tunnel effect tip when the

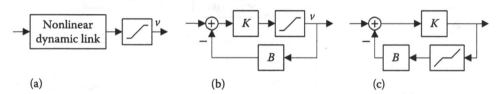

(a)   (b)   (c)

FIGURE 11.28
Dynamic saturation.

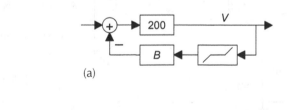

(a)

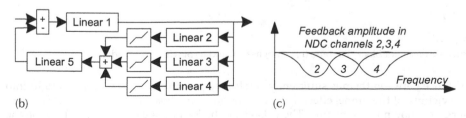

(b)   (c)

FIGURE 11.29
Dead-zone NDC in the feedback path (a), and NDC with three linear local links in (b) and (c).

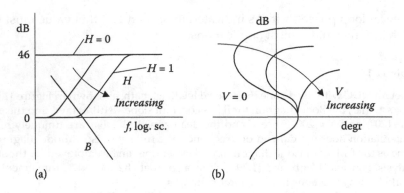

FIGURE 11.30
(a) Bode diagrams of the NDC shown in Figure 11.29a with the dead zone replaced by equivalent linear links and (b) iso-*V* Nyquist diagrams for the main loop.

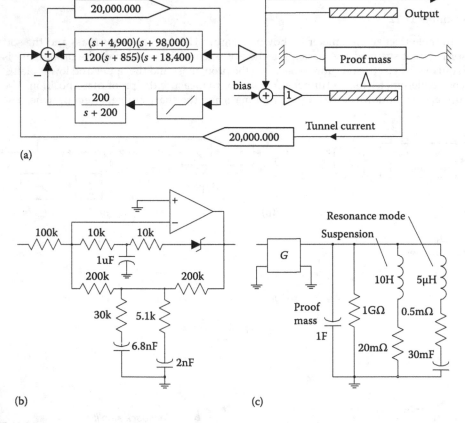

FIGURE 11.31
(a) Silicon accelerometer block diagram, (b) compensator, and (c) SPICE plant model.

distance between them is sufficiently small. The proof mass is gradually brought into the vicinity of the tunnel effect tip by an additional feedback loop using capacitive sensors, not shown in the picture. The voltage on the lower plate equals the voltage on the upper plate plus some bias. It can be shown that with proper bias voltage the upper plate voltage is proportional to the measured acceleration.

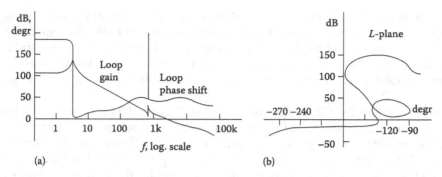

FIGURE 11.32
Accelerometer (a) Bode diagram and (b) Nyquist diagram.

To achieve the desired accuracy, the feedback in the proof mass control loop must be made larger than 100 dB at frequencies up to 5 Hz. The feedback crossover frequency f_b is limited by the dynamics (structural resonances) of the proof mass and suspension system to less than 3 kHz.

The tunnel current is the exponent of the inverse of the tunnel sensor gap. The normal value of the gap is approximately 6 Å, but when the gap is smaller, the tunnel current is exponentially larger. Since in this case the derivative of the current to the gap width (the tunnel sensor gain coefficient) increases, the loop gain becomes bigger than nominal. Without an NDC, such a system would not be globally stable. Global stability is provided by an NDC with a dead zone in the local feedback path.

The mechanical plant might have some resonance modes with uncertain frequencies over 500 Hz. The quality factor of the resonances is not higher than 20, i.e., 26 dB. The SPICE model for the plant with such a resonance is shown in Figure 11.31(c). The 1 GΩ resistor is for the SPICE algorithm to converge.

The compensator is shown in Figure 11.31(b). The dead-zone element was chosen to be nonsymmetrical (a Zener diode) since the characteristic of the tunnel effect sensor is also nonsymmetrical. For low-level signals the Zener does not conduct, and the compensator response is determined by the lower feedback path. Two series RC circuits shunting the feedback path provide two leads giving sufficient phase stability margins over the range 200–3,000 Hz. The Bode diagram and the Nyquist plot for small-amplitude signals were computed in SPICE and shown in Figure 11.32.

When the signal exceeds the Zener threshold, the diode opens and the upper feedback path, which is an RC low pass, reduces the compensator gain at lower frequencies by approximately 30 dB. This gain reduction reduces the slope of the Bode diagram, substantially increases the phase stability margin at frequencies below 200 Hz, and improves the transient response of the closed loop, which is important since the acquisition range of the tunnel effect sensor is very narrow, only about 15 Å. In experiments, the system locks rapidly into the tracking mode and remains stable whatever the initial conditions are.

Two additional examples of applications of NDCs with local feedback incorporating a dead-zone link are given in Appendix 13.

11.10 Negative Hysteresis and Clegg Integrator

Let us consider two alternatives to NDCs.

Negative hysteresis is a link with characteristics as in Figure 11.12(a), but with reversed directions of the arrows on the branches of the characteristics. The negative hysteresis effect can be achieved by switching the output at specific levels of the incident signal, as shown in Figure 11.33. Such a link introduces phase lead up to 90° for signals of certain amplitudes. However, the link does not pass signals with small amplitudes, and it is very sensitive to the signal amplitude and shape, and to the noise. Negative hysteresis links are rarely used since NDCs are simpler, more robust, and able to provide much larger phase lead.

A generalization of the *Clegg integrator* shown in Figure 11.34 consists of a splitter, two different linear links L_1 and L_2, a full-wave rectifier (i.e., absolute value link), a high-gain link with saturation realizing the *sign* operator, and a multiplier M. The instantaneous amplitude of the output signal is determined by the upper-channel output $v_1(t)$ shown in Figure 11.34(b). The sign of the output signal is defined by the sign of the lower-channel output. Therefore, the composite link output signal is $v_{out}(t) = |v_1(t)|$ sign $v_2(t)$, as seen in Figure 11.34(b).

In particular, when $L_1 = k/s$ and $L_2 = 1$, the integrator gain decreases with frequency without introducing the 90° phase lag of a conventional linear integrator, as seen in Figure 11.34(b). When this idea was first introduced, the hope was expressed that this method would work well for small-amplitude signals, thus allowing circumvention of Bode limitations and the causality principle. This, however, is not possible. A large-amplitude second harmonic is present at the output of the Clegg integrator. This harmonic's interference with the fundamental produces large phase lag for the fundamental. This and high sensitivity of H to the shape of the input signal (also the disadvantage of negative hysteresis) prevent the circuit from being used in practice. In fact, the Clegg integrator and negative hysteresis are described here only to inform the reader that such ideas have already been explored and found not particularly useful.

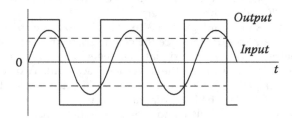

FIGURE 11.33
Negative hysteresis.

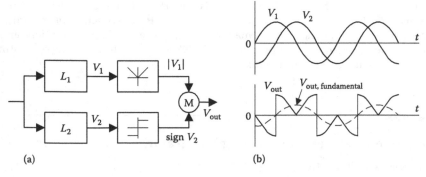

(a)   (b)

FIGURE 11.34
(a) Clegg integrator and (b) its signal histories.

11.11 Nonlinear Interaction between the Local and the Common Feedback Loops

Saturation links appear in local and common loop actuators. In addition to the common loop, a local loop is often employed in the ultimate stage of a linear amplifier to linearize its characteristic. Local feedback is employed widely in electrical and electromechanical actuators to make their characteristics more linear and stable in time, thus benefiting the main loop.

In control feedback systems, the optimal characteristic for the actuator is hard saturation, since it makes the loop gain constant up to the maximum of the output amplitude. When it is not, a predistortion memoryless nonlinear link could be installed at the input to the actuator to make the total nonlinearity hard saturation, or a local feedback about the actuator could be introduced.

The value of the local feedback varies with the signal level. Interference of the local loop with the main loop is important to understand.

Consider an example of local feedback about the actuator in the block diagram in Figure 11.35.

When the main loop is disconnected at either of the cross sections 1 or 2, the system is stable (when the actuator loop is properly designed). Either of these cross sections can be used for the stability analysis (the Bode–Nyquist criterion for successive loop closure has been described in Section 3.7). In both cases, the resulting loop includes a nonlinear dynamic loop, in one case with parallel channels and in the other case with local feedback. The nonlinear dynamic links can be analyzed as we did with NDCs. These links may introduce phase lag in the main loop that can result in a limit cycle.

For example, when L_2B_2 is $s^{-4/3}$ and B is real, the actuator's overload results in the reduction of the local feedback and in the introduction of a 120° phase lag in the main loop. To prevent an oscillation, an extra NDC with a 120° phase lead can be introduced in the loops, at either cross section 1 or 2.

Another common problem is variations in the output impedance of the driver. This impedance depends on the signal level since local current or voltage feedback loops are often employed in the driver amplifier to make its output impedance correspondingly high or low, and the feedback in these loops is affected by the actuator or driver saturation.

Example 11.13

A typical example is using current drivers for magnetic windings of actuators: voice coils, reaction wheels of spacecraft attitude control systems, flux winding for electrical power generators, solenoids, etc., as shown in Figure 11.36(a). The output current of the driver is

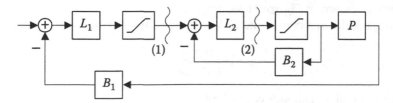

FIGURE 11.35
Local (actuator) and common loops with saturation.

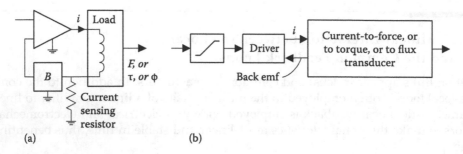

FIGURE 11.36
(a) Current feedback driver for inductive loads and (b) using extra saturation link at its input.

proportional to the driver's input voltage due to large local current feedback; without this feedback, the driver amplifier would be a voltage amplifier. Since the output force (torque) is proportional to the coil current, the transfer function of the actuator is a real number and the phase lag is 0. (Back emf does not affect the output when the signal source impedance is high.) However, when the driver is overloaded and the gain in the local feedback loop vanishes, the driver's output impedance drops, its output becomes a *voltage*, and the actuator output force becomes an integral of this voltage. In other words, overload introduces an extra integrator into the main loop, and can trigger a limit cycle. To prevent this from happening, an extra saturation link can be placed at the input of the driver, as shown in Figure 11.36(b), to limit the signal amplitude at the driver's input.

11.12 NDC in Multi-Loop Systems

As was discussed in Chapter 2, MIMO control systems most often have individual actuators for each dimension of the system output, and the plant is to a large extent decoupled; i.e., the diagonal terms are much larger than the others. Still, coupling exists and might cause instability, especially coupling due to the plant structural resonances and nonlinearities.

Let us consider the two-input, two-output feedback system shown in Figure 11.37. This system can be, for example, an x–y positioner, with some coupling between the x and y directions because of a resonant mode of the payload that is not orthogonal to x or y. Although the plant gain coefficient in the main direction is substantially larger than the coupling coefficient, the coupling is still not negligible. In this example, let us consider the main direction gain coefficients to be 1, and the coupling coefficients, k_c.

Assume that each loop is stable and robust in the absence of coupling. Then, consider the effect of coupling on the y-loop. The effect will be easier to see if we redraw the system block diagram as shown in Figure 11.38.

The coupling results in a composite link with the transfer function

$$k_c^2 \frac{B_x C_x A_x}{B_x C_x A_x + 1}$$

in parallel with the plant transfer coefficient.

When the feedback in the x-loop is large, the composite link gain coefficient is close to k_c^2. But if the feedback is positive, the gain correspondingly increases. In the nonlinear mode

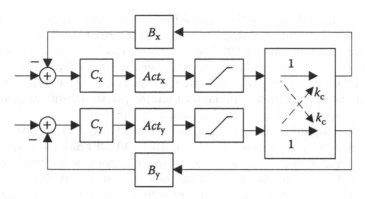

FIGURE 11.37
Two-input, two-output system.

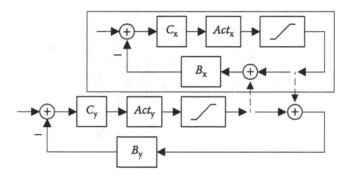

FIGURE 11.38
Equivalent block diagram.

of operation, when the x-actuator is overloaded, the feedback can become positive at any frequency below the crossover. This factor introduces additional uncertainty in the y-loop. To ensure the necessary robustness, either the stability margin must be increased with the resulting reduction in performance, or an NDC must be introduced in front of each actuator–even if the NDCs are not required for the individual loop operation (but they will certainly do no harm to the individual loops; quite the opposite, they will improve the individual loop performance). When the x-actuator becomes overloaded, the NDC introduces some phase lead in the x-loop; this eliminates positive feedback in this loop and reduces the composite link gain. Hence, with the NDC, stability margins can be reduced in the y-loop and performance can be improved without sacrificing robustness.

11.13 Harmonics and Intermodulation

11.13.1 Harmonics

By its nature, DF analysis neglects the effects of harmonics on the output fundamental. Let us analyze the resulting error and its effects.

Figure 11.39 shows the practically important case of an incident signal to a saturation link with an amplitude far exceeding the saturation levels (indicated by the dashed lines.)

The output $v(t)$ is clipped most of the time. After coming around the loop, the third harmonic in $e(t)$ delays the 0-line crossing of $e(t)$, and therefore of $v(t)$. This results in an extra phase lag for the fundamental of $v(t)$. The lag could reach $12°$ in control systems with conventional loop gain response. Harmonics higher than the third also contribute to this effect; however, their contributions are smaller. Therefore, phase stability margins for the iso-f lines should be approximately $15°$ larger than the phase stability margin accepted for the linear state of operation, i.e., for the Nyquist diagram.

When the Bode diagram for the loop gain is not monotonic, as in feedback systems with resonance modes in the plant exemplified in Figure 11.40, DF analysis is still quite satisfactory at all frequencies except for those whose third harmonic is relatively large. At these frequencies, the effect of the third harmonic must be calculated since it can produce a change of up to $30°$ or $40°$ in the phase shift for the fundamental f_o. Accordingly, some extra phase stability margin must be provided by reducing the steepness of the Bode diagram in the region close to f_o, as shown in Figure 11.40.

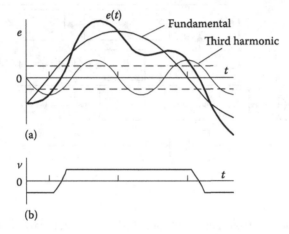

(a)

(b)

FIGURE 11.39
Effect of the third harmonic on the fundamental time delay.

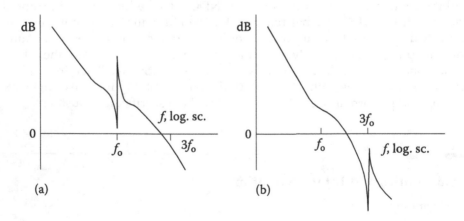

(a)

(b)

FIGURE 11.40
Open-loop Bode diagram of a system with a flexible mode (a) reducing the gain of the fundamental and (b) increasing the gain at the third harmonic frequency.

Harmonics might also be important in systems with large pure time delay in which the loop phase lag can reach π without the Bode diagram being steep.

After the DF analysis and design, the system robustness must be verified with computer simulation.

11.13.2 Intermodulation

Application of a signal having two Fourier components with different frequencies, f_1 and f_2, to the input of a nonlinear link results in the output signal $v(t)$ containing not only the harmonics of f_1 and f_2, but also intermodulation products with frequencies $m f_1 \pm n f_2$. For example, the signal shown in Figure 11.41 with large-amplitude low-frequency components and small-amplitude high-frequency components is very typical for audio signals. When the signal is clipped, the information contained in the high-frequency components over the time of clipping is lost. The information would be preserved if the lower-frequency and higher-frequency components were first separated by a fork of low-pass and high-pass filters, and then combined at the output after amplification by separate amplifiers. In this case, nonlinear distortions of the low-frequency components will not affect the high-frequency components, i.e., will not produce intermodulation. (The intermodulation in speakers can be reduced by using separate speakers for higher and lower frequencies.)

For example, antenna pointing can be affected by wind disturbances, the latter having both lower-frequency, higher-amplitude components and higher-frequency, lower-amplitude components. Similar kinds of disturbances occur in vibration isolation actuators. For large-amplitude, low-frequency vibrations not to cause the actuator, after clipping, to stop rejecting high-frequency disturbances, two different actuators for two separate bands of disturbances can be employed.

In audio recording, it is important to hide the high-frequency noise of the magnetic head and the amplifiers. When the signal high-frequency components are large, the noise is not noticeable, and the high-frequency gain should be large for better signal reproduction. When the signal high-frequency components are small, the amplifier bandwidth should be lowered to reduce the noise that would otherwise be clearly heard. This function is performed by specially designed ICs.

In control systems, similar problems can occur, and ICs designed for audio signal processing can be used as well for this application.

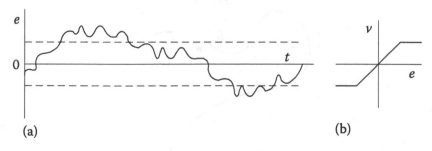

(a)   (b)

FIGURE 11.41
(a) Clipping a multicomponent signal by (b) a link with saturation.

11.14 Verification of Global Stability

The condition of harmonic balance at some frequency does not always lead to periodic self-oscillation at this frequency. For such self-oscillation to persist, the process of self-oscillation must be stable. The process is stable, by definition, if vanishing disturbances in parameters of the oscillation (in the amplitude, for example) cause deviations from the solution that exponentially decay in time. This condition signifies a limit cycle and is illustrated in Figure 11.42(a). If the deviations from the periodic solution grow exponentially, as shown in Figure 11.42(b), the process (attractor) is not stable and the limit cycle does not take place.

For example, in a system with saturation and the Nyquist diagram shown in Figure 11.43, three iso-f lines related to the frequencies f_1, f_2, and f_3 cross the critical point. From these three, the limit cycles are associated with frequencies f_1 and f_3, and the solution at frequency f_2 is unstable. This can be shown as follows.

Consider the limit cycle with frequency f_3. In this case, illustrated in Figure 11.44(a), the saturation DF reduces the loop gain such that the equivalent Nyquist diagram shrinks and passes through the –1 point. An extra increase of the signal level reduces the DF, and the Nyquist diagram shrinks further. The critical point –1 then occurs outside of the Nyquist diagram, which is the mapping of the left half-plane of s. Then, the exponent corresponding to the frequency of oscillation has a negative real component, and the signal amplitude

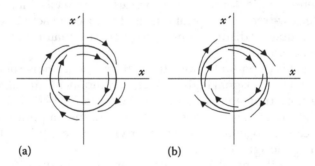

(a)   (b)

FIGURE 11.42
(a) Limit cycle and (b) unstable periodic solution.

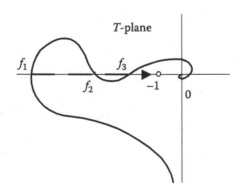

FIGURE 11.43
Nyquist diagram and iso-f lines causing two limit cycles.

reduces. On the other hand, if the signal level gets smaller, the DF increases, the Nyquist diagram expands, the critical point appears inside the Nyquist diagram, which is the mapping of the right half-plane of s, and the signal starts rising. Therefore, deviations of the signal amplitude from the equilibrium amplitude start the process of exponential adjustment of the amplitude toward the equilibrium.

The analysis for the limit cycle with frequency f_1 gives similar results, but the analysis corresponding to oscillation with frequency f_2 leads to the opposite conclusion, pointing at instability of this process. Therefore, if the oscillation at frequency f_2 is created, the signal deflects from the solution exponentially (very rapidly, i.e., practically by a jump), and enters the basin of attraction of one of the stable limit cycles which will proceed. This sort of process instability will be further studied in Chapter 12.

The simple test for global stability involves the application of a step function or a large-amplitude pulse to the command input. Such a test uncovers most of the hidden limit cycles, but not always all of them.

To discover a suspected limit cycle, initial conditions must belong to the basin of attraction, so they should be chosen close to the conditions that will exist during the limit cycle. Such a test uses bursts of oscillation of large amplitude and of various frequencies, such as those illustrated in Figure 11.45. Although, theoretically, even this test might not discover all limit cycles, and a counterexample system can be imagined where a limit cycle is triggered by only a special key signal, using a set of bursts of oscillation is good enough for testing the stability of practical control systems.

To estimate the stability margins, extra phase lag or extra gain is gradually added to the linear links of the loop until self-oscillation starts, and these extra phase and gain values are taken to be the (experimentally determined) phase and gain stability margins.

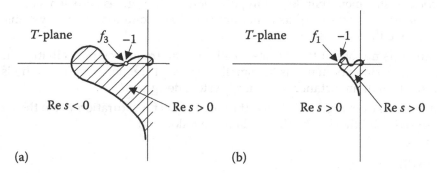

FIGURE 11.44
Nyquist diagram when the system oscillates with frequency (a) f_3 and (b) f_1.

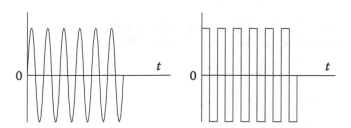

FIGURE 11.45
Signals to excite limit cycles during a global stability check.

PROBLEMS

1. The loop transfer functions are the following:

   a. $T = 200(s + 300)(s + 600) / [s(s + 20)(s + 50)(s + 100)]$

   b. $T = 1,000(s + 400)(s + 500) / [s(s + 10)(s + 40)(s + 80)]$

   c. $T = 200,000(s + 100)(s + 400) / [s(s + 10)(s + 30)(s + 600)]$

   d. $T = 1,400(s + 600)(s + 800) / [s(s + 15)(s + 20)(s + 1,200)]$

   e. $T = 180(s + 272)(s + 550) / [s(s + 12)(s + 30)(s + 80)]$

   f. $T = 2,500(s + 1,000)(s + 1,200) / [s(s + 30)(s + 80)(s + 180)]$

   g. $T = 5,000,000(s + 500)(s + 550) / [s(s + 30)(s + 32)(s + 3,000)]$

   What is the frequency of oscillation (approximately) if there is a saturation link in the loop?

   If there is a dead-zone link in the loop, is the system stable, unstable, or conditionally stable? If it is conditionally stable, how do you trigger self-oscillation?

2. What is the relative amplitude of the third, fifth, and seventh harmonics in the Π-shaped symmetrical periodic signal?

3. During oscillation in a feedback control system where the loop phase lag is 180° and the slope of the Bode diagram is constant, how much smaller are the (a) fifth and (b) seventh harmonics at the input to the saturation link relative to the fundamental?

4. Invent a nonlinear link whose output's fundamental is very sensitive to the shape (i.e., to the harmonics) of the input signal, and use it as a counterexample to the (incorrect) statement that describing function analysis gives sufficient accuracy in all situations. Are the outputs of nonlinear links of common control systems very sensitive to the signal shape?

5. DF analysis might fail to accurately estimate the frequency of oscillation in systems where stability margins are small over broad frequency bands. Why is this failure of small importance for control system designers?

6. Using the chart in Figure 11.8, find the value of DF for saturation when the ratio of the signal amplitude to the threshold is the following:

   a. 5 dB

   b. 15 dB

   c. 25 dB

   d. 4 times

   e. 8 times

7. Using the chart in Figure 11.8, find the value of DF for a dead zone when the ratio of the signal amplitude to the threshold is the following:

   a. 5 dB

   b. 15 dB

   c. 25 dB

   d. 3 times

   e. 5 times

8. Using the chart in Figure 11.8, find the value of DF for a three-level relay when the ratio of the signal amplitude to the threshold is the following:

a. 5 dB

b. 15 dB

c. 25 dB

d. 5 times

e. 10 times

9. Do Problems 6–8 using approximate formulas for describing functions.

10. The signal amplitude at the input to a saturation link with the threshold 0.12345 increases (a) from 123 to 1,230 and (b) from 314 to 628. How many times does the DF change? (Make an engineering judgment about what should be the accuracy of the answers.)

11. The signal amplitude at the input to a link with 0.12345 dead zone increases (a) from 123 to 1,230 and (b) from 2.72 to 7.4. How many times does the DF change?

12. Write an approximate expression for DF that is valid for input sinusoidal signals with amplitudes more than the saturation threshold, for the link with the following:

a. Dead zone 0.2 and saturation threshold 1.3

b. Dead zone 0.3 and saturation threshold 2.3

c. Dead zone 0.4 and saturation threshold 3.3

d. Dead zone 0.8 and saturation threshold 5

13. The linear loop links' gain is (a) 25 dB, (b) 10 dB, or (c) 15 dB, and the phase is 180° at some frequency. What is the ratio of the signal amplitude to the threshold in the saturation link during self-oscillation? If the loop contains a dead-zone link, what is the ratio of the signal amplitude at the input to the link to the dead zone?

14. Plot the transfer function on the L-plane with MATLAB and make inverse and direct DF stability analysis for a system with saturation, with threshold 1, for the following:

a. $T = 200(s + 300)(s + 650) / [s(s + 20)(s + 45)(s + 90)]$

b. $T = 10{,}000(s + 400)(s + 550) / [s(s + 10)(s + 35)(s + 80)]$

c. $T = 20{,}000(s + 100)(s + 400) / [s(s + 10)(s + 30)(s + 600)]$

d. $T = 140{,}000(s + 600)(s + 800) / [s(s + 15)(s + 20)(s + 120)]$

e. $T = 1{,}800(s + 272)(s + 700) / [s(s + 12)(s + 35)(s + 75)]$

f. $T = 2{,}500(s + 1{,}200)(s + 1{,}200) / [s(s + 22)(s + 80)(s + 160)]$

g. $T = 600{,}000(s + 650)(s + 550) / [s(s + 16)(s + 32)(s + 300)]$

15. The area of the hysteresis loop is one-half of the product of the sinusoidal input signal amplitude and the amplitude of the output signal fundamental. What is the phase lag of the link DF?

16. Find the current-voltage characteristic of the Schmitt trigger's (a) transfer function and (b) input. Discuss the stability conditions.

17. Calculate the dead beat zone and the period of oscillation in the system in Figure 11.14 where $VCC = 12$ V, the Schmitt trigger resistors are $R_1 = 100$ kΩ and $R_2 = 200$ kΩ, and the integrator elements are $R_3 = 1$ MΩ and $C = 10$ nF.

18. Calculate the oscillation period in the on–off control system shown in Figure 11.15 if the preamplifier gain is 1, the dead beat zone is 10°C, the heater power is 2 kW, the cooling losses are 200 W, and the payload thermal capacitance is 2 Btu/degree. Is the oscillation symmetrical?

19. Calculate the corner frequency (the frequency of the zero) in the plot in Figure 11.17(b) when the DF of the saturation link at all frequencies is (a) 0.5, (b) 0.1, or (c) 0.05. To what values of the signal amplitudes do these DFs correspond?

20. In the nonlinear link in Figure 11.17, what is the phase shift at frequency 10 MHz if the signal-to-threshold ratio in the saturation link is the following:

    a. 1

    b. 10

    c. 100

21. Will the saturation in the forward path of the NDC in Figure 11.25(c) substantially affect the NDC's performance?

22. Plot with MATLAB the iso-E Bode gain diagrams and phase diagrams for the link in Figure 11.17(a) for the range of $1 < E < 100$, with the threshold of the saturation link equal to 2. Mark the plots with the values of E calculated with approximate formulas for DF.

23. Produce the iso-w Bode plot shown in Figure 11.23 and the related phase responses with MATLAB or Simulink.

24. In the NDC shown in the block diagram in Figure 11.29, $B = 20/(s + 20)$, and the dead zone (for each polarity) is 1.5. Plot iso-w and iso-V Bode diagrams using MATLAB and the approximate expression for the DF of the dead-zone link.

    a. With Simulink, plot iso-w Bode diagrams for the NDC shown in Figure 11.31(a).

    b. Using SPICE, plot iso-w Bode diagrams for the NDC shown in Figure 11.31(b).

26. The composite link is a cascade connection of an integrator and a saturation link. The local feedback about this composite link makes the link gain response flat. When the saturation link becomes overloaded by a large-amplitude signal, what extra phase shift will be introduced in the main loop? Is this situation dangerous from the point of view of making the system not globally stable?

27. A link, whose gain response is flat and whose phase shift is zero over the band of interest, is enlooped by local feedback, the feedback path being an integrator. When the link is overloaded by a large-amplitude signal, what extra phase shift will be introduced in the main loop? Is this situation dangerous from the point of view of making the system not globally stable?

28. A Nyquist-stable system is made globally stable by an NDC. What initial conditions must be used to verify by computer simulations that the system is globally stable?

29. Is the system with return ratio $T = 20,000,000(s + 500)(s + 550)(s + 600)/[(s + 10)(s + 30)(s + 32)(s + 33)(s + 3,000)]$ stable in the linear mode of operation? Does the system have a limit cycle if there is a saturation link in the loop? At what frequencies is the phase stability margin equal to zero (read these frequencies from frequency responses obtained with MATLAB)? What is the loop gain at these frequencies? What is the frequency of the limit cycle? What is the value of the saturation DF if the system oscillates? What is the shape of the signal at the output of the nonlinear

link if the system oscillates? What (approximately) is the shape of the signal at the input to the saturation link? What kind of signal should be applied to the system to discover the limit cycle?

30. Research project: Design a Nyquist-stable system with an NDC with parallel channels. Compare versions with different sequences of the links in the channels.

31. Research project: Design a Nyquist-stable system with an NDC with local feedback. Compare versions with different sequences of the links in the local feedback paths.

32. Research project: Design a Nyquist-stable globally stable system with the AS approach and with the DF approach, and compare the performances.

ANSWERS TO SELECTED PROBLEMS

1a. Since the phase shift of saturation DF is zero, oscillation will occur at the frequency where $\arg T \approx -\pi$, if the loop gain at this frequency is more than $0\,\mathrm{dB}$. This happens at $f \approx 4.2\,\mathrm{Hz}$, while the loop gain is $17\,\mathrm{dB}$. If a dead-zone link is included in the loop, the system is certainly conditionally stable.

The loop response can be obtained by simulation with MATLAB, or by simulation with SPICE.

When employing SPICE, use a chain of voltage-controlled current sources. Each source should be loaded at a series RL two-pole to obtain a function zero, with $L = 1$ and R equal to the zero value, or, to implement a function pole, at a parallel RC two-pole with $C = 1$ and R equal to the inverse value of the pole. The SPICE schematic diagram is shown in Figure 11.46, and the input file for the open-loop frequency response calculation follows.

```
*ch9oc1.cir for determining oscillation condition
*T = 200(s+300)(s+600)/[s(s+20)(s+50)(s+100)], open loop
G2 2 0 0 1 200 ; gain, zero at 300
L2 2 8 1
G3 3 0 0 2 1 ; zero at 600
L3 3 9 1
R3 9 0 600
G4 4 0 0 3 1 ; pole at 20
```

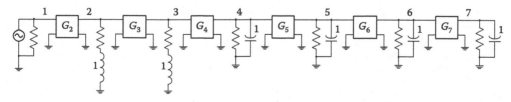

FIGURE 11.46
Schematic diagram for SPICE open-loop simulation.

```
R4  4  0  0.05
C4  4  0  1
G5  5  0  0  4  1 ; pole at 50
R5  5  0  0.02
C5  5  0  1
G6  6  0  0  5  1 ; pole at 100
R6  6  0  0.01
C6  6  0  1
G7  7  0  0  6  1 ; integrator
R7  7  0  1MEG ; de-floating resistor
C7  7  0  1
VIN 1  0  AC  1
RIN 1  0  1MEG ; source loading resistor
.AC DEC 20 1 100
.PROBE
.END
```

Use a cursor to find the frequency at which the phase shift VP(7) equals −180°. Check whether the loop gain VDB(7) is positive at this frequency.

With a dead zone in the loop, the system is conditionally stable, the limit cycle being at the frequency where the phase margin is zero. To excite the self-oscillation, a single pulse or a burst of oscillation at frequency 4.2 Hz should be applied to the closed system input. (If it is desired to perform this simulation, a dead-zone link and a summer should be added to the input file, and the loop should be closed.)

6a. −3 dB

7a. −9 dB

8a. −4.5 dB

12

Process Instability

In asymptotically process-stable (APS) systems, any infinitesimal increment to the input process causes only an infinitesimal increment of the output process. The necessary and sufficient conditions for APS require much larger stability margins than those commonly used.

The APS requirement contradicts feedback maximization, and the majority of practical control systems are not APS, since the effects of the output process instability need only to be bounded. Acceptable levels of process instability can be translated into certain boundaries on the Nyquist diagram, so they can be accounted for in a necessary and sufficient way by appropriate sizing of the stability boundaries. This is an important advantage of the frequency-domain approach to nonlinear system design.

The bounds for process instability are found for sinusoidal test signals. The relationship between the amplitude of the input sinusoidal signal and the amplitude of the fundamental error is hysteretic. As the input amplitude is gradually increased, the error amplitude increases smoothly until the error jumps up, and then continues to increase very slowly. If the input amplitude is then gradually reduced, the error and the output remain almost the same until the input amplitude is reduced to a certain value, at which point the error decreases (again, by a jump). This phenomenon is called jump resonance. Boundaries on the T-plane specify the values and the thresholds of the jumps.

The odd subharmonics may originate in control systems with saturation by a jump, when a parameter of the system is varied continuously. The second subharmonic only occurs in systems comprising nonlinear elements with asymmetric characteristics, and the subharmonic originates smoothly. With stability margins of common values, subharmonics do not show up.

12.1 Process Instability

The output process is considered unstable if an infinitesimal increment in the input triggers a finite or exponentially growing increment in the output. *Process instability* manifests itself in sudden bursts of oscillations or jumps in the output signal. These phenomena contribute to the output error and therefore need to be limited.

A system is said to be *process stable* when the output processes are stable for all conceivable inputs. In accordance with the first Lyapunov method, process stability implies stability of the locally linearized (generally, time-variable) system.

In most cases, process instability is not a critical design issue. However, during experiments with testing and troubleshooting feedback systems, process instability and the associated nonlinear phenomena such as jump resonance and subharmonic oscillation may occur and can be very confusing. For this reason, process instability conditions and manifestations need to be well understood.

12.2 Absolute Stability of the Output Process

The system in Figure 12.1(a) is said to be *absolutely process stable (APS)* if it is process stable with any characteristic of the nonlinear memoryless link $v(e)$, whose differential gain coefficient is limited by:

$$0 < \frac{dv}{de} < 1 \tag{12.1}$$

For example, the monotonic characteristic illustrated in Figure 12.2 satisfies the condition in Equation 12.1.

The system is APS if, at all frequencies,

$$\mathrm{Re}T(j\omega) > -1 \tag{12.2}$$

To prove this statement, we need to show that the feedback system linearized for deviations shown in Figure 12.3(a) is stable. Here, the nonlinear link $v(e)$ is replaced by a linear time-variable link with the gain coefficient:

$$g(t) = \frac{dv}{de}$$

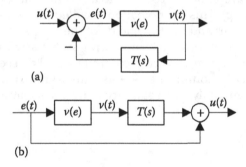

(a)

(b)

FIGURE 12.1
Feedback system and (b) its equivalent.

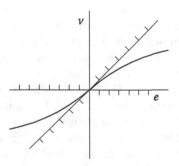

FIGURE 12.2
Characteristic of the nonlinear link with limited differential gain.

In accordance with Equation 12.1, the coefficient $g(t) \in (0,1)$, or,

$$1 / g(t) > 1 \tag{12.3}$$

The analogy between the feedback system and the two-poles connection discussed in Section 7.10.2 yields the equivalent electrical circuits, shown in Figure 12.3(b) and (c). Since $F(s) = T(s) + 1$ is positive real, the two-pole in the lower part of the circuit in Figure 12.3(c) is passive. The two-pole $1/g(t) - 1$, although time variable, is resistive and positive, i.e., energy dissipative. Hence, no self-oscillation can arise in this system. Therefore, the system in Figure 12.1 is APS if it satisfies Equations 12.1 and 12.2. (Notice that the analysis is valid even if the linear links in Figure 12.3 are time variable. Therefore, Equation 12.2 gives a sufficient stability criterion for LTV feedback systems.)

Equation 12.2 is not only sufficient, but in some sense necessary for APS. If Equation 12.2 is not satisfied, i.e., Re $T < -1$ at some frequency ω, then there can be found (1) some function $v(e)$ satisfying Equation 12.1 and (2) some periodical input signal with the fundamental ω, that together bring forth an unstable output process. Particularly, it can be shown that when $v(e)$ is saturation, and the input is a Π-shaped periodical signal with frequency ω, gradual change in the input signal amplitude renders a jump in the output signal amplitude [33].

The condition in Equation 12.2 restricts the position of the Nyquist diagram as shown in Figure 12.4 and requires the phase lag at large loop gains not to exceed 90°. This limits the slope of the Bode diagram to $-6\,dB$/octave and thereby reduces the available feedback.

In practical systems not satisfying the APS condition, typical process instability does not contribute much to the output error, and then only when the actuator is saturated and a rather unusual command is applied to the input. On the other hand, excessive process

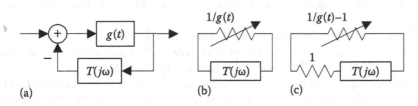

(a)   (b)   (c)

FIGURE 12.3
(a) Feedback system with time-variable real gain coefficient $g(t)$ and (b, c) the schematic diagrams described by the same equations.

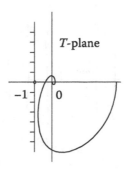

FIGURE 12.4
Restriction of the Nyquist diagram by the APS condition.

instability is not acceptable. The required feedback-system parameters should therefore lie somewhere between these two extremes. To evaluate and certify the systems, appropriate test signals should be selected among those typical for the practical system inputs. For systems where the inputs can be periodical with large amplitudes, the relative jump values of the *jump resonance* (described in the following section) are employed as the figure of merit.

12.3 Jump Resonance

The amplitude and shape of the periodic output signal are generally multivalued functions of the periodic input signal parameters, and depend on the input signal prehistory. Particularly, the output might depend on whether the current value of the input signal was arrived at by gradually increasing or gradually decreasing the input signal amplitude.

Jump resonance can be observed in the nonlinear feedback system in Figure 12.1(a) excited by sinusoidal input $u = U \sin \omega t$. While the input signal amplitude U is gradually increased, at a certain amplitude $U = U''$ after the saturation level, an infinitesimal increase in the amplitude of the input, from U'' to $U'' + 0$, causes the output signal $v(t)$ to change by a jump from time response (a) to time response (b) shown in Figure 12.5. As a result, the amplitude of the fundamental output V increases by a jump.

Next, gradually reducing the amplitude of the input starting at a value bigger than U'', down to a certain value U', causes a jump down from time response (c) to time response (d) when the input amplitude is reduced by an infinitesimal increment, to $U' - 0$.

The conditions for the jumps can be found by initially analyzing the system in Figure 12.1(b), which implements the same relationships as the feedback system in Figure 12.1(a) (as can be easily verified). Using the DF approach, we assume that $e(t) \approx E \sin \omega t$.

When the phase stability margin is rather small, say 15°, i.e., arg T is −165°, vector addition of the outputs of the two parallel channels generates a collapse of $U = 2E \sin(y\pi/2)$ in the region where $V |T| \approx E$. This is exemplified in Figure 12.6.

The dependence $U(E)$ in Figure 12.6 is single-valued. The inverse operator $E(U)$ redrawn in Figure 12.7 is three-valued. This plot comprises a *falling branch* over the interval (U', U'') between the *points of bifurcation* marked dark. When U passes U'' while being gradually increased, E jumps from before-the-jump value E_b'' to the after-the-jump value E_a'', and when U is reduced to U', E jumps down to E_a'.

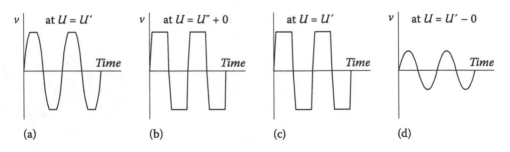

FIGURE 12.5
Shapes of the output signal $v(t)$ (a) before the jump up and (b) after the jump up, and (c) before the jump down and (d) after the jump down.

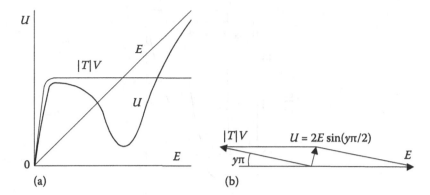

FIGURE 12.6
Output signal amplitude for equivalent parallel channel system: (a) dependence of U on E, and (b) vectors' addition at the collapse.

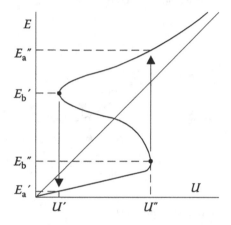

FIGURE 12.7
Three-valued dependence of the error amplitude on the input signal amplitude.

The proof that the solution on the falling branch is unstable with a real positive pole was given in Section 10.3.1. Since the system is unstable, the signal rises exponentially (jumps) between the two stable solutions, as shown by the arrows in Figure 12.7.

In the system with saturation, only the jumps down can be large. From the curves in Figures 12.6 and 12.7 it is seen that the after-the-jump amplitude E_a' depends on the loop gain and phase. The calculated dependencies are plotted in Figure 12.8. These curves allow one to specify the required stability margins from the values of the allowed jumps, and also, while experimenting, to calculate the stability margins from the observed jumps.

The jumps occur in feedback systems when arg T is such that the curve $U(E)$ in Figure 12.6 possesses a minimum. This takes place within certain frequency intervals. On the plane (U, f) in Figure 12.9, the area is traced where the dependence $E(U)$ has three solutions. At the lowest and highest frequencies, the limit case occurs if $U' = U''$ and $dU/dE = 0$ with a zero-length falling branch. The jumps can be caused by varying either U or f, or both.

Hysteretic frequency responses of the output signal amplitude similar to that shown in Figure 12.10 can be recorded while the input signal amplitude is kept constant. Such responses were first observed during the study of resonance tanks with nonlinear inductors

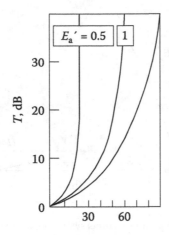

FIGURE 12.8
Lines of equal values of jumps down in a system with saturation.

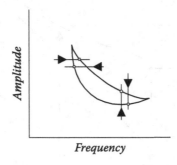

FIGURE 12.9
Amplitude-frequency areas of three-valued amplitude *U*.

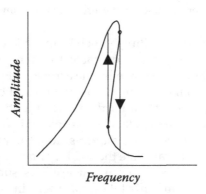

FIGURE 12.10
Jump resonance when the frequency is being changed.

having ferromagnetic cores. Because of the analogy described in Section 7.10.2, a parallel connection of an inductor and a capacitor is described by the same equations as a feedback system with the return ratio $1/(LCs^2)$. The phase stability margin is small (although it is nonzero due to the inductor losses), and therefore the jump resonance is prominent.

For feedback systems with dynamic saturation (including those with nonlinear dynamic compensators), it can be shown [33] that the after-the-jump amplitude is:

$$E_a' = \frac{1.27}{\sup |M|_{E>1}} \tag{12.4}$$

The denominator can be readily found by plotting the iso-*f* line on the Nichols chart. As shown in Figure 12.11. On the iso-*f* line, $|M|$ of 3.5 dB is maximum, i.e., $|M|$ = 1.5, and hence E_a' = 0.85.

Example 12.1

It is difficult to measure the loop responses in RF and MW feedback amplifiers since these measurements require loading the port of the open feedback loop on a two-pole with impedance imitating the impedance of the disconnected port, as shown in Figure A6.3. Instead, phase stability margins at various frequencies can be found by observing the jump resonance in the closed-loop configuration and using the plot in Figure 12.8.

Example 12.2

Jump resonance has been observed in the attitude control loop on the Mariner 10 space-craft. Nitrogen thrusters of the solar panels' attitude control excited large-amplitude periodic oscillation in the panels that have a high Q structural mode. The gyro bearing nonlinearity led to jump resonance in the gyro loop. This in turn led to the panel attitude control thrusters consuming an abnormally large amount of the propellant. The problem was rectified by reducing the gain in the control loop.

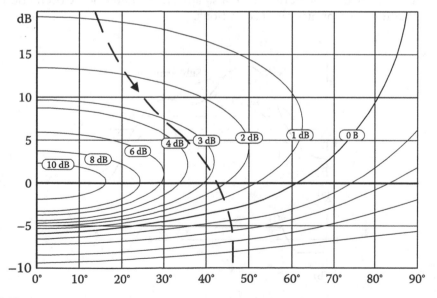

FIGURE 12.11
The jump value corresponds to the maximum of $|M|$ on the iso-*f* line.

12.4 Subharmonics

12.4.1 Odd Subharmonics

As will be shown below, subharmonics in low-pass systems with saturation can only become excited if the phase stability margin at the subharmonic frequency ω/n is rather small. This implies a steep cutoff of the Bode loop gain diagram. Therefore, the signal at the input to the nonlinear link mainly consists of the lower-frequency components: $E \sin \omega t + E_{sb} \sin(\omega t/n)$.

Let us examine whether odd subharmonic oscillation is possible with either small or large amplitudes of E and E_{sb}. We assume the saturation threshold is 1.

Evidently, with both E and E_{sb} being so small that $E + E_{sb} < 1$, the subharmonics are not observed since in this case the system behaves linearly.

If $E > 1$ and $E_{sb} \ll 1$, part of the time the nonlinear link is saturated by the signal. For small-signal increments given to its input, the link can be seen as an equivalent linear time-varying one, a pulse element with the sampling frequency ω. Considering the subharmonic as the input increment, the output increment is the product of $E_{sb} \sin(\omega t/n)$ and the Fourier expansion with the fundamental frequency ω, which characterizes the pulse element. In this product, the component with the frequency ω/n can be generated only due to the constant component of the Fourier series. Since this component is real, the nonlinear link does not introduce in the loop any phase shift for the subharmonic.

Therefore, when the parameters of the input signal are manipulated slowly and continuously, the odd subharmonics may only originate by a jump, with nonzero steady-state amplitude E (this is called *hard oscillation*).

On the other hand, if $E_{sb} \gg 1$ and $E < E_{sb}$, the input to the nonlinear link can be as shown in Figure 12.12(a). This signal is clipped at the levels shown by dashed lines, and the output of the saturation link $v(t)$ accepts the shape of trapezoidal pulses, shown in Figure 12.12(b). These subharmonic pulses are shifted in phase by some angle due to the signal $E \sin \omega t$. If this angle exceeds the phase stability margin at the frequency of the subharmonic, and the loop gain at this frequency is more than 1, a steady-state subharmonic oscillation may be excited by creating appropriate initial conditions.

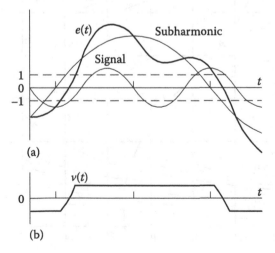

(a)

(b)

FIGURE 12.12
Third subharmonic mechanism.

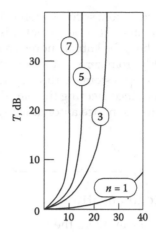

FIGURE 12.13
L-plane boundaries for odd subharmonics in a system with saturation.

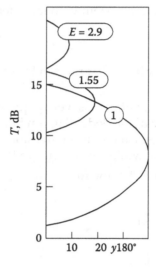

FIGURE 12.14
L-plane boundaries for second subharmonic existence in a system with single-polarity saturation.

It is seen from Figure 12.12 that the phase shift is less than $\pi/2n$. The boundaries for the third- and higher-order subharmonics are displayed in Figure 12.13 [33]. In practical control systems the phase stability margin is always bigger than $\pi/6$, and odd subharmonics are not observed.

12.4.2 Second Subharmonic

The second subharmonic can be observed only if the characteristic of the loop nonlinear link is asymmetric, and the greater the asymmetry is, the wider are the areas in which the subharmonic exists. For a system with one-sided saturation, the boundaries of the subharmonic's existence that correspond to different values of the signal amplitude are shown in Figure 12.14 [33]. As seen from the figure, the second subharmonic is not generated in systems where the phase stability margins are greater than 30°.

Unlike odd subharmonic self-excitation, second subharmonic self-excitation can be *soft*. Soft excitation means that when either U or ω, or both, is gradually changed along any trajectory entering the domain where the subharmonic can be observed, the amplitude of the subharmonic increases steadily from zero, without jumps.

We may conclude that in single-loop systems with saturation, neither the odd nor the even subharmonics present real danger, since meeting the mandatory requirement for eliminating wind-up and substantial jump resonance automatically excludes the subharmonics.

12.5 Nonlinear Dynamic Compensation

Nonlinear dynamic compensation can make the system process stable without sacrificing the available feedback. The conversion of the system with two nonlinear links to an equivalent system with one nonlinear link that was described in Section 10.7 can also be applied to designing NDCs to satisfy the process stability criteria. Design examples in Section 10.7.3 include a process-stable system with large feedback.

PROBLEMS

1. Prove that the systems in the block diagrams in Figure 12.1(a) and (b) are described by the same equations.

2. In a system with saturation, with crossover frequency $f_b = 1$, the phase stability margin is 30° and the slope of the Bode diagram below f_b is –10 dB/octave. Is this system process stable? If not, what must be the slope for the system to be process stable? Approximately, what will be the change in the feedback at 0.03 Hz?

3. Is a system with saturation APS if the following:

   a. $T = 100 / (s + 0.01)$

   b. $T = 1,000 / [(s + 0.01)(s + 2)]$

   c. $T = 123 / (s + 0.21)$

   d. $T = 500 / [(s + 0.02)(s + 3)]$

   e. $T = 272 / [(s + 2.72)(s + 27.2)]$

   f. $T = 5,000 / [(s + 0.08)(s + 0.2)(s + 0.3)]$

4. Prove the validity of the following equivalent forms for Equation 12.2 of APS:

$$\mathrm{Re}F(j\omega) > 0 \tag{12.5}$$

$$\mathrm{Re}\, 1/F(j\omega) > 0 \tag{12.6}$$

$$\cos\arg T(j\omega) > 1/|T(j\omega)| \tag{12.7}$$

and

$$\left|M(j\omega)\right| = \frac{T(j\omega)}{F(j\omega)} > \frac{1}{\sin \arg T(j\omega)} \tag{12.8}$$

5. Consider jump resonance in a system with a dead zone. Draw responses analogous to those shown in Figure 12.6 for a system with saturation.

6. Show that in a system with dead zone and saturation, the output–input amplitude dependence can be five-valued, by drawing a response analogous to that in Figure 12.6.

7. On a Nichols chart, the minimum value of $|M|$ for a system with saturation is (a) 6 dB, (b) 8 dB, or (c) 4 dB. What is the value of E after a jump down from the state of saturated output?

8. The loop gain is 20 dB in a system with saturation at some frequency. The jump down is (a) 2 times or (b) 1.5 times. What is the phase stability margin (approximately) at this frequency?

9. In a system with saturation, what must be the stability margin for the fifth subharmonic to be observable?

13

Multiwindow Controllers

Because nonlinear controllers generally perform better than linear controllers, NDC design methods are of profound interest. We already considered the design of NDCs using AS and DF approaches. In this chapter we consider NDC design from yet another angle, as multiwindow composite controllers.

Composite nonlinear controllers consist of linear high-order controllers and the means to transition between the linear controllers according to some participation rules. Each of the linear controllers is a local linear approximation to the optimal nonlinear controller. The size of the regions where a single linear controller is operational is discussed as well as the appropriate complexity of the elementary control laws.

Multiwindow control makes transitions between the elementary linear controllers on the basis of the amplitude of the error. It is relatively simple to implement and provides much better performance than linear controllers.

During the transition between the elementary controllers, it is important that the activating controller is "hot." Hot controllers are those with the input already connected to the signal. This eliminates large transients caused by the initial activation of a "cold" controller.

Wind-up is a large or long overshoot in nonlinear systems, which most often occurs in systems with a large integrating component in the compensator. The widely used anti-wind-up controllers include nonlinear links directing the signal into different paths, depending on the signal level.

The acquisition and tracking problem is that of first finding and locking onto the target, and then precisely tracking it. To perform each of these two tasks optimally, the control law cannot be the same: for acquisition, the control bandwidth must be larger but the feedback can be much smaller. The transition between the two laws must be gradual in order not to de-acquire the target. When combining linear compensators via a multiwindow nonlinear summer, it is important to guarantee that the combined transfer function remains m.p. This can necessitate using more than two parallel branches and windows.

Another typical application for multiwindow controllers is time-optimal control. This problem is related to the acquisition and tracking problem, and the solutions are similar. Several examples of nonlinear controller applications are presented.

Many issues in multiwindow controller design have not yet been investigated, and many questions remain as to how to make the best use of such controllers.

13.1 Composite Nonlinear Controllers

It was demonstrated in Chapters 10 and 11 that some nonlinear controllers perform much better than any linear controller. The optimal controller is, in general, nonlinear.

For a small region in the variable space, the nonlinear optimal controller can be approximated well by a linear controller. For an adjacent region, another linear controller can be designed that would be optimal over this region, and so on. The design problem, therefore, arises of integrating these locally optimal linear controllers into a composite nonlinear controller and providing smooth transitions between these regional linear laws.

Transitions between the control modes can be characterized by *participation rules* defining the set of *participation functions* illustrated in Figure 13.1. Over some transition interval of a variable or a condition, the controller action is the sum of the actions of the adjacent regional control modes, and at the ends of the interval, only one of the modes (or conditions, or actions) takes place.

The transition rule can be linear and expressed as:

$$action = (1-k) \times action_1 + k \times action_2 \tag{13.1}$$

where the scalar k changes from 0 to 1 over the transition interval. With these rules and when the actions are scalars or collinear vectors, the transitions between the control laws are as illustrated in Figure 13.1(a). A smoother rule is illustrated in Figure 13.1(b). Commonly, the precise shape of the participation functions does not matter much as long as it is monotonic, not too steep, and not too shallow.

In general, a monotonic shape of the participation functions does not yet guarantee the smoothness of the transition between different actions. It is also required that the adjacent control laws mix well, i.e., in particular, that the combined action of the adjacent controllers exceeds that of each individual controller. This is not always the case. For example, even if some residue that needs to be cleaned out can be removed by either an acid or a base, an acid and a base should not be used as a mixture with gradually changing content. For regulation of a reactive electrical current, a variable capacitor or a variable inductor can be used, but these elements should not be combined in series or in parallel since they might produce resonances. A low-pass link should not be carelessly mixed in parallel with a high-pass link, or else notches and n.p. shift might result.

Fuzzy-logic controllers break each smooth transition into several discrete steps. This increases the total number of regions with different control laws. Since these regions become very small, fuzzy-logic control can use low-order regional control laws. Hence, fuzzy control design can be based on phase-plane partitions, and on passivity theory expressed in state variables. In fuzzy-logic controllers, many variables need to be sensed and processed to define the boundaries of the regions.

What region size is optimal for composite controllers?

There are two advantages to making the regions small. The first is that the control laws in the adjacent regions might become very similar, which enables smooth transitions between them without taking special precautions. The second advantage is that the linear controller can be of low order, and the phase-plane can be used for the controller analysis

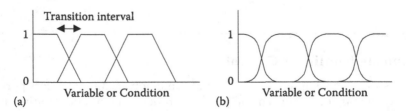

FIGURE 13.1
Participation functions of control laws in composite controllers.

and design. As claimed by some fuzzy-logic advocates, such controllers can be designed even by those ignorant of control theory. However, when the number of the regions is large, the number of boundaries between them becomes very large. Correspondingly, the number of decision-making algorithms and instruments for changing between the control modes becomes very large. This complicates both the controller design and the designed controller.

On the other hand, higher-order linear control laws can be made to remain nearly optimal over a much broader region than low-order laws. This reduces the number of regions and the number of boundaries between them.

For the design of higher-order regional control laws, frequency-domain methods should be used rather than the phase-plane. The partition between the regions should also be defined in the frequency domain. This approach requires caution and the application of the rules discussed in Chapter 4 to provide good blending of the regional control laws at the boundaries between the regions. This approach is not difficult and leads to economical and nearly optimal controllers.

For controllers designed in frequency domain, Equation 13.1 may cause non-minimum phase lag. In this case, it is worth considering the logarithmic transition rule:

$$log(action) = (1 - k)log(action_1) + klog(action_2) \tag{13.2}$$

i.e., multiplication:

$$action = action_1 \times (action_2 / action_1)^k \tag{13.3}$$

13.2 Multiwindow Control

In the following, we will consider a single-input nonlinear controller. The input of the controller is the output of the feedback summer, i.e., the feedback error. Unlike fuzzy-logic controllers, no other sensors and variables are employed to modify the control law.

As shown by N. Weiner [77], the output of a nonlinear operator can be approximated by applying the input signal to a bank of linear operators, and then combining various products of the linear operators' outputs. In other words, nonlinear dynamic links can be approximated by interconnections of linear dynamic links and nonlinear static links. The multiwindow controllers make a small subclass of such links. Although relatively simple, this subclass allows for much richer varieties of the input–output relationships than those of the linear system. Multiwindow controllers perform significantly better than linear controllers, do not require high accuracy of design and implementation (i.e., are robust), and therefore allow developing some rules of thumb suitable for the conceptual design and the design trade-offs. In spite of the subclass simplicity, rigorous, straightforward methods for the synthesis of such controllers have yet to be developed, and the initial design is further optimized in practice by repeating the computer analysis. As we concentrate on the applications, the presentations in this chapter are not rigorous and rely largely on examples and simulations.

The signal components of the error can be divided into several sets bounded by two-dimensional *windows*, shown in Figure 13.2. The windows divide the frequency spectrum (or, equivalently, the time response behavior) and the amplitude range. Within each window, a regional linear controller (compensator) is employed.

The multiwindow nonlinear controller can be implemented as follows: the error is partitioned into components falling into different windows, the components processed by the linear operators of the windows, and the results added up or combined by nonlinear functions. This architecture is referred to as *multiwindow*, and a great number of useful nonlinear control schemes can be cast in this form.

The regional linear controllers are optimized using Bode integrals as the performance bounds. The criterion for minimum-phase behavior is employed as the condition for smooth blending with the adjacent regions differing in frequency. Due to this smooth blending, the exact shape of the participation rules is not critical. The static nonlinearities used to implement the transition between the control modes can all be chosen to be of the hard saturation type, and the threshold of the saturation need not be exact.

A strong correlation exists in many systems between the error's amplitude and the error frequency content, so that the errors fall into a set of diagonal windows, as shown in Figure 13.3. This is the case, for example, when the disturbance is a stochastic force with flat spectrum density, applied to a rigid body and causing the body displacement with the spectrum density inversely proportional to the square of the frequency, and when a displacement command profile is chosen in consideration of the limited force available from the actuator. Due to this correlation, the signal components that should go to a specific window can be selected either by the amplitude or by the frequency (the order of the selection is further discussed in Section 13.5). The controllers for these systems can be

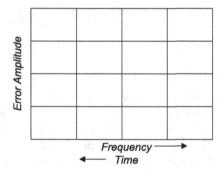

FIGURE 13.2
The multiwindow control concept: the choice of the linear controller is defined by the error amplitude and frequency content.

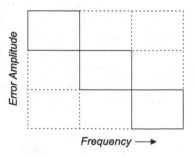

FIGURE 13.3
Diagonal windows.

composed of linear operators of the windows and nondynamic (static) nonlinear functions for pre- or postprocessing and for splitting or combining the signals. Such controllers provide good performance for a wide variety of practical problems.

The simplest multiwindow compensators are two-window compensators. The large-amplitude components are processed by the regional compensator with low low-frequency gain; the low-amplitude signals are processed with higher low-frequency gain. The two-window controllers are widely employed, in particular, in anti-wind-up schemes, in acquisition and tracking systems, and in Nyquist-stable systems for provision of global and process stability.

The multiwindow compensator is nonlinear dynamic. And conversely, the NDCs discussed in Chapters 10 and 11, whether they are made as a combination of parallel channels or as links with nonlinear local feedback, can be viewed as multiwindow controllers.

13.3 Switching from a Hot Controller to a Cold Controller

The sharpest participation rule is instant switching between controllers. Even with this switching, the transition between the regional laws can be made smooth.

Assume the inputs of several linear controllers are connected to the output of the feedback summer, as shown in Figure 13.4(a). The controllers are *hot*; i.e., they process the error all the time. A simple switching or some nonlinear law can be used to choose one of the outputs or a nonlinear function of all or some outputs to send further to the actuator. Since the controllers are hot, and if the difference between the controllers is not excessive, the output signals of the controllers are to some extent similar, and switching between them will not create large transients.

On the other hand, when a nonlinear law allocates the inputs, but the outputs are both permanently connected to the actuator input as shown in Figure 13.4(b), the controller that has been off for a long time is *cold*, and its output signal is zero or some constant. When the actuator input is switched to this controller from even a similar (but cold) controller, large transients can result. Example 13.1 illustrates the problem.

Example 13.1

Consider the system with a single-integrator plant and two switchable linear compensators, shown in Figure 13.5(a).

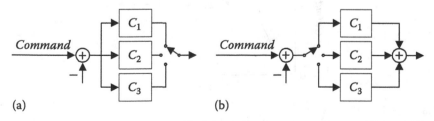

(a)   (b)

FIGURE 13.4
Switching (a) between hot controllers and (b) a cold controller.

The lower-path compensator is a lead with an integrator. The upper path provides much larger feedback bandwidth but smaller feedback at lower frequencies (these responses will be discussed in Section 13.5). The compensators are permanently connected to the output of the feedback summer so that the compensators are hot. Switching between the compensators' outputs occurs when the error magnitude achieves a certain threshold value, with some hysteresis to avoid frequent switching back and forth. In Figure 13.5(b), a similar system is shown, but here the switching occurs from a hot to a cold compensator. The saturation thresholds are –100, 100; for the hysteresis link, the thresholds are 0.1, 0.2, and the output switches between 1 and 0. The switch engages the upper path when 1 is applied to the switch input, and the lower path when 0 is applied to the switch input.

The output transient responses for both systems are shown in Figure 13.6. It is seen that in the system with hot compensators, the transients caused by the switching are small, while the transients caused by switching to the cold compensator are large.

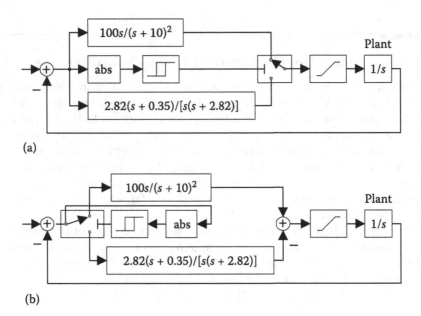

FIGURE 13.5
Switching between controllers (a) at the input and (b) at the output.

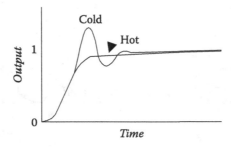

FIGURE 13.6
Transient responses with switching to hot and cold compensators.

13.4 Wind-Up and Anti-Wind-Up Controllers

The time responses of a system with saturation to small- and large-amplitude steps can be as shown in Figure 13.7. The overshoot for the large-amplitude input step is excessive and persistent—this phenomenon is called *wind-up*. The wind-up can be many times longer than the overshoot in the linear mode of operation (i.e., the overshoot for a step command of smaller amplitude).

Wind-up is typically caused by a combination of two factors: the error integration in the compensator and the actuator saturation. The saturation limits the return signal and therefore prevents the error signal accumulated in the compensator integrator from being reduced. During the initial period after the step command is applied, when the output is still low and the error is large, the compensator integrates the error. When the time comes at which this integrated error would be reduced by the return signal in the linear mode of operation, this does not happen for large commands, since the return signal is attenuated by the actuator saturation. Therefore, it might take a long time for the feedback to reduce the integrated error. The error "hangs up," and only after some time does the output signal drop to the steady-state value.

Example 13.2

We will illustrate the wind-up with an example of a simple system with a single-integrator plant, shown in Figure 13.8(a). The asymptotic loop response is shown in Figure 13.8(b).

The compensator includes a parallel connection of an integrator and a unity gain path, and a following link with a pole at 2 rad/s. The crossover frequency is 1 rad/s. As

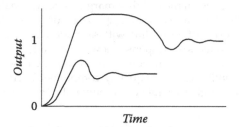

FIGURE 13.7
Transient response in linear mode (lower curve) and wind-up (upper curve).

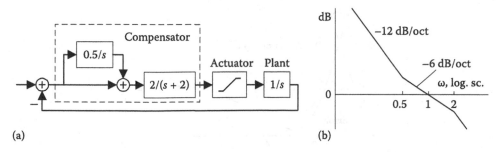

FIGURE 13.8
(a) System with saturation and (b) the loop asymptotic Bode diagram.

seen from the asymptotic Bode diagram, the system must be stable. The loop transfer function (return ratio), the closed-loop transfer function, the transfer function from the input to the error, and the transfer function to the input of the actuator are, respectively:

$$T = \frac{2s+1}{s^3 + 2s^2} = \frac{m}{n}$$

$$M = \frac{m}{n+m} = \frac{2s+1}{s^3 + 2s^2 + 2s + 1}$$

$$\frac{error}{command} = \frac{1}{F} = \frac{n}{n+m} = \frac{s^3 + 2s^2}{s^3 + 2s^2 + 2s + 1}$$

$$\frac{actuator\,input}{command} = Ms = \frac{2s^2 + s}{s^3 + 2s^2 + 2s + 1}$$

The step responses obtained with the MATLAB step command for this system without saturation are shown in Figure 13.9. It is seen that the signal at the actuator input has a large peak. The output response overshoot follows this peak with about 90° delay because of the plant integration.

Clipping the signal peak at the actuator by the actuator saturation causes wind-up. Figure 13.10 shows the output response to the step function command when the saturation level in the actuator is ±0.25. The wind-up increases the height and the length of the overshoot.

Using the DF concept, the qualitative explanation for the wind-up phenomenon goes as follows: Actuator saturation reduces the describing function loop gain, thus shifting the equivalent crossover frequency down. The resulting overshoot is long, corresponding to this low-crossover frequency. The value of the wind-up depends on the loop phase lag. When the phase stability margin is more than 70°, the wind-up is practically nonexistent, but it is large when the stability margin is 30° or smaller.

Wind-up can be reduced or eliminated by employing nonlinear dynamic compensation. For example, placing an extra saturation link with saturation level 0.1 in front of the integrator in the compensator produces the response shown in Figure 13.10 by the dashed line.

The explanation of the wind-up phenomenon can also be based on the idea of intermodulation: a large-amplitude, low-frequency component overloads the actuator and prevents it from passing higher-frequency components; the remedy suggested by this

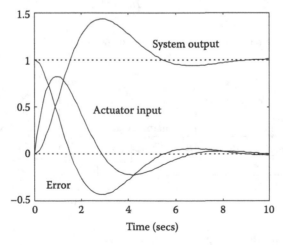

FIGURE 13.9
Time responses of a linear system.

analysis uses separate channels for processing lower-frequency and higher-frequency signal components.

Wind-up in a *PID* system is commonly reduced or eliminated by placing a saturation in front of or after the integrator, or by *resetting* the integrator, i.e., by changing its output signal to zero at the rise time.

In Figure 13.11, two approaches are shown that are widely used in analog compensators to reduce or prevent the wind-up. In Figure 13.11(a), the diodes (or Zener diodes) in parallel to the capacitor in the integrator limit the maximum voltage on the capacitor, and therefore the maximum charge in it that can be accumulated during the transients to a step command. In Figure 13.11(b), the reset option is shown: simultaneously with the application of the step command, the capacitor is shorted for a certain time (close to the rise time) to prevent the built-up charge.

Another method of the wind-up elimination is placing a rate limiter like that shown in Figure 6.19 in the command path.

A saturation link is sometimes placed as well in front of the *P* term, as shown in Figure 13.12, with a larger threshold than that of the saturation in front of the *I*-term. This makes the controller a three-window one.

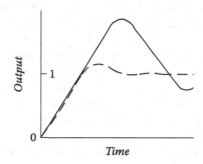

FIGURE 13.10
Time responses for the system in Figure 13.8 (solid line), with added saturation with a threshold of 0.1 in front of the integrator in the compensator (dashed line).

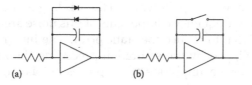

FIGURE 13.11
Analog integrators with (a) parallel diodes and (b) reset switch.

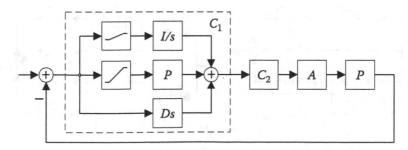

FIGURE 13.12
Saturation links in front of the *I* and *P* paths.

13.5 Selection Order

The window diagrams in Figures 13.2 and 13.3 are somewhat ambiguous since they do not indicate whether the frequency selection or amplitude selection is performed first. Often a particular order is required. This order is different in the block diagrams in Figure 13.13(a) and (b), which exemplifies two types of architectures for multiwindow compensators.

Figure 13.13(a) shows a parallel-channel compensator. In such compensators, saturation links are commonly placed in the low-frequency channel, since this channel's gain response dominates at lower frequencies. At large signal levels, the saturation link reduces the low-frequency gain, and the phase lag of the compensator decreases, thus reducing or eliminating the wind-up. Placing a saturation link with an appropriate threshold in front of the I-path commonly reduces the length and height of the overshoot. The value of the threshold is not critical. Placing a dead-zone element in front of the high-frequency channel with $k < 1$ also reduces the phase lag at large signal amplitudes, which helps to eliminate wind-up and improve the transient response.

When the saturation link follows the linear *filter fork* (a filter fork is a combination of low-pass and high-pass filters for splitting or combining low-frequency and high-frequency signals, in this case low-pass LP_1 and high-pass HP_1) as in the left side of the block diagram in Figure 13.13(b), the amplitude of the overshoot can be regulated by adjustments of the saturation threshold. However, in some cases, while the value of the overshoot substantially reduces, the time of the overshoot is still excessive. The overshoot can be reduced to say, only 1%, but the hang-off can last a long time. (In a version of the temperature controller described below, the overshoot of 1% lasted 1 h.) In certain applications, small-amplitude wind-up is acceptable.

The block diagram in Figure 13.13(b) shows a very general architecture of mixed order of nonlinear windows and filter forks.

The best architecture of a *PID* controller with anti-wind-up depends on the sort of command the controller must respond to and also on the nature of the disturbances. When the only command is a step, placing the saturation first will prevent the accumulation of integrated error that causes wind-up. However, if there is large-amplitude, short-pulse-type random noise, and the low-frequency components of this noise are substantial, the integral term needs to be functional to reduce the static error. The integration is not as effective if the peaks are being clipped by the saturation preceding the integration. In this case, it would be better to lower the amplitude of the pulses by placing a low-pass filter (or an integrator) in front of a saturation link.

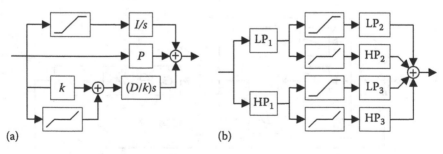

(a)   (b)

FIGURE 13.13
Multiwindow compensators with parallel channels.

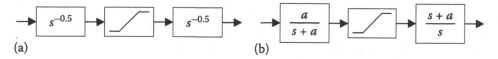

(a)   (b)

FIGURE 13.14
Possible *I*-path implementations for anti-wind-up.

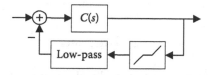

FIGURE 13.15
Anti-wind-up NDC with local feedback.

When the error signal can be arbitrary, the best performance (in a mini-max sense) might be provided by a combination of the two strategies; i.e., the saturation link could be sandwiched between two low-pass filters. Here, two options exist: (1) half of an integration can be placed before the saturation, and the other half after, or (2) using two first-order links, one can be placed before the nonlinearity, to cut off the higher frequencies using a single pole, and one after, cutting off low frequencies, with a zero compensating the pole of the first link. These implementations are indicated in Figure 13.14. The pole-zero frequency can be adjusted by a single knob. But still, counting the saturation level, two extra knobs are required compared with the linear *PID* controller. (For the half-integrator version, there is only one additional knob.)

An NDC with local nonlinear feedback is shown in Figure 13.15, which is used for the same purpose. All of these can be designed with the DF approach described in Chapter 11. The NDC shown in Figure 13.15 can also be designed as described in Chapters 10 and 12 with the absolute stability and process stability approaches.

13.6 Acquisition and Tracking

Acquisition and tracking systems are designed to operate in two modes: *acquisition mode* when the error is large, and *tracking mode* when the error is small, and the prototypical example is a homing missile intercepting a target. Another example is a pointing control system for a spacecraft-mounted camera, in which a rapid retargeting maneuver is followed by a slow, precise scanning pattern to form a mosaic image of the object. Another example is clock acquisition in the phase-locked loops of telecommunication systems and frequency synthesizers.

When the error signal is large, the system is in the acquisition regime, as indicated in Figure 13.16(a), and the controller should respond rapidly; i.e., the feedback bandwidth should be wide, as shown in Figure 13.16(b). In the acquisition mode it is not necessary that the loop gain be very large, since the error is big and even with a small gain in the compensator, the actuator applies maximum available power to the plant. In contrast, in the tracking regime, the feedback bandwidth needs to be reduced to reduce the output effects

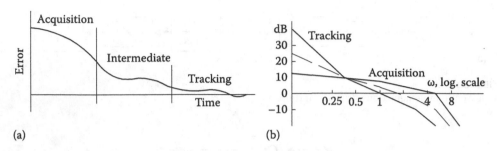

FIGURE 13.16
(a) Error time history and (b) acquisition/tracking loop responses.

of the sensor noise, but the value of the feedback at lower frequencies should be made rather large to minimize the jitter and the tracking error.

While the determination of the optimal frequency responses for the acquisition mode and tracking mode is straightforward, guaranteeing smooth transient responses during transition from acquisition to tracking is not trivial. The transition can generate large transients in the output and in the error signals. If the transients are excessively large, the target can be de-acquired.

The transition between the responses can be made by switching as shown in Figure 13.5, or by using nonlinear windows: the small errors are directed to the tracking compensator, and the large errors directed into the acquisition compensator. Special care must be taken to ensure that all intermediate frequency responses of the combined channel are acceptable. An improper intermediate response might result in an unstable system, or in a system with small stability margins that would produce large-amplitude transient responses.

As an example showing the importance of paying attention to the intermediate responses, let the total loop response be the weighted sum (Equation 13.1) of the acquisition and the tracking responses shown in Figure 13.16(b):

$$T = (1-k)T_{acq} + kT_{tr} \tag{13.4}$$

and suppose that k smoothly varies from 0 to 1. As the transition from acquisition mode to tracking mode occurs, the acquisition response gradually sinks, the tracking response rises, and the frequency at which the moduli of the two components are equal increases. When this frequency is still low, the difference in phase between the two transfer functions at this frequency exceeds 180°; therefore, the total transfer function T has a zero in the right half-plane of s. (The conditions for the transfer function $W_1 + W_2$ to become n.p. when each of the channels W_1 and W_2 is m.p. were given in Chapters 4 and 5.) The non-minimum phase lag reduces the phase stability margin and may result in self-oscillation. The transients generated while the system passes these values of k can be large and disruptive, even causing the target to be lost.

For example, in a modification of the system in Figure 13.5, where the compensators' outputs are combined via nonlinear windows, the overshoot reached 500%.

To reliably avoid the non-minimum phase lag and excessive transients, the slopes of the Bode diagrams of regional responses should not differ by more than by 9 dB/octave in two-window controllers. Hence, the two-window controller, although substantially better than a linear controller, still does not allow implementation of the best possible responses for acquisition and tracking. This can be done with a three-window controller using an intermediate frequency response like that shown by the dashed line in Figure 13.16(b).

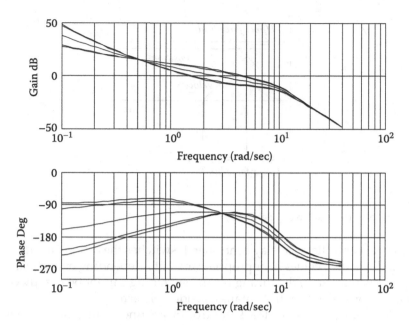

FIGURE 13.17
Responses for a smooth transition from acquisition to tracking.

The appropriate responses can also be obtained by changing the response continuously. One method for this was described in Sections 6.7.2 and 11.7.

Example 13.3

With the regulation function:

$$Q(s) = \frac{1.1(s+0.25)(s+2)(s+3)(s^2+5s+64)}{s(s+0.5)(s+7)(s^2+10s+100)}$$

and the intermediate return ratio:

$$T_0(s) = \frac{250(s+0.3)}{s(s+0.005)(s^2+10s+100)}$$

the responses of Equation 6.17 for the return ratio T are shown in Figure 13.17.

13.7 Time-Optimal Control

For *time-optimal* control the output variable must reach the commanded level in minimum time, using an actuator with limited force or power. It can be shown that for the control to be time-optimal, it is usually the case that the magnitude of the actuator action must be maximum all the time during the transition of the output variable to the commanded level. For example, shifting the position of a rigid body with a force actuator in minimum time requires the force profile shown in Figure 13.18.

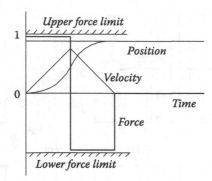

FIGURE 13.18
Time-optimal control of the position of a rigid body.

A time-optimal controller is a relay controller. It switches the action on, off, and between the positive and negative values, at appropriate instants, depending on the plant dynamics. When the plant is uncertain, the timing for the switching cannot be exactly calculated in advance, and the open-loop control entails considerable errors.

A stable, closed-loop, large-feedback control of an uncertain plant cannot employ a switch actuator since this actuator is equivalent to a saturation element with a preceding infinite gain linear link, and in practical systems the loop gain cannot be infinite at all frequencies. A good approximation to a closed-loop time-optimal controller in a system with saturation requires using large loop gain over a wide bandwidth. In order to obtain the best practical results, a proper balance must be kept between the achieved loop gain and the achieved feedback bandwidth.

Whatever the chosen feedback bandwidth, the feedback must be maximized under the limitation of keeping the system globally stable and free from wind-up. This requires using an NDC as was shown in Chapter 10, which can also be implemented as a multi-window controller. Using a multiwindow controller can also help reduce the settling time.

For most common practical problems of time-optimal control, a two-window controller suffices. More windows may be necessary when the required settling error is very small, like in beam pointing of space optical telecommunication systems.

13.8 Examples

Example 13.4: Despin Control for S/C Booster Separation

A spacecraft booster is stabilized by spinning at 85 rpm. After separation from the booster, the spacecraft shown in Figure 13.19(a) is despun by a yo-yo to about 2 rpm. (The yo-yo is a weight at the end of a cable wrapped several times around the spacecraft. When the spacecraft is released from the booster rocket, the weight is also released and begins unwrapping the cable. When all the cable length is unwrapped, the cable is separated from the spacecraft, and the yo-yo takes away most of the rotation momentum.) The remaining spacecraft spin needs to be removed by firing thrusters. Spinning of the spacecraft about the z-axis is unstable since the spacecraft is prolate, and when left uncontrolled for some time, the spacecraft will tumble. Therefore, the despin should be fast. Because of large uncertainty in the initial conditions after the separation, with various positions and spin rates and different types of coupling between the axes, the controller design for the despin function must be made very robust, and at the same time, it must perform in a nearly time-optimal fashion. After the despin is complete, the controller must be changed to provide better control accuracy in the cruise mode.

The controller shown in Figure 13.19(b) uses pulse-width-modulated (PWM) thrusters. Since each thruster produces x-, y-, and z-torques, they are combined in pairs and decoupled by the thruster logic matrix. This renders the control of each axis independent to a certain extent. The problem is, however, complicated by coupling between the x-, y-, and z-rotations due to the spacecraft dynamics, including spinning of fuel and oxidizer, initially at the rate of the booster. Due to large plant uncertainty, the despin was chosen to be proportional, providing a large phase stability margin over the entire frequency range of possible plant uncertainty and x-, y-, and z-controllers' coupling.

In the block diagram, the demultiplexer DM separates the error vector into its components. The multiplexer M does the opposite. The compensators are independent for the x-, y-, and z-rotations; i.e., the controller matrix is diagonal.

When the controllers' gains were chosen such as to despin the s/c without substantial overshoot, the z-axis response was as shown in Figure 13.20(a). It is seen that the control is not time optimal.

A better design is a nonlinear controller with two windows that switches the control law on the basis of the absolute value of the error in each channel. This was done by passing the errors via saturation/dead-zone windows with smooth transition between the control laws. The resulting control law is nearly perfect for the despin function, as well as for the cruise mode. The transient response for this controller is shown in Figure 13.20(b). The despin time was reduced by 20%.

The two-window controller performs better and is at the same time more robust than the original linear controller, with larger stability margins for the large error mode when the cross-axis coupling is the largest.

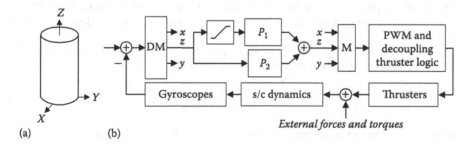

FIGURE 13.19
Spacecraft (a) local frame coordinates and (b) attitude control block diagram.

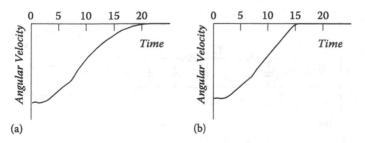

FIGURE 13.20
Time response of z-axis despinning: (a) linear controller and (b) two-window controller.

This example shows that even for complicated plants with multichannel coupled nonlinear feedback loops, a nonlinear two-window controller using only the error in individual channels for changing the control law provides nearly time-optimal performance, substantially better than that of linear controllers.

Example 13.5: Cassini Spacecraft Attitude Control with Thrusters

The plant is close to a pure double integrator; although, there are flexible modes at high frequencies. The thrusters are not throttled and not modulated, and the torque is some fixed positive or negative value, or zero (similarly to a three-position relay.) These controllers often do not include an I-channel (low-frequency disturbances are almost nonexistent), and only include a P-channel and a high-frequency channel. To avoid wind-up, they commonly use saturation in the P-channel, which is then considered to be the low-frequency channel.

Example 13.6: Temperature Controller for the Mirrors of Cassini Spacecraft's Narrow-Angle Camera

The camera is a small telescope, and the primary and secondary mirrors must be kept at approximately the same temperature in order for the mirror surfaces to match each other, and the image in the focal plane to be clear.

Figure 13.21 shows an electrical analogy to a thermal control system for a spacecraft-mounted telescope (recall Section 7.1.2). The plant is highly nonlinear because of the nonlinear law of heat radiation into free space. The heater H_2 is used to keep the temperature of the primary and secondary mirrors within 1.6 K of each other. (Another loop, which is not discussed here, drives the primary mirror heater H_1, which maintains the absolute temperature within the 263–298 K range). The heaters are pulse-width modulated with the modulation period of 6 s and the pulse-width timing resolution of 125 ms. The total heater power cannot exceed 6 W. The frequency response of the plant transfer function from the heater to the temperature differential is basically that of an integrator; however, there are also radiative losses G, which are nonlinear.

The compensator is implemented in three parallel channels. The compensation for the high-frequency (HF) channel is a complex pole pair:

$$C^{HF} = \frac{0.1}{s^2 + 0.125s + 0.1} \tag{13.5}$$

The medium-frequency (MF) and low-frequency (LF) channels are first order:

$$C^{MF} = \frac{0.5}{s + 0.1} \quad \text{and} \quad C^{LF} = \frac{0.5}{s + 0.035} \tag{13.6}$$

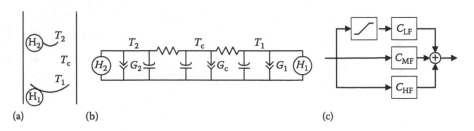

(a)   (b)   (c)

FIGURE 13.21
(a) Narrow-angle camera, (b) its thermal controller electrical analogy, and (c) the controller configuration.

and a saturation element precedes the LF compensation. The separate and combined frequency responses of the compensator channels (ignoring the nonlinearity) are shown in Figure 13.22, and the loop frequency response is shown in Figure 13.23. The parallel connection of the MF and HF channels forms a Bode step on the Bode diagram near 30 MHz. The controller was implemented digitally, and the feedback bandwidth was ultimately limited by sampling effects.

The LF compensation steepens the response in the 1–10 MHz range, providing larger feedback at low frequencies. This would result in wind-up, i.e., excessive over-shooting, for transients in which the heater saturates, unless an anti-wind-up device is provided. (For the typical power-on transient, the heater saturates immediately.) The anti-wind-up device used here is a saturation element preceding the LF path. This prevents the LF path from "integrating up" excessively when the actuator is saturated and the error is large. After a few (5–10) step response simulations were observed, the saturation threshold in the LF path was chosen to be 0.8 K. The closed-loop system transient response is notably insensitive to variations in this level, which makes a good level easy to determine. Note that placement of the saturation element *after* the LF compensation results in a transient with a wind-up error that is small but takes an excessive amount of time to decay. Industrial controllers, which often place the saturation link after the *I*-path, frequently use integrator reset features to overcome this problem.

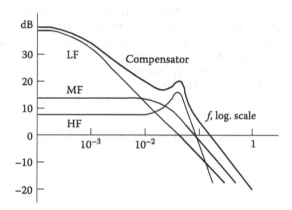

FIGURE 13.22
Parallel-channel compensator responses for a thermal controller.

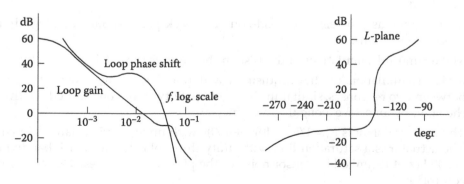

FIGURE 13.23
Loop frequency response for thermal controller.

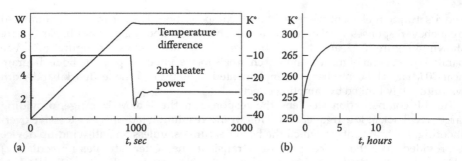

FIGURE 13.24
Step response for thermal controller.

The power-on step response for the controller is shown in Figure 13.24. The heater power is maximum most of the time while the mirror is heating up. The controller is nearly time optimal, and the overshoot is insignificant.

Example 13.7

The microgravity accelerometer that was described in Section 11.9 is another example of a two-window controller. This controller not only provides global stability with loop phase shift of π at frequencies where the loop gain is large, but it also eliminates wind-up, reduces the overshoot, and increases the acquisition band of the tunneling condition. The tunnel effect is an exponential function, and if the feedback loop were initialized when the distance in the tunnel sensor gap was much smaller than normal, then the loop gain would be much larger, and the system would become unstable if it were not for the gain reduction by the NDC.

Example 13.8

The antenna pointing controller described in Section 5.10.5 is another example of a two-window controller.

PROBLEMS

1. In a system with saturation, a double-integrator plant, and a *PID* controller, study the effect on the wind-up of the saturation links placed in front of the *I* and *P* paths. Use Simulink or SPICE, and make simulations with different saturation thresholds.

2. In the previous problem, use dead-zone feedback paths about the integral and proportional paths.

3. Make Simulink simulations of the system shown in Figure 13.5.

4. Make a simulation for the acquisition and tracking problem with switching between two responses, similar in shape (of the *PI*-type) but one shifted relative to the other by an octave along the frequency axis.

5. The nominal plant is $P_o = 1/[(s + 10)(s + 100)]$, with up to ±2 dB variations in gain. The actuator is a saturation link with unity threshold. The feedback bandwidth is 200 Hz (it is limited by sensor noise or the plant uncertainties). Design a good controller.

6. In Example 13.4, assume the plant gain is uncertain within 3 dB. Plot the transient response for plant gain deviations up and down. What is the conclusion?

7. Study command feedforward with different frequency responses of the command feedforward link in nonlinear modes of operation.

8. Study a command feedforward system for the following:
   a. Large-level commands
   b. Large deviations of the plant response from the nominal

9. Study multiwindow controllers with bounded internal variables in the plant and the actuator.

10. Study a system with multiwindow compensators, command feedforward, prefilter, and feedback path.

14

Nonlinear Multi-Loop Systems with Uncertainty

In this chapter we discuss systems whose performance is limited by uncertainties in the plant at high frequencies. After presenting the typical characteristics of these systems, we review the performance of the Bode single-loop system in this context. We also introduce some material regarding stability margins in band-pass systems, emphasizing the role of simulation to verify global stability. After a discussion of performance trade-offs in typical MIMO systems, we present nonlinear multi-loop systems, which significantly improve the feedback, and compare their performance to Bode single-loop systems. Implementation details, including the design of the internal loops, are discussed in an example. Finally, we discuss the related problem of input signal reconstruction, and how nonlinear multi-loop designs may be employed in these systems.

14.1 Systems with High-Frequency Plant Uncertainty

First mentioned in Section 4.2.2, high-frequency plant uncertainty is a common feature of electrical, thermal, and mechanical systems. A general discussion of plant tolerances was undertaken in Section 4.3.5, and the particular case of plants with structural flexibility in Sections 4.3.6 and Section 7.8. Because of the complex nature of the dynamics of real structures, and also the need to estimate controller performance prior to finalization of the design, there is typically a large amount of uncertainty regarding models of structural flexibility used for control system design. In Section 9.4, we mentioned adaptive control schemes that use frequency-domain plant identification techniques to deal with this situation, with a dedicated sensor which is accurate at the higher frequencies in the area of positive feedback. Often the mechanical resonances vary, for example, with the changing mass properties of a vehicle which is depleting propellant or reconfiguring appendages. If the modes are either known with some accuracy or identifiable, they may be compensated by creating an anti-resonance or notch filter. This block diagram is shown in Figure 14.1. With the resonances in the feedback loop significantly reduced, the loop feedback can be increased substantially.

However, the uncertainties may be more profound, especially when dealing with frequencies beyond the fundamental. It may be unclear whether the control architecture is collocated or non-collocated, and there may be significant unmodeled transport lag. The unpredictable gain variations are shown, for example, in Figure 14.2(a)—a set of narrow peaks with gain variation as much as ±20 dB, and in Figure 14.2(b)—for time delay or uncertain real exponents, as might occur in an acoustical system with varying propagation distance. We define f_u as the low-frequency boundary of this region of large gain uncertainty.

At frequencies higher than f_w, the resonance amplitudes gradually become smaller and have less of an effect on the phase at working frequencies. The average of the local maxima and minima defines the *central line*. The central line may be adjusted by the usual loop-shaping techniques, but the area encompassed by the uncertain plant boundary is formidably large.

Example 14.1: Flexible Antenna

A narrow-beam antenna on the surface of Mars is shown in Figure 14.3. The antenna's uncertain mechanical resonances are excited by the atmospheric wind. Because the mass of the antenna must be minimized, the structural designer has the first major resonances and also some additional interfering resonances only slightly higher than the control designer's specified f_u.

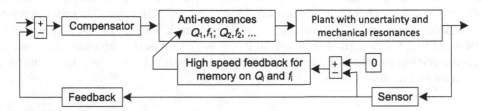

FIGURE 14.1
Anti-resonance filters with adaptive frequency adjustments.

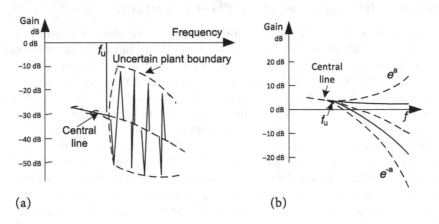

FIGURE 14.2
High-frequency area of plant uncertainty due to (a) poorly modeled lightly damped resonances and (b) uncertain time delays.

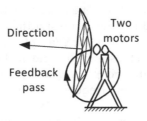

FIGURE 14.3
Antenna with uncertain structural dynamics.

In Section 14.3 we discuss the single-loop Bode response applied to this sort of system, and the ultimate single-loop limitations. Later, we present nonlinear multi-loop systems that can outperform the single-loop design. Recall that linear systems with multiple loops can always be transformed into single-loop systems using block diagram algebra with Mason's rules. Alternatively, when the different loops have significant nonlinear elements, the system is regarded as multi-loop.

Although both systems are nonlinear, we will characterize the performance of the single-loop and multi-loop designs using their frequency responses, while using nonlinear dynamic compensators to provide stability in the presence of actuator saturation. Because the internal loops of the multi-loop design are band-pass type responses with saturation-type nonlinearities, it is necessary to review adequate stability margins for these systems.

14.2 Stability and Multi-Frequency Oscillations in Band-Pass Systems

Let us consider the example of a band-pass system with frequency response depicted in Figure 14.4(a) and with the usual saturation-type nonlinearity. We have already explained in Section 10.6.2 that the Popov criterion is overly conservative for this system, so it is necessary to resort to Fourier analysis informed by experiment (refer to Section 3.4.3. of [33]).

Figure 14.4(b), redrawn from [33], shows the experimentally determined boundary of conditional instability in terms of the lower-frequency and higher-frequency phase margins. To the left and below this boundary, the system is either stable or not depending on initial conditions. The forms of the instabilities for the different numbered regions are shown in Fig. 14.4c. These oscillations are analyzed in detail in Section 3.4.3 of [33]. Using describing function type analysis, it can be shown that both the simple low-frequency instability in regions 1 and 2 as well as the multi-frequency instabilities in regions 3 and 4 are both made possible by a combination of inadequate lower- and higher-frequency phase margins.

As discussed in Section 10.6.2, lower and upper margins of 10 dB and 30° do well in practice for these sorts of systems. For general multi-loop systems, stability margins must be guaranteed for all practical situations, i.e. with any possible linear responses, nonlinear characteristics, and any nonlinear or linear loads. Stability must be checked experimentally or with detailed simulation for all conceivable cases.

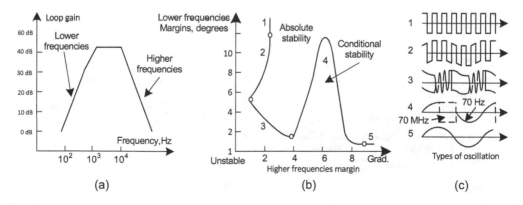

FIGURE 14.4
Band-pass system with saturation and forms of oscillation. (Reproduced by permission from Boris J. Lurie, *Feedback Maximization*, Norwood, MA: Artech House, Inc, 1986. Copyright 1986 by Artech House, Inc.)

14.3 Bode Single-Loop System

Figures 14.5(a) and (b) show a single-loop Bode system designed to accommodate the plant uncertain boundary introduced in Figure 14.2. The workable feedback area where $|T| > 0$ goes from 0 to f_t, and the feedback is typically negative up to $0.4 f_b$. (The frequency scale is normalized so that $f_b = 1$.) The stability margins are assumed to be $y = 30°$ and $x = 10$ dB, and the uncertainty gain peaks are ±15 dB. With a steep 30 dB/oct slope following the Bode step the achievable bandwidth has the ratio f_u/f_b equal to nearly 2.3 octaves. This will provide the benchmark for the nonlinear multi-loop system that we introduce in Section 14.5.

Figures 14.5(c) show a single-loop response with a narrower Bode step in which the gain is allowed to cross back up through 0 dB. Although this improves the bandwidth by 0.2 to 0.5 octaves, it requires some knowledge regarding the frequency and phase of the resonant peaks in the uncertainty envelope, as discussed in Section 4.3.6. A common example of this situation would be the control of a launch vehicle in which the first-bending mode is accurately modeled, but subsequent modes are more uncertain. This approach was taken for the Saturn V as discussed in the appendix A.13.9. Alternatively, Figure 14.5(d) shows a response similar to that depicted back in Figure 4.21, which surrounds the critical point on all sides (without a resonance). This type of design may be better than the Bode response, but the compensation in the high-frequency area may need to be high-order and noise must not be excessive.

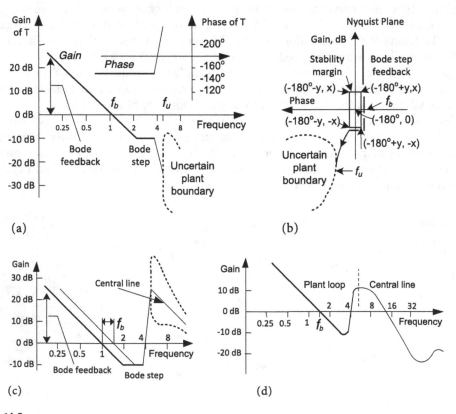

FIGURE 14.5
Single-loop Bode system with high-frequency uncertainty (a–b), and single-loop systems with improved bandwidth but requiring some additional knowledge (c–d).

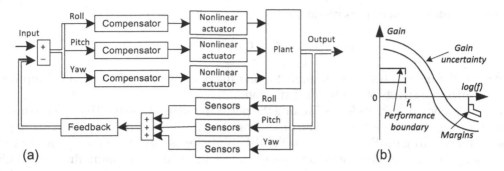

FIGURE 14.6
A typical multi-loop system (a) and (b) frequency response constraints for feedback maximization.

14.4 Multi-Input Multi-Output Systems

In Section 2.9 we introduced general multi-input multi-output (MIMO) systems and then discussed the types of MIMO systems usually encountered in practice. A typical multi-loop arrangement is shown in Figure 14.6(a), which depicts 3-axis attitude control of an aircraft. It is often best to design these systems primarily as single-loop controllers, although as we mention in Section 4.4, care must be taken to accommodate coupling, including through structural flexibilities.

In *Multivariable Feedback Control* by Skogestad and Postlethwaite [71], it is stated that the design of multi-loop feedback systems shares the same fundamental trade-offs between performance and stability margins as single-loop systems, as indicated by Figure 14.6(b) (redrawn from [71]).

The point is that we already have the tools to design these types of nonlinear multi-loop systems. In the following section we present something entirely different, which is the use of *nonlinear multi-loop feedback* for the single-input single-output (SISO) system with high-frequency plant uncertainty presented in Sections 14.1 and 14.3, with the understanding of course that the SISO system may be part of a more complex MIMO system of the type shown in Figure 14.6. Using nonlinear internal loops and with some practical limitations on high-frequency noise, we are able to improve the feedback of the Bode single-loop system, increasing the bandwidth by more than a factor of two.

14.5 Nonlinear Multi-Loop Feedback

Figure 14.7 shows the general setup for a nonlinear multi-loop system. The three *internal loop* paths with gains k, k_c, and k_a are summed, run through a nonlinear dynamic compensator, and then combined with the feedback of the main *plant loop*, which also contains an NDC. Often only one of these internal paths is used, in which case the overall system is *two-loop*.

The overall loop transmission is:

$$P = (C \times A \times FI \times P \times F \times FB) / (IL + 1)$$

where the internal loop transmission is:

$$IL = (k + k_c C + k_a A) \times ILS \times FB$$

Note that the filter FI between the actuator and the plant is most often used in amplifiers to minimize the effects of nonlinear distortion.

The frequency responses for estimating the performance of these multi-loop systems are shown in Figures 14.8(a) and (b). The plant loop crosses the internal loop at frequency f_m. The reduction in gain there is due to the different slopes (and therefore phases) of the plant loop and the internal loop and must be less than 10 dB to maintain the stability margin. The phase lag in the plant loop at low frequencies from 0 to f_m is more than 150°, and so to guarantee stability in nonlinear mode, this phase lag must be reduced with an NDC. At $f > f_m$ the internal loop gain exceeds the plant loop gain by 10–20 dB, depending on the plant phase response. The slope of the combined gain response must be such that the Nyquist diagram is stable. An additional NDC may be used in this area to reduce the frequency range of the feedback. For stability the slope of the combined response must not be steeper than –10 dB/oct at the high-frequency crossover f_{b2}.

The achieved ratio f_m / f_b, around 2.2, is the performance advantage of the multi-loop system compared to the single-loop Bode system. Figure 14.8(b) shows a system with

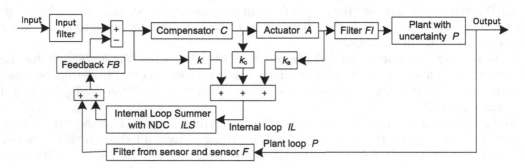

FIGURE 14.7
Nonlinear multi-loop feedback.

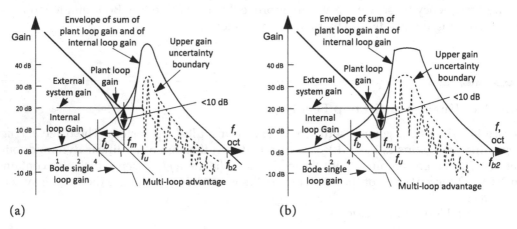

(a)   (b)

FIGURE 14.8
Frequency responses for multi-loop feedback.*

somewhat higher f_m, closer to f_u, which can accommodate greater uncertainties at higher frequencies.

In the single-loop Bode design, the feedback is positive near f_t, as it must be to maintain stability, but this exacerbates the sensitivity to uncertainties and therefore constrains the bandwidth to be well below the uncertainty region. In the nonlinear multi-loop system depicted in Figure 14.8, stability in this region is instead provided by a nonlinear dynamic compensator.

*Patent applied for by B. Lurie, Nonlinear Multi-loop Feedback Control System, Appendix 15.

Example 14.2: Nonlinear Multi-Loop System with a Three-Stage Amplifier

Figure 14.9 depicts a plant with high-frequency uncertainty driven by a three-stage amplifier. Each stage of the amplifier contains an NDC implemented as a feedback loop. The nonlinear two-loop design includes internal feedback with a fourth NDC. The achievable bandwidth is more than 1.2 octaves better than the Bode design which is constrained by the usual gain/phase relationship. Figure 14.10 compares the transient responses of the Bode system and the multi-loop system which is more than twice as fast.

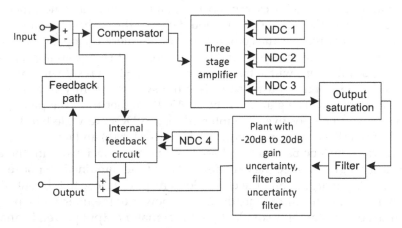

FIGURE 14.9
Two-loop system with a three-stage amplifier.

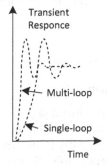

FIGURE 14.10
Transient response.

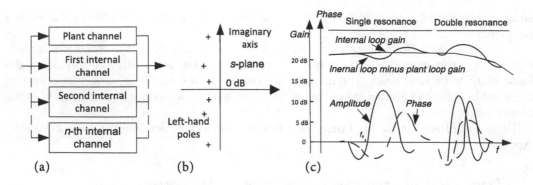

FIGURE 14.11
(a) Parallel signals keeping all the poles in the left half-plane (b). The phase variations for single and double resonances are shown in (c).

14.6 Design of the Internal Loops

In order to maximize the feedback around the plant in linear and nonlinear modes of operation, the design of the internal loops should enable the highest possible crossing frequency, f_m. In the linear regime, the combination of the internal loops with the plant loop should create a non-minimum phase function. Figure 14.11 shows an appropriate combination of these parallel loops with a frequency response which results in only LHP poles.

The internal loop response must also satisfy the system stability requirements at the frequency f_u and beyond. The gain must be 10–15 dB higher than the plant gain for the combined response to be positive. The internal-loop NDCs are wide-band, and may be implemented in parallel channels as shown in Figures 11.29(b) and (c).

Careful attention must be paid to the phase uncertainty associated with the gain peaks of the uncertainty region to keep the gain deviations of the combined response small compared to the stability margins, and the closed-loop response close to the central line.

Particularly in the nonlinear regime, the output power of the internal loops at frequencies up to f_u must be substantially smaller than the actuator output power. In analog implementations this may be inconvenient or impossible. Alternative block diagrams may be used where this occurs in the low-power feedback path. For combining the plant and internal responses it is convenient to use high voltages but lower supply current. (Obviously this is not an issue if this part of the system is implemented digitally.)

14.7 Input Signal Reconstruction

Consider the problem of estimating the input signal u_1 from the noisy output signal u_2 of the system Q. It is often not practical to invert Q to reconstruct the input signal, with difficulties arising from the uncertainty in Q, any non-minimum phase characteristics, and the potential amplification of high-frequency noise. Instead, the scheme shown in Figure 14.12 is used, which involves feedback around an approximating Q^E. The feedback reduces both the effects of the plant uncertainty and high-frequency noise (refer to Section 2.4.4. of [33]).

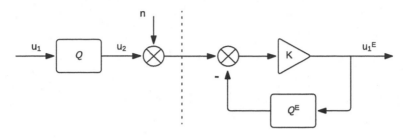

FIGURE 14.12
Input signal reconstruction using feedback.

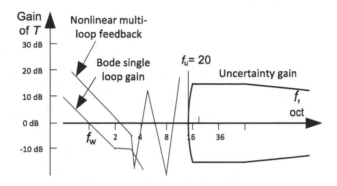

FIGURE 14.13
Alternative diagram for nonlinear multi-loop with emphasis on non-minimum phase uncertainty.

To optimize the response of this system, the feedback should certainly be maximized. The application of nonlinear multi-loop feedback to this system will outperform the equivalent Bode system by a factor of 2.2, enabling significantly faster input signal recognition. Nonlinear multi-loop methods may be applicable to other problems in filtering, estimation, and prediction.

Figure 14.13 shows an alternative sketch for the nonlinear multi-loop system which has more emphasis on non-minimum phase uncertainty rather than uncertain multiple resonances. This sort of diagram may be useful for the design of sound systems and transmitters.

Appendix 1: Feedback Control, Elementary Treatment

A1.1 Introduction

The best way to analyze a complex system is to break it into blocks which represent different physical parts or functions, with directed lines indicating their connections. Such *block diagram* analysis is used in all kinds of engineering, programming, and the sciences. Understanding how these systems function requires more than understanding how each block functions. It is also necessary to appreciate the nature of their connections, and in particular, the critical significance of feedback.

The term *feedback* was applied to closed-loop system engineering by Harold Black of the Bell Laboratories in the 1920s. It describes regulation processes found in engineering, biology, economics, social sciences, and political systems—everywhere where information about the results comes back and influences the input.

The fundamentals of feedback can be expressed in simple terms. Feedback systems can, and in our opinion should, be taught as part of a science course in high school, preceding and facilitating the teaching of physics, chemistry, biology, and the social sciences. It is important to demonstrate not only how automatic control works, but also how the system dynamics limit the speed and accuracy of control, and why and how feedback systems fail.

The word feedback is often employed in a much-simplified sense, denoting merely the obtaining of information on the results of one's action. There is much more, however, to the quantitative meaning and methods of feedback. In the modern world, feedback is employed in spacecraft and missile control, in cars, and in TVs, and is widely used to explain and quantify the processes studied in biology, economics, and social sciences. We hope the following material will provide a better perception of how the systems of this world operate.

A1.2 Feedback Control, Elementary Treatment

A1.2.1 Block Diagram

A complex system is presented as an interconnected collection of subsystems or *blocks*.

Figure A1.1 shows some *block diagrams*. If the input is a composite of several input systems, this is shown in the figure on the left with a *summer*. The output is $5 + 30 + 10 = 45$.

When the *inputs* in figure on the right are combined with the *output value*, the output value of 25 is produced. The diagram describes this arithmetic: $3 \times 10 = 30$; $30 - 5 = 25$.

The factor by which the block's output is larger than the input (10 in this case) is called the block's *gain coefficient*.

In the block diagram in Figure A1.2, the output of block 5 is fed back to the input summer forming a *feedback loop*. The arithmetic is the following: $6 - 5 = 1$; $1 \times 5 = 5$.

A1.2.2 Feedback Control

We start with examples.

Example A1.1

While a rifle is being aimed, an eye looks at the target through the rifle sights. In Figure A1.3, the rifle points down and to the left of the target. The pointing *error* is the difference between the direction to the target and the rifle pointing direction. An inscription in the block (or close to it) gives the name of a physical device represented by the block.

Using the sights, the eye registers the rifle pointing direction and communicates this information to the brain, as indicated in Figure A1.3. The brain (a) calculates the error by subtracting the pointing direction from the command, and (b) acting also as a *controller*, issues appropriate orders to the arms' muscles to correct the aim.

Example A1.2

While steering a four-wheel-drive car in a desert, the "command" can be "drive west." The block diagram in Figure A1.4 shows the process of the control system operation. The eyes estimate the actual direction of the car's motion, and the brain calculates the error and gives orders to the hands.

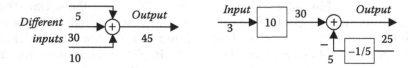

FIGURE A1.1
Block diagrams with a summer.

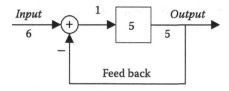

FIGURE A1.2
Block diagram of a feedback loop.

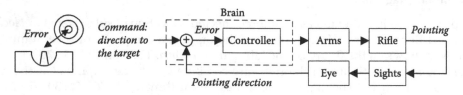

FIGURE A1.3
Block diagram describing aiming of a rifle.

Example A1.3

While pointing a small spacecraft with a hard-mounted telescope to take a good picture of a planet, the flight computer calculates the direction to the planet and sends these data as a command to the control system shown in Figure A1.5. The pointing angle sensor here can be, for example, a camera with a wider angle than the telescope; the steering means can be the jets rotating the spacecraft. The signal at the summer output is the difference between the command and the actual readings of the sensor, i.e., the error.

We may now generalize the concept of a *feedback control system* (also called a *closed-loop system*) in the form shown in Figure A1.6. The *actuator* drives the object of control, which is traditionally called the *plant*.

If the error is 0, no action is taken.

The controller's gain coefficient is large. It senses even a small error and aggressively orders the actuator to compensate for the error.

In a typical control system, the actuator is powerful, but not as accurate as the sensor. The sensor is accurate, but not powerful. The feedback control integrates the best features of both the actuator and the sensor. It is widely employed in biological and engineering systems.

We now know enough to start *designing* control systems.

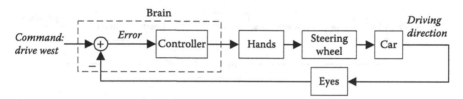

FIGURE A1.4
Block diagram describing driving a car.

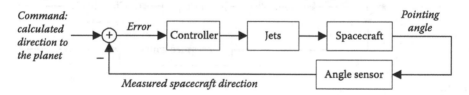

FIGURE A1.5
Block diagram describing pointing a spacecraft.

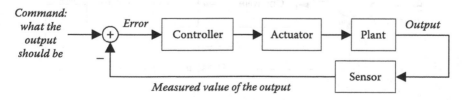

FIGURE A1.6
General block diagram describing control of a plant.

Example A1.4

Design a system to maintain the temperature at 1,206°C within an industrial furnace.

We use the general diagram of feedback control in Figure A1.6. The command here is "1,206°C." The actuator is now an electrical heater. The plant is the furnace with the payload. The sensor is an electrical thermometer. The resulting block diagram is shown in Figure A1.7.

Example A1.5

Design a system to maintain a pressure of 2.2 atm in a chamber. Now, the command is 2.2 atm, the actuator is a pump, the plant is the chamber, and the sensor is a pressure gauge, as shown in Figure A1.8.

Assume that the pressure gauge reading is 2.15 atm. This means that the command is not performed perfectly, and the error is 0.05 atm.

Example A1.6

Design a block diagram of a biological system to produce a certain amount of a specific tissue.

The block diagram in Figure A1.9 will do it. (The amount of tissue may be measured by the increasing pressure sensed by the reproducing cells, for example.)

When the feedback mechanism fails, the tissue continues to be manufactured even when there is more than enough of it. This may cause a serious health problem.

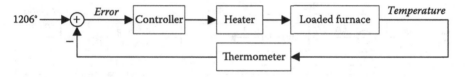

FIGURE A1.7
Block diagram of temperature control.

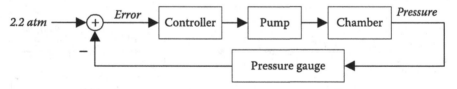

FIGURE A1.8
Block diagram of pressure control.

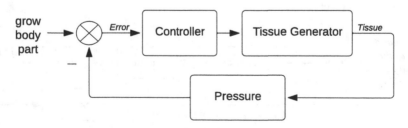

FIGURE A1.9
Block diagram of the regulation of tissue growth.

A1.2.3 Links

Feedback systems are composed of *links*, which are a particular kind of block.

An electronic thermometer, for example, produces electrical voltage proportional to the temperature. This link speaks two languages: its input understands degrees Fahrenheit, and its output speaks in volts. When the temperature is 1°F, the output is 0.01 V. At 100°F, the thermometer output is 1 V. That is, this particular thermometer generates 0.01 V per each degree as indicated in Figure A1.10.

The electronic pressure gauge displayed in Figure A1.11 produces 1 V output for each atmosphere of pressure. In other words, its transmission coefficient is 1 V/atm.

Figure A1.12 shows a connection of two links making a composite link. An electrical heater consumes electrical power from the input, in watts, and produces heat, 0.24 calories each second per each watt. The heat raises the temperature of the payload in the furnace chamber by an amount depending on the size of the payload.

Can we connect two arbitrary links—a thermometer and a pressure gauge, for example, as in Figure A1.13? No, this will not work, and not only because 13 is an unlucky number: the links speak different languages and do not understand each other.

The language must be common at the connection of the links. We can, for example, connect an electrical thermometer to the output of the links of Figure A1.12, as shown in Figure A1.14.

When several links are correctly connected in a chain, the resulting gain coefficient is the product of these links' gain coefficients.

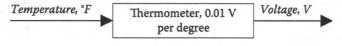

FIGURE A1.10
Thermometer link.

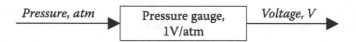

FIGURE A1.11
Pressure gauge link.

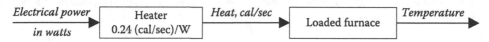

FIGURE A1.12
Composite link.

FIGURE A1.13
Links that cannot be connected.

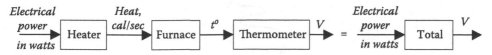

FIGURE A1.14
Equivalent composite link.

A1.3 Why Control Cannot Be Perfect

A1.3.1 Dynamic Links

We assumed before that thermometers measure the temperature instantly. This is only approximately correct. A mercury thermometer has to be kept in the mouth for several minutes for the readings to approach the mouth temperature, as shown in Figure A1.15a. Electronic thermometers settle faster, but still not instantly. The thermometer readings depend not only on the instant temperature, but also on what the temperature was seconds and minutes ago. Thus, the thermometer has *memory*.

Consider now a pendulum suspended from a frame, initially at rest, as shown in Figure A1.16 by dashed lines. Let us push the frame by some distance, considering this distance to be the input of the link. The output of the link is the pendulum position. It depends not only on the frame position at the current moment, but also on the *previous* frame position and when the position changed. The plot for the pendulum position after the frame was pushed is shown in Figure A1.15(b). The output depends on what happened in the past. The pendulum has memory.

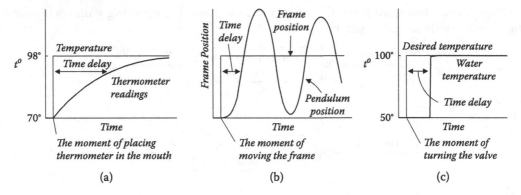

(a)   (b)   (c)

FIGURE A1.15
Dynamic links time histories: (a) time history of thermometer readings, (b) pendulum position history, (c) and temperature history after a delay.

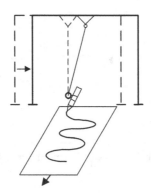

FIGURE A1.16
Pendulum.

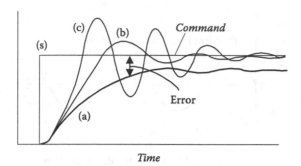

FIGURE A1.17
Control system's output time history.

Another link with memory is the shower. The input is the hot water valve position, and the output is the shower water temperature. The input and output time histories are shown in Figure A1.15(c). The output is delayed by the time it takes the water to flow through the pipe.

Devices or processes with memory are called *dynamic*. When the input is changed instantly by a certain amount, the change in the output is delayed, and the output can grow gradually and can be oscillatory.

Certainly, a feedback system composed of dynamic links itself becomes dynamic.

A1.3.2 Control Accuracy Limitations

After a step command is issued, it takes some time for the output of a dynamic control system to change, as seen in Figure A1.17(a). The error decreases with time but does not completely vanish.

A more aggressive control, with larger controller gain, reduces the error as exemplified in Figure A1.17(b). However, with bigger controller gain, the problem arises of stopping the actuator action immediately after the error is reduced to zero. Since the links in the feedback loop are dynamic, the information from the sensor that the error is already zero comes back to the actuator with some delay, and the actuator action proceeds for some time after the moment it should be terminated. Then, an error of the opposite sign will appear at the output. The process then repeats itself, and the output oscillates as seen in the Figure A1.17(b). If the controller gain is even larger, the oscillation amplitude increases and output will look more like Figure A1.17(c); with further increase of the controller gain, the oscillation becomes periodic and with large amplitude like that in Figure A1.18. A simple way to explore this process experimentally is by trying to regulate the shower temperature while being very impatient.

The larger the total delay of all dynamic links in the feedback loop, the smaller must be the controller gain for the system to remain stable, and the less accurate and the more sluggish will be the control. Thus, while a feedback control system is being designed, major attention should be paid to reduction of delays in the loop.

A1.4 More about Feedback

A1.4.1 Self-Oscillation

The time history of an oscillation is shown in Figure A1.18(a). It can be drawn by a pen bound to the pendulum on Figure A1.16, while a sheet of paper is being dragged in the direction perpendicular to oscillation. This curve is called a *sinusoid*. The number of

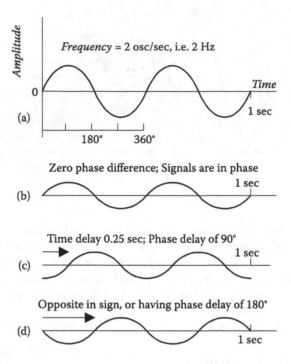

FIGURE A1.18
Time histories of motion.

oscillations per second is called the oscillation *frequency* in Hertz (Hz). A single oscillation is a *cycle*.

The cycle is 360° long. Oscillation b is *in phase* with oscillation a. Notice that if shifted by 360°, the oscillation remains in phase. Oscillation c is 90° delayed compared with oscillation a. Oscillation d is delayed by 180°, which is equivalent to changing the sign of the oscillation.

Rigorously speaking, oscillation of the pendulum is not exactly sinusoidal, and the oscillation gradually dies out. The oscillation amplitude can be maintained if the energy losses caused by friction are compensated by some actuator injecting energy into the system.

The operation of a swing shown in Figure A1.19 can be explained with the help of the block diagram in Figure A1.20. The actuator is the kid's muscles. The kid jerks his body to produce extra tension in the rope, to sustain the oscillation. The sensors he uses are in his vestibular apparatus. He detects the proper timing for his movements by feeling zero velocity in the rightmost and leftmost positions. To sustain the self-oscillation, the return signal in the feedback system must be in phase with the swing motion *and* must be strong enough.

Similar feedback systems are employed to generate radio and TV signals, and in the dynamos generating electricity at power stations.

A1.4.2 Loop Frequency Response

Links and entire systems can be tested with a set of sinusoidal inputs with different frequencies. This method is used, for example, in testing audio recording systems such as the one illustrated in Figure A1.21. This system contains a CD player, a power amplifier,

FIGURE A1.19
Swing.

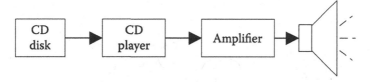

FIGURE A1.20
Block diagram for swing operation.

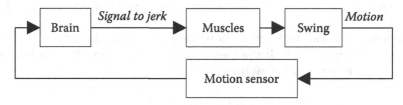

FIGURE A1.21
CD player block diagram.

and speakers. The input-to-output gain coefficient expressed in decibels (dB) is the system gain. The gain frequency responses are exemplified in Figure A1.22 for a good quality system with nearly equal gain at all frequencies from the lower frequencies of 25 Hz to the higher frequencies of 18,000 Hz, and for a portable boom box, where the lower and higher frequencies are not well reproduced, thus making the sound different from the original.

The gain frequency response is not flat (as would be desired) because the speakers resonate at many frequencies with various amplitudes. Better and more expensive speakers (bigger, with better magnets, with larger and firmer enclosures, with some special filling inside the enclosures) have wider and flatter gain responses.

Audio systems are typically characterized by only the sound amplitude responses, since our ears are to a large extent insensitive to the phase of the sound. For feedback systems, however, the phase shift in the loop is important as well—as we already know from the analysis of the swing.

A1.4.3 Control System Design Using Frequency Responses

Self-oscillation in control systems is potentially disastrous. A control system must be *stable*, i.e., self-oscillation must not occur at any frequency. To prevent self-oscillation, feedback control systems are designed such that at frequencies where the return signal is big, its phase is not such that supports oscillation, and at those frequencies where the return signal is in a phase that supports oscillation, the return signal is sufficiently small.

During control system design, the gain and the phase shift frequency responses about the feedback loop are first calculated with computers and then measured experimentally and displayed with a signal analyzer, as is shown in Figure A1.23.

A1.4.4 Some Algebra

We already know qualitatively that when the controller gain is large, the error is small. Using some algebra, we will find how small the error is, and how many times it is reduced by the feedback.

As mentioned, in a series link connection, the equivalent gain coefficient is the product of the gain coefficients of all the elementary links. Then, the feedback system in Figure A1.24 implements the following equations:

$$output = error \times CAP \tag{A1.1}$$

$$error = command - fbs \tag{A1.2}$$

$$fbs = error \times CAPS \tag{A1.3}$$

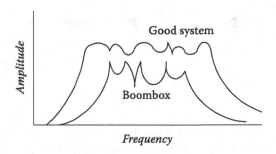

FIGURE A1.22
Frequency responses of a CD player.

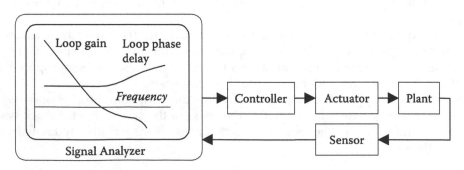

FIGURE A1.23
Measuring loop transmission frequency responses.

By substituting Equation A1.3 into Equation A1.2, we get:

$$error = command - error \times CAPS$$

and isolating the error:

$$error = command / (1 + CAPS)$$

This expression shows that the *error* is $(1 + CAPS)$ times smaller than the command. The expression $1 + CAPS$ is, numerically, the *feedback*. The larger the feedback, the smaller the error. And, we have already concluded that the feedback cannot be arbitrarily large.

After substituting this expression for the error into Equation A1.1, we find that:

$$output = command \frac{1}{S} \frac{CAPS}{1 + CAPS} \tag{A1.4}$$

When the product *CAPS* is large, much more than 1, then 1 in the denominator can be neglected, and

$$output \approx command \frac{1}{S} \tag{A1.5}$$

For example, if $S = 2$, $CA = 20$, and $P = 1$, or 10, or 100, the *output* from Equation A1.4 is correspondingly 0.488, 0.4988, and 0.49988, i.e., very close to 0.5 from Equation A1.5.

A1.4.5 Disturbance Rejection

The environmental *disturbances D* shown in Figure A1.25 add unwanted components to the plant's output, and may also represent inaccuracies of the actuator. The value of these

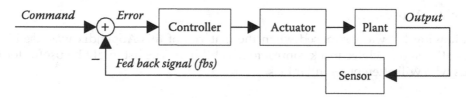

FIGURE A1.24
Feedback control system.

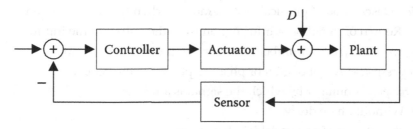

FIGURE A1.25
Disturbances at the plant input.

unwanted components at the system's output can be calculated with Equation A1.4 when using the disturbance as the command:

$$disturbances\,at\,the\,output = D\frac{P}{1+CAPS}$$

Without the feedback, the effect of disturbances at the system's output would be DP. Therefore, the feedback reduces the output effect of disturbances by $(1+CAPS)$ times. The feedback can therefore be used to reduce the effects of mechanical vibrations on some precision instruments and machinery.

For example, floor vibrations caused by passing cars, by air conditioner motors, and by people walking around disturb precision optical systems mounted on a desk. By using position sensors and piezoelectric motors to move the desktop, the amplitude of the optics' vibrations can be reduced many times.

A1.4.6 Conclusion

In this short introduction to feedback systems, we considered single-loop control with a single actuator and a single sensor. Sometimes, several sensors are employed. While operating our hands, for example, we use position and stress sensors in the muscles, tactile and temperature sensors in the skin, and the eyes. In complex systems, many loops coexist to regulate various parameters: heartbeat frequency and strength, amount of enzymes in the stomach, body temperature, and many others. It would be difficult to count all the feedback loops in a TV—there are hundreds of them.

A1.5 New Words

The following list was composed when the material of this Appendix was used by one of the authors to teach feedback control to his children. The list might be useful for those readers who will attempt a similar task.

Actuator: A device such as a motor, power amplifier, muscle.

Closed-loop control, feedback control: Using data from a sensor to correct actuator actions.

Disturbance: A source of error in the system.

Dynamics: Description of physical system motion when forces are applied.

Feedback: Return of signal or of information from the output to the input.

Frequency in Hertz (Hz): A number of oscillation periods per second.

Frequency response: A plot of gain or phase dependence on frequency.

Gain coefficient: A number by which the signal is amplified.

In phase: Without phase delay.

Link: Representation of one variable's dependence on another.

Payload: A useful (or *paid* for) load to be moved or heated.

Phase shift: A change in phase.

Plant: The object to be controlled.

Positive feedback: Feedback supporting oscillation.

Quantitative: Expressed in numbers.

Sensor: A measuring device.

Sinusoid: A curve describing periodic oscillation.

Summer: A device whose output equals the sum of its inputs.

Variable: A numerical description of a feature of a system (such as temperature or distance).

Appendix 2: Frequency Responses

A2.1 Frequency Responses

Linear systems have the property that when they are driven by a sinusoidal signal, the output variable—and, in fact, any variable of the system—is also sinusoidal with the same frequency. Thus, when signal $u_1(t) = U_{m1}\sin \omega t$ is applied to the input of a link, the output signal of the link is $u_2(t) = U_{m2}\sin(\omega t + \phi)$, where U_{m1} and U_{m2} are the signal amplitudes, ω is the signal frequency in rad/s, and ϕ is the *phase shift* between the output and input signals. The ratio U_{m2}/U_{m1} is called the *gain coefficient* of the link, and $20 \log_{10}(U_{m2}/U_{m1})$ is called the *gain* of the link in dB.

Example A2.1

If a sinusoidal signal with unity amplitude is the input, and the output sinusoidal signal is delayed in phase with respect to the input by $\pi/3$ (i.e., by 60°), and the output signal amplitude is 2, then the phase shift is $-\pi/3$, or $-60°$, the gain coefficient is 2, and the gain is approximately 6 dB.

Figure A2.1 shows how to measure the phase shift and the gain of an electrical link: a signal generator is connected to the link's input, the voltmeters read the amplitudes of the input and the output, and the phase difference between the signals can be seen with a two-beam oscilloscope or another phase difference meter.

The gain coefficient and the phase shift both are, generally, functions of the frequency. The plots of these functions are called gain and phase shift *frequency responses*. The frequency axis is commonly logarithmic.

Figure A2.2 depicts a truss structure (which might be a model of a stellar interferometer). Some relatively noisy (vibrating) equipment is placed on platform 1 (pumps, motors, reaction wheels, tape recorders), while platform 2 is the place where some sensitive optics operate. It needs to be measured to what extent the vibrations from the upper platform propagate through the truss structure to the second platform, which is supposed to be quiet. The transfer function is defined as the ratio of the measured acceleration of a specified point on platform 2 to the force applied at a specific point to platform 1. The measurements are performed in the frequency domain, and the signals are sinusoidal. For the purpose of the measurements, the force is applied by a *shaker*. Inside of the shaker there is an electromagnet that moves up and down a body called a *proof mass*. This motion generates reacting force F applied by the case of the shaker to the platform. An accelerometer (a small proof mass placed on a piezoelement, or a magnetic core of a coil suspended on a spring in the field of a permanent magnet) produces an electrical signal proportional to the acceleration a. Instead of two voltmeters and a signal generator, a *signal analyzer* is used that incorporates these three devices together with a display and a computer that sends signals of appropriate amplitude and frequency to the input of the power amplifier driving the shaker.

Example A2.2

The application of force $F = 5\sin \omega t$ to a body with mass M, as shown in Figure A2.3(a), and the calculation of the resulting acceleration, velocity, and position are reflected in the three-link block diagram shown in Figure A2.3(b).

The input to the first link is the force, and the link's output is the acceleration

$$a = (1/M)5\sin \omega t$$

This link's gain coefficient is $|a/F| = 1/M$, and the phase shift is 0. The gain in dB is plotted in Figure A2.3(c).

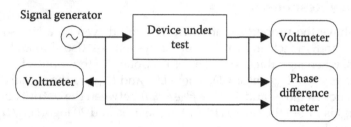

FIGURE A2.1
Measurement of a link's gain and phase shift.

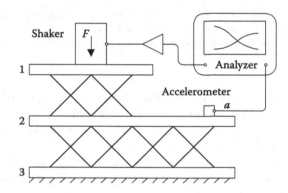

FIGURE A2.2
Measurements of a link's gain and phase with a signal analyzer.

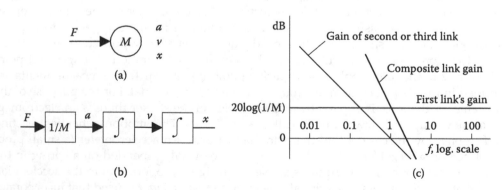

FIGURE A2.3
(a) Force acting on a rigid body, and (b) block diagram that relates the force to the body's position. (c) The links' frequency responses.

The second link is an *integrator*: its input is the acceleration and its output is the velocity

$$v = \int a\,dt = -[5/(M\omega)]\cos\omega t$$

The second link's gain coefficient is $|v/a| = 1/\omega$, and the phase is $-90°$.

The third link is also an integrator: its input is the velocity and its output is the position $x = \int v\,dt$. All three links can be integrated to form a single composite link with the gain coefficient $1/(M\omega^2)$ and the phase shift $-180°$.

Since the second and third links' gain coefficients are inversely proportional to the frequency, the gain of these links decreases with frequency with a constant slope of -20 dB per decade. The slope of the composite link gain is therefore -40 dB/decade, or, equivalently, -12 dB/octave.

A2.2 Complex Transfer Function

The two scalar variables, the gain coefficient and the phase shift, can be seen as the magnitude and the phase of a complex *transfer function*.

Example A2.3

Saying that a transfer function is $5\exp(j\pi/6)$, or $5\angle 30°$, means that the gain coefficient is 5, and the phase shift is $30°$.

Example A2.4

In Section A2.1's example of a mass driven by a force, the total transfer function from force to position is $-1/(M\omega^2)$. Notice that it can be written as $1/[(j\omega)^2 M]$.

A2.3 Laplace Transform and the s-Plane

The Laplace transform,

$$F(s) = \int_0^\infty f(t)e^{-st}\,dt$$

or

$$F(s) = L\{f(t)\}$$

renders a unique correspondence between a function of time $f(t)$ and a function $F(s)$ of the *Laplace complex variable* $s = \sigma + j\omega$. The transform is linear and exists for all practical stable functions $f(t)$. Some of the transforms are shown in Table A2.1. Here, $\delta(t)$ is the delta-function (infinitely narrow pulse whose area is 1), and $1(t)$ is the step function.

Example A2.5

For the Laplace transform $(s+a)^{-1}$, the function $f(t) = e^{-at}$. In the region of small time:

$$f(t) \approx 1 - at$$

TABLE A2.1

Laplace Transforms

F(s)	f(t)
1	$\delta(t)$
$1/s$	$1(t)$
$1/s^2$	T
$1/s^3$	$t^2/2$
$(s+a)^{-1}$	e^{-at}
$(s+a)^{-2}$	te^{-at}
$(s+a)^{-3}$	$(t^2/2)e^{-at}$
$a/[s(s+a)]$	$1 - e^{-at}$
$a/[s^2(s+a)]$	$(at - 1 + e^{-at})/a$
$a^2/[s(s+a)^2]$	$1 - e^{-at}(1+at)$
$a/(s^2+a^2)$	$\sin at$
$s/(s^2+a^2)$	$\cos at$
$\dfrac{a}{(s+b)^2 + a^2}$	$e^{-bt}\sin at$
$\dfrac{s+b}{(s+b)^2 + a^2}$	$e^{-bt}\cos at$

For transform $(s+a)^{-2}$,

$$f(t) \approx t - at^2$$

For transform $(s+a)^{-3}$,

$$f(t) \approx t^2/2 - at^3/2$$

The *initial value* and *final value theorems* are the following:

$$f(0) = \lim_{s \to \infty} sF(s)$$

and

$$\lim_{t \to \infty} f(t) = \lim_{s \to \infty} sF(s)$$

When $F(s)$ is rational, the related function $f(t)$ can be found by, first, presenting $F(s)$ as a sum of partial fractions, and then, summing the time functions that correspond to the fractions.

Figure A2.4 shows various poles of $F(s)$ in the s-plane and the related functions of time. As seen in Table A2.1, real poles make exponential signals, purely imaginary poles make sinusoidal signals, and complex poles make oscillatory time responses. As long as the pole is in the left half-plane, $\sigma < 0$, the envelope of the signal is exponentially narrowing with time.

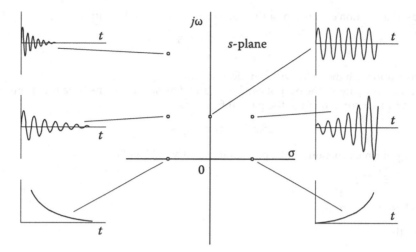

FIGURE A2.4
Poles in the *s*-plane and related time exponents.

A2.4 Laplace Transfer Function

The *Laplace transfer function* is the ratio of the Laplace transform of a link's output to that of its input.

Example A2.6

Since the step function is an integral of the delta-function, and the ratio of the step function Laplace transform to the delta-function Laplace transform according to Table A2.1 is $1/s$, evidently $1/s$ is the Laplace transfer function of an integrator. Conversely, multiplying a Laplace transform by s is analogous to differentiating the corresponding time function.

Since the Laplace transform is linear, the transfer function of several links connected in parallel equals the sum of the Laplace transforms of the links.

According to the convolution (Borel) theorem, the Laplace transfer function of a chain connection of links is the product of the Laplace transfer functions of the individual links.

Example A2.7

The step function $1(t)$ is applied to the input of a link composed of two cascaded links with transfer functions $5/(s+1)$ and $(s+1.4)/(s+2)$. Find the output time function.

Since the step function's Laplace transform is $1/s$, the Laplace transform of the output is

$$\frac{5(s+1.4)}{(s+1)(s+2)s} = \frac{5s+7}{s^3 + 3s^2 + 2s}$$

The partial fraction expansion of the Laplace transform of the output

$$\frac{-2}{s+1} + \frac{-1.5}{s+2} + \frac{3.5}{s}$$

can be found with the MATLAB function residue.

The time function of the output can be obtained by summing the time functions from Table A2.1 that correspond to the partial fractions:

$$-2e^{-t} - 1.5e^{-2t} + 3.5 \times 1(t)$$

The output time function can be plotted using MATLAB either with:

```
num = [5 7];
den = [1 3 2 0];
impulse(num, den)
```

or with:

```
num = [5 7];
den = [1 3 2];
step(num, den)
```

With the impulse command, the numerator and the denominator of the Laplace transform of the output are used (as the response to the δ-function whose Laplace transform is 1); with the step command, the numerator and the denominator of the transfer function are used.

Laplace transforms and inverse Laplace transforms (i.e., finding $f(t)$ from $F(s)$) can be directly found with MATLAB commands laplace and invlaplace.

When the input signal is an exponent $\mathrm{Re}\left(e^{-s_o t}\right)$ and s_0 is a zero of the transfer function, the output signal is zero.

For the input signal $\mathrm{Re}\left(e^{-s_p t}\right)$, where s_p is a pole of the transfer function, the transfer function is infinite; i.e., the signal is infinitely amplified and becomes infinitely large at the output.

When a pole of a transfer function is in the right half-plane of s, the output signal in response to δ-function input is an exponentially growing signal. Such a system is considered unstable since there always exist δ-function components in the input noise. In practical systems, the output grows exponentially until it becomes so large that, because of the power limitation in active devices, the output signal becomes limited, and the system no longer can be viewed as linear.

To determine whether a system with a given rational transfer function is stable, one might calculate the roots of the denominator polynomial and check whether there are roots with positive real parts. This can be done with MATLAB and many other popular software packages.

The Routh–Hurwitz criterion indicates the presence of right-sided polynomial roots when certain inequalities hold; verifying these inequalities is simpler than actually finding the roots. The criterion is not described here since software packages calculate the roots in no time.

To calculate the response to sinusoidal inputs, s should be replaced by $j\omega$. This results in a complex transfer function.

Example A2.8

If the input signal is $\mathrm{Re}[\exp(j\omega_o)]$, and at this ω_o the Laplace transfer function is $A\angle\phi$, then the output signal of the link will be $A\mathrm{Re}[\exp(j\omega_o + j\phi)]$. In other words, the signal is amplified A times and the phase is shifted (advanced) by ϕ radians.

Example A2.9

The transfer function is $100/(j\omega+5)$. At frequency $\omega=5$ rad/s, the function equals $14.1\angle-45°$; i.e., the output amplitude is 14.1 times larger than that of the input, and the output is delayed by 45°.

A2.5 Poles and Zeros of Transfer Functions

The locations of the transfer function's poles and zeros manifest themselves in two important aspects:

1. They show which signals are amplified infinitely (exponential signals with the exponents equal to the poles), and which are not transmitted at all (exponential signals with the exponents equal to the zeros). In Chapter 3, the former issue is related to system stability, and the latter to non-minimum phase shift.
2. They affect the transmission of purely sinusoidal test signals, i.e., the frequency responses.

Example A2.10

In the example shown in Figure A2.3, the Laplace transfer function is $1/(Ms^2)$. This is a rational function with a double pole at the origin. It is seen that s replaces $j\omega$.

Example A2.11

Consider the transfer function $L(s)=50/(s+5)$. The pole is at $s=-5$. The gain frequency response is the plot of the function

$$20\log|L(j\omega)| = 20\log[50/(j\omega+5)]$$

The vector $j\omega+5$ is shown in Figure A2.5(a). It is evident that this vector is lowest in magnitude at lower frequencies, where therefore the transfer function is largest. At zero

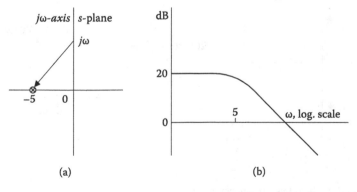

(a)　　　　　　　　　　　　(b)

FIGURE A2.5
Gain frequency response (b) corresponding to the pole at −5 on the s-plane (a).

frequency, the gain is 20 dB and the phase is 0. When the frequency increases, the vector becomes longer and eventually the gain coefficient decreases in proportion to the frequency. That is, when the frequency doubles, i.e., increases by an octave, the gain coefficient is halved, i.e., decreases by −6 dB. Therefore, the gain decreases with the constant slope of −6 dB/octave.

When ω changes from 0 to ∞, the phase changes from zero to $-\pi/2$.

Rational Laplace transform functions for physical systems always have only real coefficients. Therefore, their poles and zeros either are real or come in complex conjugate pairs. The transfer function multipliers for complex poles (or zeros) are commonly written as:

$$\left[s-\left(\sigma_o + j\omega_o\right)\right]\left[s-\left(\sigma_o - j\omega_o\right)\right] = s^2 - 2\sigma_o s + \omega_o^2 = s^2 + 2\xi\omega_o s + \omega_o^2$$

Here, σ_o is negative and $\zeta = -\sigma_o/\omega_o$ represents the damping coefficient for the pole.

Example A2.12

The transfer function

$$L(s) = \frac{c}{s^2 + 2\xi\omega_o s + \omega_o^2} = \frac{c}{\left[s-\left(\sigma_o - j\omega_o\right)\right]\left[s-\left(\sigma_o + j\omega_o\right)\right]}$$

has two complex conjugate poles, $\sigma_o \pm j\omega_o$. Figure A2.6(a) shows the poles on the s-plane, where σ_o is negative and small; the two vectors represent the multipliers in the denominator of the function. The frequency response is obtained by substituting $j\omega$ for s in the function. Then, when ω approaches ω_o, the first multiplier becomes $-\sigma_o$, the denominator becomes $2\sigma_o\omega_o$, and $|L(j\omega)|$ becomes large. With a further increase of the frequency, the denominator gets bigger and the modulus of the function decreases. Figure A2.6(b) shows the gain frequency response for the function.

The smaller the damping coefficient ζ (and therefore $|\sigma_o|$), the higher is the peak on the gain frequency response. At higher frequencies $\omega \gg \omega_o$, $|L(j\omega)|$ decreases as the square of the frequency, and the gain decreases with constant slope −12 dB/octave.

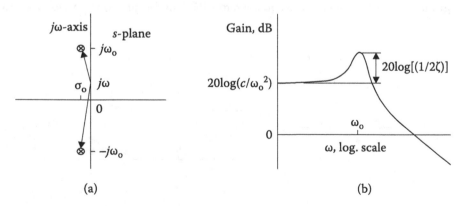

(a)   (b)

FIGURE A2.6
(a) Complex poles and (b) the related frequency response.

A2.6 Pole-Zero Cancellation, Dominant Poles, and Zeros

Next, consider a pole-zero pair with little distance between the pole and the zero, as shown in Figure A2.7. Since the vectors corresponding to the pole and to the zero are nearly the same, they compensate each other and insignificantly affect the frequency response. Adding such a pair to any transfer function will certainly change the order of the system but will have a negligible effect on the frequency response. And, since the frequency response fully characterizes the performance of a linear link, there will be no substantial difference in these links' performance in any application.

For example, a cluster of two-poles and one zero can be replaced by one pole. If the cluster is far from the frequency range of interest on the $j\omega$-axis, then the replacement is adequate even when the cluster is not very tight. But if the cluster is very close to the frequencies of interest, then the distance between the singularities must be very small for the replacement to be adequate.

In many systems, several poles and zeros mutually compensate or have a small effect on the transfer function within the frequency range of interest, and only one or a few poles and zeros have a prominent effect on the frequency response. Such poles and zeros are called dominant.

Much more about frequency responses' relation to the poles and zeros is explained in Chapter 4.

A2.7 Time Responses

In the calculation of time responses, the poles of the transfer function characterize the exponents of the solution to a system of certain linear differential equations. The closer the poles are to the $j\omega$-axis, the more oscillatory the solution is. Therefore, the closed-loop responses should not have poles too close to the $j\omega$-axis in order that the output time history tracks the command without excessive error.

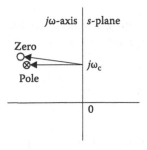

FIGURE A2.7
Pole-zero pair.

PROBLEMS

1. Can the system be linear if the input signal is 3sin(34*t*+5) and the output is the following:

    a. 3sin(5*t*+34)

    b. 3sin(34*t*+34)

    c. 5sin(5*t*–5)

    d. –13sin(34*t*+5)

    e. 2.72sin(2.72*t*+2.72)

    f. 3sin(340*t*–5)

    g. 5*t*sin(34*t*–5)

2. What is the phasor for the following functions (express the phase in radians):

    a. 22sin(ωt+12)

    b. –2sin(ωt+2)

    c. 3sin(ωt+12)

    d. –5cos(ωt+2)

    e. –2.72cos(ωt+2)

3. What is the phasor of the sum:

    a. 22sin(ωt+12) – 12sin(ωt+12)

    b. 12sin(ωt+12) – 10sin(ωt+12)

    c. $4\angle 30° - 40\angle 60°$

    d. $4\angle 30° + 40\angle 60°$

    e. $4\angle 60° - 40\angle 30°$

    f. $4\angle 60° + 40\angle 30° - 20\angle 0°$

4. What is the transfer function if the input signal phasor is $4\angle 49°$ and the output is the following:

    a. $4\angle 49°$

    b. $40\angle -59°$

    c. $40\angle 229°$

    d. $1,000\angle -30°$

    e. $2.72\angle -2.72°$

5. Find the originals of the following Laplace transforms (do so using Table A2.1 and the MATLAB function invlaplace):

    a. $1/s$

    b. s

    c. $2/(s+3)$

    d. $4/[(s+5)(s+10)]$

6. Find the Laplace transforms of the following signals (use both Table A2.1 and the MATLAB function laplace):

    a. δ-function

    b. Step function with the value of the step equal to 8

    c. $5t^2$

d. $10te^{-2t}$

e. $13(1 - e^{-4t})$

f. $6\sin2t$

g. $4e^{-4t}\cos5t$

h. $-4\cos2t$

7. Find transfer function of the following linear operators (use Table A2.1):

a. $df(t)/dt$

b. Integral of $f(t)$

c. Double integral of $f(t)$

8. The input signal is the δ-function, and the output signal Laplace transform is the following:

a. $(s^2 + 2s + 4)/(2s^3 + 4s^2 + 30)$

b. $(s^2 + 3s + 24)/(2s^3 + 12s^2 + 60)$

c. $(s^2 + 4s + 42)/(2s^3 + 12s^2 + 80)$

d. $(s^2 + 40s + 46)/(2s^3 + 2s^2 + 90)$

e. $(s^2 + 50s + 40)/(2s^3 + 22s^2 + 120)$

f. $(s^2 + 60s + 100)/(2s^3 + 20s^2 + 300)$

Find the link's transfer function.

9. Which of the following expressions can be transfer functions of stable systems:

a. $(20s^2 + 30s + 40)/[(s + 43)(s + 85)(s + 250)(s + 2{,}500)]$

b. $(s^2 + 30s + 4)/(2s^3 - 2s^2 + 30)$

c. $-(10s^2 + 10s + 40)/(2s^4 - 2s^3 + 3s^2 + 80)$

d. $(s^4 + 5s^3 + s^2 + 20s + 5)/(s^3 + 23s^2 + 200s + 300)$

e. $-60(s + 3)(s - 16)/[(s + 33)(s + 75)(s + 200)(s + 2{,}000)]$

f. $10(s + 2)(s - 22)/[(s + 40)(s + 65)(s + 150)]$

g. $-20(s + 2)(s + 26)/[(s + 43)(s + 85)(s + 250)(s + 2{,}500)]$

10. Use the MATLAB command root to calculate the poles and zeros of the following functions:

a. $(20s^2 + 30s + 40)/(2s^4 + s^3 + 5s^2 + 36)$

b. $(s^2 + 30s + 4)/(s^4 + 2s^3 + 2s^2 + 36)$

c. $(10s^2 + 10s + 40)/(2s^4 + 2s^3 + s^2 + 3)$

d. $(s^2 + 20s + 5)/(s^4 + 5s^3 + s^2 + 3)$

e. $(2.72s^2 + 27.2s + 20)/(s^4 + 2.72s^3 + 7s^2 + 2.72)$

f. $(s^2 + 10s + 8)/(s^4 + 12s^3 + 12s^2 + 33)$

11. Use the MATLAB command poly to convert the following functions to ratios of polynomials:

a. $50(s + 6)(s + 12)/[(s + 50)(s + 85)(s + 110)(s + 1{,}200)]$

b. $-60(s + 7)(s + 15)/[(s + 53)(s + 95)(s + 210)(s + 2{,}300)]$

c. $10(s + 8)(s + 62)/[(s + 50)(s + 65)(s + 150)]$

d. $-20(s + 9)(s + 66)/[(s + 63)(s + 95)(s + 240)(s + 2{,}700)]$

e. $2.72(s + 7)(s + 10)/[(s + 70)(s + 130)(s + 1{,}200)]$

12. Write the frequency response function by replacing s by $j\omega$ for the following:

   a. $(20s^2+30s+40)/(2s^4+s^3+s^2+3)$

   b. $(s^2+30s+4)/(s^4+2s^3+2s^2+3)$

   c. $(10s^2+10s+40)/(2s^4+2s^3+s^2+3)$

   d. $(s^2+20s+5)/(s^4+5s^3+s^2+3)$

   e. $(2.72s^2+27.2s+20)/(s^4+2.72s^3+7s^2+2.72)$

   f. $(s^2+10s+8)/(s^4+12s^3+12s^2+130)$

13. Three links with the following transfer functions, respectively,

$$-20(s+2)(s+26)/[(s+85)(s+250)]$$

$$(s^2+30s+4)/(s^3+2s^2+30)$$

$$(10s^2+10s+40)/(s^4+2s^3+30s^2+1{,}000)$$

are connected (a) in series (cascaded), (b) in parallel, (c) with the first and the second in parallel and the third in cascade, and (d) with the first and the second in series and with the third in parallel to this composite link. Find the transfer functions of the resulting composite links.

14. Use MATLAB to plot the frequency response for the first-, second-, and third-order functions:

   a. $10/(s+10)$

   b. $100/(s+10)^2$

   c. $1{,}000/(s+10)^3$

   Describe the correlation between the slope of the gain frequency response and the phase shift.

15. Use MATLAB to plot the frequency response and the step time response for the following first- and second-order functions:

   a. $10/(s+10)$

   b. $-10/(s+10)$

   c. $100/(s^2+4s+100)$

   d. $-100/(s^2+4s+100)$

   e. $100/(s^2+2s+100)$

   f. $100/(s^2+s+100)$

   Describe the correlation between the shapes of the frequency responses and those of the step responses.

16. Use MATLAB to convert the functions to a ratio of polynomials and plot the frequency response for the following functions:

   a. $50(s+3)(s+12)/[(s+30)(s+55)(s+100)(s+1{,}000)]$

   b. $60(s+3)(s+16)/[(s+33)(s+75)(s+200)(s+2{,}000)]$

   c. $10(s+2)(s+22)/[(s+40)(s+65)(s+150)]$

   d. $-20(s+2)(s+26)/[(s+43)(s+85)(s+250)(s+2{,}500)]$

   e. $2.72(s+7)(s+20)/[(s+10)(s+100)(s+1{,}000)]$

   f. $-25(s+2)(s+44)/[(s+55)(s+66)(s+77)(s+8{,}800)]$

   What is the value of the function at dc (i.e., when $s=0$)? What does this function degenerate into at very high frequencies?

17. Compare the frequency responses of the following:

   a. $(s+2)/(s+10)$ and $(s+2)(s+5)/[(s+10)(s+5.1)]$

   b. $(s+2)/(s+10)^2$ and $(s+2)/[(s+9)(s+11)]$

   c. $(s+2)/(s+10)^2$ and $(s+2)/[(s+7)(s+14)]$

   d. $(2.4s^2+25s+20)/(s^4+2s^3+8s^2+3)$

   Draw a conclusion.

18. Plot time responses to a step command for the functions in the previous example. Draw a conclusion.

19. Plot and compare frequency responses of the functions with complex poles and zeros:

   a. $(s+2)/(s+10)$ and $(s+2)(s+5)/[(s+10)(s+5.1)]$

   b. $(s+2)/(s+10)^2$ and $(s+2)/[(s+9)(s+11)]$

   c. $(s+2)/(s+10)^2$ and $(s+2)/[(s+7)(s+14)]$

   d. $(2s^2+22s+20)/(s^4+2.5s^3+7.5s^2+2.5)$

   Draw a conclusion.

20. Plot the frequency responses (on the same plot, using hold on feature) of the following functions:

   a. $(s+1)/(s+10)$

   b. $(s+1)/(s+20)$

   c. $(s+1)/(s+1,000)$

   d. $(s+10)/(s+1,000)$

   Over what frequency range do the functions approximate the differentiator s?

21. Make a transfer function to implement the following frequency response:

   a. 10 at lower frequencies, rolling down at higher frequencies inversely proportionally to the frequency

   b. 10 at high frequencies, rolling up at low frequencies proportionally to the frequency

   c. 10 at low frequencies, 100 at high frequencies

   d. 100 at low frequencies, 10 at high frequencies

   e. 10 at medium frequencies, rolling up proportionally to the frequency at lower frequencies, rolling down inversely proportionally to the frequency at higher frequencies

   f. Resonance peak response, rolling up proportionally to the frequency at lower frequencies, rolling down inversely proportionally to the frequency at higher frequencies

   g. Resonance peak response, flat at low frequencies and rolling down inversely proportionally to the square of the frequency at higher frequencies

   h. Resonance peak response, flat at high frequencies and rolling up proportionally to the square of the frequency at higher frequencies

   i. Notch responses of three different kinds, with different behavior at lower and higher frequencies

Appendix 3: Causal Systems, Passive Systems and Positive Real Functions, and Collocated Control

Causal systems are those systems whose output value at any t_1 does not depend on the input signal at $t > t_1$. A causal system is stable if and only if there are no poles in the right half-plane. When s is equal to a pole, the transfer function is ∞. A transfer function pole in the right half-plane means that the response to a finite input signal is an exponentially rising output signal. This cannot be a property of a *passive system*, i.e., a system without sources of energy.

A transfer function is θ *positive real* (p.r.) if the following three conditions are satisfied:

1. There are no poles or zeros in the right half-plane of s.
2. Any poles and zeros on the $j\omega$-axis are single.
3. Re $\theta(j\omega) > 0$ at all frequencies.

A system is passive if and only if its transfer function is positive real. All passive driving point impedances (a driving point impedance is the ratio of the voltage to the current at the same port) are positive real. Such an impedance has nonnegative resistance for all sinusoidal signals and positive resistance for all rising exponential signals (whether oscillating or not). A p.r. impedance does not generate but only dissipates power.

The driving point admittances of passive systems are also p.r.

The transfer impedances and admittances of passive systems can be p.r. but are not necessarily so.

The driving point impedances of active systems can be p.r. but are not necessarily so.

Any p.r. function can be *realized* (i.e., implemented) as a driving point impedance of a passive system; i.e., it is always possible to make a system composed of passive elements whose driving point impedance is the prescribed p.r. function. In some cases, the system is an arrangement of resistances, capacitances, and inductors connected in series and in parallel. For some p.r. functions, however, realization requires bridge-type circuits or transformers. A p.r. function can also be realized as a driving point impedance of an active *RC* circuit.

Collocated control is feedback control of a passive plant where the actuator and the sensor are collocated, and the ratio of the sensor readings to the actuator action is the driving point impedance or admittance, or a derivative or an integral of the impedance. This limits the range of the plant phase variations to 180°.

Appendix 4: Derivation of Bode Integrals

A4.1 Integral of the Real Part

Consider the Laurent expansion for a complex function $\theta(s)$ that is analytic in the closed right half-plane:

$$\theta(s) = A_\infty - \frac{B_1}{s} - \frac{A_2}{s^2} - \frac{B_3}{s^3} - \frac{A_4}{s^4} - \ldots$$

The series is convergent for $s \to \infty$ with $\text{Re}\{s\} > 0$. On the $j\omega$-axis, the expansion becomes:

$$\theta(j\omega) = A_\infty + j\frac{B_1}{\omega} + \frac{A_2}{\omega^2} + \ldots \tag{A4.1}$$

and it can be seen that the real and imaginary parts correspond to the A and B terms accordingly. If θ satisfies real symmetry, then the real coefficients A and B are even and odd functions, respectively, of the frequency.

The function $\theta - A_\infty$ is also analytic in the RHP so the contour integral of $\theta - A_\infty$ around it equals 0. The contour of integration may be viewed as being composed of the entire $j\omega$-axis plus a π-radian arc of infinite radius R, as shown in Figure A4.1. The integral along the arc equals πB_1; the integral along the whole $j\omega$-axis equals twice the integral of the even part of the integrand, i.e., of $A - A_\infty$, along the positive semi-axis. Therefore,

$$\int_0^\infty (A - A_\infty)\, d\omega = -\frac{\pi B_1}{2} \tag{A4.2}$$

If θ is an impedance, Z, of a parallel connection of a two-pole and a capacitance, C, then equation A4.2 becomes the resistance integral presented in 3.14.4. Alternatively, if θ is the logarithm of the feedback, and $\lim_{s \to \infty} sT(s) = 0$, then Equation A4.2 becomes Equation 3.10, which shows that the integral of the feedback is zero.

A4.2 Integral of the Imaginary Part

θ/s is also analytical in the right half-plane and on the $j\omega$-axis, except at the origin. So if the origin is avoided, as shown in Figure A4.2, the integral of θ/s around the contour equals 0. The integral along the small arc equals πA_0, where A_0 is the value of θ at zero frequency. The integral along the infinite radius arc equals $-\pi A_\infty$, where, as follows from

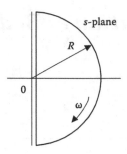

FIGURE A4.1
Contour on the s-plane.

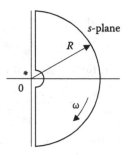

FIGURE A4.2
Contour on the s-plane avoiding a pole at the origin.

Equation A4.1, A_∞ is the value of θ at high frequencies. Along the $j\omega$-axis, the odd parts of θ/s cancel, which are the even parts of θ, leaving:

$$\int_0^\infty \frac{B}{\omega} d\omega = \frac{\pi(A_\infty - A_0)}{2}$$

i.e.,

$$\int_{-\infty}^\infty B(u)\, du = \frac{\pi(A_\infty - A_0)}{2} \tag{A4.3}$$

where $u = \ln(\omega)$. This relation is known as the *phase integral*.

Another important relation between the real and imaginary components results from integrating $(\theta - A_\infty)/\sqrt{sW}$ around the same contour. Here, W is a *reactance function*, i.e., an impedance function of a reactance of two-poles. On the $j\omega$-axis, W is purely imaginary, either positive or negative, so the function sW is purely real, positive or negative, and the function $\sqrt{sW}$ therefore alternates between being purely real or purely imaginary on adjoining sections over the $j\omega$-axis. It has branch points at the joints of these sections. The sign of the radix at the sections must be chosen so that the whole contour of integration belongs to only one of the Riemann sheets. On this contour, the function Re $W(j\omega)$ must be even and Im $W(j\omega)$, odd.

If the reactance W is proportional to s at high frequency, the integrand decreases at least as s^{-2}, and the integral along the large arc vanishes. Since the total contour integral is zero, the integral along the $j\omega$-axis equals zero as well. Its real part is certainly zero:

$$\int_{-\infty}^{\infty} Re \frac{\theta - A_\infty}{\sqrt{sW}} d\omega = 2\int_{0}^{\infty} Re \frac{\theta - A_\infty}{\sqrt{sW}} d\omega = 0 \tag{A4.4}$$

If, in particular, $W = (1 + s^2)/s$, then $\sqrt{sW} = \sqrt{1 - \omega^2}$ is real for $|\omega| < 1$ and imaginary otherwise, and Equation A4.4 yields:

$$\int_{0}^{1} \frac{A - A_\infty}{\sqrt{1 - \omega^2}} d\omega + \int_{1}^{\infty} \frac{B}{\sqrt{\omega^2 - 1}} d\omega = 0$$

i.e.,

$$\int_{\omega=0}^{\omega=1} (A - A_\infty) d \arcsin(\omega) = -\int_{1}^{\infty} \frac{B}{\sqrt{\omega^2 - 1}} d\omega \tag{A4.5}$$

A4.3 General Relation

Let's define the frequency at which the phase shift is of interest as ω_c, and also define $\theta_c \equiv A_c + jB_c \equiv \theta(j\omega_c)$, where A_c and B_c are real. Consider the function of s:

$$(\theta - A_c)\left(\frac{1}{s/j - \omega_c} - \frac{1}{s/j + \omega_c} \right) = (\theta - A_c) \frac{2\omega_c}{-s^2 - \omega_c^2} \tag{A4.6}$$

This function is analytical in the right half-plane of s and on the $j\omega$-axis except at the points $-j\omega_c$, $j\omega_c$. Therefore, the integral of the function taken around the contour shown in Figure A4.3 equals zero. The contour consists of several pieces; the sum of the integrals along these pieces is 0.

The integral along the arc of infinite radius R equals 0 because of the term s^2 in the denominator of the integrand.

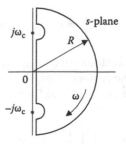

FIGURE A4.3
Contour on s-plane avoiding imaginary poles.

As s approaches $j\omega_c$, $\theta - A_c$ approaches B_c, and the second multiplier in the left side of Equation A4.6 tends to $1/(s/j - \omega_c)$. Then the integral along the infinitesimal arc centered at $j\omega_c$ equals:

$$\int \frac{B\,ds}{s/j - \omega_c} = -\pi B_c$$

The integral along the small arc centered at $-j\omega_c$ equals $-\pi B_c$ as well.

Next, equating the sum of all components of the contour integral to 0, we see that the integral along the $j\omega$-axis equals $2\pi B_c$. Neglecting the odd component of the integrand whose integral is annihilated within these symmetrical boundaries, we have:

$$\int_{-\infty}^{\infty} (A - A_c) \frac{2\omega_c}{\omega^2 - \omega_c^2} \, d\omega = 2\pi B_c$$

and finally,

$$B_c = \frac{2\omega_c}{\pi} \int_0^\infty \frac{A - A_c}{\omega^2 - \omega_c^2} \, d\omega \tag{A4.7}$$

Integrating by parts with:

$$U = \frac{A - A_c}{\pi} \tag{A4.8}$$

$$dV = \frac{2\omega_c}{\omega^2 - \omega_c^2} \, d\omega = -d \ln \coth\left(\frac{u}{2}\right) du$$

where $u = \ln(\omega/\omega_c)$, equation A4.7 becomes:

$$B_c = \frac{-1}{\pi} \left[(A - A_c) \ln \coth \frac{u}{2} \right]_{-\infty}^{\infty} + \frac{1}{\pi} \int_{-\infty}^{\infty} \frac{dA}{du} \ln \coth \frac{u}{2} \, du \tag{A4.9}$$

The left side of the equation is real. Hence, the imaginary components on the right side are annihilated after summing. We can therefore count the real components only. Since

$$\ln \coth \frac{-u}{2} = j\pi + \ln \coth \frac{u}{2}$$

replacing $\ln \coth(u/2)$ by $\ln \coth|u/2|$ does not change the real components of Equation A4.9 and is therefore permitted. After the replacement, the function in square brackets becomes even and is annihilated because of symmetrical limits. Thus,

$$B(\omega_c) = \frac{1}{\pi} \int_{-\infty}^{\infty} \frac{dA}{du} \ln \coth \frac{|u|}{2} \, du \tag{A4.10}$$

which is Equation 3.5.

A4.4 Integral Relations for Disturbance Isolations

In this section we show integral estimation for passive and active dynamic isolation.

Consider the system in Figure A4.4 of two bodies connected with an active strut. A disturbance force source F_1 is applied to the body M_1. The force F_3 is applied via the massless active strut to the body M_3.

The mobility Z_2 of the active strut is the ratio of the difference in the velocities at the end of the strut to the force (because we neglect the strut's mass, the force is the same at both ends of the strut).

To increase the disturbance rejection, the force increases the division ratio of the system $K_F = F_3/F_1$ by the feedback employed in the strut.

Using the following electromechanical analogy: mechanical power to electrical power, voltage to velocity, current to force, capacitance to mass, and inductance to inverse of the stiffness coefficient, the electrical equivalent circuit for the system is shown in the lower figure. The current division ratio I_2/I_3 is equivalent to K_F. The electrical impedance Z_2 is equivalent to the strut mobility.

The current division ratio, i.e., the force division ratio, is:

$$K_F = \frac{1/(sM_1)}{1/(sM_1) + Z_2 + 1/(sM_3)}$$

or

$$K_F = \frac{1}{1 + sM_1 Z_2 + (M_1/M_3)} \tag{A4.11}$$

At higher frequencies, the equivalent electrical impedance of the strut in Figure A4.4 degenerates into the impedance of series inductance included in Z_2, the inductance being equivalent to the inverse of the stiffness coefficient k of the strut at higher frequencies. Therefore, the force division ratio at higher frequencies turns into:

$$K_F \big|_{\omega \to \infty} = k/s^2 M_1 \tag{A4.12}$$

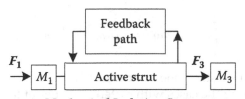

Mechanical Isolation System

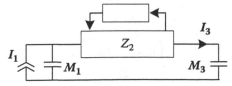

Equivalent Electrical System

FIGURE A4.4
A disturbance rejection system.

Because the value reduces as a square of the frequency, the Bode integral of the real part of a function applies. From this integral,

$$\int_0^\infty \log|K_F + 1|\, d\omega = 0 \tag{A4.13}$$

This relation remains valid with and without feedback in the active strut and allows one to estimate the effect of feedback on the disturbance isolation at higher frequencies, but only as long as the Figure A4.4 model correctly reflects the physical strut and the bodies. The model must be accurate enough over the frequency range where $\log |K_F + 1|$ is substantial. The model might become inaccurate at higher frequencies within this range because the mass of the strut cannot be neglected at high frequencies and the bodies cannot be considered rigid. Integral relations can be developed for such systems as well; however, the increased complexity of the integrand makes it more difficult to use the relations for fast performance estimation while comparing different design versions.

Another equation, which will give a better estimation of available performance at lower frequencies, can be found as follows. From Equation A4.11,

$$\frac{1}{K_F(1 + M_1 / M_3)} = 1 + sM_1Z_2 / (1 + M_1 / M_3) \tag{A4.14}$$

Consider the practical case of the feedback in the active strut to be finite at dc. At lower frequencies, the active strut degenerates into a spring. Therefore, at lower frequencies, the fraction in the right-hand side of Equation A4.14 increases with frequency as ω^2. With the frequency scale inverted, the fraction decreases at high frequencies as ω^{-2}. Then the Bode integral of the real part of the function applies, the integral of the logarithm of the expression on the right side of Equation A4.14 equals 0. Therefore,

$$\int_0^\infty \left[-\log|K_F| - \log\frac{1 + M_1}{M_3} \right] d\omega^{-1} = 0$$

or

$$\int_0^\infty \log|K_F|\, d\omega^{-1} = -\int_0^\infty \log\frac{1 + M_1}{M_3}\, d\omega^{-1} \tag{A4.15}$$

The feedback in the active strut does not affect the right-hand side of the equation. Therefore, when comparing the cases with different values of feedback in the active strut loops, the right-hand part of Equation A4.15 can be neglected. Hence, the integral of the difference in the vibration transmission between any two cases with different feedback in the active strut is 0:

$$\int_0^\infty \log|K_F|\, d\omega^{-1} = 0 \tag{A4.16}$$

Equation A4.16 is important because it places a simple fundamental restriction on what can be achieved by disturbance isolation design. Introduction of feedback in the active strut reduces the force division ratio at some frequencies, but at some other frequencies (in fact, at lower frequencies) this ratio increases, and the difference between the areas of the output force reduction and the area where the force is increased (with inverse frequency scale) is zero.

Equation A4.16 is convenient for estimation of the effects of disturbance isolation loops in complex feedback systems.

Example A4.1

In experiments with a large-scale model of an interstellar interferometer, a vibration source (representing a reaction wheel) was placed on a platform (body 1) suspended on six orthogonal active struts. Vibration propagation to the base on which a sensitive optic was installed (body 2) was reduced by band-pass feedback in the active struts by 30 dB at 20 Hz, the value gradually decreasing with frequency to nearly zero at 100 Hz. At 7 Hz frequency, however, the feedback increased the base vibrations by approximately 10 dB.

The trade-off between the vibration amplification and vibration attenuation over different frequency ranges was quite acceptable since at lower frequencies, the feedback in optical pointing loops was large (and can be made even larger, if necessary, by application of nonlinear dynamic compensation), and the total error in the optical loops was small.

Appendix 5: Program for Phase Calculation

For functions that do not have high Q resonance peaks or gaps, the phase integral (Equation A4.10) can be calculated with MATLAB program BONYQAS. The phase accuracy for a typical feedback loop shaping problem without sharp peaks or notches is better than 3°. The description and application of the program are shown in Appendix 14.

The functions are given by piece-linear segments $1(s^2 + s + 1)$, with k being the frequency of their corners. The asymptotic gains are given at the higher `hslope` and the lowest `lslope` frequencies.

The function BONYQAS calculates the Nyquist diagram and the phase characteristic.

The listing of the programs follows:

```
function [gain1,phase1] = bonyqas(u,g,lslope,hslope,nml);
% function [gain1,phase1] = bonyqas(u,g,lslope,hslope,bnm);
% BONYQAS Asymptotic Bode and Nyquist diagrams corresponding in minimum
% phase manner to a piece-linear gain response in dB specified by the
% gain g(h) at k corner frequencies u(h) in rad/sec. The lengths of
% vectors g and u must be equal. Coefficient lslope defines the
% asymptotic slope at lower frequencies in integrator units, and
% hslope defines the asymptotic slope at higher frequencies. nml is
% nonminimum phase lag in degrees at u(h); the lag is proportional
% to the frequency, default is nml = 0.
% Smoothed Bode diagram is plotted in white. Phase is plotted in greed.
% Nyquist diagram is plotted on logarithmic plane.
% The function is applicable for reasonably smooth responses. The phase
% accuracy for typical feedback loop shaping problem without sharp
% peaks or notches is better than 3 degr.
%
% Default: bonyqas without arguments plots the band-pass response:
% bonyqas([0.2 2 8 32],[-5 10 6 -12],-1,3);
% Low-pass loop Example 1: bonyqas([3 4 20 30],[16 13 -9 -10],2, 3, 0)
% Loop Example 2: bonyqas([3 3.8 22 45],[16 13 -10 -11.5],2, 3, 50)
% Copyright (c) B. Lurie 20-2-2004.

% default
  if nargin == 0,
    u = [0.2 2 8 32]; % 4 corner frequencies
    g = [-5 10 6 -12]; % gain in dB; band-pass system
    lslope = -1; % -1 integrator, i.e. 6.02 dB/oct up-slope
    hslope = 3; % 3 integrators, i.e. -18.06 dB/oct
  end
  if nargin < 5,
    nml = 0;
  end

  k = length(u); h = [1:k];

% frequency range of calculations and plots
  wmin = floor(log10(u(1)) - 0.33);
  wmax = ceil(log10(u(k)) + 0.33);
  w = logspace(wmin,wmax);
```

```
% low-frequency asymptote
  gain1 = (-lslope * (20 * log10(w/u(1))) + g(1))';
  phase1 = -lslope * 90;
% phase1 = -lslope * 90 * ones(length(w),1); % can be used instead
% semilogx(w, gain1, 'r--') % can be activated for purpose of teaching
% hold on
% segment slopes; segment no.h ends at u(h)
  for h = [2:k]
    sslope(h) = (g(h) - g(h-1))/log2(u(h)/u(h-1));
  end % [0 15 -2 -9]

% sslope % can be activated for purpose of teaching
% ray slopes
    rslope(1) = sslope(2) + 6.02 * lslope;
    rslope(k) = - 6.02 * hslope - sslope(k);
  for h = [2:(k-1)]
    rslope(h) = sslope(h + 1) - sslope(h);
  end
% rayslope_round = round(rslope) % [9 -17 -7 -9], positive means up

% frequency responses for k rays
  nray = 1; dray = [1 1 1]; % normalized ray with damping coefficient 0.5;
  for h = [1:k]
    [nrayf, drayf] = lp2lp(nray,dray,u(h)); % specifying corner frequency
    tfr = freqs(nrayf, drayf, w);
    rgain = 20*real(log10(tfr));
    rphase = (180/pi) * angle(tfr);
    rasis = (-rslope(h)/12.041) * rgain'; % minus makes positive ray go up
    % semilogx(w, rasis,'k--'); % can be activated for purpose of teaching
    % semilogx(w, rphase,'g:'); % can be activated for purpose of teaching
    gain1 = (-rslope(h)/12.041) * rgain' + gain1; % slope denormalization
    phase1 = phase1 - (rslope(h)/12.041) * rphase';
  end

% nonminimum phase lag addition
  phase1 = phase1 - nml * (w/u(h))';

% Bode diagram and phase response
  subplot(1,2,1)
  semilogx(w,gain1,'k', w,phase1,'g--', u,g,'ro')
  grid;
  xlabel('frequency, rad/sec');
  ylabel('gain, dB, and phase shift, degr');
  title('Bode diagram')
  hold off;
  zoom on;

% Nyquist diagram on log plane
  subplot(1,2,2)
  plot(phase1, gain1,'k', -180,0,'ro')
  hold on
  gain2 = (-lslope * (20 * log10(u/u(1))) + g(1))'; % marks
```

```
phase2 = -lslope * 90;
  for h = [1:k]
    [nrayf, drayf] = lp2lp(nray,dray,u(h));
    tfr = freqs(nrayf, drayf, u);
    rgain = 20*real(log10(tfr));
    rgain = real(rgain);
    rphase = (180/pi) * angle(tfr);
    gain2 = (-rslope(h)/12.041) * rgain' + gain2;
    phase2 = phase2 - (rslope(h)/12.041) * rphase';
  end
phase2 = phase2 - nml * (u/u(h))';
plot(phase2,gain2,'ro');
hold off;
set(gca,'XTick',[-270 -240 -210 -180 -150 -120 -90])
grid
axis([-270 -90 -20 70])
title('Nyquist diagram')
xlabel('phase shift, degr')
ylabel('gain in dB')
zoom on
```

Appendix 6: Generic Single-Loop Feedback System

When a feedback system cannot be easily broken into a connection of links, a more general description of the feedback system can be used as a connection of a unilateral two-port w to a passive four-port B, as displayed in Figure A6.1.

The two-port w is assumed to have zero reverse transmission and either infinite or infinitesimal input and output impedances. When the circuit diagram in Figure A6.1 is applied to the analysis of circuits with physical amplifiers, the amplifier input and output impedances may be imitated by connecting passive two-poles in parallel or in series to the amplifier's input and output. Further, these two-poles have to be integrated into the B-network.

In principle, the dimensionality of the signals at the input and output of the amplifier and at the input and output terminals of the whole system does not influence the following analysis. However, to simplify the exposition, we select one of the possible versions and characterize the signal at the input to the active element by the voltage E_3, and the signal at its output by the current I_4. The active element with transadmittance $w = I_4/E_3$ is therefore assumed to have high input and output impedances.

When an external emf E_3 is applied to the input of the amplifier disconnected from the B-circuit, as shown in Figure A6.2, the return voltage U_3 appears at port 3'. The return ratio is:

$$T = -\frac{U_3}{E_3} \tag{A6.1}$$

When the return ratio T is being measured, the emf E_1 of the source connected to the system's input 1 must be replaced by a short circuit, in accordance with the superposition principle. Thus, the source impedance Z_1 is connected to the port 1. The return ratio therefore depends on Z_1, i.e., $T = T(Z_1)$. In particular, $T(0)$ denotes the value of T measured while the input is shorted, and $T(\infty)$, while the input is open.

Figure A6.3 shows the *cross-sectioned feedback circuit*. The external two-poles Z_s and Z_r connected to the input and output of the broken loop provide appropriate loading for the disconnected parts of the B-circuit in order to keep their transfer coefficients unchanged. The emf E_6 applied to the input of the broken loop produces return signal U_5. The ratio

$$-\frac{U_5}{E_6} = -\frac{U_3}{E_6}\frac{I_4}{U_3}\frac{U_5}{I_4}$$

equals T. When T is measured this way, the two-pole Z_s need not be connected since it is shunted by the emf E_6.

Generally, the voltage input–output ratio of a linear four-pole can always be presented as the product of two ratios: the current input–output ratio, and the ratio of the load impedance to the four-pole input impedance. The latter ratio is found to be 1 when T is calculated. Therefore, T may be measured arbitrarily as the ratio of either voltages or currents.

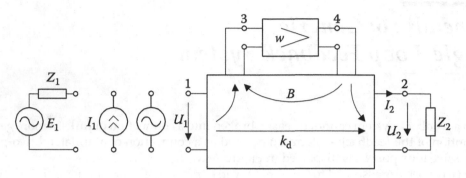

FIGURE A6.1
Single-loop feedback system.

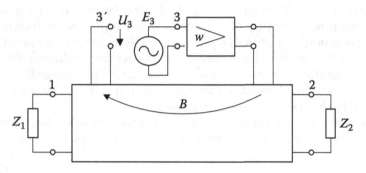

FIGURE A6.2
Disconnected feedback loop.

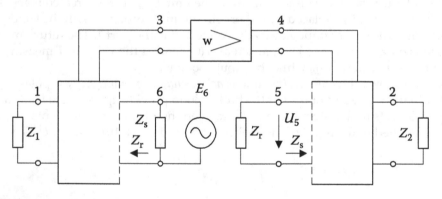

FIGURE A6.3
Disconnecting feedback loop in the feedback path.

In the closed-loop system, the signal U_3 is formed by superposition of the effects U_3° and $-U_3T$ produced, respectively, by the signal source and the output of the amplifier. Thus, $U_3 = U_3^\circ - U_3T$, whence:

$$U_3 = \frac{U_3^\circ}{F}$$

(A6.2)

Here, $F = T + 1$ is the return difference, i.e., the difference between the signals U_3 and E_3 relative to E_3.

The output of the feedback system in Figure A6.1 is a linear combination of two signal sources: the output of the amplifier and the signal source. By virtue of Equation A6.2, the signal from the amplifier output is reduced F times. Hence, the closed-loop system transfer coefficients in voltage and current, and the ratio of the output voltage to the signal emf are, respectively,

$$K_{OL} = \frac{U_2}{U_1} = \frac{K}{F(0)} + k_d \tag{A6.3}$$

$$K_{OLI} = \frac{I_2}{I_1} = \frac{K_I}{F(\infty)} + k_{Id} \tag{A6.4}$$

$$K_{OLE} = \frac{U_2}{E_1} = \frac{K_E}{F} + k_{Ed} \tag{A6.5}$$

Here, K_{OL}, K_{OLI}, and K_{OLE} are the open-loop system transmission functions, measured while the feedback path is disconnected; $F(0)$, $F(\infty)$, and $F = F(Z_1)$ are the return differences measured under the conditions of connecting zero impedance, infinite impedance, or impedance Z_1, respectively, to the system's input terminals; and k_d, k_{Id}, and k_{Ed} are the coefficients of direct signal propagation through the B-circuit and are determined under the same set of loading conditions at the input terminals.

Let Z designate the input impedance, and Z_o, the input impedance in the system without feedback (with a cross-sectioned feedback path, or with $w = 0$, i.e., with the active element zeroed out). Then,

$$K_{OL} = K_{OLI} \frac{Z_2}{Z} \tag{A6.6}$$

$$k_d = k_{Id} \frac{Z_2}{Z} \tag{A6.7}$$

$$K = K_I \frac{Z_2}{Z_o} \tag{A6.8}$$

By substituting Equations A6.7 and A6.8 into Equation A6.3, and Equation A6.4 into Equation A6.6, we get:

$$\frac{K_I Z_2 / Z_o}{F(0)} + k_{Id} \frac{Z_2}{Z} = \left(\frac{K_I}{F(\infty)} + k_{Id} \right) \frac{Z_2}{Z}$$

from which Blackman's formula follows:

$$Z = Z_o \frac{F(0)}{F(\infty)} = Z_o \frac{T(0) + 1}{T(\infty) + 1} \tag{A6.9}$$

The formula expresses Z through the three easily calculated functions: Z_o and return ratios $T(0)$ and $T(\infty)$.

Since, in principle, any two nodes of the B-circuit can be regarded as input terminals of the feedback system, Equation A6.9 can be used for calculation of the driving point impedance at any port n, provided that $F(0)$ is understood to be measured with the port n shorted, and $F(\infty)$, with the port terminals open.

If the terminals are shorted, the voltage between them vanishes, but not the current. For this reason, the feedback is called *current mode* (or *series*) with respect to the terminals n if, with respect to these terminals, $T(0) \neq 0$ and $T(\infty) = 0$. Analogously, the feedback is *voltage mode* (or *parallel*) if $T(0) = 0$ and $T(\infty) \neq 0$. If neither of them equals 0, the feedback is called *compound*.

If the feedback is infinite, the input impedance is:

$$Z = \lim_{w \to \infty} Z = Z_o \frac{wB(0)}{wB(\infty)} = Z_o \frac{B(0)}{B(\infty)} \tag{A6.10}$$

It depends exclusively on the B-circuit and not on w.

More detail along these lines is given in [6, 33].

Appendix 7: Effect of Feedback on Mobility

The following derivation of Blackman's formula follows Blackman's original proof, but in mechanical terms (see [10]).

Consider a mechanical system with an actuator accessed via a single mechanical port. That is, consider an active structural member including an actuator. To the active member, a structure is connected. The structure's mobility is Z_L. The force F is measured between the active member and the structure.

For the purpose of analysis, consider disconnecting the feedback loop at the input to the actuator and applying signal E to the actuator input. The relative velocity V across the active member and the feedback return signal E_r at the end of the disconnected feedback loop can each be expressed as a linear function of the force F and the signal E:

$$V = aE + bF, \ E_r = cE + dF \tag{A7.1}$$

Here a, b, c, and d are the constants to be determined from boundary conditions. Notice that when the feedback loop is closed, $E = E_r$, and $E = 0$ when it is open.

First, find the expressions for the mobility with and without feedback. First, the case without feedback, i.e., $E = 0$, gives $V = bF$. Thus, the active member mobility without feedback, Z_o, becomes

$$Z_o = V / F = b \tag{A7.2}$$

When the loop is closed, i.e., $E_r = E$, Equation A7.1 gives $V = [b + ad/(1-c)]F$, so that the active member mobility with feedback, Z, becomes:

$$Z = b + ad / (1 - c) \tag{A7.3}$$

Second, find the expressions for the return ratio for the two different loading conditions, when the active member is clamped and when it is free to move. The return ratio T of the active member feedback loop is defined as the ratio of the return signal to the negative of the input signal, i.e., $T = -E_r/E$. This ratio is certainly a function of the mobility of the structure Z_L, i.e., $T(Z_L)$.

When $Z_L = 0$, the active member is rigidly constrained, i.e., $V = 0$. In this case, Equation A7.1 gives $F = -a/(bE)$ and $E_r = (c - ad/b)E$. Hence, the return ratio becomes:

$$T(0) = -c + ad / b \tag{A7.4}$$

On the other hand, when $Z_L = \infty$, i.e., the active member is free to move, zero force is induced in the active member. In this case $E_r = cE$ and the return ratio is:

$$T(\infty) = -c \tag{A7.5}$$

Comparing the obtained equations (A7.2–A7.5) results in Blackman's formula as:

$$Z = Z_o \frac{T(0)+1}{T(\infty)+1}$$

The formula expresses active member mobility with feedback, Z, in terms of three other functions, Z_o, $T(0)$, and $T(\infty)$, that do not depend on the structural system to which the active member is connected.

Appendix 8: Regulation

A8.1 Dependence of a Function on a Parameter

The formulas derived in the two previous sections involve linear fractional functions, i.e., ratios of linear functions. Generally, a Laplace transform transfer function of a physical linear system can be presented as Δ/Δ_o, where the main determinant Δ and the minor Δ_o are linear functions of the value of an element of the system (e.g., for an electrical system, resistance of a resistor, capacitance of a capacitor, inductance of an inductor, gain coefficient of an amplifier). The proof can be found in [6].

For example, a return ratio is a linear fractional function of the feedback system load impedance Z_L. This function can be expressed as:

$$T(Z_L) = \frac{Z_o T(0) + Z_L T(\infty)}{Z_o + Z_L}$$

and can be used to analyze the effects of the plant uncertainty.

A8.2 Bode Symmetrical Regulation

Bode regulators are called symmetrical with respect to a certain nominal value w_0 of the variable parameter w when the maximum relative deflections of $W(w)$ from $W(w_0)$, up and down, are equal (see Section 6.7.2), i.e., when the function of regulation must be:

$$Q^2 = \frac{W(\infty)}{W(w_0)} = \frac{W(w_0)}{W(0)} \tag{A8.1}$$

In A8.1 it is known that dependence of W on w is:

$$W(w) = \frac{w_1 W(0) + w W(\infty)}{w_1 + w}$$

By combining this equation and the equation from A8.1, we have:

$$w_0 = w_1 / Q$$

and

$$W = W(w_0) \frac{1 + \dfrac{w}{w_0} Q}{\dfrac{w}{w_0} + Q} \tag{A8.2}$$

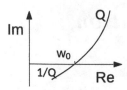

FIGURE A8.1
Frequency response of regulator.

This is illustrated in Figure A8.1. Regulators with such characteristics are called Bode variable equalizers.

The sensitivity of W to w in these regulators:

$$S = \frac{d \ln W}{d \ln w} = \frac{w_1 w}{w_1 + w} \frac{W(\infty) - W(0)}{w_1 W(0) + w W(\infty)} \tag{A8.3}$$

possesses an extreme at $w = w_0$, as can be seen by substituting (A8.1) into (A8.3).

Letting $W^* = W / W(w_0)$ and denoting the ratio:

$$\beta = \frac{w - w_0}{w + w_0} \frac{Q-1}{Q+1} = \frac{W^* - 1}{W^* + 1} \tag{A8.4}$$

the function W is expressed as:

$$W = W(w_0) \frac{1 + \beta}{1 - \beta} \tag{A8.5}$$

so

$$\ln W = \ln W(w_0) + 2 \operatorname{arctanh} \beta \tag{A8.6}$$

The Taylor expression for $\ln W$ gives:

$$\ln W = \ln W(w_0) + 2\left(\beta + \beta^3 / 3 + \beta^5 / 5 + \cdots\right) \tag{A8.7}$$

Truncated after the first term, it gives a very accurate approximation of $\ln W$ up to fairly large values of $|\beta|$. For example, with $|\beta| = 0.5$, the regulator constitutes 1 neper or 8.7 dB, and the maximum error due to omitting the second term of the series is only $(2/3)2 = 0.083$ neper or 0.72 dB.

The variable corrector comprising an auxiliary four-pole is depicted in Figure A8.2. Denoting the image parameters [24] of the four-pole as $g_c = a_c + jb_c$, z_{c1} and z_{c2}, the input impedance of it is given by:

$$W = z_{c1} \frac{1 + \dfrac{w}{z_{c2}} \coth g_c}{\dfrac{w}{z_{c2}} + \coth g_c} \tag{A8.8}$$

Equations (A8.2) and (A8.8) coincide if $w_{c1} = W(w_0)$, $w_{c2} = w_0$, and $g_c = \operatorname{arccoth} Q$. In this case,

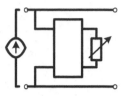

FIGURE A8.2
Four pole regulator.

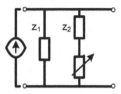

FIGURE A8.3
Simple *L*-type symmetrical corrector.

$$\frac{Q-1}{Q+1} = \exp\left(-2g_c\right) \tag{A8.9}$$

and β conforms to definitions of reflection coefficient between w and w_c. The regulation of W is then given by:

$$\ln\left|\frac{W(w)}{W(w_0)}\right| = 2\,\mathrm{Re}(\beta) = 2\exp(-2a_c)\cos(2b_c)\frac{w-w_0}{w+w_0} \tag{A8.10}$$

After selecting an appropriate structure for the auxiliary four-pole, the impedances for the four-pole branches may be found by successive approximation. The simplest *L*-type four-pole shown in Figure A8.3 can be designed in a straightforward manner. If w changes from 0 to ∞, and $w_0 = 1$, we normalize the impedances relative to intermediate impedance of the variable element, then:

$$Z_1 = Q - 1/Q$$

$$Z_2 = 1/Q$$

Most frequently, w is chosen to be real.

Appendix 9: Balanced Bridge Feedback

As is seen from Blackman's formula (Equation A6.9), each one of the following equations:

$$\left.\begin{array}{c} T(0) = T(\infty) \\ F(0) = F(\infty) \\ Z = Z_o \end{array}\right\} \tag{A9.1}$$

validates the other two. These are the conditions of a *balanced bridge*. In such a system the feedback is not dependent on the external impedance, and therefore the value of the feedback is not limited by the impedance variations.

An example of a balanced bridge circuit is shown in Figure A9.1. The Wheatstone bridge is balanced if $Z_a/Z_b = Z_c/Z_d$. In this case, the transmission from the output of the amplifier to each of the system output terminals is the same, the voltage across the bridge diagonal is zero, and connecting these terminals by any external impedance will not change the return ratio. Therefore, the first condition from Equation A9.1 is satisfied.

Zero transmission between the system's output terminals and the input of the amplifier is of similar use. An example can be drawn by inverting the direction of the amplifier in Figure A9.1. The desired output impedance of the amplifier (actuator) Z_d can be implemented by using a nested actuator feedback loop.

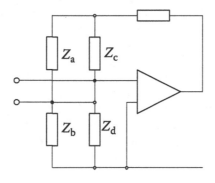

FIGURE A9.1
Balanced bridge feedback.

Appendix: Balanced Bridge Feedback

Appendix 10: Phase-Gain Relation for Describing Functions

For a system composed of linear links and a nonlinear nondynamic link, the relation between the gain and phase frequency responses $B(A)$ is the same as in linear systems if the DF (describing function) of the nonlinear link stays the same for all frequencies; i.e., the signal amplitude at the input to the nonlinear link does not depend on frequency. This is the case of the iso-w curves. When, however, the nonlinear link depends on frequency, the relation $B(A)$ becomes more complicated.

Assume that the conjunction of the filter is applicable to the problem considered. Let us determine the relationship between the gain and phase of the describing function of a nonlinear link:

$$\ln H = \ln|H| + j\phi = \ln U_2 - \ln U_1$$

Here, U_1 is the amplitude of the sinusoidal input signal, and U_2 is the complex amplitude of the output signal fundamental. The derivative is:

$$\frac{d\ln H}{d\ln U_1} = \frac{U_1}{U_2}\frac{dU_2}{dU_1} - 1 = \exp(\theta - \ln H) - 1$$

where the logarithmic transfer function for increments

$$\theta = A + jB = \ln dU_2 - \ln dU_1$$

and dU_1 is real. Therefore,

$$\frac{d\ln|H|}{d\ln U_1} = \exp(A - \ln|H|)\cos(B - \varphi) - 1$$

and, by applying the Bode relation to the differentially linearized stable m.p. link, we see that:

$$\cos\left[\varphi - \frac{1}{\pi}\int_{-\infty}^{\infty}\frac{dA(U_1, u)}{du}\ln\coth\frac{|u|}{2}d|u|\right] = \left(1 + \frac{d\ln|H|}{d\ln U_1}\right)\exp(\ln|H| - A) \qquad \text{(A10.1)}$$

Using Equation A10.1, the phase shift $\phi(U_1, \omega_1)$ can be calculated from measurements of the magnitude of the describing function, its derivative with respect to the input signal, and the frequency response of the real part of the differentially linearized circuit A. The latter must be measured while simultaneously applying to the link's input the main signal with amplitude U_1 and frequency ω_1. The magnitude A is easy to measure with a frequency-selective voltmeter if at this frequency no nonlinear product has a substantial amplitude. Since the power of the set of frequencies of these components is small in comparison with the power of all real frequencies, for the continuously differentiable function A (the differentially linearized circuit must be m.p. and not containing singularities on the $j\omega$-axis, which will be clearly seen in the results of the measurements), the integral in Equation A10.1 can be determined with sufficient accuracy from the measurement results of A on a limited number of frequencies.

Equation A10.1 is rather complicated. For practical design applications, iso-w responses are preferable and should be used whenever possible.

Appendix 11: Discussions

The function of the "Discussions" sections is to anticipate questions and objections, and also to address certain persistent misconceptions. In most cases, these are condensed transcripts of conversations that took place with the authors' colleagues and students, to whom we extend our sincere gratitude.

A11.1 Compensator Implementation

Q: It has been stated that modern technology somehow relates to high-order compensators. What does the word *modern* have to do with the compensator's order? Isn't it used here just as an advertisement gimmick?

A: It is more than a gimmick. The old way of compensator implementation used the same technology as that for the actuation. There exist mechanical, hydraulic, and pneumatic gadgets whose outputs are proportional to the integral or to the derivative of the inputs. These devices were combined to make the desired transfer function of the compensator. Since the devices were relatively expensive and rather difficult to tune, the compensators remained low order. The most complicated among them were the *PID* controllers using a proportional device, an integrator device, and a differentiator device, all connected in parallel.

The nature of the mechanical controller implementation impacted the control theory. Convenient methods were developed for low-order controller design; however, they are not suitable to handle high-order controllers.

On the other hand, feedback amplifier technology was not limited to low-order compensators since small capacitors, resistors, and chokes used in the compensators were cheap. The Bell Laboratories scientists developed synthesis theory for high-order compensators to approximate the theoretical best. These theory and design methods only partially propagated into the so-called *classical* automatic control.

The sensor technology changed completely in the 1980s and 1990s. Nowadays, in mechanical, pneumatic, and hydraulic systems the outputs of the sensors are electrical, the inputs of the actuators are electrical, and the compensators are electrical. Therefore, high-order compensators can be economically implemented in hardware or software, and the control can be made close to the best possible. Teaching control at the elementary level, however, lagged, still practicing some tools that had become not only obsolete, but plain counterproductive and confusing when used to try to make a precision controller design.

Q: The specifications on the closed-loop performance are often formulated in the time domain. Is it worthwhile to convert them into frequency-domain specifications and then design the compensator with frequency-domain methods?

A: Definitely. The benefits of frequency domain design far outweigh the trouble of converting the time domain specifications. The conversion of most practical specifications is simple and transparent.

Q: For initial design, is it advisable to optimize the performance while ignoring some details, like noise, nonlinearities, and plant parameter variations?

A: That sort of approach typically produces disastrous results in engineering practice. Even in the initial design, the system should be addressed in its entirety, or else the time and resources allocated for design will be wasted on dead ends.

A11.2 Feedback: Positive and Negative

Q: Does the minus sign at the summer mean that the feedback is negative?

A: No. At any specified frequency, the sign of the feedback also depends on the modulus and phase of the elements in the loop.

Q: From the transfer functions CAP and B, the closed-loop transfer function $CAP/(1+CAPB)$ can be calculated, and vice versa. Is that all there is to it?

A: It's not quite that simple. When the feedback is large, only the first calculation will be accurate. Even rather large errors in the knowledge of C, A, and P do not cause much uncertainty in the closed-loop transfer function. The result is that the inverse problem is ill-conditioned, with small errors in the observed or calculated closed-loop response possibly mapping into large errors in the calculation of the loop transmission. Therefore, the calculations or measurements of the closed-loop response are commonly inadequate to determine the loop transfer function.

A11.3 Tracking Systems

Q: Isn't a common definition of a tracking system that the feedback path transmission is 1?

A: $B=1$ is required for frequencies much lower than the crossover, and B can be identically 1 in a system with a prefilter. However, with $B=1$ and without a prefilter, the disturbance rejection and the compensator design need to be compromised to provide acceptable transient responses for the nominal plant. (This may be unavoidable for homing-type systems.)

A11.4 Elements (Links) of the Feedback System

Q: Should the feedback path and the compensator be analog or digital?

A: There are many factors to consider here. A digital compensator has the advantage that it can be reprogrammed in response to known changes in the plant; however, it should be pointed out that changing software is often neither trivial nor cheap. A primary disadvantage of digital compensators is that they reduce the available feedback when the computational delay in the loop is significant, as it often is. For this reason, analog compensators are the first choice for fast feedback loops. If necessary, the parameters of an analog compensator can be made "programmable" by employing multiplying D/A converters. The summing device and the feedback path can be analog as long as the required relative accuracy is not better than about 0.0001. To achieve much higher accuracy, they should probably be digital. Many other factors will affect this decision, including power requirements, cost, production quantity, and so forth.

Q: The compensator is optimized for disturbance rejection, and the feedback link (or prefilter) is designed to provide the desired closed-loop response with the nominal plant. But what happens to the closed-loop response when the plant deviates from the nominal?

A: A methodology has been developed, called quantitative feedback theory (QFT), that designs the system so as to satisfy the closed-loop performance specifications in the presence of the worst-case parameter variations. This procedure is discussed in Chapter 8. Although the methods involve a significant amount of calculation, they are useful when parameter variations are very large, i.e., more than 10 dB. For most applications, it is sufficient to first design the closed-loop response with the nominal plant, and then determine the changes caused by the plant variations and, if necessary, modify the prefilter.

Q: What is wrong with design methods that use the closed-loop response as the objective for the compensator design, such as the pole placement method?

A: With such methods, the appropriate value of feedback and the stability margins are not observed and preserved. This can cause many problems: the system stability may be only conditional, or wind-up might occur, the disturbance rejection may not be optimized, and the plant parameter variations may cause larger closed-loop response variations than what is achievable with feedback maximization methods based on the open-loop frequency response.

A11.5 Plant Transfer Function Uncertainty

Q: How accurate must the gain response approximation be for the accuracy of the related phase response to be 5°?

A: About 1/2 dB.

Q: The compensator transfer function must be high-order to approximate well the required transcendental response. But, the poles and zeros of a high-order transfer function may be very sensitive to the polynomial coefficients. Then, rounding errors will make the system not robust. Is this correct?

A: Not exactly. Variations in the values of poles and zeros are not important as long as the frequency response does not change much. The sensitivity of the frequency response to polynomial coefficients in typical cases is not high, and the compensator order is in this aspect irrelevant. For example, in analog telecommunication systems, equalizers have been routinely employed with 20th-order transfer functions, with the accuracy of implementation of the desired response of ±0.001 dB. The requirements for the control system compensators are much easier.

Q: What if the plant is non-minimum phase?

A: The non-minimum phase component of the loop phase lag must be compensated for by reducing the minimum-phase component of the phase lag. This can be done by reducing the frequency fb to increase the length of the Bode step. Simple formulas for these calculations are given in Chapter 4.

Q: The command, the prefilter, and the feedback path are known, and the sensor is measuring the plant's output. These data suffice to calculate CAP. Further, if C is known, AP can be determined exactly, and this information can be used to modify C such that the loop transmission function is as desired. Right?

A: Before deciding to do this, let's consider the errors in the calculation. The sensor is not ideal (noisy), the command is not well suited to the goal of characterizing the plant, and the problem is ill-conditioned for all frequency components where the feedback is large. This approach might be used for the frequency band where the feedback is positive, i.e., in the neighborhood of the crossover frequency f_b. However, for this application, the high-frequency components of the command need to be sufficiently large, larger than the noise, which is not usually the case.

Q: Can the effects of the plant resonances on the potentially available feedback be calculated in advance?

A: Sure. See Chapter 4, Problem 4 and the "Answer to the Problems" section.

A11.6 The Nyquist Stability Criterion

Q: Is the distance of the poles of the closed-loop transfer function from the $j\omega$-axis a better robustness measure than the stability margins on the Nyquist diagram?

A: No. A practical counterexample is an active RC notch filter. Its closed-loop poles are very close to the $j\omega$-axis, but the stability margins in this feedback system are sufficient; the system is globally stable, reliable, and widely used. Conversely, placing the poles of a closed-loop system far away from the $j\omega$-axis does not guarantee global stability, process stability, or robustness.

Q: Is it convenient to define the boundaries on the Nyquist diagram to reflect necessary and sufficient stability requirements?

A: Yes.

Q: How are the values for the stability margins determined?

A: Global and process stability must be ensured in the presence of the plant and actuator parameter variations. (Often, process stability is not strictly achieved, but the satisfaction of a certain norm on the errors in the nonlinear state of operation is acceptable.) Also, the sensitivity of the closed-loop response to plant parameter variations must be limited.

Q: Why should the area surrounding the critical point be defined by gain and phase margins, and not by a mathematically simpler circular boundary?

A: The shape of the margin boundary is defined by global and process stability requirements and by the plant parameter variations. In the majority of plants, variations of the plant gain and plant phase in the neighborhood of the crossover are not well correlated. This necessitates defining gain and phase margins independently.

Q: Is much performance lost if we simplify the design methods and use a circular stability margin boundary?

A: Changing the stability margins from those appropriate to circular would substantially reduce the integral of positive feedback in this frequency range, and therefore the integral of negative feedback in the functional frequency band.

Q: Is the definition of *Nyquist stable* the same as that of *conditionally stable*?

A: No. The Lyapunov definition for *conditional stability* is that the nonlinear system stability depends on the initial conditions. A Nyquist-stable system with the usual actuator saturation nonlinearity may be conditionally stable but can be rendered globally stable by the addition of nonlinear dynamic compensation.

A11.7 Actuator's Output Impedance

Q: Whatever the internal (output) impedance of the actuator, we can always apply to its input such a signal that the system output will be as required. Therefore, it doesn't matter what the actuator output impedance is, right?

A: Wrong. For the nominal plant, the statement is right, but the effects of the plant parameter deviations from the nominal are to a large extent dependent on the actuator output impedance.

Q: Compound feedback produces some finite output impedance for the driver. However, this can be achieved more simply by making voltage feedback and placing a resistor in series.

A: The result will be the same as long as the system remains linear. This method can be used in the laboratory for testing various control schemes. However, there will be power losses on this series resistor, and as a result, the actuator needs to be more powerful, with a larger saturation level.

Q: But, isn't the power dissipated anyway on the internal impedance of the actuator when the impedance is made finite by compound feedback?

A: No. Compound feedback causes no power losses and doesn't reduce the actuator efficiency.

A11.8 Integral of Feedback

Q: At what frequencies is the feedback positive?

A: The feedback is positive within the circle of unit radius centered at $(-1,0)$ on the T-plane. When the loop gain decreases monotonically with frequency, the major part of the area of positive feedback falls in the band $0.6f_b$ to $4f_b$. The integral of feedback at much higher frequencies is negligibly small.

Q: Well, the feedback makes the system better at some frequencies, but at others even worse than it was before. So, what is gained from the application of feedback?

A: Negative feedback is used to cover the frequency range where the plant parameter variations and the effects of disturbances are critical, and positive feedback is confined to the frequency range where the noise and the disturbances are so small that even after being increased by the positive feedback, they still will be acceptable.

A11.9 Bode Integrals

Q: How many Bode relations are there, in total?

A: There are about 30 different integral relations in Bode's book.

Q: Have new relations or expansions of these relations been made since then?

A: Yes. The most important are the R. Fano expansions of the Bode integral of the real part, which use not one but several terms in the Laurent expression. They are usable for *RF* and microwave circuits design. Also, Bode relations were expanded on unstable systems, discrete systems, multivariable systems, parallel paths of the signal propagation, and the transfer function for disturbance isolation. An example of an expansion on certain nonlinear systems is given in Appendix 10. However, these expansions are less important for practice, compared with the four fundamental Bode relations described in this book.

A11.10 The Bode Phase-Gain Relation

Q: Why is the slope of the Bode diagram expressed here in dB/octave, not in dB/decade?

A: The same reason that supermarkets sell milk by gallons, not barrels. Gain samples one octave apart typically suffice to calculate the phase response with less than 5° error as needed for sound feedback system design. And it is certainly convenient that the slope of a well-designed Bode diagram is nearly −10 dB/octave.

Q: Does it make sense to consider a loop gain response having a constant slope of −10 dB/octave? How can such a response be implemented in a physical system?

A: It can be closely approximated by a rational function of s.

Q: Are the Bode integrals applicable to transcendental plant transfer functions?

A: Yes.

Q: At a university where I have taught, the teaching relies heavily on the root locus method. Afterward, I started to work in the aircraft industry and was surprised that here the engineers prefer Bode diagrams. Why is this so?

A: Bode noticed that for an engineer striving to make the best of the design, it is difficult to cope with three scalar variables: gain, phase, and frequency. So he *reduced* the number of the variables to only *two scalars*: gain and frequency. This approach allows handling high-order systems (including MIMO and multi-loop systems). In contrast, applying the root locus method to an *n*th-order system *increases* the number of the variables to *n complex variables* (roots) crawling all over the *s*-plane in a strange, threatening, and unmanageable manner.

A11.11 What Limits the Feedback?

Q: What are the physical factors that limit the feedback?
A: The major factors are the uncertainties in the plant transfer function, the clipping of higher-frequency components of the sensor noise in the actuator, and the level of noise in the system's output.
Q: Does it pay to exploit as much as possible the available knowledge about the plant when designing the control law?
A: Certainly. This increases the available feedback. The available feedback is infinite for hypothetical full-state feedback, and no feedback is available for a completely unpredictable plant.
Q: A typical concern of the project manager: the system design trade-offs require knowledge of the available performance of its subsystems, and these subsystems often employ feedback. How can the system trade-offs be made without designing the feedback subsystems?
A: The manager should request data about available subsystem performance, i.e., the performance that can be achieved with the optimal compensator. Using the Bode approach, this can be calculated without actually designing the compensator.
Q: What kind of data do we need for this estimation? What kind of mathematics is involved?
A: The following data are generally required: the sensor noise spectral density over three octaves in the vicinity of the planned feedback bandwidth, the available (nondistorted) output signal amplitude from the actuator, ranges of frequencies where plant structural modes can fall, up to four octaves over the estimated feedback bandwidth, and sometimes more. This sort of data is typically available; although, it may not always be very precise. The available disturbance rejection can then be estimated using the Bode integrals.

A11.12 Feedback Maximization

Q: I want to develop and market a product. How can I be sure that my competitor will not enter the market, soon after me, with a superior product that uses the same actuator, plant, and sensor, but has bigger and faster feedback?
A: The Bode integral approach allows the determination of the best theoretically available system performance.

Q: Is it necessary to approximate a physical plant transfer function (which is sometimes transcendental) by a rational function or to perform plant "model reduction" when designing a feedback system with the Bode method?

A: No. The calculated or measured plant and actuator transfer functions are subtracted from the desired loop response to obtain the desired compensator response, and the rational function approximation is introduced only in the final stage of the compensator design.

Q: What is the required accuracy of the approximation of the desired loop gain response?

A: The accuracy must be such that the related phase response will approximate the desired phase response with an accuracy of about 5°.

Q: Why 5°?

A: In this case, the *average* loop phase lag must stay 5° away from the stability margin boundary. Then, the average slope of the Bode diagram will be less than the maximum acceptable by 1/3 dB/octave (from the proportion: 90° for 6 dB/octave). Therefore, over each of the three octaves of the cut-off, 1 dB of feedback is lost. Typically, these losses are marginally acceptable. The small remaining possible feedback increase may not justify a further increase in the complexity of the compensator.

Q: How accurate must the gain response approximation be for the accuracy of the related phase response to be 5°?

A: About 1/2 dB.

Q: The compensator transfer function must be high order for accurate approximation of the required transcendental response. But, poles and zeros of a high-order transfer function may be very sensitive to the polynomial coefficients. Then, the rounding errors will make the system not robust. Is this correct?

A: Not exactly. Variations in the values of poles and zeros are not important as long as the frequency response doesn't change much. The typical compensator response has no sharp peaks and notches; the sensitivity of such a function modulus to the polynomial coefficients is typically less than 1, and the compensator order is in this aspect irrelevant. Therefore, in order for the compensator to be accurate within 0.5 dB, i.e., 6% in the magnitude of the transfer function, the polynomial coefficient accuracy need not be better than 1–5%, and the effect of rounding is insignificant.

Q: What is the best available accuracy of analog compensators?

A: Equalizers for analog telecommunication systems have been routinely designed with up to a 20th-order transfer function, with an accuracy of implementation of the desired response of ±0.001 dB.

Q: How were these equalizers designed?

A: By interpolation, by cut-and-try procedures, by using asymptotic Bode diagrams, by adjustments in the element domain, by using Chebyshev polynomial series, by using the second Remez algorithm, and by using the Simplex method.

Q: What if the plant is non-minimum phase?

A: The non-minimum phase component of the loop phase lag must be compensated for by reducing the minimum phase component of the phase lag. This can be done by reducing the frequency f_b to increase the length of the Bode step. Simple formulas for these calculations are given in Chapter 4.

Q: Only analog systems have been discussed. What is the difference with regard to digital controllers?

A: Sampled-data systems, which include an analog-to-digital converter, a digital filter, and a digital-to-analog converter, are time-variable and have an inherent delay that reduces the available feedback.

Q: What about a proportional-integral-derivative (*PID*) compensator having the transfer function $C(s) = k_p + k_i/s + k_d s$?

A: With appropriate gains, the *PID* compensator roughly approximates the response of the optimal controller; however, the approximation error can be quite large. *PID* control falls short of the performance of a system with an appropriate high-order compensator.

Q: But how much better off than a *PID*'s is the optimal controller?

A: It depends on the frequency response of the plant transfer function and the disturbance spectrum density. For smooth plant responses, the expected improvement in disturbance rejection is 5–10 dB when a *PID* controller is replaced by a higher-order controller. For a plant with structural resonances, especially when it is used in conjunction with a nonlinear dynamic compensator, a 5–30 dB improvement can be expected from the high-order controller.

Q: Why aren't the Ziegler-Nichols conditions for tuning *PID* controllers presented in this book?

A: This method does not use a prefilter (or command feedforward), and is suitable only for specific plants. More general and better results can be achieved with the frequency-domain approach.

A11.13 Feedback Maximization in Multi-Loop Systems

Q: What about loop response shaping for multi-loop systems?

A: Basically, the technique is the same. Multi-loop systems can be designed one loop at a time, as long as loop coupling is taken care of by adjusting the responses and providing extra stability margins within certain frequency bands.

Q: Does this technique work well even when the number of the loops is large?

A: Yes, typically.

A11.14 Non-Minimum Phase Functions

Q: Is there a case where it is advantageous to apply compensators with non-minimum phase transfer functions?

A: Probably not. Sometimes, when the plant has high-frequency structural modes, a compensator must introduce a phase delay in the loop to make the system stable. However, this phase delay can always be achieved with a minimum-phase low-pass filter-type function—with the extra advantage of increasing the amplitude stability margin and reducing the noise effect at the actuator's input.

A11.15 Feedback Control Design Procedure

Q: Isn't it awkward to go back and forth from the Nyquist diagram on the *L*-plane to the Bode diagram? Isn't one type of diagram enough?

A: It's a case of the right tool for the right job: global and process stability characterization, stability margin definitions, and nonlinear dynamic compensation design all require the use of the *L*-plane, but the trade-off resolution and the compensator design are simplified by using the Bode diagram due to a reduced number of variables.

Q: Do the design phases discussed in Appendix 12 apply for the design of single-loop or multi-loop systems?

A: Single-loop. But multi-loop system design is a natural extension of this procedure.

Q: Still, if you have to design, say, a three-input, three-output system, will it take 3, 9, or 27 times longer than a single-loop system of comparable complexity for each channel?

A: Three times longer, typically.

A11.16 Global Stability and Absolute Stability

Q: Is it convenient to use Lyapunov functions in the design of high-performance control systems?

A: Devising appropriate Lyapunov functions is yet unmanageable for systems with a high-order linear part and several nonlinear elements.

Q: At the 1st Congress of the International Federation of Automatic Control (IFAC) in 1960 in Moscow, researchers from Russian academia presented important theoretical results using time-domain and state-variable methods. This research impressed American professors attending the conference and shifted to a large extent the direction of research in American academia. Were these methods the ones Russian engineers employed to design control laws for their rockets and satellites?

A: No. Russian rockets' control systems have been designed with frequency-domain methods in very much the same way as American rockets (see Section A13.9).

A11.17 Describing Function and Nonlinear Dynamic Compensation

Q: DF analysis fails to predict instability in certain systems. Would it be right, therefore, to discard the DF approach altogether?

A: No. It would be foolish to discard a Phillips screwdriver just because it cannot drive all screws. Similarly, although DF analysis is not a universal, foolproof tool, it is fairly accurate when employed for the analysis and synthesis of *well-designed* control loops with steep monotonic low-pass responses.

Q: What is the purpose of using DF now, when nonlinear systems can be simulated well with computers?

A: The DF advantage is seen when it is used not for the purpose of analysis only, but for the purpose of design, and especially conceptual design of control systems with several nonlinear links. In many cases, the accuracy of the analysis need not be high. The phase of the DF has an error of up to 20° compared with the phase calculated with exact analysis, but this phase is not of critical importance since NDCs can easily provide the required phase advance.

Q: DF analysis does not account for additional phase shifts created by higher harmonics interacting in the nonlinear link. So does the DF design guarantee robustness?

A: Interaction of high-order harmonics can typically yield up to 15–20° of extra phase lag for the fundamental. The NDC DF phase advance must be increased by this amount, which is rather easy to do since the typical NDC phase advance exceeds 100–200°.

Q: How much improvement in disturbance rejection can be expected from using an NDC?

A: It depends on the desired frequency shaping of the disturbance rejection—in some cases, up to 30 dB.

Q: Is it advantageous to use NDCs in MIMO systems?

A: Yes. In addition to providing the same advantages as for SISO systems, NDCs reduce the effects of nonlinear coupling between the loops on the system stability.

A11.18 Multi-Loop Systems

Q: Why is it the number of saturation elements that matters when defining what the multi-loop system is?

A: Because the DF of a saturation link changes from 1 down to the inverse of the loop gain coefficient, i.e., 100 or 1,000 times. This effect needs to be accounted for during the stability analysis and synthesis of the loop transfer function. Compared with this effect, even the effects of the plant parameter variations are small.

A11.19 MIMO Systems

Q: Where is the better place to implement the decoupling matrix—in the forward path or in the feedback path?

A: In the forward path, since in this case the decoupling matrix doesn't need to be precise. In this case, commands are formulated in the sensors' readings. When commands should be formulated in actuators' actions (and sensors are not aligned with the actuators), the decoupling matrix should be placed in the feedback path.

Q: Is the design of MIMO and multi-loop systems difficult?

A: It is not simple, but several often-ignored techniques greatly simplify the design:

The Bode technique of making trade-offs between the loops without designing the compensators.

Verifying the m.p. character of parallel signal propagation paths.

Considering the effects of loop coupling only in the narrow-frequency ranges where the coupling is critical.

Using appropriate output impedances for the actuators.

Using NDCs to provide global stability and good performance in the nonlinear mode of operation, without sacrificing the disturbance rejection.

Q: How does coupling between the loops affect the system stability and robustness?

A: The effects of loop coupling can be analyzed by considering the change in one loop transfer function caused by the plant and the actuator transfer function variations in another loop. The robustness can be provided for by correspondingly increasing the stability margins.

Q: At what frequencies is the coupling most dangerous?

A: In the nonlinear mode, at lower frequencies, where the loop gain is large and changes in it caused by saturation could affect other loops. However, this can be taken care of by using an NDC. In the linear mode, the coupling is more dangerous near the crossover frequencies where the feedback in the loops is positive.

A11.20 Bode's Book

Q: In the bibliography, the book is characterized as a book on synthesis, and not on analysis. However, Bode himself named it the book on analysis. Why?

A: The book describes the tools for analysis also, but mostly for synthesis. The choice of tools for these two purposes are, generally speaking, quite different.

Q: What about the terminology in Bode's book?

A: Bode developed several powerful approaches valid for a wide range of applications, and he optimized the terminology to serve this wide range of applications.

For example, in his definition the transfer function is the ratio of the input to the output, and the sensitivity is the ratio of the relative change of the cause to that of the result. These definitions allowed integrating feedback theory with driving point function analysis and synthesis.

Following these definitions, he defined certain functions as minimum-phase—the functions having the smallest phase shift among all transfer functions with the same gain response. Today, however, we use the inverse definition of the transfer function, which is, maybe, intuitively better for the narrower area of applications, but not as elegant and general. With the definition employed today of the transfer function as the ratio of the response to the source, m.p. function has in fact *maximum* phase shift. To bring some sense to this, an m.p. function could be called a minimum *phase lag* function, but still, in most books it is inconsistently called a minimum-phase function.

Bode didn't use the word *compensator*. Instead, as a telecommunication engineer, he used the word *corrector* for a feedback loop or for any other signal transmission channel. The meaning of this word is: we know the theoretically best response, we have measured the response (of the loop) as is, and now we have to correct the response for it to be as prescribed by the theory, to be optimal.

Bode used separate definitions for the sensitivity and the return difference, while in some contemporary works the return difference is formally defined as "sensitivity," which may be acceptable if we limit the use of this term to a very narrow class of problems, too narrow even for control engineers.

Bode formulas have been later modified and generalized for the purposes of certain research goals and for the convenience of teaching. Nonetheless, the authors of the *Classical Feedback Control* found no versions in the existing literature better or equally usable for engineering design purposes than those employed in Bode's book. This is why *Classical Feedback Control*, with very few exceptions, uses original Bode notations, definitions, and expressions.

A11.21 Plant Uncertainty in a Nonlinear Multi-Loop System

Q: While studying these systems, the frequency plant gain must be increased in the working band—but stability needs to be preserved. Is this possible, according to Nyquist margins?

A: Yes, by the estimation of internal gain, as shown in Chapter 14.

Q: Is an elementary design procedure created for this frequency maximization?

A: Yes. In a single-loop system, the frequency gain is maximized with a simple condition $f_u/f_b \approx 2.2$. Here the frequency f_b is equal to the return ratio of a single-loop system with $|T| = 0$. The frequency f_u is the lowest frequency of the loop uncertainty. The plant gain at low frequency must be limited by an NDC with nonlinear phase less than nearly 90°.

Q: What are the conditions for the feedback f_b in a multi-loop system to be maximized?

A: The plant loop gain must be increased to 20 dB to 30 dB relative to single-loop system; the internal loops must be higher than the plant loop; the frequency of crossing the internal and plant loops must be $=f_u/2$; the gain frequency response in nonlinear mode, with the help of NDC, must be higher than Bode's; the nonlinear dynamics must be stable in linear mode; the loops must guarantee their minimum-phase responses.

Appendix 12: Design Sequence

In the following, an estimate of the average time is given for completion of each of the steps by an experienced designer, and for a problem that is not unusually complex.

Phase I: Calculation of the available performance (up to 24 h)

Determine the feedback bandwidth f_b (about 30 min).

Decide what kind of nonlinear dynamic compensation (if any) will be employed, and define the stability bounds on the L-plane (5 min).

Draw the desired Nyquist diagram on the L-plane, which maximizes the feedback (5 min).

Find and draw the desired Bode diagram (5 to 30 min).

Make all trade-offs required by the frequency responses of the disturbance spectra, by other feedback loops in the system, and by the plant transfer function variation; reshape the Bode diagram (10 min to 22 h); determine the available feedback; and report the system performance to your manager.

Phase II: Design of the linear compensators and prefilters (up to 5 days)

Given the nominal plant response, or a plant model, subtract the nominal plant gain response from the desired nominal loop gain response to obtain the required compensator gain response (10 to 30 min).

Approximate the required compensator response using one of the methods discussed in Chapter 6 and obtain the compensator transfer function and specifications for the compensator hardware/software (a day or two).

Make a linear system model and calculate the closed-loop frequency response (several hours).

Decide on the type of nominal closed-loop gain frequency response that is close to the optimum (2 min to several hours).

Subtract the calculated closed-loop gain response from the desired response to obtain the prefilter gain response (10 to 30 min). Or, alternatively, find the desired response for the feedback path, B.

Find the prefilter (or feedback path link) to approximate the required response (10 min to 3 hours).

Determine the frequency and time responses for the linear system (1 to 24 h).

Phase III: Nonlinear compensator design and time response simulations

Create a nonlinear model for the system and find the responses for large-level signals (nonlinear mode) (up to 10 days).

Introduce nonlinear elements in the compensator or design a specialized nonlinear dynamic compensator, and test and tune the system performance (up to 30 days).

Subtotal time for phase III: up to 40 days

Note that only phase I is in the critical path for the global engineering system design (of which your feedback system is a subsystem). Phase I might be repeated many times for different configurations of the global system before a final decision is made, so it is important that this procedure be short, simple, and reliable.

Appendix 13: Examples

In addition to the examples in the book's chapters, several examples are given here of single-loop and multi-loop systems designed with the frequency-domain approach.

A13.1 Thrust Vector Control for the Cassini Spacecraft

Figure A13.1 depicts the Thrust Vector Control algorithm for the Cassini spacecraft, which controls the orientation during maneuvers by gimballing the engine. This is a continuous representation of a digital system with a 125 ms sample time. The sensors and actuators are flat out to a few Hz, and the 4 Hz Nyquist suggests that the system is equipped to operate up to 0.4 Hz or so. This is complicated, however, by the lightly damped magnetometer boom bending mode near 0.7 Hz.

The spacecraft model is broken out in Figure A13.2(a). In addition to the double-integrator are structural flexibility modes for the magnetometer boom and the RPWS antennas, and two slosh modes for the bipropellant. The zero frequencies are less than the poles making these modes stably-interacting, but the phase situation starts to deteriorate near the magnetometer boom bending frequency, with losses resulting from sensor-actuator dynamics, the need to derive rate by back-differencing position, and the assumed computation delay of a full sample. This situation prompted a decision to gain stabilize the magnetometer boom.

Another complication for the controller is that the maneuver execution pointing error is the attitude error and also the gimbal angle that is necessary to compensate for the center of mass offset error which is depicted in Figure A13.1. The guidance filter shown in the controller Figure A13.2(b) feeds back the combined angle with a filter, which eliminates the steady-state error. (Additional integral control was not necessary in this case.) The limiter provides global stability, similar to an anti-wind-up device used with integral control.

The compensator consists mostly of complex poles and zeros, and was digitized in stages using the bilinear transform. The L-plane Nyquist diagram is shown in Figure A13.3. (This was generated using the Linear Analysis function in Simulink with input–output indicated by the x's in Figure A13.1.) This high-order compensator provides the sharp corners around the stability boundaries and the steep gain roll-off to the magnetometer boom bending frequency. The design stability margins are ±6 dB and 30°. The design accommodates worst-case assumptions for the frequency and damping of the magnetometer boom, which was ultimately limiting the performance.

The figure also includes an additional curve for the inverse describing function that represents the small amount of freeplay in the main engine gimbal assembly as well as the hysteretic effects of the propellant flex lines. The intersection occurs near 0.03 Hz resulting in small stable limit cycle behavior during long maneuvers.

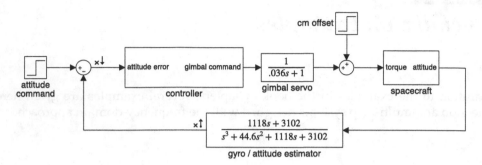

FIGURE A13.1
Cassini TVC algorithm.

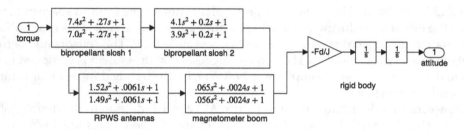

FIGURE A13.2A
Cassini TVC—Spacecraft model.

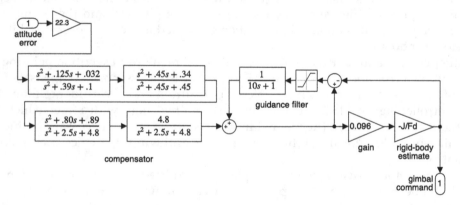

FIGURE 13.2B
Cassini TVC—Controller.

A13.2 Scanning Mirror of a Mapping Spectrometer

The mapping spectrometer of the Cassini spacecraft is an instrument producing an object image in the light of specified wavelengths. The spectrometer's scanning mirror is attached to a yoke suspended on a flexure, as shown in Figure A13.4.

The yoke is rotated by a solenoid actuator. The mirror angle is measured by a linear voltage differential transformer (LVDT). The plant's mechanical schematic diagram in Figure A13.5 shows the cores of the solenoid and the LVDT, and the mirror characterized respectively by the moments of inertia J_S, J_M, and J_{LVDT}, the stiffness and the damping in the

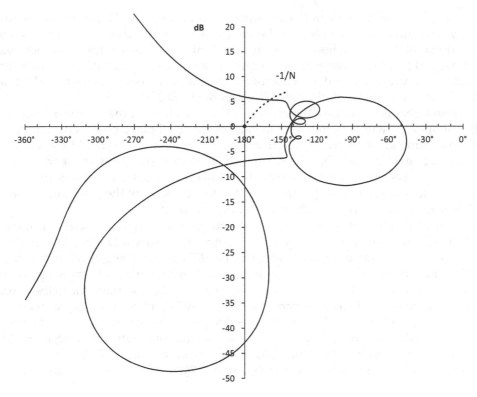

FIGURE A13.3
Cassini TVC—*L*-plane Nyquist.

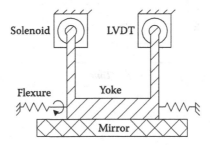

FIGURE A13.4
Scanning mirror suspension with actuator (solenoid) and sensor (LVDT).

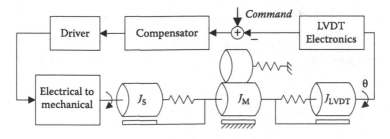

FIGURE A13.5
Mechanical schematic diagram for scanning mirror positioning with feedback control.

yoke's arms, and the suspension flexure stiffness and damping. The analytical model of the system dynamics was derived and the system was simulated as a block diagram with unidirectional blocks. Later, however, for the convenience of the analog controller design, the mechanical plant was modeled as an equivalent electrical circuit (the response for the electrical model was compared with that of the mechanical model and found to be exactly the same) and integrated with the controller's model in SPICE.

The control goal is to increase the exposure time, i.e., to scan the mirror in such a way that the light beam will remain longer in each of the focal plane photocells, with a staircase-type command as shown in Figure A13.6. The mirror angle should track the command with a rise time of less than 2 ms. Thus, the feedback bandwidth must exceed 1 kHz.

The size of a pixel, angle-wise, is 0.9 mrad. The plant parameter variations are such that for the beam to stay reliably within the pixel, the feedback about the plant must exceed 60 dB at lower frequencies. The overshoot must not exceed 30%.

The feedback bandwidth is limited by the structural resonance of the solenoid core on its arm. The nominal resonance frequency is 6 kHz with possible deviations of ±20%. The damping of the resonance is known with the accuracy of ±20% (for damping, this is excellent accuracy). Therefore, the maximum plant gain in the 6 kHz neighborhood is known pretty well, with the accuracy of ±2 dB. The plant phase, however, is quite uncertain due to the effects of the LVDT resonance (at a higher frequency) and the stray inductances and capacitances in the coils. Therefore, the loop must be gain stabilized at the frequency of the solenoid core resonance; i.e., the gain at the resonance must be below the amplitude stability margin of –10 dB.

The plant and controller simulation has been performed in SPICE. The chosen stability boundary on the *L*-plane is shown in Figure A13.7(a) and has a 45° phase margin. The

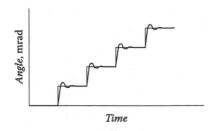

FIGURE A13.6
Closed-loop transient response of the scanning mirror.

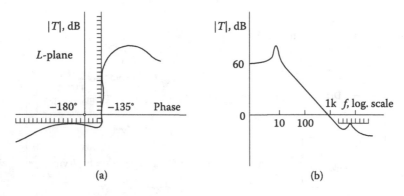

FIGURE A13.7
(a) Nyquist and (b) Bode diagrams for scanning mirror control.

Bode diagram for the design is shown in Figure A13.7(b). Here is seen the resonance of the mirror on its suspension at 8 Hz, and the solenoid core resonance on its arm at 6 kHz. The resonance frequency of the LVDT core on its arm is higher than 10 kHz.

It is seen that variations in the frequency of the plant resonances will not cause the system to oscillate. The stability margin of 45° is preserved over all frequencies where the loop gain is more than −10 dB.

The feedback path transfer coefficient (B) was chosen to be a constant. The response of a system with such a B to a small-amplitude step input is shown in Figure A13.8(a). The overshoot is about 30%, which is marginally acceptable. The overshoot can be efficiently reduced (to about 10%) by putting either a lead filter in the feedback path or an equivalent low-pass prefilter. However, this was not done since the actuator is not intended to operate in the linear mode.

It is economical to employ a small actuator whose maximum available torque is a small fraction of the maximum torque that would be seen in an idealized, linear system (without saturation).

Due to the 45° phase stability margin, large wind-up should not be expected. Therefore, application of an NDC is not required. This was confirmed by computer simulation of the output response to the step command with different actuator saturation thresholds. In Figure A13.8, curve (a) relates to the linear system and curve (b) to the threshold five times smaller than the maximum torque calculated for the linear system. Even with this small-output power actuator, the output transient response is quite good, and the overshoot is reduced.

Because of the rather large feedback bandwidth, small production quantities, and required low power consumption, it was decided to make the entire loop analog. The system block diagram and the simplified schematic diagram of the compensator are shown in Figure A13.9.

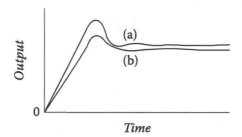

FIGURE A13.8
Closed-loop transient response (a) without and (b) with saturation in the actuator.

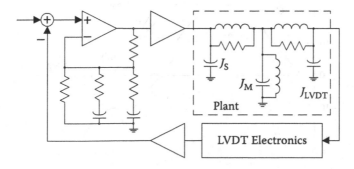

FIGURE A13.9
Analog controller and plant model for scanning mirror (simplified).

A13.3 Rocket Booster Nutation Control

It was found that the nutation angle of a rocket booster shown in Figure A13.10 increases exponentially during the main engine thrusting. To fight this, small auxiliary thrusters were employed to produce stabilizing moments. The thrust is pulse-width modulated (with the carrier frequency of 10 Hz) by a feedback loop using a gyro as a sensor. Since the nutation frequencies are in the range of 0.2–0.8 Hz, the feedback loop is a band-pass one.

It was shown that the larger the feedback, the smaller is the real part of the nutation exponent. The loop gain and phase shift frequency responses of the controller's original version are shown by curves (o) in Figure A13.11. The control loop compensator consists of a two-pole low-pass filter having a corner frequency of 4 Hz and a high-pass filter with a corner frequency of 0.1 Hz. The low-pass filter reduces the loop gain at the 10 Hz carrier frequency to the acceptable level of –27 dB. The high-pass filter attenuates dc and low-frequency components of the error to prevent them from overloading the thrusters. The filters are made analog as a cost-effective solution to low-quantity production.

The loop responses with the modified higher-order compensator, employing a notch filter to attenuate the carrier instead of the low-pass filter, are shown by the curves marked m. Both the feedback over the range of the nutation frequencies and the amplitude stability

FIGURE A13.10
Rocket booster nutation.

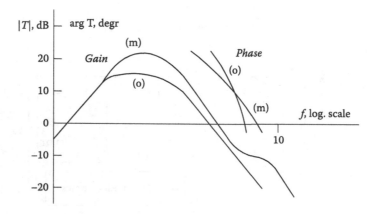

FIGURE A13.11
Bode diagrams for nutation controller.

margin are increased. The controller is globally stable; however, a substantial jump resonance could be expected because of the phase stability margin of only 30° (the same as in the older system). Although the process instability might not create a problem, it is safer to eliminate even the possibility of jump resonance and wind-up by introducing an NDC whose frequency response for small-signal levels is shown in Figure A13.12(a). The NDC schematic is shown in Figure A13.12(b). For large signals, the diodes are open and the NDC gain is 0 dB.

Computer simulations showed that after 12 seconds of thrusting, the nutation amplitude in the modified system was 50 times smaller.

A13.4 Telecommunication Repeater with an NDC

Linearity of repeaters for telecommunication systems with the frequency division multiplex should be very good to reduce the channel intermodulation to an acceptable level. The linearity of this repeater designed in Russia in the 1960s is comparable to the linearity of a concurrent American counterpart, despite the fact that the bandwidth of the transistors was smaller and the linearity of the transistor in the ultimate stage was much worse than that of the special transistors developed in the Bell Laboratories. This was achieved by the increase in the feedback obtained by using a Nyquist-stable system with the Bode diagram shown in Figure A13.13, upper curve.

The system global stability was provided by a nonlinear dynamic local feedback loop in the penultimate stage. The simplified schematic of the amplifier with the NDC is shown

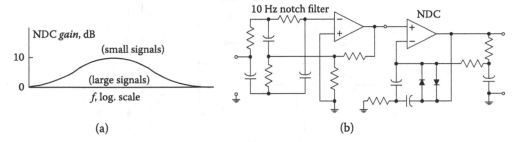

FIGURE A13.12
Nonlinear dynamic compensator: (a) gain responses and (b) schematic diagram.

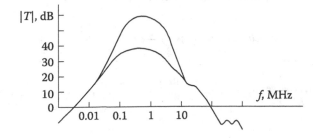

FIGURE A13.13
Bode open-loop diagrams for a telecommunication repeater feedback amplifier.

in Figure A13.14. The diodes play the role of a dead-zone link. When the signal amplitude exceeds the diodes' threshold, they start conducting and introduce negative feedback over the functional bandwidth, thus changing the gain in the common loop as shown by the lower curve in Figure A13.13. This Bode diagram is less steep and the related phase lag is smaller, so that according to the DF method, further saturation in the ultimate stage does not cause periodical oscillation.

Many initial conditions were tried during experiments and none of them elicited oscillation. The repeater was put in mass production in large quantities and no stability problems were ever observed.

A13.5 Attitude Control of a Flexible Plant

Consider the problem of pointing a spacecraft using a reaction wheel that resonates on a flexible shaft at 30 Hz in the system shown in Figure A13.15. The reaction wheel is rotated by the motor M.

We consider here only the planar rotation about the center of rotation for the system. The moment of inertia of the spacecraft is $J_{SC} = 100$ kgm^2, and of the reaction wheel, $J_R = 1$ kgm^2.

The amplitude stability margin was chosen to be 8 dB. The gyro noise limits the feedback bandwidth to less than 10 Hz. The flexible appendage with the moment of inertia $J_A = 5$ kgm^2, resonating somewhere in the 13–29 Hz region (nominally at 15 Hz) with very low damping, produces an area of uncertainty on the loop gain and phase responses, which

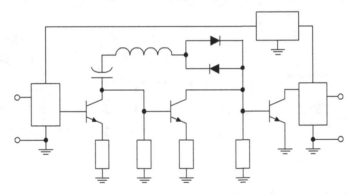

FIGURE A13.14
Nonlinear dynamic compensation in a telecommunication feedback amplifier.

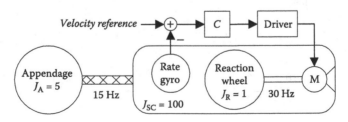

FIGURE A13.15
Attitude control of a flexible plant.

further limits the feedback bandwidth, as illustrated (not to scale) in Figure A13.16(a). The loop at the resonance must be gain stabilized. The gain peak depends on the value of the output mobility of the actuator.

A local feedback loop about the driver (not shown in Figure A13.15) makes the actuator output mobility equal to 0.03 (rad/s)/Nm. This reduces the peak-to-notch swing from 54 dB to 28 dB, compared with the case of the actuator being a pure torque source. Correspondingly, the available feedback bandwidth increases by almost an octave. The resulting Bode diagram is shown in Figure A13.16(b).

A13.6 Voltage Regulator with Main, Vernier, and Local Loops

The power supply voltage for a klystron transmitter is produced by a motor-generator whose output is rectified and passed through a low-pass filter. The −20 kV output voltage is the difference between the −22 kV at the filter's output and the 2 kV voltage drop on the bypass tube. The output voltage is regulated by (1) changing the current in the field winding of the generator (main actuator) and (2) changing the grid bias of the bypass tube (vernier actuator), as shown in Figure A13.17. The goal (the reference) for the main (slow) loop is to keep the voltage on the bypass tube at 2 kV, on average, for the tube to be operational. The goal (the reference) for the fast vernier loop is to fine-tune the voltage drop on the bypass tube for the regulator output voltage to be −20 kV (or any other specified value within the 0 to −40 kV range).

The range of the output voltage of the generator/rectifier/filter subsystem is from 0 to −42 kV. The feedback bandwidth in the main loop is limited to 10 Hz by the time constant of the generator, the maximum voltage from the power amplifier that supplies the current to the field winding of the generator, and the noise of the voltage divider and the first stage of the amplifier in the compensator.

The vernier loop bandwidth was chosen to be 60 Hz to achieve the required disturbance rejection. This regulator can only regulate within the range of 4 kV so that its average output voltage needs to be maintained at the 2 kV level (to be de-saturated) by the slower main loop.

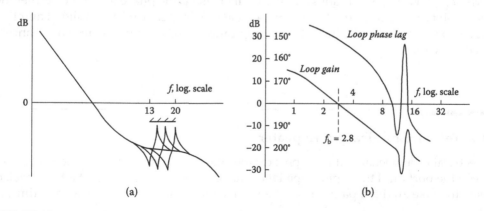

(a)   (b)

FIGURE A13.16
Bode loop gain diagrams (a) for the flexible plant control with uncertain resonance frequency, not to scale, and (b) for the nominal plant.

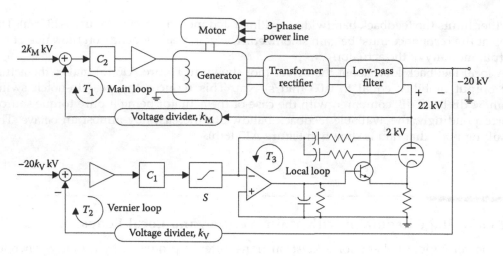

FIGURE A13.17
High-voltage power supply controller.

The plant is high order and nonlinear (because of the nonlinearity of the magnetization curve of the field winding). The loop compensator's order is rather high. The Bode diagrams and the Nyquist diagrams for the main loop return ratio T_1 and the vernier loop return ratio T_2 are presented in Figure A13.18.

Experiments showed that the system was asymptotically stable following any initial conditions that take place in practice. However, initially, the saturation link S was not included in the system, and a substantial jump resonance was observed. Analysis and experiments indicated that the reason for the jump resonance was a nonlinear interaction between the vernier loop and the local loop T_3.

This local loop is about the high-voltage transistor driving the bypass tube, to stabilize its transfer function, with 1 MHz feedback bandwidth. This loop contains an amplifier that is nearly an integrator, a power transistor that is nearly an integrator, and the feedback path, which is nearly a differentiator, so that the phase stability margin in this loop is quite big and the stand-alone loop is globally stable. However, when the transistor is overloaded by a large-amplitude input signal, the feedback in this loop reduces. This increases the closed-loop phase lag of this subsystem, thus increasing the phase lag in the vernier loop.

The stability margins in the transistor loop have been increased and a signal amplitude limiter S was installed at the input to this loop. Since then, substantial jump resonance has not been observed.

A13.7 Telecommunication Repeater

An electrical or a mechanical multiport can be characterized by transmission coefficients between the ports and the input impedances (mobilities) at the ports. MIMO feedback can be used to make all these parameters of an active multiport as desired, stable in time, and linear.

An example of such a system is a repeater designed in the Bell Laboratories in the 1930s for a 12-channel frequency division multiplex telecommunication system over open wires

[6, pp. 499–502]. The repeater amplifier must have large feedback over the operational band to reduce the intermodulation between the channels. Over the operational frequency band, the repeater's gain must be continuously adjusted by automatic level control to match the attenuation of the open wire span between the repeaters for all weather conditions, which can cause the line attenuation to vary within 40 dB at higher frequencies and within 20 dB at lower frequencies. The nominal input and output impedances of the repeater equal $Z = 150\ \Omega$. The provision of the desired impedances and the input–output gain can be illustrated with the flowchart shown in Figure A13.19. This is a MIMO system having three inputs and three outputs (not counting the slower gain response control loops).

The amplifier is a two-port described by linear relations between the two input and two output signals: I_{ain}, U_{ain}, and I_{aout}, U_{aout}. The three-port splitter and combiner connect the signal source and the load to the amplifier and the feedback path B. The ratios of the signals (variables) to be controlled are the amplifier transfer coefficient, the input impedance, and the output impedance. The input and output impedances are specified (commanded) by the transfer functions of the blocks Z. In other words, with the input current I_{in} and the outputs of the blocks Z being the command signals, the output current I_{out} is controlled using the summer S_1; the input voltage U_{in} with the summer S_2; and the output voltage U_{out} with the summer S_3.

The simplified schematic diagram including the gain control loops is shown in Figure A13.20. The system is single loop in the Bode sense since it includes only one actuator with a nonlinear link: the saturation in the ultimate stage of the amplifier. This diagram

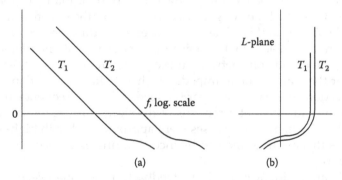

(a)   (b)

FIGURE A13.18
(a) Bode and (b) Nyquist diagrams for a high-voltage power supply controller.

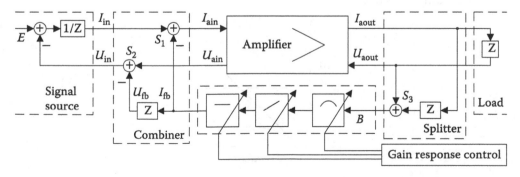

FIGURE A13.19
Flowchart of a MIMO system.

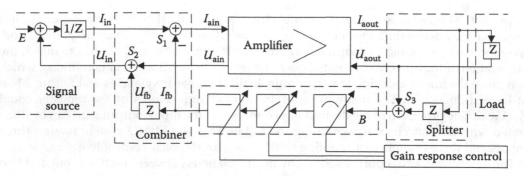

FIGURE A13.20
Telecommunication repeater simplified schematic diagram.

illustrates the benefits of structural design: over the functional frequency range, the input impedance depends exclusively on the combiner, and the output impedance, on the splitter. The repeater input–output gain depends on these two blocks and on the feedback path B. Thus, gain regulations in B do not affect the input and output impedances.

The combiner and the splitter are made with transformers having band-limiting stray reactive elements and multiple resonances. They are more complicated than the idealized ones shown in Figure A13.19: unlike in the block diagrams, splitting a signal in electrical circuits means splitting the incoming power into two (or many) loads, and there is a trade-off between how much power goes to one output at the expense of the second output. The optimal solution to this trade-off is different at different frequencies. The stray elements bound the available signal-to-noise ratio, reflection attenuations, and feedback. During the design, the trade-offs between these parameters were first resolved using the Bode integrals for the real part of an impedance that defines the available signal-to-noise ratio, for the reflection attenuation, and for the feedback. The frequency responses for the splitter and combiner in different directions of the signal propagation were chosen to be not overly complex. Then, these responses were approximated with high-order correctors and compensators that are implemented by incorporating extra electrical elements in the combiner and splitter.

The functional bandwidth over which the feedback sufficiently rejects the intermodulation was limited to 50 kHz by the quality of vacuum tubes available at the time. Today, the bandwidth of FET feedback amplifiers can reach several GHz, but the design principles remain essentially the same.

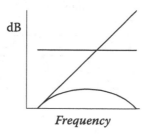

FIGURE A13.21
Gain regulation frequency responses.

The loop gain response with a crisp Bode step and maximum available feedback over the operational bandwidth was implemented with accuracy better than 0.5 dB, with the order of the compensators over 25. The feedback path was arranged in two parallel paths. The higher frequencies bypassed the feedback path B (whose transformers attenuate higher frequencies) via the high-pass filter B_{hf}. The minimum phase property of the total feedback path $B + B_{hf}$ was preserved by appropriate shaping of the frequency responses B and B_{hf}.

A simple NDC—a back-parallel pair of diodes shunting the capacitive interstage load—provided some phase advance for large-level signals and significantly reduced the jump amplitude of the jump resonance.

The slope of the Bode diagram and the available feedback over the operational band could be increased with Nyquist-stable loop design. However, in those days (and until the 1960s) the methods of providing global stability with NDCs were not yet well developed, and without such an NDC, the system would not be globally stable.

Three Bode variable equalizers in the feedback path have the following gain regulation frequency responses: flat, slanted, and convex, as depicted in Figure A13.21. Three pilot signals with frequencies at the ends and in the center of the functional frequency band were being sent continuously. An analog computer in the form of a resistor matrix decoupled the automatic gain control loops. The automatic gain control kept the pilot levels at the output of the amplifier constant, thus ensuring that the total gain of the open wire line and the repeater remained 0 dB at these frequencies, and therefore approximately 0 dB over the entire frequency band of operation.

A13.8 Distributed Regulators

The signal level along the telecommunication trunk can be controlled by sending a small-amplitude single-frequency pilot signal and changing the attenuation of the feedback paths of the repeaters such that the levels of the pilot signal selected by appropriate bandpass filters will be equal to the references, as shown in Figure A13.22.

When the number of such regulators in a long telecommunication trunk is large, there exists substantial interaction between them. When, say, the signal level on the nth repeater output suddenly becomes larger than the reference, the levels at the outputs of all repeaters starting with the nth become larger than their references, and all these automatic gain control systems start acting to correct the signal level.

The analysis of such systems shows that even a small overshoot in a single-regulator transient response causes a large overshoot at the output of the trunk, as illustrated in Figure A13.23.

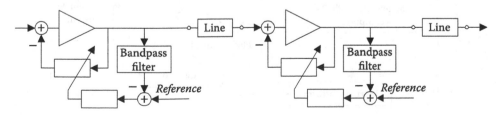

FIGURE A13.22
Cascaded signal level regulation in a telecommunication system.

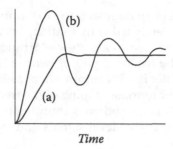

FIGURE A13.23
Transient responses for (a) a single regulator and (b) a chain of regulators.

If all regulators are identical, then to make small the overshoot of the total response, a 90° stability margin must be preserved in each regulator over the frequency band where the regulator loop gain is bigger than −60 dB, which, although feasible, presents substantial implementation difficulties.

Alternatively, NDCs can be introduced in the regulator loops, or the regulators can be made differently, with a faster regulator following several slower regulators, thus reducing the coupling between the regulators.

Chains or networks of regulation are employed for shape correction of flexible bodies (optical mirrors, for example) with multiple spatially distributed regulators.

A13.9 Saturn V S-IC Flight Control System

For the flight control system for the first stage of the Saturn V launch vehicle, the attitude is sensed using an inertially stabilized platform, and the attitude rate is available from a rate gyro package. The torque is provided by gimballing four of the five F-1 rocket engines.

The controller block diagram is shown in Figure A13.24. It is part digital and part analog. The digital computer compares the programmed attitude command history to the attitude feedback sampled from the inertial platform. The attitude error is passed through a D/A converter to the analog flight control computer. The analog feedback from the rate gyros goes directly to the flight control computer. This computer consists of the high-order *RC*

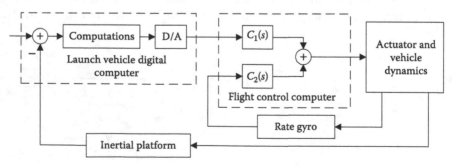

FIGURE A13.24
Control block diagram for the Saturn V.

compensators C_1 and C_2. Note that the rate feedback and the compensators are analog because they must have a fast response, while the generation of the attitude errors in the digital computer can be rather slow.

Figure A13.25 is a frequency map of the transfer function from the gimbal actuators to the sensors. It can be seen that the plant contains many resonances due to structural dynamics and propellant slosh, the frequencies of which are uncertain.

The Nyquist diagram for the pitch channel is shown in Figure A13.26. The resonance modes of the structure and propellant appear as loops. The control system has sufficient stability margins for all possible frequencies of the flexible modes within the boundaries shown in Figure A13.25.

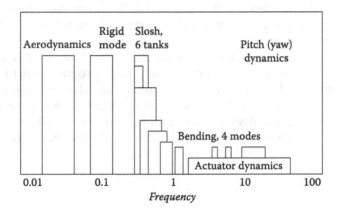

FIGURE A13.25
Frequency map of Saturn V dynamics.

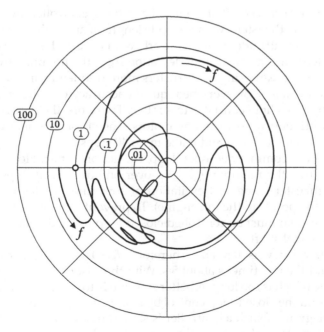

FIGURE A13.26
Nyquist diagram for Saturn V pitch control.

A13.10 PLL Computer Clock with Duty Cycle Adjustments

The 20 MHz computer clock of the Cassini spacecraft radar must have 50% duty cycle and must be synchronized with a 10 MHz ultra-stable quartz generator.

A nearly sinusoidal 20 MHz signal is generated by an *LC* VCO. The VCO is followed by a Schmitt trigger to make the clock shape rectangular. To stabilize the frequency and the duty cycle of the clock, two feedback loops are employed. The coupling between the two loops is small and can be neglected.

The first loop is a PLL. The clock frequency is halved by a D flip-flop and applied to one of the inputs of an exclusive OR gate, which is used as a phase detector. To the other input of the gate is applied the reference 10 MHz signal. The averaged phase detector output (phase error) is applied via an op-amp with *RC* compensation to the VCO tuning diode and corrects the clock frequency.

The duty cycle loop uses as a sensor a simple *RC* low-pass filter. The filter's output voltage is the average value of the clock signal and is proportional to the duty cycle. This voltage is subtracted from a half of the voltage divider, and the difference (duty cycle error) adds to the dc bias of the transistor in the VCO, via an op-amp with *RC* compensation, thus changing the degree of asymmetry of the generated signal. After being clamped by the Schmitt trigger, the output is forced by the feedback to have the required 50% duty cycle.

A13.11 Attitude Control of Solar Panels

The DS-1 (Deep Space 1) spacecraft attitude is bang-bang controlled by short bursts fired from the attitude control thrusters, with the cycle length of approximately 1,000 s. Between the firings, the spacecraft attitude changes linearly with time (if the solar wind and the outgassing are neglected) and the maximum deviation from the nominal exceeds 1°. Making the error smaller would mean firing the thrusters more frequently and using more propellant, which is unacceptable. However, the required accuracy of pointing the solar panels outfitted with Fresnel lens light concentrators is 0.5°. Therefore, the panels' attitude toward the sun must be continuously corrected by motors. The chosen stepper motors rotate the panels via a gear without backlash and make up to 40 steps/s.

When the motors are commanded open loop, the action is fast. However, the possibility is not excluded that the motors might slip over one or several steps during extreme conditions of the thruster firing and excitation of structural modes in the large panels (the panels supply the power to the ion engine). To prevent this from affecting the steady-state accuracy, the system must be controlled closed loop using the data from the encoder placed on the solar panel shaft.

The sampling frequency is 0.5 Hz. Therefore, $f_b \approx 0.05$ Hz, the closed-loop bandwidth ≈ 0.07 Hz, and the rise time is about 5 s. With this rise time, the settling time of the system is about 10 s, which is too long. A better solution is to combine the advantages of the open-loop control and the closed-loop control by using a feedback-feedforward scheme. A simplified block diagram of such a controller is shown in Figure A13.27. Here, alpha_com is the commanded angle, alpha_enc is the encoder readings, pulse_num is the number of steps the motor must make during the sampling period, fforw_num is the number of

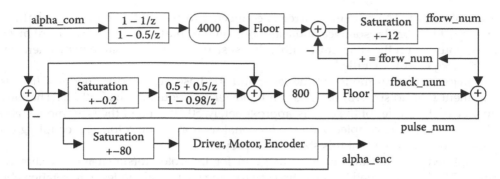

FIGURE A13.27
Simplified block diagram for solar panels' attitude control.

pulses commanded by the feedforward path, and fback_num is the number of pulses commanded by the feedback compensator.

The feedback controller has a common configuration of parallel connection of proportional control path and a low-pass link with saturation in front of it.

The upper path in the block diagram is the feedforward. It begins with a high-pass filter that does not pass dc. Therefore, in 7 or so seconds after a step command, this path becomes disabled and the feedback control takes over, but for small times it supplies up to 12 pulses to the motor to speed it up and to reduce the rise time and the settling time.

A13.12 Conceptual Design of an Antenna Attitude Control

The Microwave Limb Sounder (MLS) of the Chemistry spacecraft includes a radiometer. The radiometer antenna mirror scans the Earth horizon with 6 arcsec pointing accuracy. After each 20 s, 2° scan the antenna must be turned away by nearly 180° for calibration, and then returned to start the next scan. The required settling accuracy after the return is 10^{-5} (the ratio of 6 arcsec to 180°), and the settling time must be less than 0.8 s, which is not easy to achieve.

The radiometer pointing can be shifted by 180° with an additional mirror. In this case there will be two separate mechanisms: a scanning mirror and a switching mirror. The design of attitude controllers in the dual-mirror system is simple since the switching mirror can be small and can be switched completely off the optical path of the main mirror during the scan, and the scanning mechanism can include a gear.

An economic alternative is using a single mirror directly driven by a motor, and a single servo loop for both scanning and switching. In the attitude controller for this mechanism, the small settling time requirement necessitates using wide feedback bandwidth; the bandwidth might be limited, however, by structural resonances. The question on feasibility of the single-mirror servo has to be resolved at the initial stages of the project when accurate mathematical models of the mechanisms' dynamics have not yet been developed. Therefore, for the purposes of conceptual design and the control performance estimation, a simple specialized model of the systems dynamics has to be developed by the control system designer.

The single-mirror mechanism is shown in Figure A13.28(a). The angular encoder disk is used as the angle feedback sensor. It is placed close to the motor rotor, thus making the control collocated. The antenna mirror is connected to the motor rotor via a shaft. The frequency of torsional resonance of the antenna on the shaft is 300 Hz. The entire mechanism is mounted on the spacecraft with a thrust structure. On the same instrument base, a heavy 50 kg laser is installed.

The major structural modes of the laser and the base are rocking modes. There is a requirement that no structural mode can be lower in frequency than 50 Hz. The lowest frequency of the modes of the base is, approximately, 50 Hz, and of the laser modes, about 60 Hz. The modes are coupled with the torsional motion of the rotor. The coupling coefficients are not yet known at this stage of the project.

A simplified mechanical schematic diagram for the rotor torsional motion shown in Figure A13.28(b) is obtained by projecting the rocking structural modes onto rotation about the motor rotation axis. We will consider the worst case: the rocking mode motion aligned with the torsional modes. These projections use the moments of inertia of the rigid bodies about the rotation axis. The spring coefficients are chosen such that the torsional modes resulting from these projections have the same frequencies as the original rocking modes.

The employed models have no provisions for modal damping. The damping can be introduced in the model, but in this example there is no need for this since the modes will be already substantially damped by the dissipative output mobility of the motor driven by a voltage source.

The moment of inertia of the laser about the rotor rotation axis $J_L \approx 0.5$ kgm^2. The instrument base, the stator, and the encoder constitute a rigid body with the moment of inertia about the motor rotation axis, $J_B \approx 0.2$ kgm^2. The antenna mirror moment of inertia: $J_M = 0.01$ kgm^2. The actuator output is shown as a source of velocity Ω_S with internal mobility Z_{SM}.

The initial, rough estimation of the plant dynamics' effects on the control loop based on the diagrams in Figure A13.28 is the following. The base moment of inertia is more than an order of magnitude larger than the mirror's moment of inertia. Therefore, approximately, we can assume the base moment of inertia is infinite. This simple model is valid at frequencies up to 0.5 of the frequency of the lowest structural resonance, i.e., up to 25 Hz. The resonances at 50 and 60 Hz of the base and the laser will not affect profoundly the loop because the base mobility is rather low, relative to the antenna mobility—as long as the frequency range of interest is up to 25 Hz, i.e., the control bandwidth is below 15 Hz.

For extending the control bandwidth beyond 15 Hz, the worst-case stability analysis can be made using the plant model in Figure A13.28(b).

The equivalent electrical schematic diagram is shown in Figure A13.29(a), with a 1-to-1 ratio in the analogy of the mechanical to electrical parameters and variables, for example with J_M replaced by $C_M = 0.01$ F. The torsional stiffness coefficient of the mirror shaft k_M is replaced by $L_M = 1/k_M$. The electrical contour impedance of the driver's output is the winding resistance r_w (the motor is driven by a voltage driver); it is converted to mobility and to

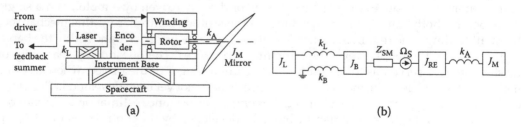

(a)   (b)

FIGURE A13.28
(a) The antenna mechanism and (b) its simplified mechanical schematic diagram.

the mobility 1-to-1 electrical impedance representation $Z_S = k^2 r_w$, where k is the electromechanical motor constant.

The schematic diagram modified by changing the order of series connection of the two-poles about the contour is shown in Figure A13.29(b). Here,

$$k = 0.85, \; r_w = 4\Omega, \; C_B = 0.2F, \; C_L = 0.5F, \; C_R = 0.0002F, \; C_M = 0.01F$$

the resonance frequencies $\omega_L = 400$ rad/s, $\omega_B = 320$ rad/s, $\omega_A = 2{,}000$ rad/s,

$$1/Z_S = 1/(k^2 r_w) = 0.346$$

$$1/Z_3 = sC_R = 0.002s$$

$$Z_4 = sL_A = s/\omega_A^2 C_A = 0.000025s$$

$$Z_M = 1/(sC_M) = 100/s$$

$$Z_2 = \frac{s^3 L_L C_L L_B + sL_B}{s^4 L_L C_L L_B C_B + s^2 (L_B C_B + L_L C_L + C_L L_B) + 1}$$

Since $L_i C_i = 1/\omega_i^2$, we can express Z_2 as:

$$Z_2 = \frac{s^3/(C_B \omega_L^2 \omega_B^2) + s/(C_B \omega_B^2)}{s^4/(\omega_L^2 \omega_B^2) + s^2 [1/\omega_B^2 + 1/\omega_L^2 + C_L/(C_B \omega_B^2)] + 1}$$

or

$$Z_2 = \frac{305 \times 10^{-6} s^3 + 48.8s}{61.035 \times 10^{-6} s^4 + 31{:}64 s^2 + 10^6}$$

This function is an impedance of a lossless two-pole, and its zeros $0; \pm j400$ and poles $\pm j184; \pm j700$ alternate along the $j\omega$-axis.

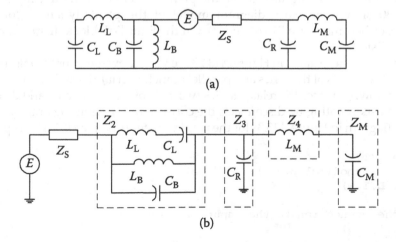

(a)

(b)

FIGURE A13.29
(a) Electrical equivalent schematic diagram for the plant dynamics of antenna attitude control and (b) the modified diagram.

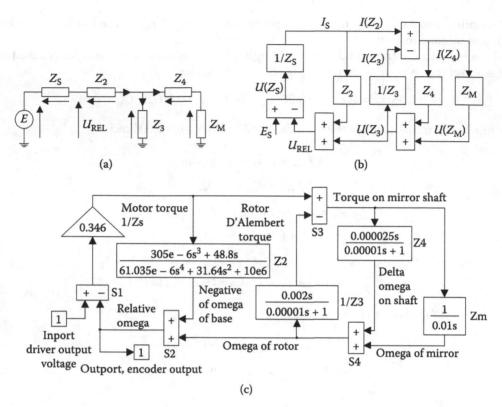

FIGURE A13.30
(a) Ladder network plant model of the antenna attitude control, (b) the block diagram model of the plant, and (c) the Simulink model.

The schematic diagram in Figure A13.29 could be analyzed with SPICE; in this case the expressions for the impedances do not need to be derived. However, for the convenience of system design and integration, the plant model is worth building in Simulink. Figure A13.30 shows the block diagram model of the plant dynamics following the approach described in Figure 7.18(d) and (f). On the Simulink block diagram, mechanical variables are listed.

Notice that diagrams b and c in Figure A13.28 are very convenient for troubleshooting and investigating the effects of the series and parallel branches in (a): disconnecting the input of a link pointing down replaces the related series two-pole by a short circuit, and disconnecting the input to a link pointing up disconnects the corresponding shunting branch.

The Bode diagram in Figure A13.31 for the plant transfer function $U(Z_M)/E_S$ is plotted with:

```
w = logspace(1,3);
[A,B,C,D] = linmod('mlsplan');
bode(A,B,C,D,1,w)
hold on
% remove the connection to the input of block Z2
bode(A,B,C,D,1,w)
```

for two cases: for the nominal values of the plant parameters, and for $Z_2 = 0$, i.e., for an infinitely massive base, as was assumed in the initial rough analysis.

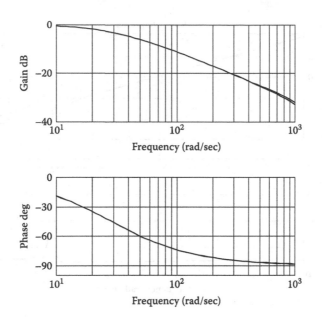

FIGURE A13.31
Bode diagram for the plant of the antenna attitude control loop.

The two gain and phase responses in the picture are very close to each other. Hence, the structural resonances of the laser and the base do not constrain the feedback bandwidth, which can therefore be 30 Hz. With an appropriate prefilter, a multiwindow controller, command profiling (see Section 5.11), and loop shaping, this feedback bandwidth ensures the required accuracy with substantial margins. Based on this analysis, a decision can be made to use the economic option of a single-mirror mechanism.

For a much larger primary mirror of the MLS GHz radiometer, single-mirror attitude-and-switching control is not feasible. This device must be made with two separate mechanisms, a scanning mirror and a switching mirror.

A13.13 Longer Antenna in the Spacecraft

The antenna for a Mars satellite must be larger to improve the information transmission; it has length $l = 15$ m and weight $M = 4$ kg. It is flexible and oriented at 90° to spacecraft main direction. The moment of inertia of the spacecraft is $J_{s/c} = 2,000$ kgm². The control loop designers must initially evaluate the new control system.

The antenna loaded at one end of the spacecraft gives the resonance of the structure, and the combination of antenna resonance with the s/c will give the second resonance. For the spacecraft rotational motion, it's a rigid body (see Example 4.19):

$$P(s) = \frac{s^2 + 2\zeta\omega_z + \omega_z{}^2}{s^2 + 2\zeta\omega_p + \omega_p{}^2} \frac{1}{J_{s/c}s}$$

Here, ω_z is the angular frequency of the zero, ω_p is the angular frequency of the pole, and ζ is the damping coefficient for these resonances. Next, we assume that the spring of the

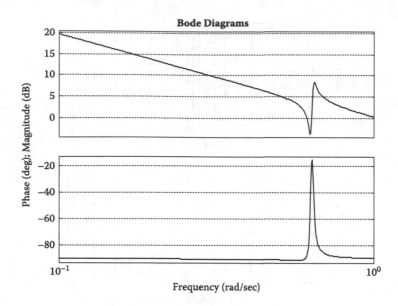

FIGURE A13.32
Bode diagram.

antenna defines its frequency to be approximately 0.1 Hz. Thus, the ratio of the pole-to-zero frequencies is the square root of 1: 0.01875.

We assume the damping coefficient for the resonances is 0.01. Each resonance by itself will make a peak or value of 40 dB. This is probably the upper estimate of the real antenna damping. Therefore, $\omega_z = 2\pi \times 0.1 = 0.63$ rad/s, and thus $\omega_p = 1.01875\omega_z = 0.6418$.

Then, the transfer function of the spacecraft is:

$$P(s) = \frac{s^2 + 0.0126s + 0.3969}{s^2 + 0.012836s + 0.41192} \frac{1}{2000s}$$

The plant Bode diagram in Figure A13.32 is plotted with MATLAB script:

```
w = logspace(-1, 0, 400);
·n = [1 0.0126 0.3969]; d =[1 0.01836 0.41192 0];
bode(n,d,w)
```

The lost feedback due to the pole-zero can be calculated using Bode phase integral. It is about 2.5 dB (the parameters of the Bode curve are very stable, and the response of the curve is also stable). The effects of the resonances on available feedback are rather small. The larger antenna is perfectly doable.

A13.14 Path Length Control of an Optical Delay Line

A13.14.1 System Description

The optical delay line of the stellar interferometer (SIM) is placed in the path of the light gathered from one of the widely separated optical elements called siderostats. The delay is regulated to provide an interference pattern on the focal plane, where this light is combined

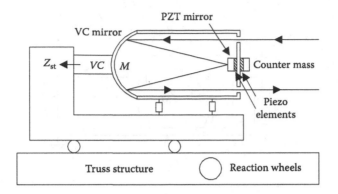

FIGURE A13.33
Optical delay line.

with the light from another siderostat. The delay line is positioned on a truss structure, as shown in Figure A13.33. A reaction wheel assembly is employed for the interferometer attitude control.

The delay line includes two mirrors. The larger, spherical mirror is actuated by a voice coil (VC) capable of a maximum displacement of 1 cm at low frequencies. The smaller, flat mirror is moved by a piezoactuator (PZT) with a maximum stroke of 30 μm. The light bouncing between the mirrors along the variable path length experiences a controllable delay. The path length must be controlled with better than 5 nm accuracy.

The PZT consists of two piezoelements and applies a force between the smaller mirror and the countermass. Since the forces applied to the supporting structure balance out, the PZT does not excite the structure, and the feedback bandwidth in this loop is not limited by the structural modes.

The VC de-saturates the PZT. The VC is, in turn, de-saturated by placing it on a cart; the cart motor control loop will not be discussed here.

The voice coil moves the mirror having mass $M = 0.5$ kg against the flexible structure with mobility Z_{st}. The equivalent electrical schematic diagrams for the VC actuation of the mirror are shown in Figure A13.34. Here, C_M reflects the mass of the mirror, and L_{susp} and R_{susp} reflect the VC suspension, with the resonance at 5 Hz. The VC mobility $Z_{VC} = (r_{vc} + Z_D)/k^2$, where r_{vc} is the coil resistance, Z_D is the driver amplifier output impedance, and k is the force/current electromechanical coefficient of the voice coil. Over the frequency range of interest, $|Z_{st}|$ is substantially smaller than the mobility of the contour it is in and does not much affect the velocity of the mirror (the voltage on C_M).

A13.14.2 Numerical Design Constraints

The PZT maximum displacement is 30 μm over the entire frequency range of interest.

The VC maximum displacement amplitude is:

$$D_{max} = (2\pi f)^{-2} k I_{max} / M$$

With the VC constant $k = 0.3$ N/A and the current saturation threshold of the VC driver amplifier $I_{max} = 3$ A, the frequency at which the VC maximum displacement amplitude equals that of the PZT, $D_{max} = 0.00003$ m, is:

$$f_{cross} = \sqrt{k I_{max} / (M D_{max})} / (2\pi) \approx 40 \, \text{Hz}$$

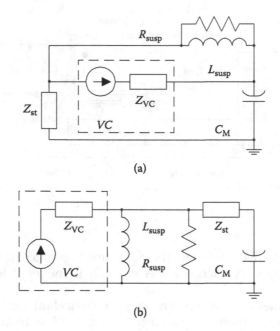

FIGURE A13.34
Equivalent electrical circuit for the voice coil drive: (a) following the structure and (b) modified.

The feedback in the VC and PZT loops must be sufficiently large to reject the vibrational disturbances over the bandwidth up to 500 Hz, caused by the reaction wheels. The feedback must be at least 60 dB at 16 Hz. The disturbance forces' spectral density responses are assumed to be flat over the 500 Hz bandwidth, so that the vibration amplitudes are, on average, inversely proportional to the square of the frequency. Therefore, both the feedback and the maximum actuator output amplitude can decrease inversely proportional to the square of the frequency. Hence, the loop gain slope must be close to −12 dB/octave.

The available feedback in the VC loop is affected by the mobility of the structure Z_{st}, as seen in Figure A13.34. As can be estimated using an asymptotic Bode step response, the structural modes and their uncertainties prevent the VC loop bandwidth from exceeding 100 Hz—if the loop is designed as stand-alone stable. In this case, the gain in the VC loop at 16 Hz is only 26 dB, less than the required 60 dB.

Therefore, the VC loop has to be designed as stand-alone unstable; in this case, the gain and the slope of the loop Bode diagram can be substantially increased. The loop can be designed as a self-oscillating dithering system, as described in Section 9.7.

The sampling frequency for the PZT loop is 5 kHz. For the VC loop, the sampling frequency can be chosen to be 1 kHz.

A13.14.3 Higher-Level Design Objectives

The control system performance index is the mean square error in the delay line path length. The principal design objective is to keep the mean square error below 5 nm.

In addition to the principal design objective, the following characteristics are required or highly desirable:

System robustness.

Good output responses to commands of different shapes and amplitudes.

Transient responses to large-amplitude vanishing disturbances that are not excessively large in amplitude or duration.

A large disturbance/command triggering threshold of nonlinear phenomena, if the nonlinear phenomena responses are prolonged and violent and cannot be excluded by the design, in order that these nonlinear phenomena happen infrequently.

A13.14.4 Design Approach

The controller for this nonlinear, flexible, and uncertain plant must be reasonably close to the best achievable, but not overly complicated. We choose the controller to be multi-loop, high order, and nonlinear, but not time variable (i.e., not adaptive).

The conceptual design employing Bode integrals and the Bode asymptotic diagrams should produce a solution in the vicinity of the globally optimal time-invariable controller (the "optimality" here means best satisfaction of the higher-level design objectives, according to the opinion of the customers, i.e., system engineers).

The design begins with making some reasonable assumptions and translating the higher-level objectives into a set of lower-level guidelines. The latter consists of the design objectives and the design considerations. If possible, these objectives and considerations should be formulated in a mutually decoupled (orthogonal) form to simplify the system trade-offs and speed up the design. This is more easily accomplished using frequency-domain specifications.

A13.14.5 Lower-Level Design Objectives

The lower-level design objectives for the feedback system under consideration can be formulated as follows:

To effectively reject the vibrational disturbances of rather large amplitudes, the VC loop gain must exceed 50 dB at 40 Hz and the feedback must increase toward the lower frequencies.

To effectively reject vibrations at higher frequencies, however with smaller amplitudes, the feedback in the PZT loop must have wide bandwidth; it has been calculated that 600 Hz bandwidth will suffice.

The transient response to the commands of different amplitudes must be good.

The system must be globally stable.

If the VC loop is made unstable when stand-alone, the frequency of oscillation in this loop should be as high as possible in order for the amplitude of the limit cycle oscillation in this loop to be as small as possible and not overload the PZT.

The *design considerations* are the following:

The VC and PZT loops are nearly m.p., and the combined loop must be m.p. For this, the parallel channels must be shaped appropriately.

At frequencies below $f_{\text{cross}} \approx 40$ Hz, the VC provides a larger stroke and is therefore the main actuator; at higher frequencies, PZT is the main actuator. Therefore, rejection of the large-amplitude disturbances depends at $f < f_{\text{cross}}$ on the loop gain in the VC loop and at $f > f_{\text{cross}}$ on the PZT loop. Outside of these ranges, the only requirement for the VC and PZT loop gain shaping is the provision of stability and robustness.

(Generally, making the loop cross at f_{cross} is not necessary. Extra gain in the loop that is not the main one does not hurt the system performance, but it unnecessarily consumes a part of the area specified by a related Bode integral; therefore, this is not worth doing.)

At the frequency of 5 Hz, the VC driver output impedance should be small to damp the VC suspension resonance. At frequencies over 50 Hz, the impedance can be made higher, thus making $|Z_{\text{VC}}|$ higher in order for the loop transfer function at these frequencies to be less affected by the system structural modes.

A13.14.6 Conventional Design Approach

Most commonly, industrial main/vernier systems do not include NDCs or only include simple NDCs for improving the responses to the commands. Such systems are made globally stable by making all stand-alone loops globally stable.

When the feedback and the disturbance rejection that are achievable in such systems are not sufficient, the disturbance rejection can be improved at the price of making the main/vernier system only conditionally stable. This, however, causes the command generator to be complicated and the reaction to the commands sluggish. It also necessitates a procedure to recover the system from the limit cycle, which can be quite violent and damaging to the hardware. For this reason, this option is normally not used in industrial systems.

A13.14.7 The Chosen Design Options

The chosen design options for the system with an NDC follow:

The VC loop is stand-alone unstable. In this case the slope of the Bode diagram of the VC loop can be made steeper, thus improving the disturbance rejection and the handling of large-amplitude disturbances.

An NDC provides global stability.

The PZT loop bandwidth is 600 Hz wide, with a Bode step response, with the Bode diagram slope −10 dB/octave at frequencies 150–1,200 Hz, and −12 dB/octave over the range 40–150 Hz. At 40 Hz the gain in the PZT loop is therefore 12 dB/octave $\times \log_2(150/40) + 20 = 43$ dB. The gain may gradually roll down at lower frequencies.

We choose the asymptotic Bode diagrams shown in Figure A13.35(a), since for this particular design task the advantages of using the better-shaped diagrams exemplified in (b) do not justify higher complexity of the design.

The VC loop Bode diagram crosses the PZT loop Bode diagram at 40 Hz.

The PZT loop phase margin at 40 Hz is approximately 40°.

The VC loop Bode diagram slope is −18 dB/octave down to 20 rad/s, and −6 dB/octave below this frequency. The gain at 40 Hz must be 43 dB, i.e., 141 times. From here, $T_{\text{VC}} = a/[s(s+20)^2]$, where $a = 141(2\pi40)^3 = 2.24 \times 10^9$.

At 40 Hz, the phase difference between the loops is $270° - 140° = 130°$. This ensures minimum phase character of the total loop with a 50° safety margin.

To ensure the system's global stability and good transient responses to the commands, an NDC will be included in the combined loop, with the dead zone equal to the threshold of the PZT actuator. As the linear block of the NDC, the PZT loop transfer function is employed.

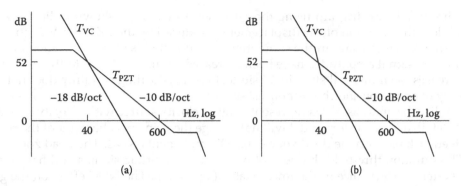

FIGURE A13.35
Asymptotic Bode diagrams for stand-alone loop transfer functions: (a) simpler and (b) higher performance.

A13.14.8 Block Diagram with Rational Transfer Functions

The asymptotic (transcendental) responses shown in Figure A13.35(a) need to be approximated with rational function responses.

The frequency-normalized Bode step-type loop response (including certain n.p. lag of sampling) is obtained with MATLAB function `bostep` (from the bode step toolbox) and shifted in frequency with `1p21p` for the crossover frequency to be $2\pi600$ rad/s. This forms the stand-alone PZT loop transfer function. The stand-alone VC loop transfer function has already been defined as $2.24 \times 10^9/[s(s+20)^2]$.

The Bode diagrams for the stand-alone VC and PZT loops and the summed loop Bode diagram are obtained with the following script:

```
npzt = [2.33e-10 6.7e-6 -5.14e11 1.26e17 1.5e21 9.5e24 1.2e28];
dpzt = [1 2.83e5 7e9 9.18e13 6.5e17 1.684e21 0 0];
nvc = [2.5e9]; dvc = [1 40 400 0];
ntot1 = conv(npzt,dvc);
ntot2 = conv(dpzt,nvc);
adz = zeros(1,length(ntot1)-length(ntot2));
ntot = ntot1 + [adz ntot2];
dtot = conv(dvc,dpzt);
w = logspace(1,4.5);
bbode(npzt,dpzt,w);
hold on
bode(nvc,dvc,w);
bode(ntot,dtot,w);
hold off
zoom on
```

The logarithmic plane Nyquist diagrams are plotted with `nyqlog` from the bode step toolbox:

```
nyqlog(3700, nvc,dvc); hold on
nyqlog(3700, npzt,dpzt); hold on
nyqlog(3700, ntot,dtot); hold off
```

The diagrams shown in Figure A13.36 indicate that the system in the linear mode of operation is stable, robust, and provides 69 dB disturbance rejection at 16 Hz.

The Simulink block diagram using these transfer functions is shown in Figure A13.37. The path length is the sum of the displacements produced by the PZT and VC actuators.

In this block diagram and in the simulations, the thresholds of the saturation, the dead zone, and the signal amplitudes have been increased 10 times relative to the real system. This increases the numerical stability of simulation. This scaling certainly does not affect the theory of operation and the responses' shapes.

To simplify the analysis, VC is represented by a velocity source with velocity saturation; using a detailed VC actuator model will not change the principal character of the results.

The feedback path via the dead zone dz0.0003 implements NDC. The dead zone equals the PZT saturation threshold. It does not pass small-amplitude signals and has no effect on the system performance in the linear state of operation. The NDC effect on the global stability and the transient responses to large-amplitude commands will be discussed later.

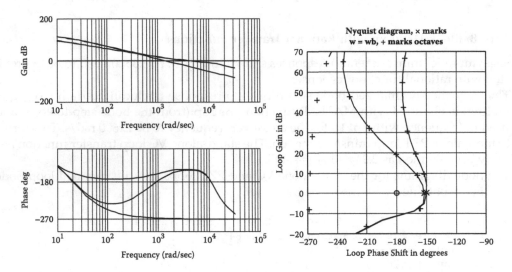

FIGURE A13.36

Bode and Nyquist diagrams for the rational transfer functions approximating the asymptotic diagrams for the VC, the PZT, and the combined loop. (The combined response is the top one on the gain response, and the intermediate one, on the phase response and the Nyquist diagram.)

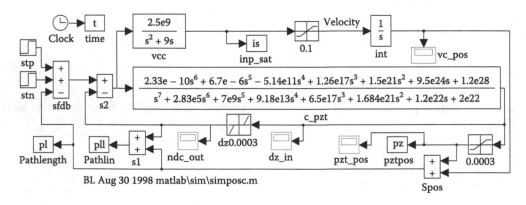

FIGURE A13.37

Simplified Simulink block diagram of path length control.

A13.14.9 Self-Oscillation in the System without an NDC

With frequency responses shown in Figure A13.36, the system is stable in the linear mode of operation when both the PZT loop and the VC loop are closed. The stand-alone VC loop is unstable.

Without an NDC, the system is not globally stable. Figure A13.38 shows the path length oscillation triggered by a 30 ms, 3 mm pulse command. The oscillation approaches a limit cycle with the period longer than 5 s.

Compared with the amplitude of this oscillation, the PZT output is negligibly small. Therefore, the system can be analyzed as having a single nonlinear link, the saturation in the VC actuator, and the analysis can be performed with the describing function method. The accuracy of such analysis is sufficient since the VC loop is low pass and the signal at the input to the saturation link becomes nearly sinusoidal when approaching the limit cycle, as seen in Figure A13.38(b).

In other words, when the vibrations' amplitudes are large enough to saturate the PZT, the VC loop is left alone and the system bursts into the limit cycle self-oscillation at a low frequency with large signal amplitude. When the high-amplitude, high-frequency vibrations vanish and no longer saturate the PZT, the limit cycle oscillation does not change substantially since the amplitude of the output of the PZT is negligibly small compared with the displacement generated by the VC. Thus, the limit cycle of the stand-alone VC loop is, in fact, also the limit cycle of the system as a whole. Large-amplitude vanishing signals with substantial frequency components of the limit cycle oscillation belong to the basin of attraction for the limit cycle. The system is conditionally stable.

A13.14.10 Limit Cycle of the VC Loop with the NDC

Since the stand-alone VC loop is unstable and the NDC dead zone does not pass the signals of small amplitudes, the stand-alone VC loop is unstable even with the NDC. It is important to emphasize, however, that the NDC radically changes the amplitude and the frequency of the limit cycle oscillation.

The amplitude of the limit cycle oscillation is determined by the dead zone in the NDC, since for signals much larger than the dead zone, the dead-zone-describing function approaches 1, the system becomes close to linear, and such a system, as has been shown

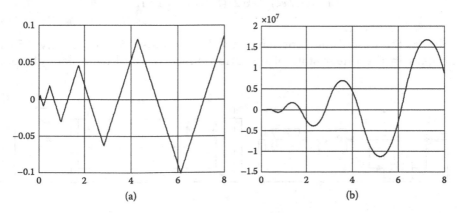

(a)   (b)

FIGURE A13.38
Oscillation triggered by a short pulse command in the system without an NDC (a) of the VC output path length component and (b) at the input to the saturation link.

before, is stable with sufficient stability margins. Therefore, oscillation can only take place with amplitudes not much higher than the dead zone.

Since the oscillation amplitude is small, during the describing function stability analysis the saturation link in the VC actuator can be replaced by 1, and the system can be viewed as that shown in Figure A13.39(a). This system's linear part is, however, unstable; for the conventional describing function methods to be applicable, diagram a is transformed equivalently to diagram b by replacing the dead zone with parallel connection of a unity link and a saturation link with the threshold equal to the dead zone. The stand-alone linear part of the latter (the dashed box) is stable since this is a feedback system with parallel connection of VC and PZT loops. The transfer function of the box is:

$$\frac{T_{\text{PZT}}}{1 + T_{\text{VC}} + T_{\text{PZT}}}$$

Notice that the signal about the loop via the saturation is applied in phase.

The Bode diagram for this transfer function, shown in Figure A13.40, is obtained with the script:

```
npzt = [2.33e-10 6.7e-6 -5.14e11 1.26e17 1.5e21 9.5e24        1.2e28];
dpzt = [1 2.83e5 7e9 9.18e13 6.5e17 1.684e21 0 0];
nvc = [2.5e9]; dvc = [1 40 400 0];
n1 = conv(npzt,dvc);
d1 = conv(dvc,dpzt);
d11 = [0 n1] + d1;
d2 = conv(dpzt,nvc);
deq = d11 + [0 0 0 d2];
w = logspace(0,4);
% nyqlog(1260,-n1,deq)
bbode(n1,deq,w)
```

It is seen that the condition of oscillation (zero loop phase shift while the loop gain is positive) occurs at approximately 1,600 rad/s, or approximately 250 Hz. The loop gain at this frequency is 3.5 dB; i.e., the gain coefficient is 1.5. The oscillation amplitude E can be found by equating the describing function to the inverse of the loop gain coefficient, i.e., using Equation 11.8,

$$1.27(E/e_{\text{S}})^{-1} - 0.27(E/e_{\text{S}})^{-4} \approx 1/1.5$$

where $E \approx 1.8e_{\text{S}} \approx 0.00054$.

With this E, the describing function of the PZT equals 0.67, i.e., is only 3.5 dB less than the gain in the linear state of operation. Since in the linear state of operation

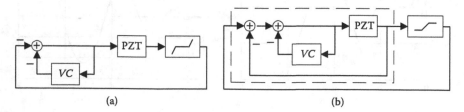

(a)   (b)

FIGURE A13.39

Equivalent block diagrams for describing function analysis, (a) with the unstable linear part and (b) with the stable linear part.

stability and safety margins on the order of 10 dB are provided, the system is asymptotically globally stable.

Therefore, when the PZT actuator recovers from being saturated by disturbances or commands, it is capable of delivering an output signal comparable to the self-oscillation in the VC loop. Application of this signal makes the system stable, and the oscillation in the VC asymptotically dies down.

The time history of the path length in this case is shown in Figure A13.41, (a) without and (b) with saturation in the VC that limits the slew rate.

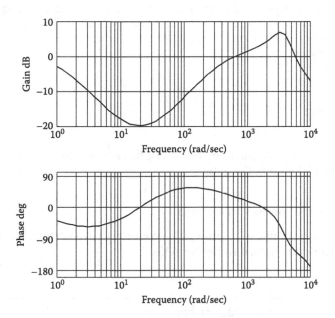

FIGURE A13.40
Bode diagram for the oscillation describing function analysis.

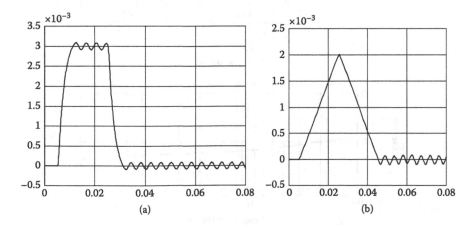

FIGURE A13.41
Path length time history after a pulse command in a system with the NDC and without a PZT actuator, (a) without limiting the VC velocity and (b) with a ±0.1 m/s limiter of the VC velocity.

A13.14.11 Global Stability of a System with the NDC

The dead-zone branch is in parallel with the PZT saturation branch. These two parallel branches equal a linear branch with the same transfer function. (Augmenting a nonlinear system with an extra nonlinear subsystem, in our case by the dead zone, so that the total system properties are those of a linear system, is sometimes called *exact linearization*.)

The system therefore includes only one nonlinear link: the VC saturation. The transfer function of the equivalent linear link from the VC saturation output to the VC saturation input is $T_{VC}/(1 + T_{PZT})$. The Bode diagram of this transfer function found with the script:

```
npzt = [2.33e-10 6.7e-6 -5.14e11 1.26e17 1.5e21 9.5e24 1.2e28];
dpzt = [1 2.83e5 7e9 9.18e13 6.5e17 1.684e21 0 0];
nvc = [2.5e9]; dvc = [1 40 400 0];
n1 = conv(nvc,dpzt);
d1 = conv(dvc,dpzt);
d2 = conv(npzt,dvc);
deq = d1 + [0 d2];
w = logspace(1,4);
bode(n1,deq,w)
```

is plotted in Figure A13.42. Application of the Popov criterion to this system shows that the system is asymptotically globally stable.

A13.14.12 Transient Responses to Commands

The output transient response to a 3 mm pulse command in a system, assuming no saturation in the VC actuator, is shown in Figure A13.43.

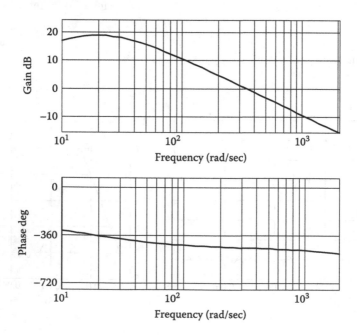

FIGURE A13.42
Bode diagram for transfer function $T_{VC}/(1 + T_{PZT})$.

The zoomed-on parts of the responses are shown in Figure A13.44. After the PZT becomes de-saturated, the NDC dead zone stops passing the signal, the path length response coincides with the response of the equivalent linear system, and the error dies down rapidly.

The output transient response to a 50 ms pulse command in a system with saturation in the VC actuator with threshold 0.01 is shown in Figure A13.45. The saturation limits the slew rate of the output, and in other aspects the responses are similar to those in Figure A13.43.

Settling time to high accuracy is that of the linear system. The tail of the response is that of a linear system. Figure A13.46 shows the responses of the equivalent linear system (dotted line) and the response of the real system (solid line).

These figures demonstrate the function of an NDC: not only does it ensure global stability, but it also provides good transient response to large signals.

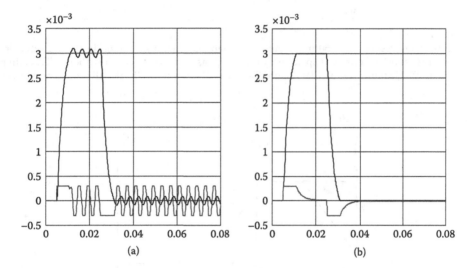

FIGURE A13.43
Closed-loop transient responses to 20 ms, 3 mm pulse command, without saturation in the VC actuator; path length, upper curves, and PZT mirror displacement time response, lower curves; (a) without PZT, ending in a small-amplitude 250 Hz limit cycle oscillation, and (b) with PZT, asymptotically globally stable, rapidly settling.

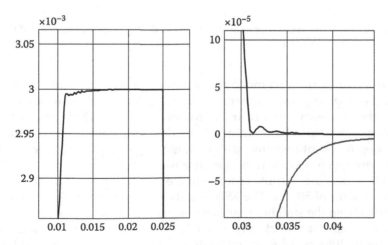

FIGURE A13.44
Zoomed-in pieces of the closed-loop transient response to a 20 ms, 3 mm pulse command.

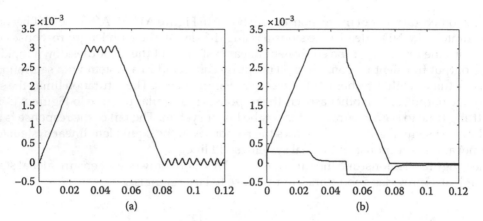

FIGURE A13.45
Closed-loop transient responses to 50 ms, 3 mm pulse command, with saturation in the VC actuator; path length, solid lines, and PZT mirror displacement time response, dashed lines; (a) without PZT, ending in a small-amplitude 250 Hz limit cycle oscillation, and (b) with PZT, asymptotically globally stable, rapidly settling.

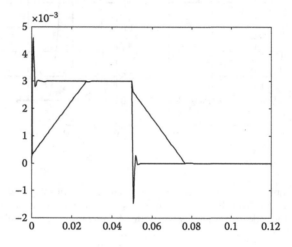

FIGURE A13.46
Transient responses to step commands of the linear system (Π shaped with overshoots) and the main/vernier nonlinear system (with finite slew rate).

In the large-signal (nonlinear) mode of operation, the loop response is, approximately, $T_{vc}/(1+T_{pzt})$. This loop response, shown in Figure A13.42, is close to a single-integrator response over the frequency range near the crossover, from 1 to 15 Hz. Such a loop response results in a single-pole closed-loop transfer function, which is known to produce a rather good closed-loop transient response (although not as good as a higher-order Bessel filter response, so there remains some room for improvement).

Figure A13.47 shows the output response to the (a) small and (b) large step commands applied at the instant of 50 ms. The 75% overshoot in the small-signal (linear) mode of operation results from the stability margins that are justifiably chosen reasonably narrow (since the plant parameter uncertainty is small). If needed, the small-signal overshoot can be reduced by a prefilter or a command feedforward, but these are not needed because in absolute values the overshoot is small.

A13.14.13 Robustness

As long as the threshold in the PZT saturation is exactly equal to the dead zone in the NDC, the system is equivalent to a system with only one nonlinear element, the VC actuator. This system satisfies the Popov criterion with conventional stability margins, and is therefore robust against variations in the linear links' transfer functions and in the nonlinear characteristic of the VC actuator.

However, it would be desirable to make the dead zone wider than the threshold of the PZT saturation by 10–30% so as not to impair the PZT output stroke. Therefore, the system must also be robust against variations in the threshold of the saturation in the PZT. To such systems that include more than one nonlinear link, the Popov criterion is not directly applicable. Instead, we rely on computer simulations. When the PZT saturation threshold is changed from 0.0003 to 0.00015, the system remains stable and the response to the pulse command shown in Figure A13.48 remains acceptable. Further reduction of the threshold to 0.0001 makes the system unstable and causes a small-amplitude oscillation similar to that shown in Figure A13.43.

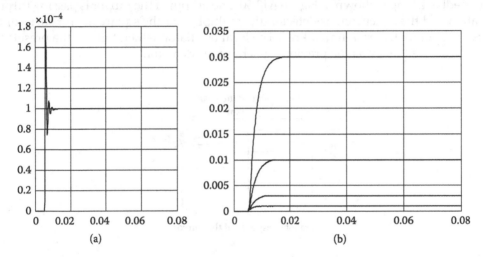

FIGURE A13.47
Transient responses to step commands of (a) 0.001 m and (b) larger amplitudes' commands.

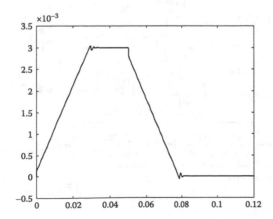

FIGURE A13.48
Transient responses to the pulse command in the system with PZT saturation threshold 0.00015.

A13.15 Optical Receiver

The measure of a space optical system is made by 10–12 laser beams launched from the beam launchers. The control systems must regulate the length and the attitude of the beams.

The accuracy of each system is much narrower than the beam. The deviation in all directions from the central line, which goes through the center of a parabolic mirror, is less than 50 nm, as shown in Figure A13.49.

The beam delay from the central point to touching the mirror equals 50 nm.

The vibrations lower the control accuracy. The frequency band in such a system is higher than 5 Hz since there lie the solar panel main resonance and harmonics. It is also higher at frequencies above 10 Hz. The usable range is therefore from 5 to 10 Hz.

The feedback for single-beam variations covers the bandwidth of 0.3 Hz. With the additional frequency interval needed for the frequency separation between the channels, the bandwidth should be made 0.5 Hz. The total bandwidth for 10 channels is therefore 5 Hz.

The feedback loop is shown in Figure A13.50. The output of the mirror is passing through saturation and the synchronous detector (the multiplier, or the square mirror surface) and enters the photodetector in nm/mkrad. The local oscillator, which follows the preselector, has the same frequency as the previous synchronous detector.

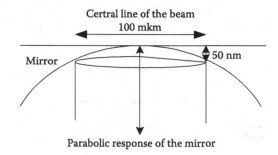

FIGURE A13.49
Beam attitude system.

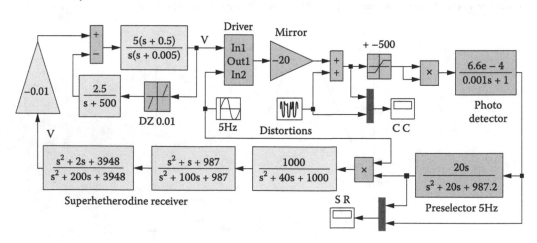

FIGURE A13.50
Optical receiver.

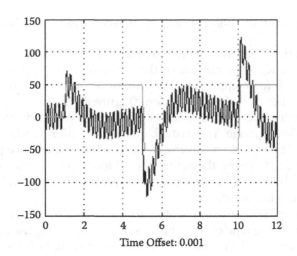

FIGURE A13.51
Application of temporary shift of bias.

The nonlinear dynamic compensator with a dead zone goes after the intermediate frequency filter and serves to make the feedback larger.

The feedback system can then be used for reducing the detector outputs. The intermediate filter changes the 0.3 Hz frequency band, depending on the input channel.

The Simulink oscilloscope output signals are shown in Figure A13.51. The distortion signal caused by mirror jerks is shown by the thin line. It changes at the arbitrary intervals at the output of the mirror. The 5 Hz filled signal shows the conversion of the signal to the initial signal level in less than 3 s, which is quite sufficient.

A13.16 Six-Axis Active Vibration Isolation

The platform on which the vibration source is placed is suspended on six identical active struts in an orthogonal (hexapod) configuration. Each active strut contains a diaphragm that allows motion only in the direction. For this direction, the diaphragm represents a reasonably soft spring. In parallel with this spring is a voice coil actuator. The strut is connected to the structure in series with an axial force sensor and two axially stiff, cross-blade flexures serving to reduce coupling between struts. The stiffness of the diaphragm was chosen such that the vertical mount mode is roughly 10 Hz, small enough to provide sufficient passive vibration isolation above the controller bandwidth. At the same time, the diaphragm is stiff enough to pass reaction wheel attitude control torques and accommodate gravity sag. Ideally, each strut transmits force only in the axial direction. However due to imperfect cross-blade flexure, each strut passes nonaxial vibration as well, creating coupling between the struts.

Disturbance isolation must be especially large over the range of 20–60 Hz, where feedback in the optical control loop is limited and therefore fails to sufficiently reduce the disturbances.

In this frequency range, the isolation provided by the strut must be increased by application of feedback from the force sensor to the voice coil in each strut. The feedback is single-input single-output (SISO), and each of the six loops is largely decoupled from the other five loops.

Feedback in the active strut changes the logarithmic force transfer function log $|K_F|$ by Δlog $|K_F|$. Specifically, K_F is the ratio of (1) the vibration force applied by the strut to the base structure to (2) the force applied by the vibration source. The integral of this change along the inverse frequency scale must be zero (Figure A13.52).

In other words, when the disturbance propagation is reduced over some range, it is increased over a different range. The disturbance propagation needs to be reduced in the range of 10–100 Hz. At frequencies lower than 10 Hz, the disturbance propagation will be increased by feedback. However, the active optical loop has large feedback in that range. At high frequency, sufficient disturbance isolation is provided passively, by the softness of the diaphragm of the active strut.

We apply the disturbance isolation control to the base structure having many flexible modes with uncertain frequencies. For this system to be stable, the active-strut mobility function must be positive real. For the disturbance isolation to be large, the modulus of the mobility needs to be paid attention to in order to reduce transmission at the frequencies of the lightly damped structural modes.

Generally, the strut mobility is $Z = Z_o(1 + T_F)$, where Z_o is the strut mobility without feedback and T_F is the force feedback loop transfer function. The mobility phase angle must be within the interval $-90°$ to $90°$. Since Z_o degenerates at low frequency into the diaphragm mobility whose phase angle is $90°$, the phase angle of $(1 + T_F)$ at low frequency must approach zero. In other words, the force feedback must degenerate into a constant.

An amplifier with high-output impedance (i.e., a current driver) drives the voice coil. In this case the back electromotive force has no effect and the vibration force transmission without feedback (and with amplifier power on) is only that caused by the diagram stiffness. Compared to the case of using a voltage driver, this substantially reduces $|K_F|$ without feedback over the range up to 60 Hz. At high frequency, where the structural mobility increases, the choice of current or voltage drivers makes no difference since the mobility of the voice coil is already much larger than the mobility of the load. As an additional

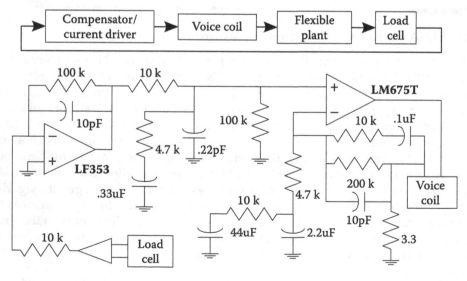

FIGURE A13.52
Block diagram in one-direction isolation loop and related circuit diagram.

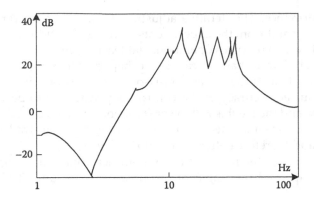

FIGURE A13.53
Isolator loop gain for a single strut with all other loops closed.

advantage, using current drivers reduces both the effects of the flexible structure on the feedback loop and the coupling between the loops.

Since no commercial current driver was readily available at a reasonable price, the amplifiers were custom designed around power operational amplifiers. (*Warning*: power operational amplifiers can self-oscillate with a frequency many times higher than the nominal when the loads are reactive. To prevent the oscillation, filter capacitors need to be appropriately placed and wire connections need to be short as governed by the radio frequency design rules.) Since the actuator inputs and the load cell inputs are all analog, the feedback loop compensators were also implemented as analog active resistor-capacitor (RC) circuits, resulting in a stand-alone system. Although active isolation modifies the plant for both optical control and attitude control, the plant for each remains passive.

The Bode diagram for a feedback loop measured with the other five loops closed is shown in Figure A13.53. All individual loop responses are similar. The coupling between the individual loops is smaller than 130 dB for frequencies up to 100 Hz.

A13.17 MIMO Motor Control Having Loop Responses with Bode Steps

An approximation to the theoretical loop response with Bode step described in Section 4.2.3 has been employed in precision SISO motor control systems for retroreflector carriage motion control and for gyroscope attitude control of an instrument on the Chemistry spacecraft.

It was also employed for several less precise attitude, position, and motion controls of robot-mounted drills: for a robot designed to take samples of concrete at the Chernobyl nuclear station, for a rover to take samples of Mars rocks, and for a device to take material samples of a comet or an asteroid in a nearly zero-gravity environment.

In these MIMO control systems, the feedback loops' bandwidths are limited by the rover and drill structural resonances. The same nominal loop response was used in most of the critical feedback loops. The ith loop response was therefore fully defined by the crossover frequency f_{bi}. The trade-offs between the different performance requirements for the loops of these MIMO systems were reduced to finding the set of scalar numbers f_{bi} that provide good system performance. The iterative adjustments of the crossover frequencies with computer simulations and from the experiments converged rapidly.

After the control system becomes operational (at least in simulations), the loop response at frequencies much lower than f_{bi} can be reshaped following the Bode relation in Equation 3.12. In most cases there was no reason to do this and it was not done, but in some cases the required accuracy of the control loop was able to be well satisfied with much less feedback, and the feedback at lower frequencies was reduced. Within the limitations imposed by the Bode integrals, this loop gain reduction was traded for bigger phase stability margins, and therefore better transient responses.

The last design step was the introduction of nonlinear dynamic compensation in some of the loops that improved the transient responses to large-amplitude commands and disturbances.

A13.18 Mechanical Snake Control

A gigantic animated mechanical snake employed in the movie *Anaconda* was constructed of chain-connected identical mechanical links, each with the ability to bend in pitch and yaw directions. Position feedback allowed achieving good transient response of a single link. However, transient responses of several such links in a chain were very oscillatory. The oscillation was caused by the interference of mechanical waves propagating along the snake.

To correct the problem, force feedback was added and, using the methods detailed in Chapter 7, the output mobilities of the actuators were made dissipative. The damping introduced into the system reduced the effects of the waves' interference, the transient response of the chain of several links was drastically improved as exemplified in Figure A13.54 for the five-link chain, and the movements of the entire mechanical snake became agile and very impressive.

This example was contributed by *JAS Company*.

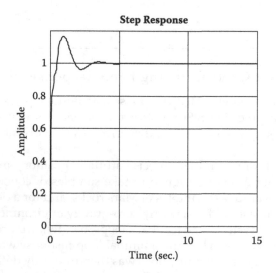

FIGURE A13.54
Transient responses of five links in a chain with position and force feedback.

A13.19 Optical Control with Simulink Multi-loop Systems

The simplified optical controller diagram is shown in Figure A13.55. The feedback loops and their bandwidth are limited by the masses of the mirrors in the delay lines. The slow (marked by the index S) and fast (index F) optical delay lines are combined by the filter at the output of the diagram to produce close to the best estimate of the fringe position.

The mirror mass for FDL is 80 g, and for SDL, 120 g. The stroke for the SDL mirror is 10 μm, for FDM, only 1 μm. The structural mode frequency for the SDL is within the range of 200–500 Hz.

In the block diagram (Figure A13.56) the plant models reflect the mirror mass, the stiffness of the PZT, and the stiffness of the support structure. The output impedance of the driver is considered small (voltage driver).

The driver models are shown in Figure A13.57.

Analog compensation allows for reducing the computer calculation load and increasing the sampling rate. The compensators are to be placed at the driver input, after the D/A converter.

The compensator models are shown in Figure A13.58.

The Bode diagrams for the main and vernier loops are shown in Figure A13.59.

Transient responses obtained with Simulink simulation are shown in Figure A13.60. The oscillation on the responses is caused by the structural mode in the slow DL.

A13.20 Industrial Furnace Temperature Control

Temperature control of a precision industrial furnace (Figure A13.61) is implemented as a single-loop SISO system depicted in Figure A13.62. The plant includes a sink representing heat dissipation into the environment, two thermal capacitances, the furnace, and the payload, connected by the heat flux conductance as illustrated in Figure A13.61(a). The plant equivalent electrical circuit is shown in Figure A13.61(b), and the plant transfer function is the ratio of the payload temperature to the difference of the power of the heater

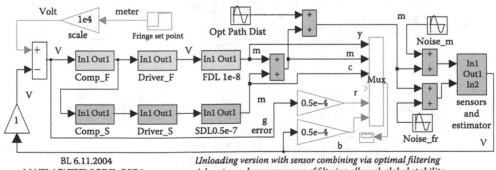

FIGURE A13.55
Block diagram for the control and dither circuitry.

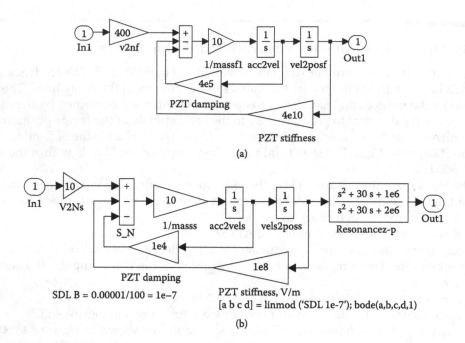

FIGURE A13.56
Delay line models: (a) fast and (b) slow.

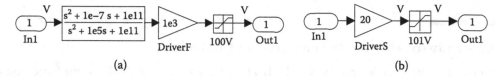

FIGURE A13.57
Driver models: (a) fast and (b) slow.

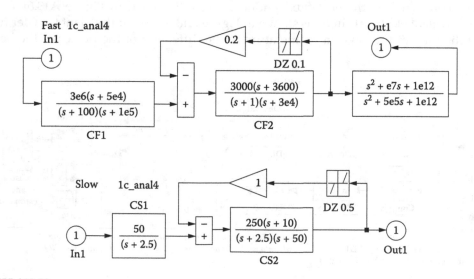

FIGURE A13.58
Compensator models: fast (top) and slow (bottom).

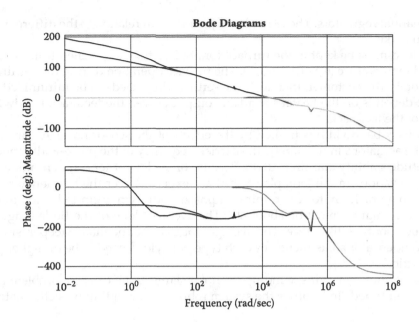

FIGURE A13.59
Control system Bode diagrams for both actuators on (highest) and only the fast actuator on, and the closed loop (lowest curve) with both actuators on.

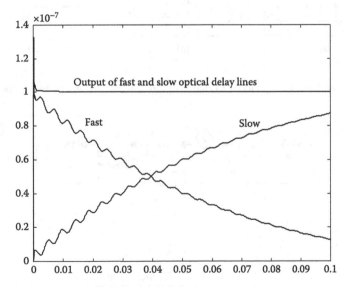

FIGURE A13.60
Transient responses for the fast and slow delay lines.

and that dissipated to the environment. The plant gain frequency response is shown in Figure A13.61c.

The actuator is a pulse-width modulated electrical heater H, the carrier being 60 Hz. The signal amplitude of the thermocouple sensor TC is rather small, so that the error amplifier noise limits the system's accuracy. The response of the PID compensator can be adjusted by three manual regulators. The regulator adjustments in relation to the different payloads are easy to make.

The payload must be kept in the furnace for not less than a specified time (say, 3 h) at a prescribed temperature (e.g., 1,150° ± 2°). The furnace operation during the settling time is not paid for by the customer; therefore, the settling time needs to be minimized. The settling time depends on the heater's available output power, the feedback bandwidth, and the value of the feedback.

The feedback bandwidth is limited by the effects of the sensor noise and the amplifier noise on the actuator's input, and by the carrier frequency of the pulse-width modulation. The amplitude stability margin is 10 dB. Because of the low-order compensator employed, rather big variations in the loop phase shift are expected over the frequency range with positive loop gain. Then, for the minimum phase stability margin to be 40°, the average phase stability margin must be about 60°. The average slope of the Bode diagram must therefore be about −8 dB/octave. The average slope cannot be made steeper, since in this case the compensator adjustments for each type of payload need to be of higher precision than is practical.

With such stability margins, wind-up cannot be large, but it can be a problem if the system is not well tuned. To improve the system robustness, wind-up was eliminated with a

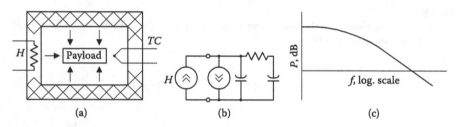

FIGURE A13.61
(a) Industrial furnace with payload, (b) its equivalent thermal schematic diagram, and (c) the plant frequency.

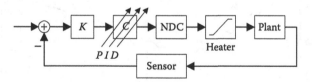

FIGURE A13.62
Temperature control system.

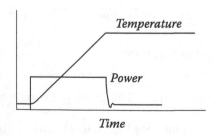

FIGURE A13.63
Closed-loop transient response.

simple nonlinear dynamic compensation (NDC) that reduces the loop phase lag for large-level signals.

The system transient response is shown in Figure A13.63. The control looks nearly time optimal because the settling time in the linear mode of operation is many times shorter than the full-power preheating time.

Digital implementation of the controller simplified the interfacing and made the controller cheaper in mass production.

Appendix 14: Bode Step Toolbox

The toolbox includes the following MATLAB functions:

1. NYQLOG (Nyquist plot on logarithmic scale with frequency in octave marks)
2. BOLAGNYQ (Bode diagram for the gain and the lag margin, and the logarithmic Nyquist plot)
3. TFSHIFT (frequency transform, frequency de-normalization, similar to lp2lp)
4. BONYQAS (asymptotic Bode diagram, phase plot, and logarithmic Nyquist plot); this program is stated in Appendix 5
5. BOSTEP (rational function approximation to optimal Bode step response)
6. BOCLOS (prefilter design for closed-loop low overshoot and fast settling)
7. BOINTEGR (breaking compensator into two parallel paths, one of which is dominant at low frequencies)
8. BOCOMP (calculation of the compensator function for a dc motor control)
9. NDCP (Nyquist plot for describing functions of NDC with parallel paths)
10. BNDCP (Bode plot for describing functions of NDC with parallel paths)
11. NDCB (Nyquist plot for describing functions of NDC with feedback path)
12. BNDCB (Bode plot for describing functions of NDC with feedback path)

These MATLAB functions simplify designing linear/nonlinear servo loops with Bode optimal loop responses and prefilters. The m-functions (and MATLAB scripts for the examples and the problems listed in the book) are available from the author's Web page: www.luriecontrol.com. The functions do the following:

A. Plotting routines
   1. NYQLOG plots a Nyquist diagram on the logarithmic plane with octave marks. The marks allow reading the slope of the Bode diagram without plotting it. The loop shaping can be performed iteratively using this plot instead of the Bode diagrams.
   2. BOLAGNYQ plots frequency responses of the gain and the phase lag stability margin, and plots the Nyquist diagram on the logarithmic plane with octave marks. This plot's arrangement is convenient for loop shaping. Using the plot of the lag margin allows one to pay attention to only a small area of the plot where the gain response crosses the 0 dB level, and to read there the value of the guard-point phase lag margin.
B. Calculations
   3. TFSHIFT de-normalizes an initially normalized frequency response similarly to the MATLAB function lp2lp but does this by a polynomial transform in the s-domain instead of matrix manipulation in the t-domain (lp2lp gives complex numbers as the answers for the gain and, sometimes, produces inaccurate answers). It can be used, for example, in problems like that described in Example 4.8.

C. Plant modeling, asymptotic diagrams, and conceptual design

    4. BONYQAS simplifies the system conceptual design. In particular, it helps to accomplish the following two tasks:

        I. Calculating and plotting the phase shift of the plant when only the plant gain has been estimated or measured (since the measurements of the plant gain are often less time-consuming than the measurements of the phase shift).

        II. Completing the conceptual design of the feedback loop using asymptotic piece-linear Bode diagrams. The function generates the asymptotic Bode diagram and calculates and plots the phase and Nyquist diagrams. It allows for easy shaping of the responses by changing the corner frequencies, the gain at the corner frequencies, the linear non-minimum phase lag coefficient, and the asymptotic slopes at low frequencies (the type of loop) and at high frequencies. The obtained parameters of the Bode step can be further used as the input file parameters for the BOSTEP function.

D. Feedback loop design

    5. BOSTEP generates a rational function approximation to the theoretical transcendental (or irrational, obtained with BONYQAS) loop response with a Bode step for specified margins, asymptotic slope, non-minimum phase, and the feedback type.

    6. BOCLOS generates a normalized closed-loop transfer function without and with a prefilter, and the frequency response of a fourth-order Bessel filter. Also, it generates the step response. It is usable for designing a prefilter rendering a good closed-loop transient response.

    7. BOINTEGR converts the linear compensator transfer function into a sum of two transfer functions, the low-pass one (a generalization of an integrator of a PID controller) and the rest. Then, saturation can be introduced into the low-frequency path to improve the transient response and provide global stability.

E. Compensator design

    8. BOCOMP calculates the compensator transfer function for the loop response with a Bode step of a servo with a dc permanent magnet motor. The input file includes the loop transfer function generated by BOSTEP and the dc motor and the inertial load parameters.

F. Nonlinear dynamic compensator design

    9. NDCP plots iso-*e* describing functions on the logarithmic Nyquist plane for an NDC with parallel paths, one of which includes a variable or a nonlinear (typically, a saturation) link. A series link is also included that can imitate the rest of the feedback loop so that the loop Nyquist diagrams can be plotted, with the values of the describing function of the nonlinear element from 0 to 1.

   10. BNDCP plots iso-*e* describing functions Bode diagrams for an NDC with parallel paths, one of which includes a variable or a nonlinear link (typically, a saturation). A series link is also included that can imitate the rest of the feedback loop so that the loop Bode diagrams can be plotted, with the values of the describing function of the nonlinear element from 0 to 1.

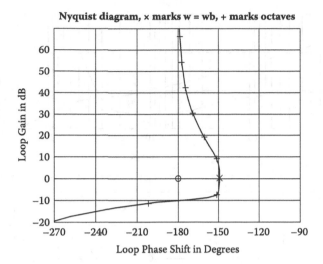

Nyquist diagram, × marks w = wb, + marks octaves

FIGURE A14.1
Default/demo for nyqlog and bostep.

11. NDCB plots iso-*e* describing functions on the logarithmic Nyquist plane for an NDC with a feedback path that includes a variable or a nonlinear link (typically, a dead zone). A series link is also included that can imitate the rest of the feedback loop so that the loop Nyquist diagrams can be plotted, with the values of the describing function of the nonlinear element from 0 to 1.

12. BNDCB plots iso-*e* describing functions Bode diagrams for an NDC with a feedback path that includes a variable or a nonlinear link (typically, a dead zone). A series link is also included that can imitate the rest of the feedback loop so that the loop Bode diagrams can be plotted, with the values of the describing function of the nonlinear element from 0 to 1.

The functions' help files include default/demo that generates plot when the name of the function is typed in without arguments. The plots are shown in Figures A14.1–A14.7 and A14.9–A14.12.

A14.1 DC Motor Servo Design

A dc motor control system (like that described in Example 4.7) can be designed using the following sequence of the toolbox functions.

First, with bonyqas, an asymptotic Bode diagram with a Bode step is chosen such that the feedback satisfies the disturbance rejection requirements, and the attenuation in the feedback loop is sufficient at the frequencies of structural modes to guarantee system stability with the chosen stability margins. An example of the diagram is shown in Figure A14.4. The stability margins, the frequencies of the beginning and the end of the Bode step, the asymptotic slope, the non-minimum phase lag, and the system type are determined during this conceptual feedback loop design.

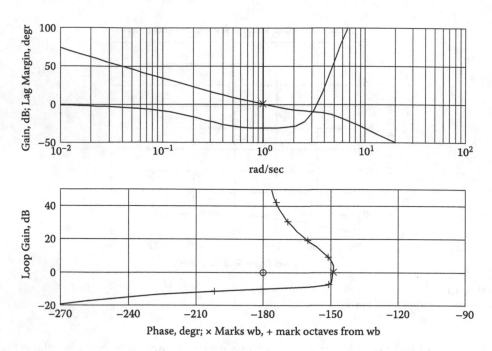

FIGURE A14.2
Default/demo for bolagnyq.

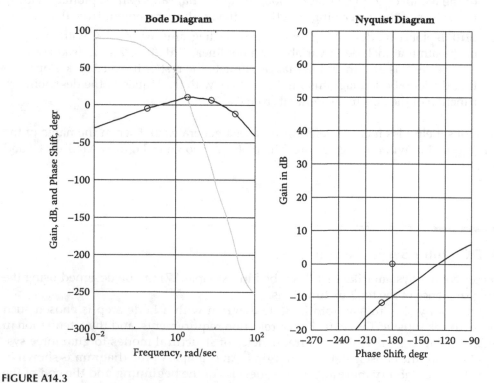

FIGURE A14.3
Default/demo 1 for bonyqas, calculating and plotting a band-pass plant phase; circles on the gain response mark the corner frequencies of the piece-linear approximation.

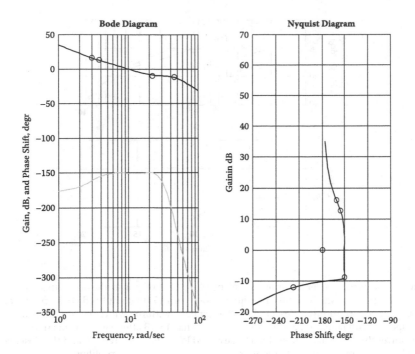

FIGURE A14.4
Default/demo 2 for bonyqas, asymptotic loop response; circles on the gain response mark the corner frequencies of the piece-linear asymptotic Bode diagram for loop response with Bode step.

Second, with the function `bostep` a rational function approximation for the loop response is obtained, and a compensator function for the nominal single-integrator plant with a small non-minimum phase lag is calculated (as was done in Example 4.6). The responses at this point are normalized in frequency; i.e., the loop gain response has unity crossover frequency, like that shown in Figure A14.1.

Third, with the function `boclos` the closed-loop frequency response is plotted, the prefilter is calculated, and the closed-loop with the prefilter response is plotted, like that shown in Figure A14.5. The MATLAB function can be used to calculate the closed-loop response with a prefilter comprising two notches, one peak, and a low pass—or with any subset of these.

The goal for the closed-loop response with the prefilter is a response close to a de-normalized Bessel filter response. The de-normalized filter response can be obtained by shifting the nominal filter response along the logarithmic frequency axis.

Fourth, using the motor and load parameters, the compensator and the plant transfer functions are calculated and plotted with `bocomp`, as were those in Figure A14.6.

Fifth, with the function `bointegr`, the compensator transfer function is split into a parallel connection of the low-frequency path and the second path (for the medium and higher frequencies), as exemplified by the plots in Figure A14.7.

Next, a Simulink model of the system should be built like that shown in Figure 7.26. The model must include flexible modes of the load and the nonlinear elements in the actuator (saturation in current and voltage), in the friction model, and in the compensator (saturation is introduced in front of the low-frequency path of the compensator). With the model, the controller is fine-tuned to perform well over the specified range of the plant parameter variations, and the system performance and robustness are evaluated.

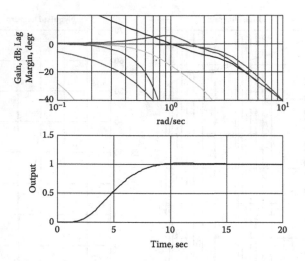

FIGURE A14.5

Default/demo for boclos plots several responses: an open-loop gain response (the one with large gain at low frequencies); closed-loop responses, without a prefilter (the one with a 7 dB hump) and with a prefilter that consists of a notch, a peak, and a third-order Bessel filter (the response with the widest bandwidth); and fourth-order Bessel filter response with nominal bandwidth 1 rad/s that has 15 dB attenuation at this frequency. The phase responses (the lower three) are of the Bessel filter, the closed loop with the prefilter, and the closed loop without a prefilter; the lower plot shows the closed-loop transient response with the prefilter.

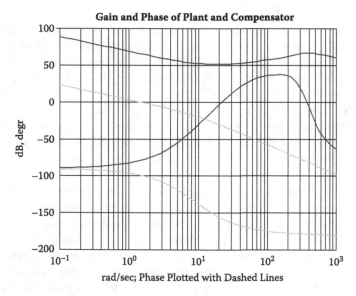

FIGURE A14.6

Default/demo for bocomp, the gain and phase responses for the compensator and the plant of a dc motor control system.

A14.2 NDC Iso-E Describing Functions

The iso-E responses for signal paths shown in Figure A14.8 are exemplified in Figures A14.9– A14.12 with the DF values 0, 0.2, 0.4, 0.6, 0.8, and 1. These values correspond to the signal levels $E/e_s = \infty$, 6.3, 3, 2, 1.5, and < 1 for the saturation and $E/e_{dz} < 1$, 1.5, 2, 3.5, 8, and ∞ for the dead zone.

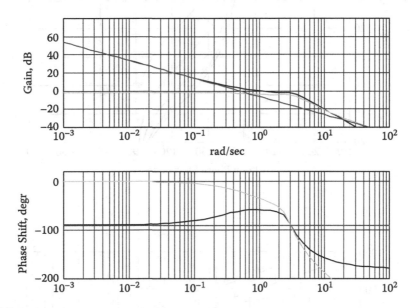

FIGURE A14.7
Default/demo for bointegr, the gain and phase responses for the entire compensator and its two parallel paths.

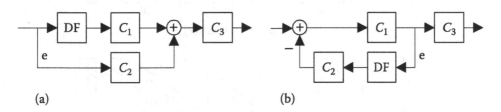

(a)   (b)

FIGURE A14.8
NDCs with (a) parallel paths and (b) with a feedback path.

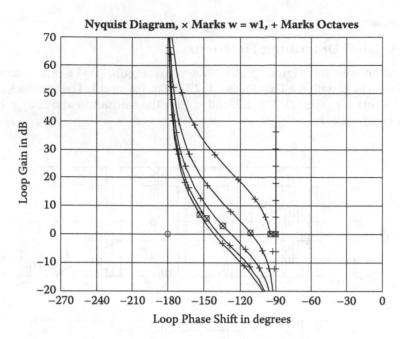

FIGURE A14.9
Default/demo for ndcp.

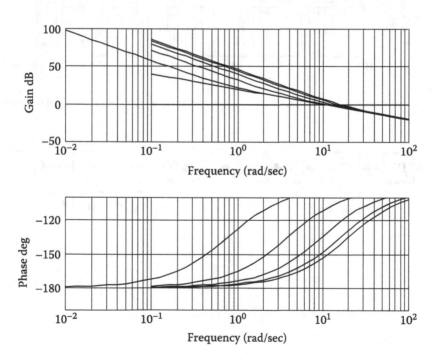

FIGURE A14.10
Default/demo for bndcp.

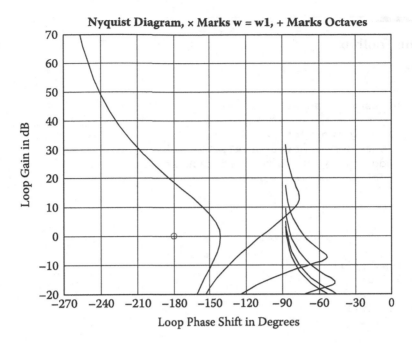

FIGURE A14.11
Default/demo for ndcb.

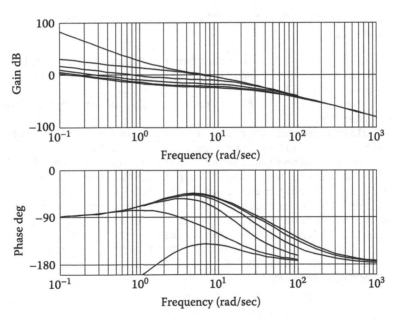

FIGURE A14.12
Default/demo for bndcb.

A14.3 The Toolbox

ADVECTOR

```
function [vectsum] = advector(v1,v2)

% ADVECTOR Adding two vectors
% function [vectsum] = advector(v1,v2)
% ADVECTOR adds two vectors of different length
% Copyright (c) B.J.Lurie 7-22-98.

  format short e

  ll = length(v1) - length(v2);
if ll < 0
  adz = zeros(1,-ll);
  vectsum = [adz v1] + v2;
else
  adz = zeros(1,ll);
  vectsum = v1 + [adz v2];
end
```

BOCLOS

```
function [dclos, nprf1, dprf1, nprf2, dprf2, numtot, dentot] =
boclos(num1,den1,y)

% BOCLOS Normalized closed loop transfer function without and with
% prefilter, and frequency response of 4th order Bessel filter. Also,
% step response, optional.
% function [dclos, nprf1, dprf1, nprf2, dprf2, numtot, dentot] =
% boclos(num1,den1,y)
% BOCLOS calculates denominator dclos of the closed loop transfer
% function without prefilter (the numerator is num1);
% prefilter1 and prefilter2 transfer functions nprf1(s)/dprf1(s)
% and nprf2(s)/dprf2(s); and closed loop transfer functions:
% num1(s)/dclos(s) without prefilter and numtot(s)/dentot(s)
% with prefilter; the coefficients are in descending powers of s.
% The defaults follow the default outputs of BOSTEP:
% num1 = [-9.6082 626.39 1947.8 3326.0 1093]
% den1 = [1 75.102 492.80 1716.3 3222.6 2217 0 0]; y = 0.1667;
% Copyright (c) B.J.Lurie 3-15-98. See also BOSTEP, BOCOMP, BONYQAS,
% NYQLOG, BOLAGNYQ, TFSHIFT

% default:
  if nargin < 3
    y = 0.1667;
  end
  if nargin < 2
```

```
   denl = [1 75.102 492.80 1716.3 3222.6 2217 0 0];
  end
  if nargin == 0
    numl = [-9.6082 626.39 1947.8 3326.0 1093];
  end
  format short e

% Normalized closed loop function denominator (without prefilter)
  ll = length(denl) - length(numl);
if ll < 0
    adz = zeros(1,-ll);
    dclos = [adz denl] + numl;
  else
    adz = zeros(1,ll);
    dclos = denl + [adz numl];
  end
% First and second prefilter notches
  nprf1 = [1 y/0.1667 1]; dprf1 = [1 2 1]; % prefilter 1
  nprf2 = [1 0.47*y/0.1667 1]; dprf2 = [1 0.6 1]; % prefilter 2

% Prefilter 3 provides gain hump
  nprf3 = [1 2.5 3.6]; dprf3 = [1 1.5 3.6]; % prefilter 3

% Closed loop transfer function with prefilters 2 and 3
  numtot = conv(conv(nprf3,numl),conv(nprf1,nprf2));
  dentot = conv(conv(dprf3,dclos),conv(dprf1,dprf2));

% Closed loop function with extra 3rd order Bessel lowpass
% prefilter,w = 0.25
  numtot = conv([15],numtot);
  dentot = conv([64 96 60 15],dentot);

% Open- and closed-loop and Bessel filter frequency responses
  w = logspace(-1,1);
  [magop,phaseop] = bode(numl, denl, w) % open-loop
  [magcl,phasecl] = bode(numl, dclos, w) % clos-loop without prefilter
  [magtot,phasetot] = bode(numtot, dentot, w) % clos-loop with prefilter
  [magbf,phasebf] = bode(105, [1 10 45 105 105], w) % Bessel 4th order
% [magbf,phasebf] = bode(15, [1 6 15 15], w) % Bessel 3th, option

% plotting
  subplot(2,1,1)
  semilogx(w, 20*log10(magop),'k', w,(phaseop),'k--');
  hold on
  semilogx(w, 20*log10(magcl),'b', w,(phasecl),'b--');
  semilogx(w, 20*log10(magtot),'g', w,(phasetot),'g--');
  semilogx(w, 20*log10(magbf),'r', w,(phasebf),'r--');
  wmin = min(w); wmax = max(w);
  axis([wmin wmax -40 20])
  xlabel('rad/sec')
  ylabel('gain, dB; lag margin, degr')
  grid
```

```
hold off
subplot(2,1,2)
step(numtot, dentot)
xlabel('time, sec')
ylabel('output')
grid
zoom on
```

BOCOMP

```
function [numc_pl,denc_pl] = bocomp(wb,numc,denc,kmot,resist,mom_inert);
% BOCOMP Compensator transfer function numc_pl(s)/denc_pl(s)for a dc
% motor system with loop response generated by bostep. The vectors
% NUMC_PL and DENC_PL contain polynomial coefficients in descending
% powers of s,
% The driver voltage gain is 1. The sum of its output resistance with
% the motor winding resistance is resist.The motor constant is kmot.
% The rigid body load moment of inertia is mom_inert.
% The loop crossover frequency in rad/sec is wb.
% The function numc(s)/denc(s) is the normalized compensator obtained
% with function bostep for the nominal plant (1/s)(-s+wc/bnc)/(s+wc/bnc).
% BOCOMP calculates: the compensator polynomial coefficients,
% the gain coefficient, the poles and the zeros,
% the biquad coefficients for complex poles and zeros,
% and plots: Bode diagrams for the plant (in green) and
% compensator (in read).
% Defaults/demo initiated by typing bocomp:
% mom_inert = 0.027; resist = 2; kmot = 0.7; wb = 100;
% denc = [ 1 6.8631 24.466 46.749 32.488 0];
% numc = [9.6082 29.262 48.975 16.016];
% [numc_pl,denc_pl] = bocomp(wb,numc,denc,kmot,resist,mom_inert)
% Copyright (c) B.J.Lurie 6-18-98. See also BOSTEP, BOCLOS, NYQLOG,
% TFSCHIFT, BOLAGNYQ, BONYQAS, BOINTEGR

  if nargin < 6,
    mom_inert = 0.027; ;
  end
  if nargin < 5,
    resist = 2;
  end
  if nargin < 4,
    kmot = 0.7;
  end
  if nargin < 3,
    denc = [1 7.9547 33.536 74.518 58.275 0];
  end
  if nargin < 2,
    numc = [13.046 45.855 76.234 22.827]; % ratio of 4th to 6th
  end
  if nargin == 0,
    wb = 100;
  end
```

```
% plant transfer function kmot/(resist*mom_inert)/[s(s+a)] where
  a = kmot^2/(resist*mom_inert);    %real pole of the plant
  npm = kmot/(resist*mom_inert); dpm = [1 a 0]; % minimum phase plant
  % without flex modes

% denormalized m.p. compensator for n.p. single integrator plant
  [numc_wb, denc_wb] = tfshift(numc,denc,wb);

% denormalized m.p. compensator for motor plant
  numc_pl = wb*deconv(conv(numc_wb,dpm),[1 0]);
  denc_pl = conv(denc_wb,npm);
  format short e

  gain_coeff = numc_pl(1)/denc_pl(1);
  roots_num = roots(numc_pl);
  biquad_num = [1 -2*real(roots_num(2)) abs(roots_num(2))^2];
  roots_den = roots(denc_pl);
  biquad_den = [1 -2*real(roots_den(2)) abs(roots_den(2))^2];

% printout
  gain_coeff
  roots_num
  biquad_num
  roots_den
  biquad_den
  numc_pl
  denc_pl

% plotting
  wmin = floor(log10(wb)-3);
  wmax = ceil(log10(wb)+1);
  w = logspace(wmin,wmax);
  [magc,phasc] = bode(numc_pl,denc_pl,w);
  [magnpm, phasenpm] = bode(npm,dpm,w);
  semilogx(w,20*log10(magc),'r', w,phasc,'r--'); % compensator
  hold on
  semilogx(w,20*log10(magnpm),'g', w,phasenpm,'g--') % plant
  hold off
  xlabel('rad/sec; phase plotted with dashed lines')
  ylabel('dB, degr')
  title('gain and phase of plant and compensator')
  grid
  zoom on
```

BOINTEGR

```
function [ao, p1, numr, denr] = bointegr(numc,denc,wb)
% BOINTEGR expresses the compensator transfer function as a sum of the
% term of the partial fraction expansion ao/(s + p1) which is dominant
% a lower frequencies and the rest of the terms combined into transfer
% function numr/denr, where the coefficients in the vectors of
% coefficients are in the descending powers of s. numc is up to
% 9th order. wb is the crossover frequency or anything close enough,
```

```
% it is used here only to scale the frequency axis; default is 1.
% Bode diagrams for the compensator and its two parallel paths are
% plotted.
% The default/demo is
% numc = [9.6082 29.262 48.975 16.016];
% denc = [1 6.8631 24.466 46.749 32.488 0]; wb = 1;
% function [ao, p1, numr, denr] = bointegr(numc,denc);
% Copyright (c) B. Lurie 6-19-98. See also BOLAGNYQ, BONYQAS, BOSTEP,
% BOCLOS, BOCOMP, NYQLOG, TFSHIFT

% default
  if nargin < 3
    wb = 1;
  end
  if nargin < 2
    numc = [9.6082 29.262 48.975 16.016];
    denc = [1 6.8631 24.466 46.749 32.488 0];
  end
  format short e

% Partial fraction expansion
  ld = length(denc);
  [r,p,k] = residue(numc,denc);
  ao = r(ld-1); p1 = p(ld-1);

% Combining the rest of the terms
  if ld == 2
    rr = 0;
    pr = 1;
  end
    if ld == 3
    rr = [r(ld-2)];
    pr = [p(ld-2)];
  end
  if ld == 4
    rr = [r(ld-3) r(ld-2)];
    pr = [p(ld-3) p(ld-2)];
  end
  if ld == 5
    rr = [r(ld-4) r(ld-3) r(ld-2)];
    pr = [p(ld-4) p(ld-3) p(ld-2)];
  end
  if ld == 6
    rr = [r(ld-5) r(ld-4) r(ld-3) r(ld-2)];
    pr = [p(ld-5) p(ld-4) p(ld-3) p(ld-2)];
  end
  if ld == 7
    rr = [r(ld-6) r(ld-5) r(ld-4) r(ld-3) r(ld-2)];
    pr = [p(ld-6) p(ld-5) p(ld-4) p(ld-3) p(ld-2)];
  end
  if ld == 8
    rr = [r(ld-7) r(ld-6) r(ld-5) r(ld-4) r(ld-3) r(ld-2)];
    pr = [p(ld-7) p(ld-6) p(ld-5) p(ld-4) p(ld-3) p(ld-2)];
  end
```

```
  if ld == 9
    rr = [r(ld-8) r(ld-7) r(ld-6) r(ld-5) r(ld-4) r(ld-3) r(ld-2)];
    pr = [p(ld-8) p(ld-7) p(ld-6) p(ld-5) p(ld-4) p(ld-3) p(ld-2)];
  end
  [numr,denr] = residue(rr,pr,k);
  numr = real(numr); denr = real(denr); % elimination of imaginary
  % parts caused by numerical errors

% Print the answers
    ao
    p1
    numr
    denr
    % if further breaking into parallel channels needed, rr and pr

    wmin = floor(log10(wb)-3);
    wmax = ceil(log10(wb)+2);
    w = logspace(wmin,wmax);
  [mc,pc] = bode(numc,denc,w);
  [mr,pr] = bode(numr,denr,w);
  [mi,pi] = bode(ao,[1 p1],w);

% plotting
  wmin = min(w); wmax = max(w);
  subplot(211)
  semilogx(w, 20*log10(mc),'k', w, 20*log10(mr),'g', w,
20*log10(mi),'r');
  axis([wmin wmax -40 80])
  xlabel('rad/sec')
  ylabel('gain, dB')
  grid
  subplot(212)
  semilogx(w,(pc),'k', w,(pr),'g', w,(pi),'r');
  axis([wmin wmax -200 20])
  ylabel('phase shift, degr')
  grid
  hold off
  zoom on
```

BOLAGNYQ

```
function BOLAGNYQ(wb,numl,denl)
%BOLAGNYQ Gain and -(phase+180) (i.e. lag margin) responses, and Nyquist
% diagram on logarithmic plane
%function BOLAGNYQ(wb,numl,denl)
% The loop transfer function is numl(s)/denl(s), where numl and
% denl contain the polynomial coefficients in descending powers of s.
% Frequency wb in rad/sec specifies the position of cursor x; it is
% convenient to put it at the frequency where the loop gain is 0 dB.

% Default/demo: wb = 1, numl = [-26 178 654 1269 429],
% denl = [1 18.8 132 543 1164 860 0 0]
% Copyright (c) B.J.Lurie 6-16-98, La Crescenta. See also BOSTEP,
```

```
% BOCLOS, BONYQAS, BOINTEGR, TFSHIFT, STEPLOG, NYQLOG

  if nargin < 3,
    den1 = [1 18.8 132 543 1164 860 0 0];
  end
  if nargin < 2,
    num1 = [-26 178 654 1269 429];
  end
  if nargin == 0,
    wb = 1;
  end

  [mag, phase, w] = bode(num1, den1);

  subplot(2,1,1)
  semilogx(w, 20*log10(mag),'k', w,(-180-phase),'k--');
  hold on
  [mmb,ppb] = bode(num1, den1,wb);
  semilogx(wb,mmb,'kx');
  wmin = min(w); wmax = max(w);
  axis([wmin wmax -50 100])
  xlabel('rad/sec')
  ylabel('gain, dB; lag margin, degr')
  grid
  hold off

  subplot(2,1,2)
  plot(phase, 20*log10(mag),'k')
  set(gca,'XTick',[-270 -240 -210 -180 -150 -120 -90])
  grid
  axis([-270 -90 -20 50])
  hold on
  [a,b] = bode (num1,den1,wb); plot(b,20*log10(a),'kx')
    for w = [0.0156 0.03125 0.0625 0.125 0.25 0.5 2 4],
      [a,b] = bode(num1,den1,wb*w); plot(b,20*log10(a),'r+');
    end
  plot(-180, 0, 'ro');
  hold off
  xlabel('phase, degr; x marks wb, + mark octaves from wb')
  ylabel('loop gain, dB')
  zoom on

  wb
  gain_at_wb = mmb
  lag_margin = 180 + ppb
```

BONYQAS

```
function [gain1,phase1] = bonyqas(u,g,lslope,hslope,nml);
% function [gain1,phase1] = bonyqas(u,g,lslope,hslope,bnm);
% BONYQAS Asymptotic Bode and Nyquist diagrams corresponding in minimum
% phase manner to a piece-linear gain response in dB specified by the
% gain g(h) at k corner frequencies u(h) in rad/sec. The lengths of
```

```
% vectors g and u must be equal. Coefficient lslope defines the
% asymptotic slope at lower frequencies in integrator units, and
% hslope defines the asymptotic slope at higher frequencies. nml is
% nonminimum phase lag in degrees at u(h); the lag is proportional
% to the frequency, default is nml = 0.
% Smoothed Bode diagram is plotted in white. Phase is plotted in greed.
% Nyquist diagram is plotted on logarithmic plane.
% The function is applicable for reasonably smooth responses. The phase
% accuracy for typical feedback loop shaping problem without sharp
% peaks or notches is better than 3 degr.
%
% Default: bonyqas without arguments plots the band-pass response:
% bonyqas([0.2 2 8 32],[-5 10 6 -12],-1,3);
% Low-pass loop Example 1: bonyqas([3 4 20 30],[16 13 -9 -10],2, 3, 0)
% Loop Example 2: bonyqas([3 3.8 22 45],[16 13 -10 -11.5],2, 3, 50)
% Copyright (c) B. Lurie 6-13-98 See also BOSTEP, BOCLOS, BOCOMP,
% BOINTEGR, NYQLOG, TFSHIFT from Bode toolbox. Further explanations
% are given in Advanced Nonlinear Classical Control by B.J.Lurie and
% P.J.Enright.

% default
  if nargin == 0,
    u = [0.2 2 8 32]; % 4 corner frequencies
    g = [-5 10 6 -12]; % gain in dB; band-pass system
    lslope = -1; % -1 integrator, i.e. 6.02 dB/oct up-slope
    hslope = 3; % 3 integrators, i.e. -18.06 dB/oct
  end
  if nargin < 5,
    nml = 0;
  end

  k = length(u); h = [1:k];

% frequency range of calculations and plots
  wmin = floor(log10(u(1)) - 0.33);
  wmax = ceil(log10(u(k)) + 0.33);
  w = logspace(wmin,wmax);

% low-frequency asymptote
  gain1 = (-lslope * (20 * log10(w/u(1))) + g(1))';
  phase1 = -lslope * 90;
% phase1 = -lslope * 90 * ones(length(w),1); % can be used instead
% semilogx(w, gain1, 'r--') % can be activated for purpose of teaching
% hold on

% segment slopes; segment no.h ends at u(h)
  for h = [2:k]
    sslope(h) = (g(h) - g(h-1))/log2(u(h)/u(h-1));
  end % [0 15 -2 -9]
% sslope % can be activated for purpose of teaching

% ray slopes
    rslope(1) = sslope(2) + 6.02 * lslope;
    rslope(k) = - 6.02 * hslope - sslope(k);
```

```
  for h = [2:(k-1)]
    rslope(h) = sslope(h + 1) - sslope(h);
  end
% rayslope_round = round(rslope) % [9 -17 -7 -9], positive means up

% frequency responses for k rays
  nray = 1; dray = [1 1 1]; % normalized ray with damping coefficient
0.5;
  for h = [1:k]
    [nrayf, drayf] = lp2lp(nray,dray,u(h)); % specifying corner frequency
    tfr = freqs(nrayf, drayf, w);
    rgain = 20*real(log10(tfr));
    rphase = (180/pi) * angle(tfr);
    rasis = (-rslope(h)/12.041) * rgain'; % minus makes positive ray go up
    % semilogx(w, rasis,'k--'); % can be activated for purpose of teaching
    % semilogx(w, rphase,'g:'); % can be activated for purpose of teaching
    gain1 = (-rslope(h)/12.041) * rgain' + gain1; % slope denormalization
    phase1 = phase1 - (rslope(h)/12.041) * rphase';
  end

% nonminimum phase lag addition
  phase1 = phase1 - nml * (w/u(h))';

% Bode diagram and phase response
  subplot(1,2,1)
  semilogx(w,gain1,'k', w,phase1,'g--', u,g,'ro')
  grid;
  xlabel('frequency, rad/sec');
  ylabel('gain, dB, and phase shift, degr');
  title('Bode diagram')
  hold off;
  zoom on;

% Nyquist diagram on log plane
  subplot(1,2,2)
  plot(phase1, gain1,'k', -180,0,'ro')
  hold on
  gain2 = (-lslope * (20 * log10(u/u(1))) + g(1))'; % marks
  phase2 = -lslope * 90;
    for h = [1:k]
      [nrayf, drayf] = lp2lp(nray,dray,u(h));
      tfr = freqs(nrayf, drayf, u);
      rgain = 20*real(log10(tfr));
      rgain = real(rgain);
      rphase = (180/pi) * angle(tfr);
      gain2 = (-rslope(h)/12.041) * rgain' + gain2;
      phase2 = phase2 - (rslope(h)/12.041) * rphase';
    end
  phase2 = phase2 - nml * (u/u(h))';
  plot(phase2,gain2,'ro');
  hold off;
  set(gca,'XTick',[-270 -240 -210 -180 -150 -120 -90])
  grid
  axis([-270 -90 -20 70])
```

```
title('Nyquist diagram')
xlabel('phase shift, degr')
ylabel('gain in dB')
zoom on
```

BOSTEP

```
function [wd,wc,width,numc,denc,numl,denl] = bostep(x,y,n,bnc,
zetaz,zetap,typ)
% BOSTEP Normalized loop transfer function with Bode step.
% function [wd,wc,width,numc,denc,numl,denl] = bostep(x,y,n,bnc,zetaz,zet
ap,typ)
% Defaults: [wd,wc,width,numc,denc,numl,denl] = bos
tep(10,0.1667,3,0.1,0.6,0.4,2),
% x is gain stability margin in dB, default x = 10;
% y is phase stability margin y*pi rad, default y = 0.1667;
% n defines asymptotic slope -6n dB/oct, default n = 3;
% bnc (0 to 1) is n.p. shift in rad at the end of Bode step (at wc),
% default bnc = 1;
% zetaz is damping of zeros at beginning of Bode step (at wd),
% default zetaz = 0.6;
% zetap is damping of poles at the end of Bode step (at wc), default
% zetap = 0.4;
% typ (1 or 2) is the number of poles at the origin, default typ = 2.
% BOSTEP calculates the following parameters of the asymptotic Bode
diagram
% withcrossover frequency 1 rad/sec: frequencies of beginning and end of
% the step wd, wc, and the width of the step width.
% BOSTEP also produces numerator and denominator of rational loop
transfer
% function numl(s)/denl(s) approximating the asymptotic loop Bode
diagram,
% where numl and denl contain the polynomial coefficients in descending
% powers of s. The typ is the number, 1 or 2, of the poles at the origin.
% BOSTEP also calculates the minimum phase compensator transfer function
% numc(s)/denc(s) for a single integrator plant with n.p. lag
% (-s + 2wc/bnc)/(s + 2wc/bnc).
% Copyright (c) B.J.Lurie 3-1-98, La Crescenta. See also BOCLOS, BONYQAS,
% BOCOMP, NYQLOG, TFSHIFT

  if nargin < 7,
    typ = 2;
  end
  if nargin < 6,
    zetap = 0.4;
  end
  if nargin < 5,
    zetaz = 0.6;
  end
  if nargin < 4,
    bnc = 0.1;
  end
  if nargin < 3,
```

```
   n = 3;
 end
 if nargin < 2,
   y = 0.167;
 end
 if nargin == 0,
   x = 10;
 end

% ** Asymptotic diagram parameters
 width = 0.9*(n + (pi/2)*bnc)/(2*(1 - y));
 wd = 2^(x/12/(1 - y));
 wc = wd*width;

% ** Poles and zeros for typ = 1
 z1 = 0.5*(1-y);
 z23 = [1 2*zetaz*wd wd^2]; % a pair of complex zeros
 k = 0.35*wc^2*sqrt(0.17/y);
 p1 = 0.07*(1-y); p2 = 0.7*wd; p3 = (wd + wc)/2.7;
 p45 = [1 2*zetap*wc 0.81*wc^2]; ;% a pair of complex poles

% ** Poles and zeros for typ = 2
 if typ == 2,
   p1 = 0;
   p3 = (wd + wc)/2.2;
   z23 = 0.95*[1 2.2*zetaz*wd*sqrt(sqrt(y/0.17)) wd^2];
   p45 = [1 2.2*zetap*wc 0.81*wc^2];
 end

% ** Minimum phase compensator
 numc = k * p3 * conv([1 z1], z23);
 denc = conv(conv([1 p1],[1 p2]),conv([1 p3],p45));

% ** Loop with 1/s plant and n.p. lag (-s + 2wc/bnc)/(s + 2wc/bnc):
 numl = conv(numc,[-1 2*wc/bnc]);
 denl = conv(conv(denc,[1 0]),[1 2*wc/bnc]);
 format short e
 numl
 denl

% numl_mp = numc; ne nada!
% denl_mp = conv(denc,[1 0]);ne nada!

% ** L-plane Nyquist diagram
 [mag, phase] = bode(numl, denl); plot(phase, 20*log10(mag),'k')
 set(gca,'XTick',[-240 -210 -180 -150 -120 -90])
 grid
 axis([-240 -90 -20 70])
 hold on
 nl = numl;
 dl = denl;
 [a,b] = bode (nl,dl,1); plot(b,20*log10(a),'bx')
   for w = [0.0156 0.03125 0.0625 0.125 0.25 0.5 2 4],
     [a,b] = bode (nl,dl,w); plot(b,20*log10(a),'r+');
   end
```

```
plot(-180, 0, 'ro')
hold off
title('Nyquist diagram, x marks w = 1, + marks octaves')
xlabel('loop phase shift in degrees')
ylabel('loop gain in dB')
zoom on
```

NIQLOG

```
function NYQLOG(wb,numl,denl)

%NYQLOG Nyquist diagram on logarithmic plane
%function NYQLOG(wb,numl,denl)
% The loop transfer function is numl(s)/denl(s), where numl and
% denl contain the polynomial coefficients in descending powers of s.
% wb is the crossover frequency in rad/sec, i.e. frequency at which
% loop gain is 0 dB.
% Default: wb = 1, numl = [-26 178 654 1269 429],
% denl = [1 18.8 132 543 1164 860 0 0]
% Copyright (c) B.J.Lurie 4-1-98, La Crescenta. See also BOSTEP,
% BOCLOS, BONYQAS, BOINTEGR, TFSHIFT, STEPLOG

  if nargin < 3,
    denl = [1 18.8 132 543 1164 860 0 0];
  end
  if nargin < 2,
    numl = [-26 178 654 1269 429];
  end
  if nargin == 0,
    wb = 1;
  end
  [mag, phase] = bode(numl, denl); plot(phase, 20*log10(mag),'k')
  set(gca,'XTick',[-270 -240 -210 -180 -150 -120 -90])
  grid
  axis([-270 -90 -20 70])
  hold on
  [a,b] = bode (numl,denl,wb); plot(b,20*log10(a),'bx')
    for w = [0.0156 0.03125 0.0625 0.125 0.25 0.5 2 4],
      [a,b] = bode(numl,denl,wb*w); plot(b,20*log10(a),'r+');
    end
  plot(-180, 0, 'ro');
  hold off
  title('Nyquist diagram, x marks w = wb, + marks octaves')
  xlabel('loop phase shift in degrees')
  ylabel('loop gain in dB')
  zoom on
```

STEPLOG

```
function [yout,x,t] = steplog(a,b,c,d,iu,t)
%STEPLOG Step response of continuous-time linear systems.
% steplog(A,B,C,D,IU) plots the time response of the linear system:
```

```
%  .
%  x = Ax + Bu
%  y = Cx + Du
%  to a step applied to the input IU. The time vector is auto-
%  matically determined. sstelog(A,B,C,D,IU,T) allows the specification
%  of a regularly spaced time vector T.
%
%  [Y,X] = steplog(A,B,C,D,IU,T) or [Y.X,T] = steplog(A,B,C,D,IU) returns
%  the output and state time response in the matrices Y and X
%  respectively. No plot is drawn on the screen. The matrix Y has
%  as many columns as there are outputs, and LENGTH(T) rows. The
%  matrix X has as many columns as there are states. If the time
%  vector is not specified, then the automatically determined time
%  vector is returned in T.
%
%  [Y,X] = steplog(NUM,DEN,T) or [Y,X,T] = steplog(NUM,DEN) calculates the
%  step response from the transfer function description
%  G(s) = NUM(s)/DEN(s) where NUM and DEN contain the polynomial
%  coefficients in descending powers of s. If no arguments, plots the
%  absolute value of the responses with logarithmic scale versus time.
%  See also: LSIM and DSTEP.

%  J.N. Little 4-21-85
%  Revised A.C.W.Grace 9-7-89, 5-21-92
%  Copyright (c) 1986-93 by the MathWorks, Inc.
%  Modified from STEP B.J. Lurie 4-5-98 (scale; color; precision)

nargs = nargin;
if nargs==0, eval('exresp(''step'')'), return, end

error(nargchk(2,6,nargs));

if (nargs < 4) % Convert to state space
  [num,den] = tfchk(a,b);
  if nargs==3, t = c; end
  [a,b,c,d] = tf2ss(num,den);
  iu = 1;
  nargs = nargs+3;
else
  error(abcdchk(a,b,c,d));
end

[ny,nu] = size(d);
if nu*ny==0, x = []; t = []; if nargout~=0, yout=[]; end, return, end

if nargs>4
  if ~isempty(b), b=b(:,iu); end
  d=d(:,iu);
end

% Workout time vector if not supplied.
if (nargs==5 | nargs==4),
  if isempty(a),
```

```
    t = 0:.1:1;
  else
    % The next two constants control the precision of the plot
    % and the time interval of the plot.
      st=0.00001; % Set settling time bound = 0.5%, (changed,bl)
      precision=100; % Show approx 30 points for simple graph (changed,bl)
    % Step response is effectively equal to placing initial conditions
    % on the plant as follows:
      [n,m]=size(b);
      if abs(rcond(a)) > eps
  x0 = -a\(b*ones(m,1));
    % Cater for pure integrator case
      else
  x0 = ones(n,1);
      end
      t=timvec(a,b,c,x0,st,precision);
  end
end

% Multivariable systems
if nargs==4
  [iu,nargs,y]=mulresp('step',a,b,c,d,t,nargout,0);
  if ~iu, if nargout, yout = y; end, return, end
end

% Simulation
dt = t(2)-t(1);
[aa,bb] = c2d(a,b,dt);
n = length(t);
[nb,mb] = size(b);
x = ltitr(aa,bb,ones(n,1),zeros(nb,mb));
if isempty(a),
  x = [];
  y = ones(n,1)*d.';
else
  y=x*c.'+ ones(n,1)*d.';
end
if nargout==0, % If no output arguments, plot graph
    dcgain = 0;
    if abs(rcond(a)) > eps
  dcgain=-c/a*b+d;
    end
    ni = find(y<0);
    pi = find(y>=0);
    yabs = abs(y);
    semilogy(t(ni),yabs(ni),'b.', t(pi), yabs(pi),'g.')
    xlabel('Time (secs)'), ylabel('Amplitude')
    grid
    return % Suppress output
end
yout = y;
zoom on
```

TFSHIFT

```
function [numsh, densh] = tfshift(num,den,q,p)
%TFSHIFT Transfer function with shifted frequency response.
% [numsh, densh] = tfshift(num,den,q,p)
% TFSHIFT produces numerator and denominator of transfer function
% NUMSH(s)/DENSH(s) shifted in frequency q times from NUM(s)/DEN(s)
% where NUM and DEN contain the polynomial coefficients in
% descending powers of s. NUMSH(s)/DENSH(s) has the same value
% at frequency qw as the function NUM(s)/DEN(s) at frequency w.
% The numerator and denumerator are both multiplied by 10^p for
% the coefficients to have convenient values. Default is p = 0.
% Copyright (c) 1998 A.Ahmed and B.J.Lurie 2-27-97

  ln = length(num)-1;
   ld = length(den)-1;
   pn = [ln: -1: 0];
   pd = [ld: -1: 0];
   numsh = num./(q.^pn);
   densh = den./(q.^pd);
   if nargin == 4,
     numsh = numsh*10^p;
     densh = densh*10^p;
   end
```

Appendix 15: Nonlinear Multi-Loop Feedback Control (Patent Application)

Although the following patent application of B. Lurie was rejected by the US patent office, we include it here because the claims or a part of it may still be correct. Claims 1, 8, 12, 13, 15, 19 and 20 are slightly changed here, and most of the figures have already appeared in Chapter 14.

[0001] The patent application claims the benefit of United States Provisional Patent Application Serial Number 62/165,860, entitled "nonlinear Multi-Loop Feedback Control," filed May 22, 2015, with application is incorporated in its entirely here by this reference.

Technical field

[0002] This invention relates to a multi-loop feedback control system with nonlinear dynamic compensators that increases a plant's feedback in comparison to a single-loop plant feedback system with plant uncertainty.

Background

[0003] A multi-loop feedback system employs the existing subsystems (sensors, motors, actuators and other devices with set parameters) automatically, without the necessity of opening the system loops and adjusting them from inside the system. The multi-loop feedback system adds compensators and system structures to an actuator control. A nonlinear or linear compensators can appear anywhere in the multi-loop system but mostly is placed on inputs of actuators. A loop gain is a multiplication of all gains of a loop, and a loop phase is a sum of link cases.

[0004] The term "plant" means (often) a load on the actuators that changes its position, direction, or chemical analysis et cetera. A plant without feedback does not meet the desired specification due to its time parameters' uncertainty. An actuator beyond the normal working conditions becomes nonlinear.

[0005] The plant gain ends with uncertainty at higher frequencies, as shown on Fig. PA.1 (Numbers on pictures 1 to 6 are local). An uncertainty boundary, f_u, is shown and extends to a dashed line, wherein the plant gain is unpredictable within the uncertain boundary, such that it may vary from –15 dB to 15 dB. A middle line of uncertainty is called a gain central line.

[0006] The gain and phase of return ratio T for linear single-loop feedback system by Bode theory are shown in 14a and 14b in Figure PA1.2. The plant has the workable bandwidth that ends at f_w. The plant boundary is marked by thicker lines. The Figure PA1.2 assumes $|T(f_w)| = 1$. The feedback is $|F|| = |T| + 1|$.

[0007] As can be seen from Figure PA1.2, rather sharp corners occur on an amplitude of a Bode step. The amplitude is shown as a dark line at lower frequencies up to

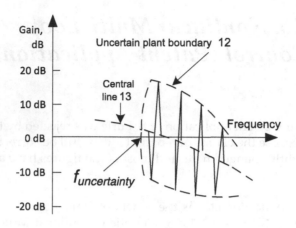

FIGURE PA1.1

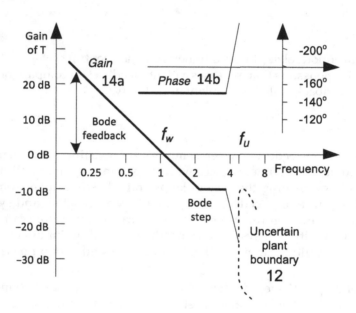

FIGURE PA1.2

the Bode step. The extension of this line goes with steep slope down and enters a plant uncertainty. All methods and/or designs relating to a single-loop feedback system achieve inferior stability and narrower bandwidth as compared to the present invention comprising a multi-loop feedback system.

[0008] A phase is a dark line over the same frequency range as the amplitude. Any smoothing of the Bode step corners, caused by an improper rational function approximation, reduces the available feedback.

[0009] A band-pass system is demonstrated in Figure PA1.3(a). Saturation exist in the output of the system in Figure PA1.3(b) with different instability forms seen in Figure PA1.3(c). The oscillations reduce the stability margin, as shown in Figure PA1.3(a), to nearly 7°.

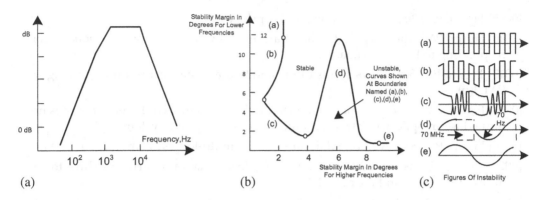

FIGURE PA1.3

[0010] The stability margin for band pass and low pass feedback system must be $y \sim 30°$ in phase and $x \sim 10$ dB in gain. The phase of the Bode diagram is typically $-150°$.

[0011] The range between f_w and f_u per Figure PA1.2 with these stability margins and the boundary of uncertain plant is nearly 2.3 octaves.

Summary

[0012] The present invention is directed to a multi-loop feedback control system that focuses on increasing the frequency range of control and more specifically controlling a nonlinear actuator that moves a mechanical plant. The feedback control system comprises an actuator having an electrical motor used to drive the mechanical plant through a first filter.

An output sensor measures movement of the mechanical plant and provides plant output to the feedback pass summer through a second filter.

The feedback pass summer adds the plant output though the second filter and a summer of feedback passes outputting to a feedback pass.

The summer of feedback passes adds output pass of a compensator, the actuator, and the feedback summer. The feedback summer subtracts the signal from the feedback path from the input signal through an input filter and provides a signal to the compensator. The feedback control system increases the range of control, especially between frequencies with positive gain and frequencies in the uncertain plant boundary. More specifically, the multi-loop feedback control system increases the workable plant frequency by at last about one or two octaves in comparison to the workable plant frequency with any existing linear single-loop feedback control system.

Brief description of drawings

[0013] Figure PA1.1 shows a frequency uncertainty area in plant gain for a prior art.

[0014] Figure PA1.2 shows a Bode gain plot and a Bode phase plot for a prior art.

[0015] Figure PA1.3(a) shows a Bode plot for prior art.

[0016] Figure PA1.3(b) shows a stability margin for lower and higher frequencies corresponding to Figure PA1.3a.

[0017] Figure PA1.3(c) shows forms of instability diagrams corresponding to Figure PA1.3(b).

[0018] Figure PA1.4 shows a multi-loop feedback system that could me for. Nut is not limited to, an optical, electromechanical or electro-optical system.

[0019] Figure PA1.5 shows Bode diagrams for the multi-loop system in Figure PA1.4.

[0020] Figure PA1.6(a) shows a MATLAB feedback diagram corresponding to the multi-loop system in Figure PA1.4.

[0021] Figure PA1.6(b) shows a MATLAB transient response corresponding to Figure PA1.6(b).

Detailed description of the invention

[0022] The detailed description set forth below in connection with the appended drawings is intended as a description of presently preferred embodiments of the invention and is not intended to represent the only form in which the present invention

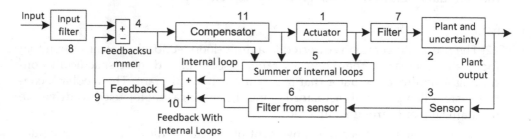

FIGURE PA1.4

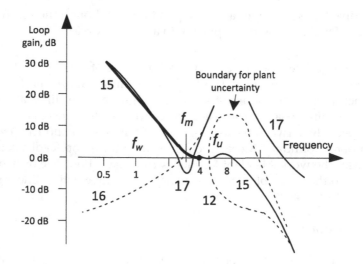

FIGURE PA1.5

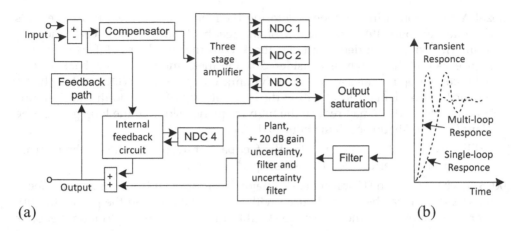

(a)   (b)

FIGURE PA1.6

may be constructed or utilized. The decryption sets forth the functions and the sequence of steps for constructing and operating the invention in connection with the illustrated embodiments. It is to be understood, however, that the same or equivalent function and sequences may be accomplished by different embodiments that are also intended to be encompassed within the spirit and scope of the invention.

[0023] The present invention, as shown in Figure PA1.4, is a nonlinear multi-loop feedback system (18) that comprises a plant (2) with uncertainty, a compensator (11), an actuator (1), a filter (7), a feedback summer (4), and an input filter (8) on a forward pass.

The multi-loop feedback system (18) further comprises an output sensor (3), a summer for the internal feedback (5) with an output added to a feedback with internal loops (10), which adds a signal from a filter from sensor (6) to the summer for internal feedback (5), and then applies a signal to feedback (9) and to the feedback summer (4). Transmission of the multi-loop feedback system (18) in a forward direction includes starting from the feedback summer (4) and ending with plant output.

[0024] An increase in feedback in a multi-loop feedback system must involve (a) changing the plant loops gain, (b) non-minimum phase connection of the multi-loop feedback system to plant uncertainty loop, and (c) achieving the maximum value of plant feedback. Also, the design must verify and check: d) uncertainty of the plant, (e) links with nonlinear dynamics to be stable in linear mode, (f) the frequency band-pass dynamics with asymmetrical nonlinear responses, (d) the connection of minimum-phase responses, (g) satisfying the contradictory between the amplitude and the gain responses, and (h) achieving positive stability analysis with a nonlinear mode of instability.

[0025] Figure PA1.5 shows a diagram portraying a forward transmission of gain, where a dark gain (15) has the same frequency band as Bode amplitude diagram 14a in Figure PA1.2. Figure PA1.5 also contains a feedback transmission (16) of an internal loop (19) and its effect gain (17) of connecting the dark gain (15) to the feedback transmission (16) of the internal loop (19). The extension of the effect gain (17) at higher frequencies follows a central line of uncertainty (13) in Figure PA1.1.

[0026] A collection of internal loops entering the feedback summer (4) equal equals the internal loop (19) at an output of the feedback summer (4). The collection of internal loops is added by the feedback with internal loops (19), as shown in Figure PA1.3. The system is considered a nonlinear multi-loop feedback system if a second loop includes a nonlinear dynamic compensator (NDC), rather than a sensor, but the plant (2) is not within the internal loop (19). The nonlinear dynamic compensator (NDC) may reduce the high frequency slope of the loop gain of the plant (2), to be shallower than about 10 dB/oct.

[0027] For example, the dark gain (15), the feedback transmission (16) of the internal loop (19), and the effect gain (17) are shaped as follows:

[0028] 1. The dark gain (15) presents the major form for front transmission of a feedback system. It can be typically used with slope 10 dB/oct and the phase stability margin of about 30°, since these are stability margins of the multi-loop feedback system (18).

[0029] 2. The phases of the dark gain (15) and the internal loop (16) must forbid, as the frequency f_m, reducing phase of the effect gain (17) by more than 150°, or else due to the effect gain (17), the multi-loop feedback system (18) will have a smaller stability margin.

[0030] 3. At frequencies higher than f_m, the effect gain (17) must be higher than the boundary of uncertainty by at least the stability margin of about 10 dB, for which the summer of internal loop (5) and the feedback (9) of Fig.PA.4 will make composite resonances at approximately 1 octave higher than f_u.

[0031] 4. The effect gain (17) is chosen approximately as shown on Figure PA1.5. The advantage of a multi-loop system over a single-loop system is the appearance of f_m near 1.5 oct (this is the point of crossing the dark gain (15) and the internal loop (16) serving as plant feedback) to f_w (in our case $f_w| = 1$), or 1.5 octaves.

[0032] In a single system with similar parameters in loop uncertainty, $\log_2 f_u/fw| = 2.3$ oct. in the multi-loop system, $\log_2 f_m/f_m$ is, by Figure PA1.4, nearly 0.7 octaves. Then $\log_2| = f_m/f_w| = 1.5$ octaves is a maximum increase in workable bandwidth in the multi-loop (18) compared to a single-loop Bode system.

[0033] On the boundary for plant uncertainty, the effect gain (17) must exceed the boundary for plant uncertainty by about 10 or 15 dB. The power related into the effect gain (17), this might be a problem. However, the amplitude of the effect gain (17) also may be measured as load impedance multiplied by initial impedance of feedback (9), which is zero since this part of the multi-loop feedback system (18) is digital.

[0034] For some analog versions of the multi-loop feedback system (18), a distance between the effect gain (17) and the uncertainty boundary (12) may be made lower by summing the peaks of uncertainty with Bode non-minimal phase additional of internal loops.

[0035] In place of a curve of the effect gain (17), a digital curve may be used with jumps up and down, or a curve with changing positive and negative slopes.

[0036] For the system with reducing nonlinear distortion in linear amplifiers, the filter (7) can be installed at the plant input to prevent high order nonlinear product to make the plant output nonlinear.

[0037] In a preferred embodiment, the nonlinear multi-loop feedback system (18) comprises a plant loop with lower than about 40 dB gain uncertainty on higher frequencies, wherein one or several forward links are considered to the plant (2) and are in parallel for a working feedback range, such that obtained loops are internal, and internal loops of the nonlinear multi-loop feedback system (18) increase a plant feedback by about 0.7 to about 2 octaves, depending on plant linearity and feedback requirements of the nonlinear multi-loop system (18), as compared to a single-loop feedback system with the same plant, uncertainty, and stability margin.

[0038] Alternatively, in another preferred embodiment, the nonlinear multi-loop feedback system (18) comprises a plant loop with lower than about 40 dB gain uncertainty on higher frequencies, wherein one or several forward links are connected to a plant and are parallel to the plant at lower frequencies.

[0039] In another preferred embodiment, the nonlinear multi-loop feedback system (18) comprises a plant loop with lower than about 40 dB gain uncertainty on higher frequencies, wherein one or several forward links are connected to the plant (2) and are in parallel for the working feedback range, such that obtained loops are internal, and a transient response is twice as fast compared with single-loop system with the same plant, uncertainty and stability margins.

[0040] In addition, bridge or compound feedback together with internal feedback may be used to increase the plant feedback. A plant loop's gain at a boundary for plant uncertainty may be positive and increase a plant bandwidth as compared to a negative gain of the plant loop with the single-loop feedback system.

Realization of invention in MATLAB

[0041] The plant uncertainty in this software example is shown in Figure PA1.6(a). The plant is modeled in MATLAB with a tree stage amplifier. Each stage contains an NDC. Another NDC contains in internal feedback loop. The feedback with multiple resonances in filter was placed in cascade with the plant.

[0042] The increase in frequency of plant loop gain by more than about 1.2 octaves is achieved using the Bode gain/phase relation and Bode minimum phase stability methods.

[0043] On Figure PA1.6(b), two transient responses are provided, for the single loop and for the signal for both systems remain the same. The transient response for double loop system is shown at Figure PA1.6(b).

Application of the invention

[0044] The description of a preferred embodiment of the invention has been presented for the purpose of illustration and description. It is not intended to be exhaustive or to limit the invention to precise forms disclosed. Obviously, many modifications and variations will be apparent to a practitioner skilled in this art. It is intended that the scope of the invention be defined by the following claims and their equivalents. The present invention for the nonlinear multi-loop feedback control is not presently used for problems relating to large increase of feedback systems with high frequency uncertainties.

[0045] For example, this control method may be used for a device with a compound feedback and an antenna on an optical transmitter having an uncertainty in heterodyne frequency, or acoustical high bandwidth headphones, or a laser with minimal performance time for a conveyer line, or landing a large rocket, or using rockets in rare atmosphere on the Mars landing.

[0046] The foregoing description on the preferred embodiment of the invention has been presented for the purpose of illustration and description. It is not intended to be exhaustive or to limit the invention to the precise from disclosed. Many modifications and variations are possible in light of the above teaching. It is intended that the scope of the invention not be limited by this detailed description, but by the claims and the equivalents to the claims appended hereto.

Claims

What is claimed is:

1. A nonlinear multi-loop feedback system having a plant loop with lower than about 40 dB gain uncertainty on higher frequencies, wherein one or several forward links are connected to a plant and are in parallel for a working feedback range, such that obtained loops are internal, and internal loops of the nonlinear multi-loop feedback system increase a plant feedback up to 2 ± 1 octaves, depending on plant linearity and feedback requirements of the nonlinear multi-loop feedback system, as compared to a single-loop feedback system with the same plant, uncertainty, and stability margins.

2. The nonlinear multi-loop feedback system of claim 1, wherein bridge or compound feedback together with internal feedback is used to increase the plant feedback.

3. The nonlinear multi-loop feedback system of claim 1, wherein the plant's loop gain at a boundary for plant uncertainty is positive and increases a plant bandwidth as compared to a negative gain of the plant loop with the single-loop feedback system.

4. The nonlinear multi-loop feedback system of claim 3, wherein bridge or compound feedback together with internal feedback is used to increase the plant feedback.

5. The nonlinear multi-loop feedback system of claim 3, further comprising a nonlinear dynamic compensator (NDC) that can reduce a plant response to one octave or lower in frequency.

6. The nonlinear multi-loop feedback system of claim 1, further comprising a nonlinear dynamic compensator (NDC) that can reduce a plant response to one octave or lower in frequency.

7. The nonlinear multi-loop feedback system of claim 6, wherein bridge or compound feedback together with internal feedback is used to increase the plant feedback.

8. A nonlinear multi-loop feedback system having a plant loop with less than 40 dB gain uncertainty on higher frequencies, wherein one or several forward links are connected to a plant and are parallel to the plant at lower frequencies, such that obtained loops are internal, and internal loops of the nonlinear multi-loop feedback system increase a plant feedback by 0.7 to 2 octaves, depending on plant linearity and feedback requirements of the nonlinear multi-loop feedback system, as compared to a single-loop feedback system with the same plant, uncertainty, and stability margins.

9. The nonlinear multi-loop feedback system of claim 8, wherein bridge or compound feedback together with internal feedback is used to increase the slopes of plant feedback gain going down steeper than 10 dB/oct.

10. The nonlinear multi-loop feedback system of claim 8, wherein the plant loop's gain at a boundary for plant uncertainty is positive and increases a plant bandwidth as compared to a negative gain of the plant loop with the single-loop system.

11. The nonlinear multi-loop feedback system of claim 10, wherein bridge or compound feedback together with internal feedback is used to increase the plant feedback.

12. The nonlinear multi-loop feedback system of claim 10, further comprising a nonlinear dynamic compensator (NDC) that allows the slope of the plant's loop gain to be steeper with up to 10 dB/oct.

13. The nonlinear multi-loop feedback system of claim 8, further comprising a nonlinear dynamic compensator (NDC) that allows the slope of the plant's loop gain to be steeper up to 10 dB/oct.

14. The nonlinear multi-loop feedback system of claim 13, wherein bridge or compound feedback together with internal feedback is used to increase the plant feedback.

15. A nonlinear multi-loop feedback system having a plant loop with less than 40 dB gain uncertainty on higher frequencies, wherein one or several forward links are connected to a plant and are in parallel for working feedback range, such that obtained loops are internal, and a transient response is twice as fast compared with a single-loop system with the same plant, uncertainty, and stability margins.

16. The nonlinear multi-loop feedback system of claim 15 wherein bridge or compound feedback together with internal feedback is used to increase a plant feedback.

17. The nonlinear multi-loop feedback system of claim 15, wherein the plant loop's gain at a boundary for plant uncertainty is positive and increase a plant bandwidth as compared to a negative gain of the plant loop with the single-loop system.

18. The nonlinear multi-loop feedback system of claim 17, wherein bridge or compound feedback together with internal feedback is used to increase a plant feedback.

19. The nonlinear multi-loop feedback system of claim 17, further comprising a nonlinear dynamic compensator (NDC) that allows the slope of the plant's loop gain to be steeper up to 10 dB/oct.

20. The nonlinear multi-loop feedback system of claims 15, further comprising a nonlinear compensator (NDC) that allows the slope of the plant's loop gain to be steeper up to 10 dB/oct.

Bibliography

1. Abramovici, A., and J. Chapsky, *A Fast-Track Guide for Scientists and Engineers*, Kluwer, Academic Publishers, Norwell, MA, 2000.
2. Bayard, D.S., "An Algorithm for State-Space Frequency Domain Identification without Windowing Distortions," *IEEE Transactions on Automatic Control*, 39(9), 1994.
3. Bayard, D.S., "Necessary and Sufficient Conditions for LTI Representation of Adaptive Systems with Sinusoidal Regressors," paper presented at Proceedings of the American Control Conference, Albuquerque, NM, June 1997.
4. Bernstein, D.S., "Four and a Half Control Experiments and What I Learned from Them: A Personal Journey," paper presented at Proceedings of the American Control Conference, Albuquerque, NM, June 1997.
5. Biernson, G., *Principles of Feedback Control*, vols. 1 and 2, John Wiley and Sons, New York, 1988.
6. Bode, H.W., *Network Analysis and Feedback Amplifier Design*, Van Nostrand, New York, 1945.
7. Bode, H.W., and C. Shannon, "A Simplified Derivation of Linear Least Square Smoothing and Prediction Theory," *Proceedings of the IRE*, 38, 1950, 4.
8. Brugarolas, P., and B. Kang, "Instrument Pointing Control System for the Stellar Interferometry Mission—Planet Quest," paper presented at SPIE, Astronomical Telescopes and Instrumentation, 2006.
9. Bryan, H.K., D. Boussalis, and N. Fathpour, "SIM Planet Planet Quest Lite Interferometer Guide 2 Telescope Pointing Control System," SPIE, vol. 7064, 2008.
10. Chen, G.S., and B.J. Lurie, "Active Member Bridge Feedback Control for Damping Augmentation," *Journal of Guidance, Control and Dynamics*, 15(5), 1992.
11. Chen, G.S., and B.J. Lurie, "Bridge Feedback for Active Damping Augmentation," AIAA paper 901243, 1989.
12. Chen, G.S., C.R. Lawrence, and B.J. Lurie, "Active Member Vibration Control Experiment in a KC135 Reduced Gravity Environment," paper presented at First U.S./Japan Conference on Adaptive Structures, Maui, Hawaii, November 1990.
13. Crafton, P.A., *Shock and Vibration in Linear Systems*, Harper and Brothers, New York, 1961.
14. Dolgin, B.P., F.T. Hartley, B.J. Lurie, and P.M. Zavracky, "Electrostatic Actuation of a Microgravity Accelerometer," paper presented at 43rd National Symposium of American Vacuum Society, Microelectronical Mechanical Systems Topical Conference, Philadelphia, PA, October 14–18, 1996.
15. Dorf, R.C., and R.H. Bishop, *Modern Control Systems*, Addison-Wesley, Menlo Park, CA, 1997.
16. Doyle, J., and G. Stein, "Robustness with Observers," *IEEE Transactions on Automatic Control*, 24(4), August 1979.
17. Enright, P.J., F.Y. Hadaegh, and B.J. Lurie, "Nonlinear Multi-Window Controllers," paper presented at AIAA Guidance, Navigation and Control Conference, San Diego, July 1996.
18. Fang, S., J. Chen, and H. Ishii, *Towards Integrating Control and Information Theories*, Lecture Notes in Control and Information Sciences 465, Springer, 2017.
19. Fanson, J.L., Cheng-Chih Chu, and B.J. Lurie, "Damping and Structural Control of the JPL Phase 0 Testbed Structure," *Journal of Intelligent Material Systems and Structures*, 2(3), 1991.
20. Freudenberg, J.S., and D.P. Looze, *Frequency Domain Properties of Scalar and Multivariable Feedback Systems*, lecture notes in control and information science, vol. 104, Springer Verlag, Berlin, 1987.
21. Garnell, P., *Guided Weapon Control Systems*, 2nd ed., Pergamon Press, New York, 1980.
22. Goldfarb, L.C., "On Some Nonlinear Phenomena in Regulatory Systems," *Automatika i Telemekhanika*, 8(5), 1947; translated in U. Oldenburger (ed.), *Frequency Response*, Macmillan, New York, 1956.

23. Grogan, R.L., G.H. Blackwood, and R.J. Calvet, "Optical Delay Line Nanometer Level Pathlength Control Law Design for Space-Based Interferometry," Proc. SPIE, Koha, Hawaii, 1998.

24. Guillemin, E.A., *Communication Networks*, John Wiley and Sons, New York, vol. 1, 1931, vol. 2, 1935.

25. Hench, J.J., B.J. Lurie, R. Grogan, and R. Johnson, "Implementation of Nonlinear Control Laws for the RICST Optical Delay Line," paper presented at IEEE Aerospace Conference, Big Sky, MT, March 2000.

26. Horowitz, I.M., *Synthesis of Feedback Systems*, Academic Press, New York, 1963.

27. Kalman, R.E., "When Is a Linear Control System Optimal?" *Journal of Basic Engineering*, 86(1), March 1964.

28. Kang, B.H., D.S. Bayard, and G. Macala, "SIM-Planet Interferometer Real-Time Control System Architecture," *Proceedings of SPIE*, 6675, 667507-1, 2007.

29. Kochenburger, R.J., *Analysis and Synthesis of Contactor Servomechanisms*, Sc.D. thesis, Massachesetts Institute of Technology, Department of Electrical Engineering, Cambridge, MA, 1949.

30. Kuo, B.C., *Automatic Control Systems*, 5th ed., Prentice Hall, Upper Saddle River, NJ, 1996.

31. Lin, Y.H., and B.J. Lurie, "Nonlinear Dynamic Compensation for Control System with Digital Sensor," U.S. Patent 5,119,003, June 2, 1992.

32. Lur'e, A.I., *Certain Nonlinear Problems in the Automatic Regulating Theory*, Gostekhizdat, 1951 (In Russian, English Translation: London, HMSO, 1975).

33. Lurie, B.J., *Feedback Maximization*, Artech House, Dedham, MA, 1986.

34. Lurie, B.J., "Three Loop Balanced Bridge Feedback Pointing Control," paper presented at Proceedings of the American Control Conference, Atlanta, GA, 1988.

35. Lurie, B.J., "Global Stability of Balanced Bridge Feedback," *ICCON '89*, paper presented at IEEE International Conference on Control and Applications, Jerusalem, April 1989.

36. Lurie, B.J., "Balanced Bridge Feedback Control System," U.S. Patent 4,912,386, March 27, 1990a.

37. Lurie, B.J., "Multiloop Balanced Bridge Feedback in Application to Precision Pointing," *International Journal of Control*, 51(4), 1990b.

38. Lurie, B.J., "Three-Parameter Tunable TID Controller," U.S. Patent 5,371,670, December 6, 1994.

39. Lurie, B.J., a letter to Bob's Mailbox, *Electronic Design*, March 4, 1996.

40. Lurie, B.J., "Integral Relations for Disturbance Isolation," *Journal of Guidance Control Dynamics*, 20(3), 1997.

41. Lurie, B.J., "Multi-Mode Synchronized Control for Formation Flying Interferometer," paper presented at AIAA Conference on Guidance, Navigation and Control, Austin, TX, August 2003.

42. Lurie, B.J., A. Ahmed, and F.Y. Hadaegh, "Asymptotically Globally Stable Multiwindow Controllers," paper presented at AIAA Conference on Guidance, Navigation and Control, New Orleans, LA, August 1997.

43. Lurie, B.J., A. Ahmed, and F.Y. Hadaegh, "Acquisition and Tracking with High-Order Plant and Nonlinear Regulation of Bode Diagram," paper presented at Control Conference, Pullman, Washington, September 1998.

44. Lurie, B.J., and J. Daegas, "An Improved High-Voltage DC Regulator for a Radar and Communication Transmitter," paper presented at Proceedings of the 18th Power Modulator Symposium, Hilton Head, SC, 1988.

45. Lurie, B.J., B. Dolgin, and F.Y. Hadaegh, "Motor Control with Active Impedance of the Driver," paper presented at Space 2000, Albuquerque, NM, January–February 2000.

46. Lurie, B.J., J.L. Fanson, and R.A. Laskin, "Active Suspensions for Vibration Isolation," paper presented at 32nd SDM Conference, Baltimore, April 1991.

47. Lurie, B.J., A. Ghavimi, F.Y. Hadaegh, and E. Mettler, "System Trades with Bode Step Control Design," paper presented at AIAA Conference on Guidance, Navigation and Control, Boston, 1998.

48. Lurie, B.J., A. Ghavimi, F.Y. Hadaegh, and E. Mettler, "System Architecture Trades Using Bode Step Control Design," *Journal of Guidance Control Dynamics*, 25, 2002.

49. Lurie, B.J., and F.Y. Hadaegh, "Applications of Multiwindow Control in Spacecraft Systems," paper presented at NASA URC Tech. Conference, Albuquerque, February 1997.

50. Lurie, B.J., J.J. Hench, A. Ahmed, and F.Y. Hadaegh, "Nonlinear Control of the Optical Delay Line Pathlength," paper presented at AeroSense '99, SPIE Conference, Orlando, FL, April 5–7, 1999.

51. Lurie, B.J., J. O'Brien, S. Sirlin, and J. Fanson, "The Dial-a-Strut Controller for Structural Damping," paper presented at ADPA/AIAA/ASME/SPIE Conference on Active Materials and Adaptive Structures, Alexandria, VA, November 1991.

52. Lurie, B.J., J.A. Schier, and M.M. Socha, "Torque Sensor Having a Spoked Sensor Element Support Structure," U.S. Patent 4,932,270, June 12, 1990.

53. Lurie, B.J., and Carol Smidts, "Classical and Reliable," *IEEE Control Systems Magazine*, 2, 2007.

54. Macala, G.A., "Design of the Reaction Wheel Attitude Control System for the Cassini Spacecraft," paper presented at AAS/AIAA Space Flight Mechanics Meeting, San Antonio, TX, January 2002.

55. MacFarlane, A.G.J. (ed.), *Frequency-Response Methods in Control Systems*, IEEE Press, New York, 1979.

56. Melody, J.W., and G.W. Neat, "Integrated Modeling Methodology Validation Using the Micro-Precision Interferometer Testbed: Assessment of Closed-Loop Performance Prediction Capability," paper presented at American Control Conference, Albuquerque, NM, June 1997.

57. Mitter, S., and A. Tannenbaum, "The Legacy of George Zames," *IEEE Transactions on Automatic Control*, 43(5), 1998.

58. Murray, A., "DSP Motor Control Boosts Efficiency in Home Appliances," *Electronic Design*, May 25, 1998, pp. 102–110.

59. Neat, G.W., A. Abramovichi, J.W. Melody, R.J. Calvet, N.M. Nerheim, and J.F. O'Brien, "Control Technology Readiness for Spaceborn Optical Interferometer Missions," paper presented at SMACS 2, Toulouse, France, May 1997.

60. Neat, G.W., J.W. Melody, and B.J. Lurie, "Vibration Attenuation Approach for Spaceborn Optical Interferometers," *IEEE Transactions on Control Systems Technology*, 6(6), 1998.

61. Neat, G.W., J. O'Brien, B. Lurie, A. Garnica, W. Belvin, J. Sulla, and J. Won, "Joint Langley Research Center/Jet Propulsion Laboratory CSI Experiment," paper presented at 15th Annual AAS Guidance and Control Conference, Keyston, CO, February 1992.

62. O'Brien, J., "Rudder Roll Stabilization by Nonlinear Dynamic Compensation," US Patent Application Publications US20100147204.

63. O'Brien, J.F., and B. Lurie, "Integral Force Feedback for Structural Damping," paper presented at First International Conference on Adaptive Structures, San Diego, November 1992.

64. O'Brien, J.F., and G.W. Neat, "Micro-Precision Interferometer: Pointing Control System," paper presented at 4th Conference on Control Applications, Albany, NY, September 1995.

65. Phillips, C.L., and J. M. Parr, *Feedback Control Systems*, 5th ed., Pearson Education, 2011.

66. Popov, V.M., "Absolute Stability of Nonlinear Systems of Automatic Control," *Automatika i Telemekhanika*, 22, 1961.

67. Rahman, Z., J. Spanos, and D. Bayard, "Multi-Tone Adaptive Vibration Isolation of Engineering Structures," paper presented at 36th AIAA/ASME/ASCE/AHS/ASC Structures, Structural Dynamics and Material Conference, New Orleans, LA, April 1995.

68. Rohrs, C.E., J.L. Melsa, and D.G. Schultz, *Linear Control Systems*, McGraw-Hill, New York, 1993.

69. Shields, J., B. Metz, R. Bartos, A. Morfopoulos, C. Bergh, J. Keim, D. Scharf, and A. Ahmed, "Design, Modeling and Control of a Pointing Sensor for the Formation Control Testbed (FCT)," AIAA Guidance, Navigation, and Control Conference, Chicago, IL, 2009.

70. Sidi, M.J., *Spacecraft Dynamics and Control*, Cambridge University Press, New York, 1997.

71. Skogestad, S., and I. Postlethwaite, *Multivariable Feedback Control*, Wiley, Southern Gate, Chichester, West Sussex, England, 2009.

72. Spanos, J.T., and M.C. O'Neal, "Nanometer Level Optical Control of the JPL Phase B Testbed," paper presented at ADPA/AIAA/ASME/SPIE Conference on Active Materials and Adaptive Structures, 1992.

73. Spanos, J.T., Z. Rahman, and A. von Flotov, "Active Vibration Isolation on Experimental Flexible Structure: Smart Structures and Intelligent Systems," paper presented at SPIE Proceedings 1917-60, Albuquerque, NM, 1993.

74. Spanos, J.T., and Z. Rahman, "Narrow-Band Control Experiments in Active Vibration Isolation: Vibration Monitoring and Control," paper presented at SPIE Proceedings 2264-01, San Diego, CA, July 1994.

75. Talwar, A., "Noise and Distortion Reduction in Amplifiers Using Adaptive Cancellation," *Microwave Journal*, No. 8, p. 98, August 1997.

76. Tustin, A., "The Effects of Backlash and of Speed-Dependent Friction on the Stability of Closed-cycle Control Systems," *Journal IEE*, London, pt. II, vol. 94, May 1947, pp. 143–151.

77. Wiener, N., *Extrapolation, Interpolation, and Smoothing of Stationary Time Series, with Engineering Applications*, John Wiley and Sons, New York, 1949.

78. Zames, G., "On the Input-Output Stability of Time-Varying Nonlinear Feedback Systems," Parts I and II, *IEEE Transactions on Automatic Control*, vol. AC-11, April 1966.

Notation

A, $A(s)$	actuator transfer function, 1.1
$A(\omega)$	log magnitude of the loop transfer function, 3.9.2
A	system matrix, 8.3
A_o	maximum available feedback, 4.2.5
A_2, A_4	even coefficients in Laurent expansion of logarithmic transfer function, A.4.1
A_∞	value of A at $s = \infty$, 3.14.2
B, $B(s)$	feedback path transfer function, 1.1
$B(\omega)$	phase of the loop transfer function 3.9.2
B	control-input matrix, 8.3
$B(0)$	feedback path transfer function with two specified nodes connected, Appendix 6
$B(\infty)$	feedback path transfer function with two specified nodes disconnected, Appendix 6
B_1, B_3	odd coefficients in Laurent expansion of logarithmic transfer function, A.4.1
B_n	non-minimum phase lag, 3.12
C	capacitance, 3.14.4
C, $C(s)$	compensator transfer function, 1.1
C	output matrix, 8.3
E	emf of the signal source, 7.2.2
E	amplitude of the fundamental at the input to a nonlinear link, 11.2
E_1	amplitude of the fundamental at the input to NDC, 11.6
E_2	amplitude of the fundamental at the input to the actuator, 11.6
E_a'	amplitude of the fundamental after the jump down, 12.3
E_a''	amplitude of the fundamental after the jump up, 12.3
E_b'	amplitude of the fundamental before the jump down, 12.3
E_b''	amplitude of the fundamental before the jump up, 12.3
E_{1C}, E_{2C}	values of E_1, E_2 causing $|T| =$, 11.6
E_{sb}	amplitude of a subharmonic, 12.4.1
F	force, 7.1.1
$F = T + 1$	return difference, 1.1
$F(0)$	value of return difference when two specified terminals shorted, Appendix 6
$F(\infty)$	value of return difference when two specified terminals open, Appendix 6
$|F|$, or $20 \log|F|$	feedback, 1.1
$G(s)$, G	return ratio of linear links in NDC, 10.7.1
G	plant noise distribution matrix, 8.4
H	measurement matrix, 8.4
$H(E, j\omega)$	describing function, 11.2
H_∞	norm, 8.5
J	moment of inertia, 1.6, 7.1.1
J	quadratic cost functional, 8.4

K	closed-loop transmission coefficient in voltage, Appendix 6
K	gain matrix, 8.3
$K(s)$	coupling transfer function, 4.4
K_E	closed-loop ratio of the output voltage to the source emf, Appendix 6
K_E	estimator gain matrix, 8.4
K_I	closed-loop transmission coefficient in current, Appendix 6
K_{OL}	open-loop system transmission coefficient in voltage, Appendix 6
$K_{OL\,I}$	open-loop system transmission coefficient in current, Appendix 6
$K_{OL\,E}$	open-loop system ratio of the output voltage to the source emf, Appendix 6
L	inductance, 3.14.4, 7.1.1
M	mass of a rigid body, 4.3.6, 7.1.1
M	T/F, 1.3
M, N	coefficients in linear fractional functions in an NDC, 11.7
N_A	noise at the actuator input, 4.3.3
$P, P(s)$	plant transfer function, 1.1
$P_o, P_o(s)$	transfer function of nominal plant, 1.9
Q	regulation of a symmetrical regulator, 6.7.2
$Q = 1/(2\zeta)$	quality factor of a resonance, 5.5
Q	weighting matrix, 8.4
$R, R(s)$	prefilter transfer function, 2.2
R	weighting matrix, 8.4
R_L	load resistance, 7.2.3
R_T	thermal resistance, 7.1.2
S	sensitivity, 1.8
S_H	Horowitz sensitivity, 1.9
T	return ratio, 1.1
T	absolute temperature, 7.1.2
$T(0)$	return ratio in a system with the two specified nodes connected, 7.4.2, Appendix 6
$T(\infty)$	return ratio when two specified nodes are disconnected, 7.4.2, Appendix 6
T_P	return ratio about the plant, 10.7.1
T_E	return ratio in an equivalent system, 10.7.1
T_S	sampling period, 5.10.1
U	voltage, 7.1.1
U	amplitude of sinusoidal signal at the system's input, 12.3
U', U''	threshold values of U causing jumps in E, 12.3
$U_{2(n)}$	amplitude of nth harmonic at a system's output, 1.7
V	amplitude of fundamental at the output of nonlinear link, 12.3
V	velocity, 7.1.1
V	voltage, 7.1.1
W	reactance function, A4.2
$W(w)$	transfer function (immitance) of a regulator, 6.7.1, A8.2
$W(w_0)$	nominal transfer function (immitance) of a regulator, 6.7.1, A8.2
$W(0), W(\infty)$	regulator functions with zero or infinite parameter values w, 6.7.1, A8.2
Y	admittance, 3.14.4
Y'	admittance of a two-pole that is not zero at infinite frequency, 3.14.4
Z	impedance, 3.14.4
Z'	impedance of a two-pole that is not zero at infinite frequency, 3.14.4
Z_L	load impedance, 7.2.2

Z_o	impedance (mobility) of a system without feedback, 7.4.2, Appendices 6 and 7
Z_S	source impedance, 7.2.2
a_1, a_2, etc.	polynomial coefficients, 5.10.4
b_1, b_2, etc.	polynomial coefficients, 5.10.4
$e, e(t)$	signal at the input to a nonlinear link, 12.2
e_d	dead zone, 11.3
e_s	saturation threshold, 11.3
f	frequency in Hz, 1.4.1
f_b	frequency at which loop gain is 0 dB, i.e. $|T| = 1$, 1.4.1
f_c	lowest frequency at which loop gain is $-x$ dB, 4.2.2
f_c	central frequency of a segment of constant slope of a Bode diagram, 3.10
f_d	highest frequency at which loop gain is $-x$ dB, 4.2.2
f_g	lowest frequency at which loop gain is x_1 dB, 4.2.7
f_h	highest frequency at which loop gain is x_1 dB, 4.2.7
f_p	frequency of a pole, 5.2
f_s	sampling frequency, 5.10.1
f_T	unity gain bandwidth, 6.1.1
f_z	frequency of a zero, 5.2
f_m	frequency where integral and plant feedbacks intercept, 14.5
f_u	frequency where uncertainty amplitude becomes big, 14.1
k	gain coefficient, 11.7
k	motor constant, 7.6.2
k	spring stiffness coefficient, 7.1.1
k_i	piecewise-linear gain coefficients, 11.3
k_o	coefficient of forward propagation in voltage, Appendix 6
k_{oE}	coefficient of forward signal propagation relative to the source emf, Appendix 6
k_{oI}	coefficient of forward signal propagation, Appendix 6
k_c	coupling coefficient, 11.12
n	slope coefficient of an asymptotic Bode diagram, 4.2.2, 5.2
n	transformation coefficient, 7.6.5
$p_1, p_2, \ldots$	polynomial coefficients, 5.10.1
q	Popov's criterion coefficient, 10.5.1
q_1, q_2	coefficients in modified Popov criterion, 10.6.1
$q_1, q_2, \ldots$	polynomial coefficients, 5.10.1
r	reference vector, 8.3
$s = \sigma + j\omega$	Laplace variable, 1.4, A2.3
t_d	delay time, 3.1.1
t_r	rise time, 3.1.1
t_s	settling time, 3.1.1
u	$\ln \omega/\omega_c$, 3.9.2
$u, u(t)$	signal at system's input, 3.8
u	control vector, 8.3
u_{th}	dead-beat threshold, 11.4
u_n	signal sample at the input to trapezoid integrator, 5.10.1
$v, v(e)$	signal at the output of a nonlinear link, 12.2
v	vector of process noise, 8.4
v_n	signal sample at the output of trapezoid integrator, 5.10.1
w	scalar parameter for variable immittance, 6.7.1, A8.2

w	transadmittance, Appendix 6
w	width in octaves of a trapezium gain response segment, 3.10
w	vector of sensor noise, 8.4
w_o	nominal value of variable element in Bode symmetrical regulator, 6.7.1, A8.2
x	displacement, 7.1.1, 11.12
x	lower amplitude stability margin in dB, 3.5
x	vector of state variables, 8.3
x_1	upper amplitude stability margin in dB, 3.5
x_E	state estimate, 8.4
y, or $y\pi$	phase stability margin, 3.5
y	output vector, 8.3
y	displacement along y-axis, 11.12
z	measurements, 8.4
z^{-1}	delay operator, 5.10.1
Δ	difference, 3.14.2
Δ	main determinant, A8.1
Δ_o	minor, A8.1
Ω	angular velocity, rad/s, 7.1.1
Ω_f	free-run angular velocity, rad/s, 7.2.3
θ, $\theta(s)$, $\theta(j\omega)$	transfer function, A4.1
θ	angle of rotation, 7.1.1
φ	nonlinear function, 10.5.1
ρ	wave resistance (mobility), 7.8.2
τ	torque, 1.6
τ_b	brake torque, 7.2.3
$\zeta = 1/(2Q)$	damping coefficient, 5.5
ω	frequency, rad/s, 1.4.1
ω_c	frequency at which phase shift is calculated, 3.9.2

Index